Modern earth buildi

Related titles:

Materials for energy efficiency and thermal comfort in buildings
(ISBN 978-1-84569-526-7)
Achieving a sustainable level of energy efficiency in buildings, while maintaining occupant comfort, would both substantially reduce energy demand and improve energy security, as well as improving the environmental impact, including the carbon footprint, of building stock worldwide. Energy efficiency and thermal comfort can be achieved in both old and new buildings through application of advanced building materials and technology. This book critically reviews the development and utilisation of advanced building materials and applications of this technology in a range of building types and climates.

Sustainability of construction materials
(ISBN 978-1-84569-349-3)
Major concerns have been raised on the sustainability of construction activities, in particular the sustainability of construction materials. A tremendous amount of research has been carried out to quantify the ecological and environmental impact caused during the life cycle of construction materials. This book compiles all the information on the sustainability of construction materials together into one complete volume. *Sustainability of construction materials* is an invaluable source of reference to all those professions involved in the process of infrastructure development, construction and design. It presents current research from a leading team of international contributors.

Building materials in civil engineering
(ISBN 978-1-84569-955-6)
The construction of buildings and structures relies on having a thorough understanding of building materials. *Building materials in civil engineering* provides an overview of the complete range of building materials available to civil engineers and all those involved in the building and construction industries. Chapters cover the basic properties of building materials including cement, concrete, mortar, construction steel and wood. The book also examines materials used for insulation and finishing. It serves as an essential reference for civil and construction engineering students and professionals alike.

Details of these and other Woodhead Publishing materials books can be obtained by:

- visiting our web site at www.woodheadpublishing.com
- contacting Customer Services (e-mail: sales@woodheadpublishing.com; fax: +44 (0) 1223 832819; tel.: +44 (0) 1223 499140 ext. 130; address: Woodhead Publishing Limited, 80 High Street, Sawston, Cambridge CB22 3HJ, UK)
- in North America, contacting our US office (e-mail: usmarketing@ woodheadpublishing.com; tel.: (215) 928 9112; address: Woodhead Publishing, 1518 Walnut Street, Suite 1100, Philadelphia, PA 19102-3406, USA)

If you would like e-versions of our content, please visit our e-platform: www. woodhead publishingonline.com.Please recommend it to your librarian so that everyone in your institution can benefit from the wealth of content on the site.

Woodhead Publishing Series in Energy: Number 33

Modern earth buildings

Materials, engineering, construction and applications

Edited by

Matthew R. Hall, Rick Lindsay and
Meror Krayenhoff

WP

WOODHEAD
PUBLISHING

Oxford   Cambridge   Philadelphia   New Delhi

Published by Woodhead Publishing Limited,
80 High Street, Sawston, Cambridge CB22 3HJ, UK
www.woodheadpublishing.com
www.woodheadpublishingonline.com

Woodhead Publishing, 1518 Walnut Street, Suite 1100, Philadelphia,
PA 19102-3406, USA

Woodhead Publishing India Private Limited, G-2, Vardaan House,
7/28 Ansari Road, Daryaganj, New Delhi – 110002, India
www.woodheadpublishingindia.com

First published 2012, Woodhead Publishing Limited
© Woodhead Publishing Limited, 2012
The authors have asserted their moral rights.

British Library Cataloguing in Publication Data
A catalogue record for this book is available from the British Library.

Library of Congress Control Number: 2012934737

ISBN 978-0-85709-026-3 (print)
ISBN 978-0-85709-616-6 (online)
ISSN 2044-9364 Woodhead Publishing Series in Energy (print)
ISSN 2044-9372 Woodhead Publishing Series in Energy (online)

The publisher's policy is to use permanent paper from mills that operate a sustainable
forestry policy, and which has been manufactured from pulp which is processed using acid-
free and elemental chlorine-free practices. Furthermore, the publisher ensures that the text
paper and cover board used have met acceptable environmental accreditation standards.

Typeset by Replika Press Pvt Ltd, India
Printed by TJ International, Padstow, Cornwall, UK

10 0670643 3

Contents

(* = main contact)

Editors

Preface and Chapter 1

Dr Matthew R. Hall*
Nottingham Centre for
 Geomechanics
Division of Materials, Mechanics
 and Structures
Faculty of Engineering
University of Nottingham
University Park
Nottingham
NG7 2RD
UK

E-mail: matthew.hall@nottingham.ac.uk

Rick Lindsay
Earth Structures (Australia) Pty Ltd
PO Box 653
Mansfield 3724
Victoria
Australia

E-mail: rick.earthstructures@gmail.com;
 lindsay@mansfield.net.au

Meror Krayenhoff
SIREWALL Inc
212 Cusheon Lake Road
Salt Spring Island
BC
V8K 2B9
Canada

E-mail: meror@sirewall.com

Chapter 2

Dr Matthew R. Hall* and S. Casey
Nottingham Centre for
 Geomechanics
Division of Materials, Mechanics
 and Structures
Faculty of Engineering
University of Nottingham
University Park
Nottingham
NG7 2RD
UK

E-mail: matthew.hall@nottingham.ac.uk

Chapter 3

Dr Christina J. Hopfe*
BRE Institute of Sustainable
   Engineering
School of Engineering
Cardiff University
The Parade
Cardiff
CF24 3AA
UK

E-mail: hopfec@cardiff.ac.uk

Dr Matthew R. Hall
Nottingham Centre for
   Geomechanics
Division of Materials, Mechanics
   and Structures
Faculty of Engineering
University of Nottingham
University Park
Nottingham
NG7 2RD
UK

E-mail: matthew.hall@nottingham.ac.uk

Chapter 4

Dr Eng. Horst Schroeder
Faculty of Architecture
Dept. Ecological Building
Bauhaus University Weimar
99421 Weimar
Germany

E-mail: horst.schroeder@uni-weimar.de;
   dvl@dachverband-lehm.de

Chapter 5

Prof. Ludwig Rongen
University of Applied Sciences
Erfurt and Rongen Architects
Propsteigasse 2
D 41849
Wassenberg
Germany

E-mail: l.rongen@rongen-architekten.de

Chapter 6

Lakshmi N. Reddi*
University Graduate School
Florida International University
Miami
Florida
33199
USA

E-mail: lreddi@fiu.edu

Arun K. Jain and Hae-Bum Yun
Department of Civil
Environmental and Construction
   Engineering
University of Central Florida
Orlando
Florida
32816
USA

E-mail: arkujain@gmail.com;
   Hae-Bum.Yun@ucf.edu

Chapter 7

Andrew Dawson
Faculty of Engineering
University of Nottingham
University Park
Nottingham
NG7 2RD
UK

E-mail: Andrew.Dawson@nottingham.
ac.uk

Chapter 8

Dr Charles E. Augarde
School of Engineering and
   Computing Sciences
Durham University
South Road
Durham
DH1 3LE
UK

E-mail: charles.augarde@dur.ac.uk

Chapter 9

Dr M. R. Hall* and K. B. Najim
Nottingham Centre for
   Geomechanics
Faculty of Engineering
University of Nottingham
University Park
Nottingham
NG7 2RD
UK

E-mail: matthew.hall@nottingham.ac.uk

P. Keikha Dehdezi
Nottingham Transportation
   Engineering Centre
Faculty of Engineering
University of Nottingham
University Park
Nottingham
NG7 2RD
UK

Chapter 10

Dr Ren Kebao* and Dr Douglas
   Kagi
Tech-Dry Building Protection
   Systems Pty Ltd
177-179 Coventry Street
South Melbourne
Victoria 3205
Australia

E-mail: ren@techdry.com.au

Chapter 11

Dr Jean-Claude Morel
Université de Lyon
Ecole Nationale des Travaux
   Publics de l'Etat
Département de Génie Civil et
   Bâtiment
CNRS FRE-3237
3 rue Maurice Audin
69518 Vaulx-en-Velin Cedex
France

E-mail: Jean-Claude.Morel@entpe.fr

Dr Quoc-Bao Bui
Université de Savoie
Polytech Annecy-chambery
Laboratoire Optimisation de la
    Conception et Ingénierie de
    l'Environnement (LOCIE)
CNRS UMR-5271
73376 Le Bourget du Lac Cedex
France

E-mail: Quoc-Bao.BUI@univ-savoie.fr

Erwan Hamard
IFSTTAR
MAT6-GPEM
Route de Bouaye CS4
44344 Bouguenais cedex
France

E-mail: Erwan.Hamard@IFSTTAR.fr

Chapter 12

Dr Paul Jaquin
Integral Engineering
Bath
UK

E-mail: pauljaquin@gmail.com

Chapter 13

Prof. B. V. Venkatarama Reddy
Department of Civil Engineering
Indian Institute of Science
    Bangalore
Bangalore 560 012
India

E-mail: venkat@civil.iisc.ernet.in

Chapters 14 and 15

David Easton and T. Easton
Rammed Earth Works
4024 Hagen Road
Napa, CA 94558

E-mail: easton@rammedearthworks.com

Chapter 16

Giuseppe Calabrese
Unit 48/21-27 Meadow Crescent
Meadowbank
2114 NSW
Sydney
Australia

E-mail: giuseppe@architetto.com.au;
    jseppec@yahoo.com

Chapter 17

J. F. D. Dahmen*
Assistant Professor
University of British Columbia
School of Architecture and
    Landscape Architecture
Room 402
6333 Memorial Road
Vancouver
BC V6T 1Z2
Canada

E-mail: joe.dahmen@gmail.com

Professor John A. Ochsendorf
MIT
Room 5-418C
77 Massachusetts Avenue
Cambridge
MA 02139
USA

E-mail: jao@mit.edu

Chapters 18, 22 and Appendix 1

Rick Lindsay
Earth Structures (Australia) Pty Ltd
PO Box 653
Mansfield 3724
Victoria
Australia

E-mail: rick.earthstructures@gmail.com;
        lindsay@mansfield.net.au

Chapter 19

Hugh W. Morris
Department of Civil and
    Environmental Engineering,
The University of Auckland
Private Bag 92019
Auckland 1142
New Zealand

E-mail: hw.morris@auckland.ac.nz

Chapter 20

Prof. Wei Wu,* Tensay G. Berhe
    and Taha Ashour
Institute of Geotechnical
    Engineering (IGE)
University of Natural Resources
    and Applied Life Sciences
Feistmantel Strasse 4
A-1180
Vienna
Austria

E-mail: wei.wu@boku.ac.at;
        tensay@engineer.com;
        taha.ashour@boku.ac.at

Chapter 21 and Appendix 2

Meror Krayenhoff
SIREWALL Inc
212 Cusheon Lake Road
Salt Spring Island
BC
V8K 2B9
Canada

E-mail: meror@sirewall.com

Chapter 23

Dr Matthew R. Hall*
Nottingham Centre for
    Geomechanics
Division of Materials, Mechanics
    and Structures
Faculty of Engineering
University of Nottingham
University Park
Nottingham
NG7 2RD
UK

E-mail: matthew.hall@nottingham.ac.uk

W. Swaney
Earth Structures (Europe) Ltd
The Manor
Ashley
Nr Market Harborough
Northamptonshire
UK

Chapter 24

Raefer K. Wallis
SIREWALL China/GIGA/A00
    Architecture
Yongjia Road
Lane 310/356
House 34
Shanghai
200031
People's Republic of China

E-mail: rkw@azerozero.com

Woodhead Publishing Series in Energy

Preface

Traditional earth building materials were the mainstay of a significant proportion of past societies, and today it is estimated that around one-third of the world's population live in buildings made from unfired earth. Since these materials and techniques are the most prolific, both historically and in modern times, and have one of the greatest proven track records in terms of longevity (some buildings being several hundred years old) it might well seem reasonable to argue that by definition they are the most conventional construction material of all. Ironically, they are commonly referred to with terms such as 'non-conventional', 'alternative' and 'low environmental impact' within the modern construction industries of western society. Whilst the latter point is not intended to be derogatory, it is also implicit in these terms that the material must be at a low level of technological advancement, in-situ performance, and with little capacity for refinement or integration with 'conventional' materials and systems of building. However, these conclusions rest on one significant assumption; that earth construction materials are already well-understood and that our rationale for not using them instead of fired brick masonry, concrete, timber or pre-fabricated composites must be based on this knowledge. The main purpose in creating this book is to test this hypothesis and to challenge the assumption on which it is based. The editors assert that modern earth construction represents high-quality materials and finishes, standardised approaches and accurate placement systems, significant flexibility in terms of advanced structural engineering, the ability to passively regulate indoor temperature and humidity, enhance thermal comfort and indoor air quality, and present a viable alternative to load-bearing masonry or in-situ concrete walling with numerous functional advantages. It is hoped that the evidence for these assertions, presented throughout this book, will validate this perspective and encourage others to explore the opportunities that modern earth construction can bring.

Modern earth building could make significant contributions towards the current priorities of the construction materials sector by restoring a carbon neutral and economically sustainable existence. Firstly and most obviously, we know that sustainability *is* durability and that earth buildings have demonstrated that they can last for hundreds of years and still remain structurally sound, they can be made using less energy than fired materials and require less operational energy during their lifetime to retain comfortable, healthy indoor environments. Secondly, modern earth buildings have the capacity to refresh humanity's delight in the simple and the modest without compromising comfort, style and desirability. Indeed a properly built modern earth house adds immeasurably to the comfort of inhabitants with less cost to the environment.

Part I: Introduction to modern earth buildings begins with an overview of modern earth building including its definition and global extent as an industry, with discussion relating to the changing demands in terms of aesthetics, economics, sustainability and continually advancing structural solutions (Hall, Lindsay and Krayenhoff). This is complemented by two successive chapters that consider the behaviour and performance of earth structures in terms of building physics fundamentals. The first of these chapters focuses on hygrothermal behaviour (coupled heat and moisture transport/storage), which is central to the way earth walls passively regulate indoor air temperature and relative humidity, as well as controlling mould growth and enhancing air quality. The chapter presents materials data, approaches to modelling hygrothermal behaviour, and occupant thermal comfort and health (Hall and Casey). The next chapter relates to fabric insulation materials (including properties), and also includes a detailed review of thermal bridging phenomena and current approaches to 2D and 3D computer modelling. Within this context, cavity earth wall construction detailing is explained and illustrated along with acoustics performance and isolation (Hopfe and Hall). This is followed by a global review of building codes, standards, regulations and compliance relating to modern earth construction. It also includes details of regional and national level standards and normative documents covering aspects such as material selection, production methods and various material types (Schroeder). The section concludes with a discussion of energy efficient design in buildings with particular emphasis on the applications for high-mass construction materials such as earth. This is framed within the context of Passive House design principles as an aspirational standard of operational efficiency for future earth building projects (Rongen).

Part II: Earth materials engineering and earth construction begins with a technical review of soil properties, classification and test methods that can be applied in order to determine suitability for a given application or technique plus details of clay mineralogy, Atterberg limits and compaction (Jain, Reddi and Yun). The next chapter reviews the types, origins and properties of alternative and recycled materials as alternatives to quarried (primary) aggregates for potential use in earth construction. It also includes the relevance of life cycle assessment and future recycling, as well as age-related performance and leaching (Dawson). Strength and mechanical behaviour are dealt with in the next chapter in a review of soil mechanics, which primarily relates to soils without the addition of binders. The topics covered include effective stress, experimental and numerical models for shear stress, and unsaturated behaviour in relation to earth construction materials (Augarde). This is complemented by a detailed technical review of soil stabilisation including lime, cement and pozzolanic reactions, bituminous binders and emulsions, synthetic, polymeric, adhesives and fibre reinforcement including a selection tool for modern earth construction materials (Hall, Najim and Dehdezi). The next chapter explains the state-of-the-art technology in admixtures and surface treatments for modern earth materials. This includes hydrophobic additives with associated test data for durability and moisture transport, stabiliser set enhancers/retarders, workability enhancers and surface coatings (Kebao and Kagi). Weathering and durability of earthen materials and structures is addressed with a technical discussion of current test methodologies along with discussion of the principal erosion mechanisms and case studies of the efficacy of surface coatings/renders (Morel, Bui and Hamard).

Part III: Earth building technologies and earth construction techniques opens with a view to the past covering the historic origins and progression of different earth construction techniques throughout the world dating from ancient times to modern history (Jaquin). The other chapters in Part III review the modern-day forms of these traditional techniques, many of which have become highly commercialised enabling them to compete with conventional materials in the global construction industry. Stabilised compressed earth blocks for structural masonry is a prolific modern earth technology and is given a comprehensive review that includes soil grading/suitability, block production, approaches to stabilisation, physical-mechanical properties and extensive details of results from experimental testing and case studies (Reddy). The following chapter provides detailed and insightful discussion about modern rammed earth construction techniques. Detailed coverage includes material sourcing and proportioning, mixing and formwork technology, along with installation (including compaction) and a forecast of future trends, technological challenges and developments (Easton and Easton). Next is a technical review and summary of pneumatically impacted stabilized earth (PISE); a novel modern earth construction technique that has only recently been developed. Specific details include proportioning, mixing and delivery along with details of formwork and placement methodologies (Easton). Finally, techniques and methodologies for the conservation of historic earth buildings are explained, along with detailed case studies and principles relating to erosion mechanisms and the factors that affect them (Calabrese).

Part IV: Modern earth structural engineering is intended to provide both an understanding and an appreciation of the current capacity for structural engineering and innovation in modern earth construction. It opens with a comprehensive chapter that explains the structural theory behind arches, vaults and domes, with specific examples and case studies relating to both traditional and modern earth buildings, as well as providing a set of design criteria (Dahmen and Oschendorf). In relation to modern-day stabilised rammed earth (SRE) wall construction, the next chapter provides structural design guidelines for the use of steel reinforcement. This includes technical drawings and detailed specifications from a range of real-life examples (Lindsay). Design and construction guidelines for modern earth buildings in seismic regions, as well as regions subject to high wind loading/ tropical storms, flooding, landslides and volcanic activity are reviewed in a comprehensive chapter that also gives details of structural reinforcement and risk mitigation (Morris). Finally, the application of modern earth materials in embankments and earthfill dam structures is reviewed in terms of materials and techniques, design, maintenance and stability assessment tools (Wu, Berhe and Ashour).

Part V: Application of modern earth construction: international case studies provides a series of region-specific reviews demonstrating the current state-of-the-art in contemporary earth materials across the world. It begins with North America (including USA and Canada) including many case studies and the development of stabilised insulated rammed earth (SIRE), design variables to suit the wide range of climatic conditions, seismic design, embodied energy and energy efficiency. It also includes significant development in mixing, formwork and delivery technology as well as admixtures and surface treatments, and several advances in surface details and texturing effects (Krayenhoff). Next, we move over to Australasia where numerous case studies are used to explain the significant developments in material selection and standardisation of formwork technology, along with insulated SRE, design for thermal comfort, costs and business models for the industry, plus a summary of future trends (Lindsay). Since the 1970s modern earth construction in India is prolific, perhaps much more so than any other single country, and is almost exclusively represented by stabilised compressed earth block technology which is already covered in great detail in Part III (Reddy). The next chapter in Part V summarises the current developments and case studies in Europe, which consists of both stabilised and unstabilised rammed earth and compressed earth block, plus non-structural clay-based renders and in-fill materials. It explains how heritage conservation and the revival of traditional techniques has led to modest yet active growth in the application of modern earth materials in many parts of Europe and a high level of innovation and research (Hall and Swaney). Finally, modern rammed earth construction has entered the rapidly expanding nation of China and has already resulted in significant innovation and expansion of this technique as demonstrated by the numerous case studies cited, which look set to expand even further (Wallis).

The main text of the book is supported by two appendices that are intended to provide the reader with supporting information in relation to techno-economic analyses coupled with environmental impact assessment. This has been limited to the two regions of the world (Australasia and North America) where modern earth construction has been developed and proven to the extent that it now forms a part of mainstream construction industry, and represents a viable technology that can be selected or indeed exported to other developed regions of the world where the expertise/industry does not currently exist. Clearly, the techno-economic criteria for these two regions are quite different, and so have been presented as separate appendices for greater clarity whilst incorporating the relevance of environmental impact in each case.

The editors sincerely hope that this book will serve not only to educate and inspire students of this subject around the world, but to provide a useful reference source to professionals including designers, planners and potential end users who may seek to utilise the benefits of modern earth construction materials in their future projects. The editors are also very aware that technological developments and the current state-of-the-art in one region of the world may not be well known in other parts, particularly for diverse fields such as earth construction. Therefore, this book is also intended to provide knowledge, understanding and an appreciation of the global context in terms of achievement and development in this field, whilst giving the relevant technical information needed at local level for all readers.

M. R. Hall
University of Nottingham, UK
R. Lindsay,
Earth Structures Group, Australia, and
M. Krayenhoff
SIREWALL Inc, Canada

Part I

Introduction to modern earth buildings

1

Overview of modern earth building

M. R. HALL, University of Nottingham, UK, R. LINDSAY,
Earth Structures Group, Australia and M. KRAYENHOFF,
SIREWALL Inc, Canada

Abstract: This chapter explains the current position of modern earth
building within the construction sector, its appeal and the role it might play
in the future. The definition of modern earth building is discussed along with
the ways in which this has evolved along with technological advancement
and industrialisation. The significance of modern earth building, both for
our current and future construction industries, is tackled from developed and
developing country perspectives. The chapter then discusses the changing
aesthetic of modern earth building including the role this can play in terms
of material functionality, before moving on to its position with regard to
the changing social morality towards climate change and the construction
industry's response. The chapter ends with a discussion of how earth
building approaches have been developed to meet and complement the
efficiency requirements of the modern construction industry.

Key words: modern construction sector, definitions, present and future
industry, aesthetic, functionality, climate change, efficiency.

1.1   Introduction

Earth building is simple by its very nature, but 'simple is hard to do' as the
late Steve Jobs noted on his retirement as CEO of Apple. In modern earth
building we are blessed with something simple and worthwhile. Imagine
a world where everyone is living and working in insulated or uninsulated
rammed earth buildings. That is the norm all over the planet. Everyone
has grown up in rammed earth buildings and, because they work so well,
no other alternative building materials are ever considered. Then one day
an adventurous builder decides that he is going to try building with wood.
People think he is crazy. He endures the ridicule, but is determined to give
the idea of building with wood a try. His first challenge is to try to figure
out how to cut the tree down and then determine which part of the tree is
best for building with. He has to invent tools to cut the tree down. He thinks
fire might be useful somehow. He is not sure whether to cut the tree into
rounds or into lengths. He is not sure how to do this either. He is not sure
if there is a difference between the wood of the oak and the wood of the
willow. He is a very long way from planed 2 × 4s and plywood. In the same

3

way, the rammed earth industry is in relatively early stages of development compared to its ultimate potential as a mass-market technique. There is so much innovation beckoning.

The size of the insulated and uninsulated rammed earth market is vast and the technology is ahead of its time. There are still trees to cut down and land with enough topsoil left to grow new trees. We have not yet accomplished on a planetary scale what Lebanon did to their great cedar forests or Easter Island did by cutting down its last tree, but we are working on it. At some point in the future we may not have the option of building with wood and will need to build with inorganic building materials. Of course it is preferable that we make that move while there are still forests and topsoil. As a building culture, we will need to move from veneers to substance, from disposable building and thinking to sustainable communities and healthy environments, and move to the broad use of local materials and labour. The emerging rammed earth industry will need to move from a small capacity cottage industry and levels of building to a much larger economic capability to allow an expanded range and complexity of projects.

Earth building has an honest appeal. Formwork and compaction lines are all there to see – nothing is hidden. Few construction consumables are so simple and honest. Yet it takes courage for many people today to build an earth house. Modern humanity appears to have great faith in highly manufactured and sophisticated elements. Gadgetry that can accompany a simple bushwalk, for example, is sometimes more empowering for a modern-day person than the walk itself. For the modern occupier of an earth house, the walls serve to relax our obsession with neatness, uniformity and linear perfection. While a symmetry of form is retained, the 'cleanliness' anxiety created by smooth painted surfaces is excused and somehow deleted. Some primal happiness is restored. It takes courage for many modern building contractors to embrace what can appear to be a step backwards in technology. Earth walling is assumed by many builders to be something that used to happen before the Industrial Age.

There is a real challenge, however, for people to take modern earth building seriously. Modern earth building has the capacity to reduce substantially some of the damaging effects that humanity has on the planet. Earth buildings, in conjunction with good design and clever marketing, could take us towards an acceptable status of smaller and more sensible buildings and will produce a positive and lasting environmental legacy for future generations. The human condition is capable of huge technical ability to overcome some of our most drastic environmental problems. The same condition is also capable of rationalising or justifying extravagant lifestyle choices that have created the problems in the first place. This anomaly in our intelligence is nowhere more evident than in the construction industry. We seek to create environmental solutions with clever low-emission materials and ideas. With these 'solutions' under our belt we then build even more extravagant houses for ourselves. This is particularly the case in Australia and the USA where the building space per occupant has increased fourfold over the past 50 years. Earth building has a role to play in creating an awareness of what humanity is capable of in simplifying our housing needs. It serves as a reminder that simple is possible and acceptable. Modern earth building lends status to simplicity, and good earthen architecture gives simple buildings huge credibility.

1.2   Definition of modern earth building

An earth building is one where a significant part of the structure and/or building fabric comprises graded soil (i.e. earth) that has been prepared using one or more techniques, e.g. rammed earth or compressed earth block. However, what characteristics can make an earth building 'modern'? One of the requirements must surely be that the techniques applied should build on what has been learnt from the past whilst benefiting from the advancements and improvement that state-of-the-art innovation has brought. Since our learning is ongoing it also follows that what constitutes 'state-of-the-art' is relative to the era in which the new building is set. A faithful reproduction of a historic form of earth building from the 19th century, for example, that does not benefit from any of the technological or material advancements of the last 200 years cannot be considered as a modern earth building if it is being constructed in the 21st century. A further consideration must be that 'modern' earth buildings have a high level of quality, dimensional tolerances and accuracy, compatibility with required/expected fixtures and fittings, compliance with building regulations/codes and scope to excel in best practice and future design standards, e.g. Passivhaus or Cradle to Cradle. This implies that, as our understanding of mix design, material selection and grading, formwork and delivery technology, and the development of new admixtures, binders and coatings continues to progress, then so too will the minimum standard and overall quality of 'modern' earth materials. Lastly, the ability of a 'modern' earth building to meet the ever-increasing expectations of occupants, in terms of comfort, health/well-being and shelter, should at the very least be maintained and preferably improved. In order to minimise impact and maximise sustainability, this should be achieved with a primary goal of reducing the operational energy towards zero carbon, and a secondary goal of reducing embodied and future demolition/disposal embodied energy.

1.3   The significance of modern earth building in the current and future construction industries

1.3.1   Modern earth building in developed countries

Presently there are few building materials or technologies that offer relief from the world of veneer and artifice. Though there are many reasons that a consumer might select modern rammed earth walling, the beauty and the visceral impact of massive walls made from lightly processed materials trumps them all. The fact that these walls offer healthy indoor environments coupled with energy efficiency, long-life durability with low maintenance, as well as improved acoustics, improved seismic stability and fire barriers, are seldom the determining factors in selecting rammed earth. There is little doubt that consumer patterns of the developed world will rapidly adjust to accommodate the realities of the effects on the planet caused by our lifestyles. In this context modern earth building should be in some demand. There will be current restraints on the modern earth building industry due to its relatively high labour cost. The challenge for the industry in developed countries will be technical: increasing efficiencies and construction safety while ensuring the product retains a low embodied energy. There are endless ways of mechanising the production of earth walls. The challenge will be to engage mechanisation without abandoning the environmental and aesthetic worth of the material.

1.3.2   Modern earth building in developing countries

In developing countries, most of whom have significant earth-walling industries already, the challenge will be one of marketing. A previously impoverished family who have spent generations living in an earth-walled shelter will aspire to live in a 'modern' building made from concrete, steel and glass. It may not be enough for the modern earth building to perform extremely well; there will need to be an obvious visual differentiation and one that is aspirational for the end user. In any case, modern earth building in developing countries will be an environmental necessity as a burgeoning middle-class population is likely to place impossible demands on resources for 'modern' housing. The role of social marketing in these countries to increase the status of earth building will be vitally important. In many ways the development and increased sophistication of modern earth building in developed countries is likely to significantly enhance the aspirational qualities of these structures in the eyes of developing nations.

1.4 Changes in the modern earth building industry

1.4.1 Aesthetic change

The past two decades have seen a slow uptake in the status attached to environmentally responsible lifestyle choices. In North America, this is the Lifestyles of Health and Sustainability (LOHAS) market segment, and represents approximately 50 million people willing to vote with their wallet. The size, shape and economy of cars, for example, have changed dramatically in recent years. The increase in eco-friendly leisure activities as a market sector is high. Construction materials are changing to reflect the demand for greener credentials, though ironically house size (i.e. m^2 floor area) is still increasing in many parts of the world.

The changing aesthetic of modern earth building (in developed countries) as a consequence of our social aspirations has been exciting. Modern earth building has developed within a period of change for the overall construction industry. Earth walling contractors must work with other established building trades to allow for efficient and safe outcomes. Many modern earth buildings have similar shapes to houses made using 'conventional' materials, though increasingly architects are making the most of the inherent functional properties that these walls can offer to the operation of a building and interaction with its occupants. In Australia, for example, there is an increasing use of rammed earth on spine walls to bring significant thermal and hygric mass to the core of the building interior whilst allowing lighter, well-insulated elements to clad much of the external fabric. In North America, the potential to use stabilised insulated rammed earth (SIRE) to achieve net zero energy buildings is being applied both in very cold and very hot climates. The humble presence of earth walls in living areas is useful to bring a tactile element to minimalist architecture. Good earthen architecture uses these qualities to create a sense of calm refuge for the occupants. Children love earth walls; watch a child walk up and hug the protruding end of an earth wall. Adults tend to slap the surface of a huge earth wall and feel heartened by the immensity of them. Art works sit comfortably on earth walls; if good art makes us relax into a moment of reflection, then a simple earth wall offers a peaceful place in which to do it.

In all things, good design is about more than just a sense of appearance – it is also about functionality. Humans are smart enough to enjoy the completeness of both the beauty and the function of good buildings. With well-designed earth buildings, this sense of completeness is made apparent. Frank Lloyd Wright was reputed to have said that he felt confident he could design a building that would cause a happily married couple to break up within a year. The acoustics, the smells, the temperature, the humidity, the draughts, the way the eye is moved around the space, the colours and the textures all contribute to (or indeed undermine) our sense of well-being.

We are deeply affected by our surroundings, whether we realise it or not. Well-designed modern earth buildings can enhance all of these functions with one material.

As the functionality of modern earth building has been dramatically enhanced over the last few decades, so has there been progress in making available wider visual options. Now it is possible to have rammed earth with or without forming seams, and with or without through-tie holes. There is a level of control now possible on the finished texture of the wall surface (for improved acoustics) and whether or not a surface finish is applied. The colour of the wall can be almost anything desired, although there remains the variability that comes with using a raw material. The use of multi-colour schemes is becoming more popular, despite its expense. Embedding shells and stones in the rammed earth, or carving formwork to create art within the walls has broadened what is available for either the designer or the consumer to express themselves.

1.4.2   Social morality and the response to climate change

Within the context of our current social and economic circumstances, the consumer drives social morality. The construction industry builds to fulfil the demands of that consumer. If the consumer demands small, functionally well-designed, environmentally responsible buildings, the construction industry will build them. If the consumer demand is for larger houses of a particular aesthetic with a high level of energy-consuming technological features, then the building industry will build them. A property developer who builds houses ranging from those with large floor areas and no garden to those with small floor areas and larger gardens, may discover that the large ones sell easily while the smaller ones linger on the market. The market-driven developer learns to build bigger, less environmentally smart houses since they are simply serving a demand. The source of social morality in this case lies with the consumer. Given this predicament, the changing social morality within the context of modern earth building rests less with the construction industry and more with the building consumer.

For the consumer to embrace modern earth buildings they will need opportunities to learn about and experience these options. Opportunities exist for this to happen and many people who understand this point of view are prepared to take the risk by building something different. Well-designed modern earth buildings inevitably make their way into glossy housing magazines, and this serves to educate consumers. This creates some recognition for the consumer to feel confident when their architect suggests, for example, stabilised rammed earth. Carefully placed, understated advertisements for earth buildings in the same magazines will further serve to embolden the earth wall consumer. Once consumers make headway into taking alternative options with their housing, then the industry has a responsibility to provide for the demand efficiently, safely and in an environmentally responsible manner.

The Australian experience towards this transition has been one of incremental progression. The eco-morality of the Australian consumer is changing for the better, and the next generation of consumers have been highly educated about environmentally responsible behaviour. Advertising agencies also appear to be increasingly aware of a changing eco-morality in Australia; advertising writer Michael Mulchay (2011) of DDB Group, Melbourne wrote:

> Young consumers are acutely aware of their role in playing their part in protecting our fragile environment. They increasingly feel linked to a bigger system than their own immediate needs.

However the speed with which this morality will affect consumer patterns may be hastened by the financial imperative. Michael Mulchay again stated:

> Advertising inevitably speaks to the self interest of the consumer. If modest, well designed buildings prove to substantially reduce living costs, then the market could adapt more frugal and sensible housing choices accordingly. Given the realities of our inevitable ecological challenges, the next generation will look beyond the financial bottom line to consider the ecological bottom line.

From a North American perspective, social morality is seldom the real driver of the modern earth building marketplace. There are, indeed, those who select modern earth buildings for their environmental appropriateness. However, especially in the commercial domain, being able to 'fly the green flag' has high commercial importance. Designers or clients want to let the LOHAS market segment know that they are green, and spending money on fibreboard that uses less toxic glue than formaldehyde provides them with little to show off. There are very few visible ways to be green in construction and modern earth excels in flying that flag. As far as the response of the North American construction industry to climate change, most of the big players have repackaged themselves and their products by now. What they make has changed very little. It is hard to think of a single product in a timber frame wall assembly that does not contain formaldehyde, fungicide, volatile organic compounds or phthalates. The focus in the greenbuilding world has been on energy efficiency and recycling. Positive strides have indeed been made in energy efficiency but recycling has not been good for the American home. A recent ruling by the EPA (US Environmental Protection Agency) now allows fly ash to be recycled as a drywall component. This increases the potential risk of exposure to contaminants such as mercury, arsenic and other heavy metals, which fly ash may contain. Most importantly, as the

American consumer continues to focus on first cost for big ticket items, the durability of homes and buildings continues to decline. The concept and advantages of the disposable styrofoam cup are now being seen in the disposable American home that is intended to only last the span of the mortgage. It is against this backdrop of 'normalcy' that the modern earth building industry competes.

While it may be true that the consumer has a big say in what's being built, the impacts of those choices will be felt by future generations in the form of toxic waste and resource depletion. If those decisions are not based on complete information and if product manufacturers are permitted to continue to 'greenwash', then it falls on the government to ensure complete product information is disclosed. If governments required all structural building products to have a lifespan of 100 years, then modern earth building would overnight become a major industry. Obviously that would be politically very difficult. It is useful to highlight that the current building paradigm is dependent on disposability for its competitiveness. The consumers that we will depend upon to change the market appear to be unaware of this at the moment.

1.5   Managing the demands of the modern construction industry

1.5.1   Structural demands

Modern earth building is still a small part of the overall construction industry in the developed world, though it remains a major player in the developing world. The requirements of modern building codes have meant many traditional earth building methods have fallen behind as they cannot fulfil basic structural requirements such as being load bearing or having high levels of insulation. In his recent book *Earth Architecture*, Ronald Rael (2009) suggests a political motive behind the demise of some modern earth building cultures:

> Increasingly, it is illegal to build with earth because of building codes that are enforced by municipalities. While these decisions are made in the name of safety, it is more likely that manufacturers of industrialised products have lobbied to prevent the use of a free and versatile material such as earth.

Many current Australian building codes make building with uninsulated earth walls almost impossible. This is due to computer modelling based entirely on the steady-state thermal resistance (R-value) of cladding materials. This extraordinary situation exists despite many hundreds of long-term occupants of earth buildings proving they live very comfortably without mechanical air conditioning or heating, and have very low fluctuations in indoor air temperature and relative humidity. Indeed they are, in many cases, proving they live more comfortably and frugally than the occupants of the highly insulated buildings deemed suitable by the current building codes. The technical challenge for modern earth builders will be to heed increasingly rigorous structural building codes and adapt building methods to maintain a competitive edge in the construction market. This needs to be done in a way that does not result in earth building becoming another form of concrete thinly disguised as 'earth'.

The realities of a diminishing supply of cheap energy-rich building materials such as steel and fired bricks will lend some credence to the future of proper earth building. Currently anything can be built, including modern earth walling, if structural engineers incorporate enough steel and concrete in it. In an increasingly ecologically responsible market place, however, designs will need to be less extravagant as the availability of these structurally useful elements falls away and their cost increases. Without the use of structural steel or concrete bonding systems, earth building can be used extensively by the more modest building market.

1.5.2   Efficiency requirements

It would not be uncommon for a modern earth builder to find themselves on a large commercial building site surrounded by super-fast concrete forming systems, highly mechanised concrete delivery, pre-fabricated structural steel being craned through the air, and observing that the efficiency is far greater compared to current methods in modern earth building. In the context of these efficiencies, the earth building industry has a long way to go. In all developed countries the earth building industry is relatively small. Often it is large commercial projects requiring stabilised rammed earth walling constructed at high delivery rates that have pushed the boundaries and efficiencies of the material to its present limits.

In Australia, commercial projects such as the current RACV (Royal Automobile Club Victoria) Resort at Torquay (see Figs 1.1 and 1.2), the Port Phillip Estate Winery in Red Hill (Fig. 1.3), the Tarrawarra Art Museum in Healesville (Fig. 1.4), and the Charles Sturt University Campus at Thurgoona have contributed to the advancement of construction efficiencies of the local industry. Precast transportable stabilised rammed earth panels, the development of telescopic handling systems, formwork systems compatible with cladding portal frame steel structures and many other advances are incrementally nudging the industry along.

In North America, commercial projects in very cold climates such as the Southeast Wyoming Welcome Center Wyoming (see Figs 1.5 and 1.6), and Grand Beach (see case study, Chapter 24) have required adoption of

1.1 RACV Project, Torquay, Australia. Insulated SRE cladding a steel portal frame to 11.2 m height using telescopic handler access (© Earth Structures Pty Ltd, 2011).

1.2 RACV Project, Torquay, Australia. Precast SRE panels ready for hoisting to Level Three. Cloister walls in background (© Earth Structures Pty Ltd, 2011).

1.3 Port Phillip Estate, Red Hill, Australia. Facet SRE panels to 12.6 m height (© Earth Structures Pty Ltd, 2011).

1.4 Tarrawarra Art Museum, Healesville, Australia. Curving panels to 5.4 m height (© Earth Structures Pty Ltd, 2011).

1.5 Formwork stripping at Southeast Wyoming Welcome Center, Wyoming: in order to meet tight timelines, larger panel sizes are useful (© SIREWALL Inc., 2011).

1.6 Multiple gang forms at Southeast Wyoming Welcome Center, Wyoming: in order to meet tight schedule timelines, it is useful to have multiple walls underway at different stages of the process, thereby keeping the forming team, the ramming team and the mixing teams busy at all times (© SIREWALL Inc., 2011).

previously untried techniques for heating and hoarding, e.g. building inside a tent in the winter (see Figs 1.7 and 1.8). Also, the development of soil design blends that survive extreme temperatures (+40°C to –40°C) and quick changes in temperature and humidity. The Nk'Mip Desert Cultural Center required the suspension of almost 100 000 pounds (45 359 kilos) of

1.7 Sublette County Library, Wyoming: building the heating and hoarding tent using scaffolding (© SIREWALL Inc., 2011).

1.8 Sublette County Library, Wyoming: the heating and hoarding tent in the snow with air temperatures well below −30°C (© SIREWALL Inc., 2011).

rammed earth over a glass window opening (see case study, Chapter 24). Due to building seasons that are predicated around winter, the available time to build rammed earth walls can determine if the product can be used.

It has happened that the rammed earth price for a project is acceptable but the time frame is not. For that reason commercial projects like Van Dusen Gardens and the South East Wyoming Welcome Center push the industry to adopt and refine different gang forming methods.

The grand challenge for global modern earth building will be to increase construction efficiencies without abandoning the essential elements that create the difference between earth building and, for example, a highly efficient but ecologically unsustainable concrete block and render industry. The core values that draw people to earth building need to be upheld as the quest for greater efficiencies continues. In a world that is beginning the adjustment to climate change, modern earth building offers resilience to unprecedented changing temperatures, flooding, high winds, fires and rising energy costs. This book offers an insight into some of the sophisticated technologies that are currently available with the intention of broadening the technical credibility of the emerging modern earth building industry.

1.6   References

Mulchay, M. 2011. 'Environmental Marketing'. Melbourne, DDB Group Advertising Agency

Rael, R. 2009. *Earth Architecture*. New Jersey, Princeton Architectural Press

2

Hygrothermal behaviour and occupant comfort in modern earth buildings

M. R. HALL and S. CASEY, University of Nottingham, UK

Abstract: This chapter starts by introducing the subject of hygrothermal behaviour and some of the key scientific concepts underpinning this. This includes the transport and storage of both heat and moisture in porous construction materials and, more specifically, the hygrothermal behaviour of these materials including all relevant functional properties. The approach to numerically modelling this behaviour is explained, along with the key commercial packages currently in use. Detailed discussion of the hygrothermal behaviour of rammed earth materials is presented along with results from several recent and current research projects. Finally, the correlation between hygrothermal behaviour, the resultant quality of indoor air and thermal comfort are summarised. Details of further sources of information are presented for those wishing to explore some of the concepts discussed in this chapter in more detail.

Key words: hygrothermal behaviour, humidity buffering, thermal buffering.

2.1 Introduction

In order to comply with the latest Building Codes/Regulations, all buildings need to provide adequate comfort for the occupants and to achieve this in the most energy efficient manner possible. Building physics includes the study of the interactions between heat, moisture and air movement between indoor and outdoor environments, and also the fabric of the building. As with all buildings, earth walls must provide suitable levels of thermal insulation in order to maintain a comfortable indoor environment by using as little energy as possible. The study of comfort is partly scientific, in terms of a person's physical and mental health, and partly subjective in terms of a person's perceptions. Compared to many conventional construction materials such as brick, concrete and plaster that is covered with paint or wallpaper, earth buildings can offer very high levels of thermal comfort combined with improved energy efficiency through their ability to regulate the amount of change in indoor air temperature and relative humidity. This chapter explains the scientific concepts underpinning all of these themes. It also combines this with the findings of recent and current research showing the latest advances in our understanding of how and why earth materials behave as they do, and what we can do to design and specify those properties.

17

One of the most environmentally damaging trends in terms of the way we use energy in buildings is the widespread use of air conditioning. In addition, if poorly designed or maintained, air conditioned environments can be uncomfortable for the occupants and sometimes result in respiratory health risks. It is often claimed that rammed earth walls (both unstabilised and stabilised) can be used as a building-integrated form of 'passive' air conditioning. The walls are 'building integrated' because they form part of the building structure and are load bearing. Critically, they offer the ability to passively cool a building by absorbing heat from the adjacent air. They can store this heat and re-radiate it when the ambient temperature falls, for example, at night time. This property of temperature buffering is referred to as the thermal flywheel effect and can greatly increase indoor comfort levels. The porous nature of rammed earth materials is coupled with the presence of hygroscopic clay minerals that also enable the walls to passively buffer the relative humidity (RH) of a building's internal environment, i.e. hygric mass. There have been relatively few comprehensive research investigations aimed at quantifying the 'passive air conditioning' properties of earth materials, despite the fact that this behaviour has been exploited and monitored in many case study buildings.

Previously the thermal properties of rammed earth have been studied in isolation using steady-state heat gain/loss models, while the main advantage of the material is in fact its dynamic behaviour and the utilisation of this with passive design. Some publications give values for the thermal resistance of rammed earth materials, which is their resistance to conduction of heat under steady-state conditions defined by d/λ (m^2 K/W). According to Standards New Zealand (1998), in the absence of any laboratory test data the thermal resistivity (R) of a stabilised rammed earth (SRE) wall can be approximated as $R = 2.04d + 0.12$. The CSIRO have published some experimental data, along with Walker *et al.* (2005) in the *Rammed Earth Design and Construction Guidelines* book showing the significant relationship that exists between the bulk density of earth materials and their dry-state thermal conductivity as shown in Table 2.1.

Table 2.1 The correlation between bulk density and dry-state thermal conductivity for rammed earth materials (Walker *et al.*, 2005)

Bulk density (kg/m^3)	Thermal conductivity (W/m K)
1400	0.60
1600	0.80
1800	1.00
1900	1.30
2000	1.60

2.2   Hygrothermal loads and modelling

The hygrothermal behaviour of a material can be defined as the change in a material's physical properties as a result of the simultaneous absorption, storage and release of both heat and moisture (liquid and/or vapour phase). In a building envelope these loads may be thermal or hygric and are experienced on both internal and external aspects. The thermal and hygric loads commonly experienced by a building are illustrated in Fig. 2.1. A building envelope protects the indoor environment from the outside, i.e. there will be large fluctuations in both temperature and humidity outside, but because of the envelope, only a minor part of these fluctuations is transmitted to the indoor environment. The heat and moisture fluxes that generate these loads may also change direction as the air temperature and relative humidity change throughout the day and also across the seasons. The principal hygrothermal loads experienced by the envelope can be summarised as:

- short wave radiation (solar) loads
- long wave radiation (objects) loads
- thermal convection loads
- thermal conduction loads
- drying/evaporation loads
- vapour diffusion loads.

The hygrothermal behaviour of a material results from the fluxes of (i) heat transfer to and from the material, (ii) heat storage within the material, (iii) liquid and vapour moisture transfer to and from the material and (iv) moisture storage within the material. The possible directions of these fluxes

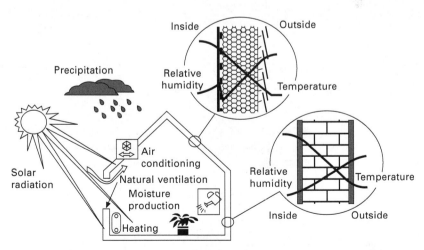

2.1 Hygrothermal loads affecting a typical building (Künzel *et al.*, 2005).

are illustrated in Fig. 2.2. Each hygrothermal flux does not act independently, but instead all are inter-dependent and occur simultaneously in a complex manner.

When a dry porous material is moved to a high(er) moisture content environment it may start to fill with water, assuming the porous structure is permeable, which can have the effect of dramatically increasing the thermal conductivity and thus the wall's thermal resistance will be reduced. This is partially due to the capillaries within the material filling with water, which, in the case of soils, leads to the formation of menisci at points of inter-particle contact (Hall and Allinson, 2009a). These menisci form a 'bridge' increasing inter-particle contact area and allowing greater heat flux within the material as depicted by Fig. 2.3. A further consideration must be the transport of heat by convection within the pores which, in a partially saturated material, includes either liquid or water vapour. Where it is possible for the water to evaporate from larger pores, transport in the form of vapour, and then condense at another location within the material the heat flux increases dramatically due to the latent energy of vaporisation–condensation. This is a complex topic that can be studied in further detail by referring to ISO 10456 (1999) and Hall and Allinson (2010a), or from dedicated texts on transport phenomena such as Bird *et al.* (2001).

Alongside the hygrothermal building loads, knowledge of the hygrothermal functional properties of all construction materials used in the building envelope is also required for any predictive modelling of its behaviour. These properties include:

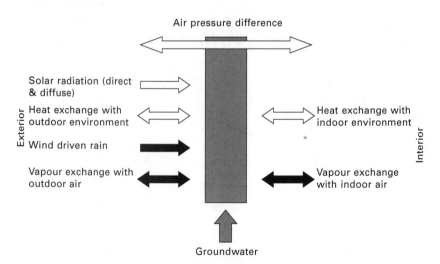

2.2 Hygrothermal fluxes and their alternating directions across an earth wall (adapted from ASHRAE, 2009).

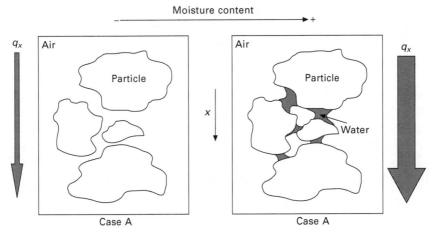

2.3 The influence of moisture content on the thermal conductivity of a partially saturated soil.

- Specific heat capacity (c_p) – the amount of energy required to raise one unit of mass by one Kelvin (J/kg K) and at constant pressure, which can also be converted to volumetric heat capacity by multiplying by the material bulk density
- Moisture content-dependent heat capacity (c_p^*) – the heat capacity of a porous material increases significantly with the degree of saturation, since the specific heat capacity of water is comparably high, and so is a function of moisture content for partially- or fully saturated materials
- Bulk porosity (n) – the ratio of the volume of pore space to the total volume of a material, where the ratio of dry mass to total volume is bulk density (ρ_d)
- Thermal conductivity (λ) – the coefficient for thermal conduction through matter due to a temperature gradient
- Moisture content-dependent thermal conductivity (λ^*) – thermal conductivity increases significantly with degree of saturation and so is a function of moisture content for partially or fully saturated materials
- Water vapour diffusion resistance factor (μ) – the ratio of the vapour diffusion resistance of air to the vapour diffusion resistance of the porous material
- Moisture storage function – the isothermal relationship between the equilibrium moisture content (EMC) of a porous material at each value of relative humidity from 0–100%, also referred to as the 'sorption isotherm'
- Water absorption coefficient (A_w) – the rate of water (mass) absorption against the square root of elapsed time due to capillarity in a porous material, also known as the sorptivity (S).

The development and significance of numerical modelling in building physics has been growing for many years. However, until the advent of fully transient hygrothermal models, modelling of building fabric traditionally considered heat transfer/storage and moisture transfer/storage in isolation, to some degree. Two previously accepted standard building physics models are the Glaser Model and the Admittance Model, although other more recent models also exist such as the Heat Balance Method (ASHRAE, 2009). These have been the basis of many computer-based modelling packages such as EnergyPlus (USDE, 2007), IES Virtual Environment and TAS (EDSL, 2010) amongst others. According to BS EN 15026: 2007 (BSi, 2007) all hygrothermal models must include the following transport and storage phenomena:

- heat storage in dry building materials and any absorbed water
- heat transport by moisture-dependent thermal conduction
- latent heat transfer by vapour diffusion
- moisture storage by vapour sorption and capillary forces
- moisture transport by vapour diffusion
- moisture transport by liquid transport (surface diffusion and capillary flow).

Some of the hygrothermal numerical models that have been developed are listed in Table 2.2. For further information refer to *Materials for Energy Efficiency and Thermal Comfort in Buildings* (Hall, 2010).

2.2.1   Calculation and results

The outputs from a hygrothermal numerical model can be categorised into two distinct sections:

1   temperature and heat flux distributions and their evolution
2   water content, RH and moisture flux distributions and their evolution.

The format of these results can be as numerical data for further input/analysis, or can be displayed graphically as a time-dependent simulation of the fluxes and distribution through a building envelope cross-section. Hygrothermal models allow the user to monitor many points throughout the envelope, including critical boundaries between differing materials. An example of this is the graphic output from the WUFI Pro v4.1 model shown in Fig. 2.4.

Interpretation of modelling results can be addressed in three stages:

1   The model requires validation to ensure all predicted results are accurate and consistent with physical modelling of the same/similar scenarios
2   Interpretation of heat and moisture levels on and within the building envelope
3   Post-processing of the results to predict corrosion, ageing, energy consumption/saving, susceptibility to rot and mould growth, etc.

Table 2.2 Source and application of existing hygrothermal numerical models in published literature

Model	Developer	Published use
1D HAM	Chalmers University, Sweden & MIT, Cambridge, United States	(Kalamees and Vinha, 2003)
BSim	Building Research Institute, Denmark	(Rode *et al.*, 2004)
Clim2000	EDF, France	(Woloszyn *et al.*, 2009)
DELPHIN	Bauklimatik, Dresden University	(Nofal *et al.*, 2001)
HAM-BE	Concordia University, Canada	(Li *et al.*, 2009)
HAM-FitPlus	British Columbia IT, Canada	(Yoshino *et al.*, 2009)
HAM-Tools	IBPT, Sweden & Denmark	(Woloszyn *et al.*, 2009)
IDA-ICE	EQUA, Sweden	(Woloszyn *et al.*, 2009)
LATENITE-VTT	VTT, Finland	(Simonson *et al.*, 2002) (Salonvaara *et al.*, 2004)
MATCH	Technical University of Denmark	(Kalamees and Vinha, 2003)
MATLAB	Mathworks Inc., United States	(van Schijndel, 2008)
TRANSYS (v16)	University of Wisconsin, United States	(Woloszyn *et al.*, 2009)
WUFI Group (Plus, Pro & 2D)	IBP Fraunhofer, Germany	(Allinson and Hall, 2010) (Charoenvai *et al.*, 2005) (Hall and Allinson, 2010) (Holm and Künzel, 2002) (Kalamees and Vinha, 2003) (Künzel and Kiessl, 1996) (Künzel *et al.*, 2005)
SIMULINK	Mathworks Inc., United States	(van Schijndel, 2008)

The results, once validated, can be used to predict moisture levels in the envelope over a specified time scale (months, years, etc.). Specific materials in the envelope will have service limits for temperature and moisture content, and using the results one can predict whether these limits are being exceeded. It is also possible to predict the drying out of initial moisture from construction and performance of specialised envelope components such as selective vapour barriers and insulation elements. Post-processing of the results can be beneficial in predicting the formation of mould in critical areas (e.g. cold bridges), rotting and premature ageing of the envelope materials due to increased moisture levels, and corrosion of ferrous envelope components (e.g. steel wall ties). These issues are discussed in further detail in *Moisture Control in Buildings* by Trechsel (1994).

2.2.2   Validation and applications

The WUFI group of hygrothermal models have been extensively validated against physical models and field work in many studies. Both the 1-D

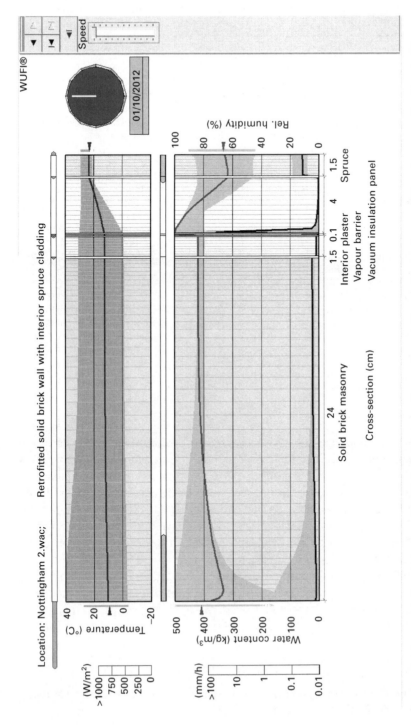

2.4 A graphical output from WUFI Pro v4.1 hygrothermal modelling software for a solid masonry wall with retrofit internal insulation based in Nottingham, UK, showing the temperature and moisture content profiles along with indicators of relative magnitude and direction of flux at the boundary between each material layer.

version (WUFI Pro v4.1) (Hall and Allinson, 2010b) and the whole building energetic model version (WUFI Plus v1.2) have been used for validation of SRE walls and buildings by Allinson and Hall (2010). The research showed that SRE has very good potential for passive indoor climatic control by significantly reducing the amplitude of RH fluctuations, when compared to standard building materials (i.e. unpainted and painted plaster board), as well as having high thermal mass comparable to that of dense concrete. In another study, the drying of an autoclaved aerated concrete (AAC) roof was successfully simulated by Holm and Künzel (2002) to determine whether an unvented flat AAC roof would perform correctly without a vapour retarder, in order to avoid the issue of construction moisture being prevented from drying, which could lead to a net loss in the insulation value of the roof. The model was also successfully validated against field experiments on natural stone facades by Künzel and Kiessl (1996).

WUFI 2-D was validated against experimental results for assessing the performance of three different wall fabrics with different insulation and vapour barriers by Kalamees and Vinha (2003). The study was conducted by simulating wall behaviour under ambient Nordic climatic conditions from autumn to spring using three different models (1D-HAM, MATCH and WUFI 2D) and then validating these results against laboratory experiments. The study concluded that, whilst the experimental results validated all three of the models in general, some inaccuracies were found due to simplifications, limitations and/or approximations of the material properties in the model and also the limiting accuracy of the experimental measurements themselves.

The full energetic building (3D) version of the programme, WUFI Plus, has been accurately validated against field results for an unoccupied SRE test building that was monitored for a continuous period of 10 months by Allinson and Hall (2010). Künzel et al. (2005) validated the model against the humidity buffering performance of wood-lined rooms. Testing has also been performed in two custom-built test rooms at the IBP Fraunhofer facility in Germany (Künzel et al., 2004). One room was constructed using aluminium foil as the interior lining, leaving the room with negligible vapour permeability and hence sorption capacity, while the other room was lined with plasterboard to simulate a standard room. The addition of wood linings to the aluminium room allowed for assessment of the humidity buffering performance of the wood, and the results from the model were in close agreement with those from the experiments.

2.3   Thermal and hygric properties of earth materials

The material functional properties required as inputs for hygrothermal modelling are described in section 2.1. Experimental measurement of these properties is more involved than for steady-state properties as the majority are functions as opposed to single value coefficients. The thermal conductivity, for example, cannot be specified as a single value rather as a function of moisture content and so has a range of values from when the material is completely dry to when it is completely saturated and anywhere in between. The difference in these functional properties between one type of material and another is often significant. Before their hygrothermal behaviour can be accurately studied, these functional properties have to be experimentally determined. In the case of rammed earth, the specific heat capacity of the soil can be determined across a specified temperature range using Differential Scanning Calorimetry (DSC). However, this technique requires small samples of the order of 25 mg for testing and the soil is composed of a range of mineral types. A more common approach for non-homogeneous materials such as soil is to measure their bulk heat capacity by determining the dominant mineral type(s) and then multiplying the specific heat capacity for that mineral by the bulk density of the material ($\rho\,C_p$), which gives the volumetric heat capacity. If the 'soil' is a blend of more than one soil/mineral type, and contains cement or another stabiliser, then the bulk heat capacity can be calculated in terms of the mass proportion of each phase. This approach can be further extended to include the relative water content, w, so that the moisture content-dependent heat capacity can be calculated using (Hall and Allinson, 2008):

$$c_p^* = \frac{\rho_{dry}}{\rho_T}\left(\varpi_{sand}c_{sand} + \varpi_{gravel}c_{gravel} + \varpi_{clay}c_{clay} + \varpi_{cem}c_{cem} + wc_w\right)$$

[2.1]

The thermal conductivity of earth materials can be experimentally determined using the heat flow meter method, either in the dry state as described in ISO 8301 (1991), or in the partially or fully saturated state in accordance with ISO 10051 (1996). The degree of saturation, S, in soils is defined in Chapter 8 (Augarde) and ranges from 0 when dry up to 1 when fully saturated. Thermal conductivity in soils, concrete and similar materials are often observed to increase linearly with S. The slope of this line can obviously be assessed by linear regression and gives a value called the moisture factor, m_f, which is a measure of the material's sensitivity to moisture content in terms of its thermal conductivity. The moisture content-dependent thermal conductivity can therefore be determined using $\lambda^* = \lambda(1 + S \times m_f)$. The vapour permeability of earth materials can be assessed using the wet cup/dry cup method as detailed in BS EN ISO 12572:2001 'Hygrothermal performance of building materials and products – determination of water vapour transmission properties' (BSi, 2001). In this case the motivating potential for vapour to transport through the pore structure of the material is the partial vapour pressure (relative humidity) differential. Liquid water transport is commonly assessed by determining the sorptivity (also known as the water absorption coefficient) by the gravimetric partial immersion method as described in Hall *et al.* (1984). This applies to partially saturated materials where the motivating potential for water absorption (and transport within the material) is capillary potential, sometimes referred to as 'suction'. In the case of a fully saturated specimen the capillary potential is zero and hence the liquid will only transport within the pore network if there is an applied pressure differential. In this case the hydraulic conductivity is assessed with reference to Darcy's Law. Clearly, the geometry of the pore network inside the earth materials determines the hygrothermal functional properties as illustrated in Fig. 2.5:

A range of hygrothermal functional properties for a limited number of stabilised rammed earth materials have been published by Hall and Allinson (2009b), as summarised in Table 2.3.

An important point is that the hygroscopic moisture storage function (ξ) refers to the gradient of the middle portion of the sorption or desorption isotherm. The actual input for a hygrothermal model is the entire sorption isotherm, which represents the equilibrium moisture content of the material at every point between 0 and 100% relative humidity. Further details of the heat and mass transport and storage process discussed in this section can be found in Hall and Allinson (2010a), while further details of the experimental data presented above and the test apparatus can be found in Hall and Allinson (2008). Significantly, the hygrothermal properties of SRE materials are thought to be 'tunable' through careful modification of the internal pore structure, either by changing the particle size distribution of the soil, or through addition of clay minerals.

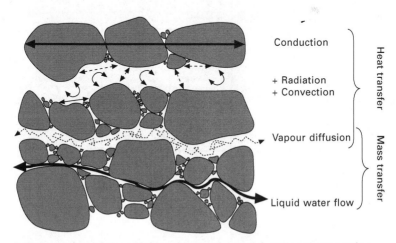

2.5 Illustration of the origins of hygrothermal functional properties within the micro-structure of rammed earth (Hall and Allinson, 2009b).

Table 2.3 Hygrothermal functional properties for three contrasting stabilised rammed earth materials (Hall and Allinson, 2009)

Rammed earth mix type	Thermal properties							Hygric properties			
	ρ_{dry}	n	C	λ	m_f	λ*		S	W	ξ_a	ξ_d
						$S_r = 0$	$S_r = 1$				
	kg/m³	–	MJ/m³ K	W/m K	–	W/m K		mm/min$^{0.5}$	kg/m²s Pa	kg/kg	kg/kg
433	2120[a]	0.239[a]	1.754	1.010[a]	0.802	1.010	1.820	1.487[b]	1.56E^{-10}	23.20	31.56
613	2020[a]	0.273[a]	1.728	0.833[a]	0.643	0.833	1.369	2.117[b]	3.23E^{-10}	28.71	19.19
703	1980[a]	0.302[a]	1.719	0.866[a]	0.955	0.866	1.693	2.700[b]	4.79E^{-10}	13.93	21.30

2.4   Hygrothermal behaviour and passive air conditioning

At the outside surface interface, rammed earth walls exchange heat by convection and radiation, and moisture by vapour exchange or by absorbing rain water. In a similar fashion, the inside wall surfaces interact with the indoor environment by exchanging both heat and water vapour and, as a result, can greatly influence the indoor air temperature and relative humidity. Obviously the hygrothermal functional properties of the earth wall material determine its behaviour under changing hygrothermal loads. Some earth materials have higher density and/or volumetric heat capacity and so are able to absorb, store and release larger quantities of heat energy, which, depending upon the rate of thermal diffusion can help to buffer variation of indoor air temperatures. In a similar way, some earth materials are able to absorb, store and release different amounts of water vapour, and depending on the quantity and rate of diffusion, can be used to buffer indoor relative humidity variation. In general terms, published data suggest that SRE materials can be identified as having both high thermal mass and a high hygric mass so can be used for both air temperature and relative humidity buffering. Unlike brick, timber or plastered walls, due to their natural aesthetics, rammed earth walls are almost always presented without any internal rendering, paint or wallpaper and so are in direct contact with the indoor air.

The passive temperature and humidity-buffering behaviour of rammed earth materials have been monitored, and to some extent exploited, in several modern earth buildings before. Few published studies have been conducted to determine which micro-structural material properties control this behaviour, or indeed how this understanding could be further utilised to optimise the material behaviour in a selective manner. Some case studies have been used to both measure and demonstrate the increased energy efficiency that can be gained when SRE walls are used for temperature buffering in domestic buildings. An electronically monitored SRE house in Sydney, Australia, for example, was left unoccupied and with no curtains for over a year and yet internal sensor readings recorded over that period showed an average annual indoor temperature range of between 18°C and 27°C, compared with an outdoor temperature range of between 7°C and 42°C (Mortensen, 2000). Similarly, a two-storey SRE office building in New South Wales, Australia, was monitored and displayed excellent thermal performance throughout a hot summer period due to peak temperature reduction by passive cooling and also dehumidification (Taylor and Luther, 2004). As a, result the amount of energy required to meet the heating and cooling loads in each of these buildings was significantly lower than that of conventional buildings. Ironically, under the CSIRO energy efficiency rating scheme, these buildings were incapable of attaining a high score because the steady state thermal resistance (R, m^2 K/W) of the SRE walls was too low. However, this factor is simply a measure of the wall's resistance to heat conduction.

An SRE test building, located in Leicestershire, UK, was constructed by Earth Structures (Europe) Ltd using the Australian-style formwork and pneumatic compaction techniques described in Chapter 22 (Lindsay). The air temperature and relative humidity was monitored in one of the rooms at 30-minute intervals for the period 4 July 2008 to 1 April 2009 as part of a published study conducted by Dr Hall and Dr Allinson at the University of Nottingham. The room measured 3.2 × 2.5 m internally and the SRE walls were of the insulated cavity design comprising inner and outer leaves 175 mm thick plus internal wall ties and 50 mm extruded polystyrene cavity insulation. Further details of the material properties and instrumentation used for monitoring can be obtained from Allinson and Hall (2010). Detailed local weather data were also obtained over the monitoring period. The energetic hygrothermal building energy modelling package WUFI Plus v1.2 was used to predict the indoor air temperature and humidity for the SRE test room over the same period. This was validated to a high degree of accuracy against the data obtained from monitoring. This enabled the model to be expanded to include, for example, different hypothetical wall materials and occupancy patterns to compare and contrast the behaviour of the SRE walls with conventional materials, and then assess what effect they had on passively buffering air temperature and relative humidity. The example chosen for simulation and analysis was a bedroom occupied by two adults between 22:00 and 06:00 each day. The data showed that the SRE walls significantly reduced the amplitude of RH fluctuations, during both summer and winter periods, when compared to walls that were covered with conventional materials, e.g. painted plaster board and aluminium foil vapour barriers (as commonly applied beneath some plasterboard coatings). This is illustrated by Fig. 2.6, which shows the simulated RH variation across a typical week in summer. These results were obtained assuming that the room was completely unconditioned by mechanical heating, ventilation or air conditioning (HVAC) and so may not always provide adequate thermal comfort, for example on very hot or very cold days, where an additional 'top up' by mechanical systems may be required.

The results of the model were further analysed taking into account mixed mode operation comprising the 'active' temperature and humidity regulation by a mechanical HVAC system in the room, and also the 'passive' temperature and humidity regulation of the SRE walls. The mixed mode operation was considered both for continuous operation in parallel with the passive system, and in intermittent operation where the mechanical HVAC operation was governed by thermostat and humidistat set point values. These set points can be varied and provide precise control over the indoor thermal

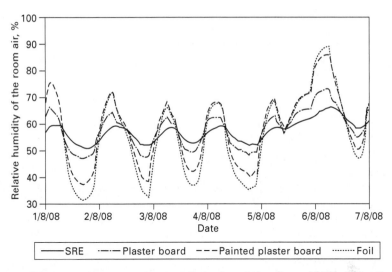

2.6 The simulation variation in relative humidity, due to different wall materials, in a bedroom over a typical week in summer (Allinson and Hall, 2010).

comfort. In this study they were set to an air temperature of between 18 and 20°C, and between 40 and 50% RH, which represents a very closely controlled indoor climate. Adequate thermal comfort may be achieved with much wider parameters in the set points, which obviously reduces overall energy consumption in its own right, for example 18–23°C and 40–70% RH. The intention, however, is that the 'energy free' passive system should be used in preference to minimise the amount of energy needed to operate the mechanical system. Further details of the modelling parameters can be obtained from Allinson and Hall (2010). The results showed that the SRE walls were very responsive to periods of high internal humidity loads and responded by absorbing moisture vapour for later release when the humidity decreased. When used in conjunction with constant mode mechanical HVAC, there was a significant reduction in the energy required for humidification and dehumidification energy demand when compared to the conventional wall materials, as shown by the data in Table 2.4.

The heating and cooling energy loads were not significantly different, as predicted, since the test room had a very simple geometry with minimal glazing and therefore could not capitalise on the benefits of passive solar design discussed in Chapter 5 (Rongen). Further studies would be required to quantify this behaviour. When the intermittent mode HVAC was simulated, there was still an overall reduction in humidification and dehumidification energy use. However, the reductions for the SRE wall were less than that for the impermeable case because the SRE absorbed significant amounts of

Table 2.4 Predicted daily HVAC energy consumption for constant mixed mode operation with SRE and conventional wall materials (Allinson and Hall, 2010

HVAC operated all day			Plasterboard		
		SRE	Unpainted	Painted	Foil
Heating power	kW	1313.7	1313.7	1313.8	1314.1
Cooling power	kW	−135.0	−135.0	−135.0	−135.0
Latent heat humidification	kW	−1.9	−7.0	−26.2	−35.0
Latent heat dehumidification	kW	41.2	58.7	86.2	89.5
Total humidifying and dehumidifying	kW	43.1	65.7	112.4	124.5
Total energy	kW	1491.8	1514.4	1561.2	1573.5
Saving (cooling)	%	0.0	0.0	0.0	0.0
Saving (humidification)	%	94.5	79.9	25.3	0.0
Saving (dehumidification)	%	54.0	34.4	3.6	0.0
Saving (humidifying and dehumidifying)	%	65.4	47.2	9.7	0.0
Saving (heating)	%	0.0	0.0	0.0	0.0
Saving (total)	%	5.2	3.8	0.8	0.0

Table 2.5 Predicted daily HVAC energy consumption for intermittent mixed mode operation with SRE and conventional wall materials (Allinson and Hall, 2010)

HVAC operated between			Plasterboard		
		SRE	Unpainted	Painted	Foil
0600–0800 and 1600–2200					
Heating power	kW	1060.6	1060.8	1061.0	1061.2
Cooling power	kW	−89.9	−90.0	−90.0	−90.0
Latent heat humidification	kW	−1.3	−4.6	−16.5	−22.3
Latent heat dehumidification	kW	23.3	14.0	19.0	22.0
Total humidifying and dehumidifying	kW	24.6	18.6	35.5	44.3
Total energy	kW	1175.2	1169.4	1186.5	1195.4
Saving (cooling)	%	0.1	0.0	0.0	0.0
Saving (humidification)	%	94.1	79.3	25.8	0.0
Saving (dehumidification)	%	−5.8	36.3	13.7	0.0
Saving (humidifying and dehumidifying)	%	44.4	57.9	19.8	0.0
Saving (heating)	%	0.1	0.0	0.0	0.0
Saving (total)	%	1.7	2.2	0.7	0.0

water vapour while the HVAC system was inoperative. This is highlighted by the data shown in Table 2.5.

When the HVAC became operative it attempted to remove this excess moisture from the wall causing a slight increase in energy use. This highlights the need for a detailed understanding of building physics at the design stage. The potential exists to make significant energy savings and increased thermal comfort using SRE walls over conventional materials, but if used inappropriately in conjunction with mechanical HVAC systems can actually have the reverse effect.

2.5   Indoor health and air quality

The indoor air quality is the most significant thing to affect occupants' thermal comfort and it is widely believed to cause a variety of issues. Physiologically, poor air quality decreases work productivity especially under a high-humidity environment (Tsutsumi *et al.*, 2007) whilst economically it can lead to significant energy waste (Osanyintola and Simonson, 2006). The study of 'thermal comfort' is often focused on the air temperature range, whilst the consideration of RH is often neglected. As previously discussed, the enthalpy (latent energy) associated with high RH in air cannot be overlooked. It has been suggested that a range of between 30 and 80% RH may not produce significant problems for human health, but that a range of 30 to 60% RH and an air temperature range of 18 to 24°C is preferable (Crump *et al.*, 2002). Another study on the effect of humidity on human comfort in office spaces in a hot/humid climate suggests that the appropriate range may be 40 to 70% RH (Tsutsumi *et al.*, 2007). A hot–humid environment can cause sweating, whilst a low-humidity environment can cause dryness of the eyes and skin (Crump *et al.*, 2002). Hines *et al.* (1993) concluded that the most advisable humidity range for thermal comfort and occupant health in buildings is between 40 and 60%.

In terms of the currently available data, the DTi-sponsored study of rammed earth for UK housing, led by the University of Bath, alluded to the importance of thermal mass for increasing energy efficiency and thermal comfort, but only provided generic values for the thermal conductivity and specific heat capacity of a limited range of earth materials (Walker *et al.*, 2005). The product information provided by the Affiliated Stabilised Earth Group (ASEG, 2003) of SRE contractors specifies an estimated thermal time lag of 6–8 hours in a 300 mm-thick wall. It is evident that materials which can rapidly absorb significant quantities of moisture vapour can buffer RH changes (i.e. hygric mass) in the same way that materials which absorb large quantities of heat can buffer air temperature changes (i.e. thermal mass). Covering these materials with low vapour permeability layers (e.g. paints) dramatically reduces the moisture buffering behaviour (Kalamees *et al.*, 2009). The interrelation between indoor air quality and RH should be considered as having a direct influence on occupants' comfort and health. Wolkoff and Kjærgaard (2007) state that interior spaces with low humidity can result in some clinical symptoms such as dry eyes and sensory irritation, whilst the occurrence of mould growth can increase when surfaces are in contact with air that is very humid. Previous research has shown that SRE materials can reduce the frequency of high humidity periods at the wall surface and were therefore judged to be highly beneficial in reducing mould growth in buildings (Allinson and Hall, 2010). SRE materials are good moisture-buffering materials and are normally left uncovered inside buildings, therefore ideally suited to passive humidity control. It has been concluded from previous research that SRE walls have the potential to improve thermal comfort, improve indoor air quality and reduce the energy demand in buildings, however care should be taken if used in conjunction with mechanical HVAC system design including the conditioning strategy and ventilation rate.

2.5.1   Measuring indoor health and air quality

There are six basic parameters that define an occupant's comfort environment (Parsons, 2003):

1   air temperature
2   mean radiant temperature
3   relative humidity
4   air velocity
5   metabolic heat generation
6   clothing.

The first four are referred to as the 'environmental factors', while the final two are 'personal factors'. The four environmental factors can be measured directly. In reference to the personal factors, however, the human body is constantly subjected to the influences of the environmental ones. In an attempt to regulate itself at a constant temperature, it uses counter processes to act against these influences, e.g. evaporation of sweat for skin surface cooling, respiratory evaporation, conduction, convection via the blood, radiation and metabolic storage. This constant loop process is called thermoregulation, and clothing and level of activity also contribute to this. The thermoregulation loop is essentially a heat-balance loop and can therefore be represented by a heat-balance equation (ASHRAE, 2009). In his text Fanger also proposed a similar equation (Fanger, 1972):

$$M - W = Q_{sk} + Q_{res} = (C + R + E_{sk}) + (C_{res} + E_{res}) \qquad [2.2]$$

While it is possible to easily measure and monitor the environmental and personal factors contributing towards thermal comfort, it is less simple to measure the physiological phenomena. As thermal comfort is subjective, it follows that the most efficient method of measuring the physiological elements of thermal comfort is to ask that person. These subjective techniques fall within two categories: personal and environmental, i.e. how the person feels (hot/cold) and what is their acceptance and/or tolerance of the thermal environment. In both techniques, scales are used to give a qualitative indication of the person's response. The Predicted Mean Vote (PMV) and Percentage of People Dissatisfied (PPD) are primarily used to measure thermal comfort. PMV and PPD can be calculated using the standard BS EN ISO 7730 (BSi, 2005), while subjective judgement scales can be found in BS EN ISO 10551 (BSi, 2001).

Pollution of the indoor air can also affect occupants' comfort due to poor air quality. The causes of this pollution will generally stem from one of (or a combination of) the following factors (Maroni *et al.*, 1995):

- poorly maintained environments
- poor ventilation or air exchange
- biological contamination from poor moisture control strategies
- internal pollutant generation
- external pollution ingress.

This means that the quality of indoor air can be affected by sources within the building and from sources outside through air infiltration into the space. Typical pollutants found in the indoor environment can be seen in Table 2.6.

With respect to the RH, the indoor air quality (IAQ) and the biological problems associated with it are summarised in Figure 2.7. The 'optimum zone' is the acceptable range of RH that will minimise pollution within the indoor air and improve the IAQ, quantified as 30–55% RH. Measuring

Table 2.6 Types and sources of indoor pollutants (adapted from Maroni *et al.*, 1995)

Type	Name	Source
Organic	Volatile organic compounds (VOC's)	Cleaning agents, cosmetics, furnishings, tobacco
	Pesticides	Agricultural spraying
	Formaldehyde	Tobacco, lacquers, varnish
Inorganic	Carbon dioxide (CO_2)	Breathing, combustion
	Carbon monoxide (CO)	Combustion, tobacco
	Nitrogen dioxide (NO_2)	Combustion, tobacco
Physical	Particulate matter (PM)	Combustion, biological, tobacco, aerosols
	Man made mineral fibre (MMMF)	Furnishings, insulation
Biological	Dust mites	Bedding, furnishings
	Allergens	Pets, vermin, pollen
	Bacteria	HVAC, humans, animals, dust aerosols
	Fungi/mould	Damp organic matter

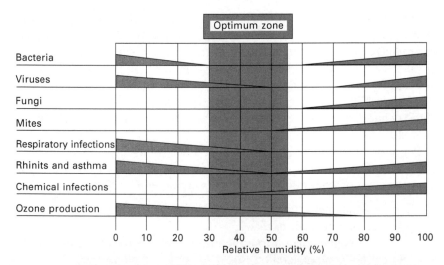

2.7 The humidity-related effects on health and indoor air quality (Simonson *et al.*, 2001).

the factors affecting IAQ involves similar measures to those employed in analysing thermal comfort, i.e. temperatures, RH levels, air exchange rates, pollutant concentrations, occupant behaviour and perception, but may also involve medical examinations to determine health issues.

Guidelines for IAQ measurement can be found in the international standard BS ISO 16814:2008 (BSi, 2008). The perceived air quality (PAQ) method of expressing IAQ uses subjective scales derived from questionnaires similar to those used in thermal comfort surveys, i.e. PPD. According to the standard, one of two methods can be used to express IAQ:

• in relation to the health risk to the occupants
• in relation to the acceptability, based on their PAQ.

2.6   Sources of further information

The following professional organisations offer membership, annual conference events and workshops, and relevant publications including official guidance documents:

• The Chartered Institute of Building Services Engineers (CIBSE), 222 Balham High Road, London, SW12 9BS. Tel +44 (0)20 8675 5211, Fax +44 (0)20 8675 5449
• American Society of Heating, Refrigeration and Air Conditioning Engineers (ASHRAE), 1791 Tullie Circle, N.E., Atlanta, GA 30329, USA.

An excellent source of further reference material is international peer-reviewed journals, which publish high-quality scholarly articles of direct relevance to the topics discussed in this chapter. Strongly recommended journals for the topics covered in this chapter include *Building and Environment*, *Energy and Buildings*, *Applied Thermal Engineering*, *Construction and Building Materials*, *Cement and Concrete Composites*, *Journal of Building Physics*, The American Society of Civil Engineers (ASCE) family of journals and the Proceedings of the Institute of Civil Engineers (ICE) family of journals.

2.7   References

Allinson D and Hall M, 2010, 'Hygrothermal analysis of a stabilised rammed earth test building in the UK', *Energy and Buildings*, 42, pp. 845–852

ASEG, 2003, *New Earth Structures* [brochure], available from Affiliated Stabilised Earth Group, PO Box 161, North Fremantle, 6159, Australia

ASHRAE, 2009, *ASHRAE Handbook – Fundamentals (SI Edition)*, American Society of Heating, Refrigerating and Air-Conditioning Engineers, Atlanta, Ga., USA

Bird R B, Stewart W E, and Lightfoot E N, 2001, *Transport Phenomena – 2nd Edition*, Wiley, Chichester

BSi, 2001, 'BS EN ISO 10551:2001: Ergonomics of the thermal environment – Assessment of the influence of the thermal environment using subjective judgement scales', British Standards Institute, London

BS EN ISO 12572, 2001, 'Hygrothermal performance of building materials and products – Determination of water vapour transmission properties', British Standards Institute, London

BSi, 2005, 'BS EN ISO 7730:2005: Ergonomics of the thermal environment – Analytical determination and interpretation of thermal comfort using calculation of the PMV and PPD indices and local thermal comfort criteria', British Standards Institute, London

BSi, 2007, 'BS EN 15026 2007: Hygrothermal performance of building components and building elements – Assessment of moisture transfer by numerical simulation', British Standards Institute, London

BSi, 2008, 'BS EN ISO 16814: 2008: Building environment design – Indoor air quality – Methods of expressing the quality of indoor air for human occupancy', British Standards Institute, London

Charoenvai S, Khedari J, Hirunlabh J, Asasutjarit C, Zeghmati B, Quenard D and Pratintong N, 2005, 'Heat and moisture transport in durian fiber based lightweight construction materials', *Solar Energy*, 78, pp. 543–553

Crump D, Raw G, Upton S, Scivyer C, Hunter C and Hartless R, 2002, 'A protocol for the assessment of indoor air quality in homes and office buildings'. BRE Report BR 450, BRE Bookshop, Watford

EDSL, 2010, EDSL TAS industry-leading building modelling and simulation, available at www.edsl.net/main/Software.aspx (accessed 14 July 2010)

Fanger P O, 1972, *Thermal Comfort*, McGraw-Hill, New York

Hall M R (ed.), 2010, *Materials for Energy Efficiency and Thermal Comfort in Buildings*, Woodhead Publishing, Cambridge

Hall M and Allinson D, 2008, 'Assessing the moisture-content dependent parameters of stabilised earth materials using the cyclic-response admittance method', *Energy and Buildings*, 40 11, pp. 2044–2051

Hall M R and Allinson D, 2009a, 'Assessing the effects of soil grading on the moisture content-dependent thermal conductivity of stabilised rammed earth materials', *Applied Thermal Engineering*, 29, pp. 740–747

Hall M and Allinson D, 2009b, 'Analysis of the hygrothermal functional properties of stabilised rammed earth materials', *Building and Environment*, 44 9, pp. 1935–1942

Hall M R and Allinson D, 2010a, 'Heat and mass transport processes in building materials', in Hall M R (ed.), *Materials for Energy Efficiency and Thermal Comfort in Buildings*, Woodhead Publishing, Cambridge

Hall M R and Allinson D, 2010b, 'Transient numerical and physical modelling of temperature profile evolution in stabilised rammed earth walls', *Applied Thermal Engineering*, 30, pp. 433–441

Hall C Hoff W D, and Nixon M R, 1984, 'Water movement in porous building materials – VI: Evaporation and drying in brick and block materials', *Building and Environment*, 19, 1, pp. 13–20

Hines A L, Ghosh T K, Loyalka S K and Warder R C (eds), 1993, *Indoor Air-Quality and Control*, Prentice-Hall, Englewood Cliffs, NJ, USA

Holm A H and Künzel H M, 2002, 'Practical application of an uncertainty approach for hygrothermal building simulations-drying of an AAC flat roof', *Building and Environment*, 37, pp. 883–889

ISO 8301, 1991, 'Thermal insulation – Determination of steady-state thermal resistance and related properties – Heat flow meter apparatus', International Organization for Standardization, Genève, Switzerland

ISO 10051, 1996, 'Thermal insulation – Moisture effects on heat transfer – Determination of thermal transmissivity of a moist material', International Organization for Standardization, Genève, Switzerland

ISO 10456, 1999, 'Building materials and products: Procedures for determining declared and design thermal values', International Organization for Standardization, Genève, Switzerland

Kalamees T, Korpi M, Vinha J and Kurnitski J, 2009, 'The effects of ventilation systems and building fabric on the stability of indoor temperature and humidity in Finnish detached houses', *Building and Environment*, 44, 8, pp. 1643–1650

Kalamees T and Vinha J, 2003, 'Hygrothermal calculations and laboratory tests on timber-framed wall structures', *Building and Environment*, 38, pp. 689–697

Künzel H M and Kiessl K, 1996, 'Calculation of heat and moisture transfer in exposed building components', *International Journal of Heat and Mass Transfer*, 40, pp. 159–167

Künzel H M, Holm A, Sedlbauer K, Antretter F and Ellinger M, 2004, 'Moisture buffering effects of interior linings made from wood or wood based products: IBP Report HTB-04/2004/e', Technical investigation commissioned by Food Focus Oy and the German Federal Ministry of Economics and Labour, IBP, Institutsteil Holzkirchen

Künzel H M, Holm A, Zirkelbach D and Karagiozis A N, 2005, 'Simulation of indoor temperature and humidity conditions including hygrothermal interactions with the building envelope', *Solar Energy*, 78, pp. 554–561

Li Q, Rao J and Fazio P, 2009, 'Development of HAM tool for building envelope analysis', *Building and Environment*, 44, pp. 1065–1073

Maroni M, Seifert B and Lindvall T, 1995, *Indoor Air Quality: A Comprehensive Reference Book*, Elsevier, Amsterdam

Mortensen N, 2000, 'The naturally air conditioned house', [online] Earth Building Research

Forum, University of Technology Sydney, Australia, available at www.dab.uts.edu.au/ebi/index (last accessed 25 March 2004)

Nofal M, Straver M and Kumaran M K, 2001, 'Comparison of four hygrothermal models in terms of long-term performance assessment of wood-frame constructions', *Eighth Conference on Building Science & Technology, Solutions to Moisture Problems in Building Enclosures*, Toronto, Canada, pp. 118–138

Osanyintola O F and Simonson C J, 2006, 'Moisture buffering of hygroscopic building materials: Experimental facilities and energy impact', *Energy and Buildings*, 38, pp. 1270–1282

Parsons K C, 2003, *Human Thermal Environments: The Effects of Hot, Moderate, and Cold Environments on Human Health, Comfort, and Performance*, Taylor & Francis, London

Rode C, Mendes N and Karl G, 2004, 'Evaluation of moisture buffer effects by performing whole-building simulations', *ASHRAE Transactions*, 110, pp. 783–794

Salonvaara M, Ojanen T, Holm T, Kunzel H and Karagiozis A, 2004, 'Moisture buffering effects on indoor air quality – experimental and simulation results', *Proceedings (CD) of the Performance of Exterior Envelopes of Whole Buildings IX International Conference*, Florida, USA

Simonson C J, Ojanen T and Salonvaara M, 2001, *Improving Indoor Climate and Comfort with Wooden Structures*, Technical Research Centre of Finland, Espoo, Finland

Simonson C J, Salonvaara M and Ojanen T, 2002, 'The effect of structures on indoor humidity – possibility to improve comfort and perceived air quality', *Indoor Air*, 12, pp. 243–251

Standards New Zealand, 1998, *NZS 4298: 1998 Materials and Workmanship for Earth Buildings*, Wellington, New Zealand

Taylor P and Luther M B, 2004, 'Evaluating rammed earth walls: a case study', *Solar Energy*, 76 pp. 79–84

Trechsel H R, 1994, *Moisture Control in Buildings*, Astm Intl, Philadelphia, USA

Tsutsumi H, Tanabe S, Harigaya J, Iguchi Y and Nakamura G, 2007, 'Effect of humidity on human comfort and productivity after step changes from warm and humid environment', *Building and Environment*, 42, 12, pp. 4034–4042

USDE, 2007, '*US* Department of Energy, EnergyPlus 2.0 – Major release – New functionality', PDF document available at: http://apps1.eere.energy.gov/buildings/energyplus/pdfs/energyplus_v2_brochure.pdf (accessed 14 July 2010)

Van Schijndel A W M, 2008, 'Estimating values for the moisture source load and buffering capacities from indoor climate measurements', *Journal of Building Physics*, 31, pp. 319–331

Walker P, Keable R, Martin J and Maniatidis V, 2005, *Rammed Earth: Design and Construction Guidelines*, BRE Bookshop, Watford

Wolkoff P and Kjærgaard S K, 2007, 'The dichotomy of relative humidity on indoor air quality', *Environment International*, 33, 6, pp. 850–857

Woloszyn M, Kalamees T, Abadie M, Steeman M and Kalagasidis A, 2009, 'The effect of combining a relative-humidity-sensitive ventilation system with the moisture-buffering capacity of materials on indoor climate and energy efficiency of buildings', *Building and Environment*, 44, pp. 515–524

Yoshino H, Mitamura T and Hasegawa K, 2009, 'Moisture buffering and effect of ventilation rate and volume rate of hygrothermal materials in a single room under steady state exterior conditions', *Building and Environment*, 44, pp. 1418–1425

2.8   Appendix: nomenclature

A	area	m^2
A_w	water absorption coefficient	$kg/m^2 \, s^{1/2}$
b_m	moisture effusivity	$kg/(m^2 \, Pa \, s^{1/2})$
c_p	constant pressure specific heat capacity	J/kg K
E_{res}	rate of evaporative heat loss from respiration	W/m^2
E_{sk}	rate of evaporative heat loss from skin	W/m^2
H	enthalpy	J
M	rate of metabolic heat production	W/m^2
m	mass	kg
n	bulk porosity	m^3/m^3
P	fluid pressure	Pa
p_{sat}	saturation vapour pressure	Pa
Q_{sk}	total rate of heat loss from skin	W/m^2
Q_{res}	total rate of heat loss through respiration	W/m^2
R	thermal resistance	$m^2 \, K/W$
RH	relative humidity	%
t_p	period	s
U	internal energy	J
V	volume	m^3
W	rate of mechanical work accomplished	W/m^2
w_m	moisture content	kg/kg
α	thermal diffusivity	m^2/s
β	thermal effusivity	$W \, s^{1/2}/m^2 \, K$
δ_p	water vapour permeability	kg/m s Pa
ρ	density	kg/m^3
ρ_d	dry density	kg/m^3
μ	water vapor diffusion resistance factor	–
λ	dry state thermal conductivity	W/m K
φ	relative humidity (decimal)	–

3

Fabric insulation, thermal bridging and acoustics in modern earth buildings

C. J. HOPFE, Cardiff University, UK and M. R. HALL,
University of Nottingham, UK

Abstract: Buildings account for nearly 40% of the energy consumption in the UK, producing almost half of the CO_2 emissions at a national level. Using appropriate insulation is a cost-effective way to substantially improve the energy efficiency of any building and reduce the overall contribution to global CO_2 emissions. This chapter focuses on cavity insulation with an overview of insulation material types and physical properties. Examples of cross-sectional construction detailing for typical stabilised rammed earth (SRE) cavity walls are given. It also provides an insight into established research of 2D and 3D modelling of thermal bridges, and the relevance of acoustic properties in relation to thermal mass. Finally, the more widely available software products for both acoustic and thermal bridge simulation are evaluated and summarised.

Key words: hygrothermal and acoustic behaviour, stabilised rammed earth, properties insulation materials, software tools

3.1 Introduction

In a global market where future energy may be substantially more expensive, guidance for the Architecture, Engineering and Construction industry (AEC) will be increasingly important both for refurbishment and new build. Appropriate insulation for instance will significantly decrease the energy demand of a building and will thereby play a central role in reducing global CO_2 emissions. This chapter summarises different approaches to the fabric insulation of modern earth buildings by addressing the different issues one has to face in terms of acoustic and thermal behaviour. Acoustic isolation of a building is very important in terms of achieving occupant comfort, in addition to the temperature, humidity and air quality. Research addressing thermal bridges is also particularly relevant in this context, as these account for a significant proportion of heat loss in buildings; providing the potential to dramatically decrease the energy demand.

Current environmental issues call for urgent changes to the way in which buildings are thermally designed. At present, commercial, residential and industrial buildings use nearly 40% of the energy in the UK, producing almost half of the CO_2 emissions at a national scale; 55–60% of this energy is used

41

for heating and cooling of the premises; the rest is consumed by electrical appliances, lighting and other uses (Department of Trade and Industry, 2001). If change is to be undertaken, a reasonable focus should lie on the thermal insulation and reduction of unwanted thermal losses and gains through the outer envelope of buildings. Proper thermal design is an essential step to ensure significant improvements in a building's overall energy performance. Using appropriate insulation is one of the most straightforward options for substantially improving the operational energy efficiency of any building and thereby reducing its contribution to global CO_2 emissions. Targeting near-zero heat losses can dramatically decrease the energy demand and contribute to a greener and more sustainable environment, whilst simultaneously improving the thermal comfort of the occupants. Although material selection is important, achieving the correct level of construction detailing is perhaps the most essential factor during the design and installation of building fabric insulation. Without careful consideration of the positioning, thickness and continuity of insulation a building can inherit one or more performance defects such as interstitial condensation (i.e. within the wall fabric), thermal bridging and isolation of thermal and/or hygric mass. These can impact on both the operational energy efficiency and also the health and comfort of the occupants. A responsible approach in planning new buildings might be the key to guaranteeing the future of upcoming generations (Hipworth, 2011).

3.2   Approaches to fabric insulation

Conventionally, insulation is placed in one of three different locations: (i) internally towards the building's interior; (ii) externally, on the outside of the building shell; or (iii) in between the building's outer and inner layer known as cavity insulation. Less commonly, insulation may be located in multiple layers throughout a construction, or comprise the entirety of the construction, however these rarely used solutions will not be discussed here as they do not apply to earth wall construction techniques. Obviously, there are a number of advantages and drawbacks associated with each of the insulation techniques. For example, in the context of heritage buildings where one has to protect the outer appearance of the building, most probably internal insulation will provide the only acceptable solution to planning authorities.

Regardless of whether the building is insulated during the building process or after the building is built, the insulation specification is ultimately a compromise between what is actually possible (e.g. feasibility, investment costs, legislation and norms), what is comfortable (i.e. heat and moisture, acoustics behaviour, thermal comfort and health-related issues, etc.) and the actual savings in terms of operational costs. According to Standards New Zealand (1998), in the absence of laboratory test data the total thermal resistivity (R_T) of a stabilised rammed earth (SRE) wall, of thickness d (m), can be calculated as shown by Equation 3.1.

$$R_T = 2.04d + 0.12 \hspace{4cm} [3.1]$$

The Building Regulations and minimum value of thermal resistance that is required vary according explicitly to the country/region in which planning consent is being sought, and implicitly by the ambient climate. The client/ designer may wish to exceed these minimum requirements if they are seeking to achieve a greater level of thermal resistance and associated reductions in building energy consumption. This could simply be an aspirational R-value, one determined as the product of building performance simulation, or it could be set by a voluntary 'best practice' scheme of accreditation, e.g. BREEAM in UK, LEED in US, CASBEE in Japan, MINERGIE in Switzerland or the PassivHaus standard in Germany, etc. In England and Wales (UK), for example, wall fabric may comply with the Building Regulations if it has a total R-value of the order 2.85 m² K/W. By means of comparison for a warmer climate, to achieve a '5 Star' energy efficiency rating (i.e. 'high', on a scale from 2 to 6 stars) under the non-compulsory Australian system (equivalent to BREEAM in the UK), an R-value of 2.2 m² K/W would be required, whilst under the German PassivHaus standard (a non-compulsory, best-practice design code – Chapter 5) an R-value of between 5.0 and 10.0 m² K/W is needed (Schnieders, 2009). Clearly, in climates where there is a significant heating load and a cold season, even the minimum R-value requirements cannot easily be met by an earth wall without insulation. The placement and installation detailing of insulation materials is discussed in the following sections, as well as details of the insulation material types that could be used with SRE walls along with details of their physical properties.

3.2.1   External wall insulation

Insulating the building externally decreases the possibility of thermal bridges. These are local regions of the earth wall (or adjoined structure/feature) in which heat transport is comparatively high and any applied insulation is ineffective. Thermal bridges can either be physical (e.g. where an uninsulated steel lintel conducts heat and bypasses the insulation) or geometric (e.g. on wall corners where the external surface area is greater than the internal). Thermal bridging is discussed in detail in Section 3.4. Furthermore when the insulation is located to the outside of a building's load-bearing wall, the structural components are better protected from extreme temperature variations and condensation risk. This reduces the possibility of damage due to thermal stress and water vapour saturation (Krus *et al.*, 2005).

The thermal storage capacity is also improved by external insulation in heavyweight structures, especially in modern earth buildings or in reinforced concrete structures where stored heat is less able to transport through the wall and escape to the outdoor environment. One effect of this is that external insulation will limit the overheating frequency during hot summer months, since the wall fabric is not heated by external gains and is therefore more able to absorb excess internal heat gains. Where the insulation is external, it needs to also protect the building fabric from the influences of rain ingress and wind; unless the structure is overclad with a rain screen. This leads to the necessity of using water-resistant renders over external insulation and normally comprises a proprietary system.

Due to the external exposure of the water-repellent plaster, which is thermally decoupled from the load-bearing wall by the insulating material, it must resist high temperature variations. These variations are a consequence of diurnal temperature variation, wind exposure and long wave radiant heat gains and losses. On clear nights, the external boundary layer will typically be below the dewpoint temperature which leads to a high risk of condensation and potentially mould growth. In terms of applying external insulation to an existing building, one will face relatively high installation costs where scaffolding is necessary. External insulation can change the building's appearance, to the point where it may become unsympathetic to the local character, whilst in some cases also increasing the building's footprint. SRE walls do not require protection from the weather and, due to their attractive sedimentary appearance and natural earthen colour, the external aesthetics form part of the appeal to designers and clients. For this reason, external insulation is generally not favoured and cavity insulation is used instead (see below). Unstabilised rammed earth, however, does need to be protected from the outdoor environment and so, despite the significant extra costs, protective waterproof external render systems can be applied that also incorporate the desired level of insulation. Similarly, rain skin cladding or double skin façades can also be applied in this situation.

3.2.2   Internal wall insulation

Internal insulation is often the most viable in terms of costs when insulating a building retrospectively, because this method is relatively easy to apply and does not require additional weather protective coverings. It is often the best thermal solution for buildings that are of masonry construction in colder climates and only used intermittently as it reduces the re-warming time by reducing the thermal admittance of the internal surfaces. Using this approach a room can be heated fairly quickly as the external wall layer does not need to be heated in order to achieve a satisfactory operative temperature. Compared to external insulation or cavity wall insulation (Section 3.2.3), this method is least desirable in modern earth buildings or in reinforced concrete structures in hot climates where the internal insulation will decouple the thermal mass of the building and therefore not contribute to lowering the overheating frequency. An exception to this occurs where adequate thermally massive internal (e.g. partition) walls, or dense large floor slabs and exposed concrete soffits/upper floors, are available to dampen the diurnal temperature swing.

The usage of internal insulation is less appropriate in terms of fire protection where flammable and toxic insulation materials are used, such as those that are based on organic foams such as polystyrene or polyurethane (Krus *et al.*, 2005). In high-density urban locations, where ground is expensive, the reduction of the treated floor area through the use of internal insulation is another drawback. The practical problems associated with relocating occupants are another issue that is often overlooked during the retrospective installation of internal insulation. However, the biggest technical problems one will face with internal insulation are related to the occurrence of thermal bridges, condensation risks and the reduced possibilities for drying out of the load-bearing wall material beneath. Internal insulation is rarely specified for modern earth building designs because it removes one of the principle advantages offered by the material, i.e. passive buffering of indoor air temperature and relative humidity.

3.2.3   Cavity wall insulation

Cavity wall insulation refers to insulation that is placed between the outer and the inner shell of an external wall; typically in masonry constructions but also in situ or pre-cast materials such as concrete and stabilised earth. A cavity wall presents advantages similar to externally insulated structures that often benefit from reduced overheating hours in summer due to the retention of thermal mass in the inner leaf (Krus *et al.*, 2005).

Amongst the disadvantages of cavity insulation is the fact that the air space in between the outer and inner leaves of the wall may be filled with a porous insulation material. The air (or gas) trapped inside the small pores remains the actual insulator, significantly reducing natural convection, and thus reduces heat losses and the associated heating costs significantly (Langdon, 2010). The material used can either be loose fill (e.g. glass fibre, rock wool or a fibrous material such as cellulose insulation or glass wool (see Table 3.1)) or it can be solid fill – either full or partial (e.g. expanded urethane or polystyrene).

Cavities present technical challenges when retrofitting as moisture penetration due to bridging and/or interstitial condensation can sometimes occur. In the case of minor defects in the construction process, or bad workmanship, the external wall layer may not be sufficient in terms of resistance to rainwater penetration – a situation that leads consequently to the danger of capillary moisturisation and humidification of the insulation.

Table 3.1 Summary types and physical properties of insulation materials available for use in wall cavity applications[1]

	Raw material	Thermal conductivity W/(m·K)	Heat capacity J/(kg·K)	Water vapour diffusion resistance factor μ	Density kg/m³	Primary energy content kWh/m³	Water repellent	Number of drills/m² and diameter (d)	Advantages	Disadvantages
Two-shell masonry										
1 Synthetic resin melamine foam	Synthetic resin, urea, formaldehyde	0.035–0.040	1500	1–4	8–15	/	Yes	2–3 per m², d: 12–16mm	Small to medium size drill holes which close automatically after injection, no preparation necessary	Small percentage of formaldehyde after injection for short time
2 Polyurethane foam	Oil, flame retardants, also herbal	0.027	1400–1500	30–100	55	800–1500	Yes	1 per m²	Small to medium size drill holes which close automatically after injection, no preparation necessary	Blowing agent which has a GWP of 7–13 and also end of life disposal
Mineral fibre cavity insulation										
3 Mineral wool	Natural stone, binder, defibration waste	0.04	840	1–2	70–150	150–400	Yes	4–5 m², d:20–24mm	Small size drill holes, close automatically	High formation of dust during installation, sealing is time consuming
Mineral injected cavity insulation										
4 Nanogel	Silica moulding	0.021	700–1150	2–3	85–95	/	Yes	Dependent on geometry		

No.	Material	Raw material								
5	Perlite	Rock	0.042–0.045	1000	1–3	75–85	160–260	Yes	10–15 per m², d: 20–24mm	Possible for small cavities, from ca. 2cm onwards	Preparation/preliminary work is necessary, as particles of sand can flow out, high formation of dust
6	Silicate light foam	Waste glass, lime glass, soda lime glass, silica glass	0.035	1000	3	20–30	/	Yes	10–15 per m², d: 20–24mm	Possible for small cavities, from ca. 3cm onwards	High formation of dust, sealing is time consuming

Organic injected cavity insulation

| 7 | Expanded polystyrene | Oil, coal | 0.033–0.034 | 1300 | 3–5 | 16–26 | / | Yes, very | 10–15 per m², d: 20–24mm | Material could be removed completely if required, dust free installation, low moisture absorption | Preparation/preliminary work is necessary, as particles of sand can flow out |
| 8 | Polyurethane foam recycled granulate | Polyurethane foam recycled granulate | 0.036 | 1200–1500 | 30–200 | 40–50 | / | Yes | Dependent on geometry | | Preparation/preliminary work is necessary, as particles of sand can flow out |

Table 3.1 Continued

	Raw material	Thermal conductivity W/(m·K)	Heat capacity J/(kg·K)	Water vapour diffusion resistance factor μ	Density kg/m³	Primary energy content kWh/m³	Water repellent	Number of drills/ m² and diameter (d)	Advantages	Disadvantages
Insulation board/panel										
9 Expanded polystyrene	Oil, coal	0.033–0.034	1300	3–5	16–26	/	Yes, very	n/a		
10 Mineral wool	Natural stone, binder, defibration waste	0.04	840	1–2	70–150	150–400	Yes	n/a		
11 Polyurethane foam	Oil, flame retardants, also herbal	0.027	1400–1500	30–100	55	800–1500	Yes	n/a		n-pentane blowing agent has a GWP of 7–13 and is known to leak in to the atmosphere over time. No end of life disposal
12 polyisocyanurate foam		0.023						n/a		
13 Extruded Polystyrene		0.035–0.38						n/a		
14 VIPS		0.002–						n/a		
15 Hemp batts		0.04–						n/a		

16	Wood fibre board	0.038–0.042	n/a	
17	Stramit board	0.102	n/a	
18	Cork	0.038–0.050	n/a	Manufactured from renewable resources (subject to management). Reusable if not adhesive or render-coated. Recyclable as loose fill. Can be used in energy recovery. Cork forests support indigenous wildlife (but detailed research of impacts on biodiversity is still lacking). Cork production helps to sustain communities in poorer agricultural areas. Sequesters CO_2 during tree growth. Waterproof. Naturally resistant to insect and rodent attack (except wasps). Resistant to compression. Dimensionally stable. Very high embodied energy. Cork dust

Table 3.1 Continued

Raw material	Thermal conductivity W/(m·K)	Heat capacity J/(kg·K)	Water vapour diffusion resistance factor μ	Density kg/m³	Primary energy content kWh/m³	Water repellent	Number of drills/ m² and diameter (d)	Advantages	Disadvantages
								may be a health issue – avoid inhalation. Wet cork boards usually harbour mould which lead to allergic reactions. Small emissions of naturally occurring formaldehyde	
19 Loose fill cellulose	0.038–0.4						n/a		Highly sorbtive material – moisture in cavity can lead to slumping
20 Sheep's wool	0.04						n/a		

[1] Data and information are taken from (IPEG, 2011; Jansen, 2010; CIBSE, 2006; Green Spec, 2010)

In this case it is essential to use a hydrophobic or water-repellent insulation material. This situation can equally apply to stabilised compressed earth block walls, where an unfilled air cavity has been left in place as part of the construction, or if the earth wall forms part of a composite cavity wall construction, e.g. with internal/external timber frame and cladding.

In the UK, cavity wall construction was introduced as a method to stop wind-driven rain from penetrating to the inside surfaces of masonry walls (Howell, 2008). The Energy Savings Trust started to financially support the injection of insulation material in cavities. This funding continues today in many parts of the UK. The current scheme, which is the third of its kind, is driven by the UK carbon emission target and has been in place since 2008. In this process the partially ventilated cavity air layer in between the outer and inner leaves that originally gave only limited thermal insulation is now filled with an insulating material (Table 3.1). The question that remains is if cavities were intentionally built as air voids, but subsequently filled at a later stage in the buildings life, how does this alter the dew point and hygroscopic characteristics of the wall?

The thermal resistance of the post-filled cavity wall is of course limited by the thermal conductivity of the blown loose fill insulation medium (e.g. ~0.04 W/m K for glass fibre) and the thickness of the air cavity (typically around 50 mm) and so any ambitious retrofit U-value targets are likely to require additional internal/external insulation.

The initial idea of the cavity is to stop water penetrating through the wall to the inside of the house. Since 1974, the Building Regulations have required new build houses to contain insulation material in the cavity (Building Regulation, Part L), and in theory this should give an ideal way of insulating the building as long as it is correctly detailed. As a result a number of different insulation products have been developed that can be fixed on to the inner leaf of a cavity wall, leaving a small air gap to prevent rainwater from being directly wicked across the wall. For this reason the (often hydrophobic) insulation material allows the rainwater and any condensate build up to drain downwards, under the influence of gravity and the presence of vertical aligned fibres. However, most problems occur in cases where insulation was retrospectively added to create a fully filled cavity resulting in damp problems due to a non-water-repellent material being used such as mineral or glass fibre. What happens as a result is that moisture is transferred back to the inner leaf of the construction and houses become damp and mouldy on their inside, and consequently the buildings become structurally damaged and uninhabitable.

The problem of rainwater crossing the cavity wall insulation is well documented (BRE Building Guide). It is stated that:

'there can be an increased risk of rain penetration if a cavity is fully filled with insulation, i.e. moisture is able to transfer from the outer to the inner leaves resulting in areas of dampness on internal finishes. Rainwater, under certain driving rain conditions, can penetrate the outer leaf of masonry leading to wetting of the cavity insulation, a reduced thermal performance and damage to internal finishes'.

It has also been revealed that 'single-leaf brick walls always leak when exposed to wind-driven rain' (Stirling, 2000). The leakage occurs at the vertical joints between adjacent bricks, because of the drying shrinkage in the mortar (Stirling, 2000). This can be a particular problem for clay brick or concrete block masonry walls, as the BRE studies have shown, since the hydraulic conductivity (i.e. pressure-driven moisture ingress) and sorptivity (i.e. capillary-driven moisture ingress) are typically very high for these materials, especially in the case of decorative facing bricks as used on the external leaf of the cavity wall. This problem can be overcome if the materials are treated with waterproofing agents, however these are typically high volatile organic compound (VOC) solvent-based surface treatments (e.g. silicone emulsions) that penetrate and fill the material's pore network. High-quality engineering brick, polished stone or walls of considerable thickness do not suffer the same level of moisture penetration. In the case of insulated SRE walls, the insulation used is a solid fill of interlocking boards usually made from extruded polystyrene (XPS), polyisocyanurate (PI) or expanded polyurethane (PU) (see Table 3.1). All are suited/designed for solid wall applications where the use of cavities is not possible and the insulation has to resist compression (e.g. during compaction of the adjacent earth material) and be resistant to moisture ingress or damage. SRE is high density (typically > 95% Proctor compaction) and so has very low permeability and sorptivity (Walker and Stace, 1996; Hall and Djerbib, 2004). Particularly for cavity rammed earth walls, where the inner and external leaves are usually a minimum of 175 mm thick in a load-bearing wall, the mix design must also include a hydrophobic admixture. This has the effect of significantly reducing liquid transport and effectively rainproofing the wall material, but leaving the macropore network open such that vapour permeability is largely unaffected. For more details of hydrophobic admixtures refer to Chapter 10. Note that masonry cavity walls typically have an inner/outer leaf thickness of 102.5 mm for fired clay brick or 100 mm for concrete block.

Condensation will always form on the coldest surface in a wall construction, which is typically found at the window/wall connection or areas of thermal bridges. Cold spots resulting from geometric or construction-related thermal bridges reduce the localised moisture-holding capacity of the air resulting in water vapour condensing, a phenomenon commonly referred to as 'interstitial condensation'.

A number of different materials used for cavity insulation are shown in Table 3.1. Commonly used in the UK are mineral wool fibre, bonded polystrene beads or insulation foams.

Figures 3.1, 3.2 and 3.3 show typical SRE solid cavity wall cross-sections during construction of Straightways Farm, Devon, UK in 2007 by Earth Structures (Europe) Ltd. Note that proprietary formwork systems have to be modified to allow the central insulation board to be mounted securely and allow compaction of the earth on either side. The two separate wall leaves are mechanically tied as for masonry walls but instead with specially designed stainless steel staples that become bonded with the compacted earth during compaction. This ties the two leaves together (both of which are supported by a common foundation) in order to prevent buckling, i.e. vertical bending moments. In cases of high wind loading or multi-storey walls, additional vertical restraint can be added, for example through the incorporation of stanchions. Where structural steel stanchions or similar

3.1 SRE cavity wall with 175 mm inner/outer leaves and 50 mm interlocking extruded polystyrene insulation core used in conjunction with Stabilform™ formwork system (© Earth Structures Europe Ltd 2007).

3.2 Insulation detail around a steel stanchion to avoid cold bridging across a solid external SRE wall (© Earth Structures Europe Ltd, 2007).

3.3 Interlocking extruded polystyrene insulation positioned in conjunction with stainless steel reinforcement in a structural SRE cavity wall (© Earth Structures Europe Ltd, 2007).

framework members are included in the external wall fabric, continuity of the insulation can be achieved with careful detailing to avoid any risk of thermal bridging. Construction of stabilised insulated (cavity) earth walls in seismic regions (e.g. North West America) requires additional detailing and support in order to reduce 'mass mobilisation' and the likelihood of cavity tie failure/buckling. Figure 3.4 shows a stablised insulated rammed earth (SIREWALL) construction detail which addresses all of these concerns by incorporating a rigid cross-braced steel framework to tie the two leaves

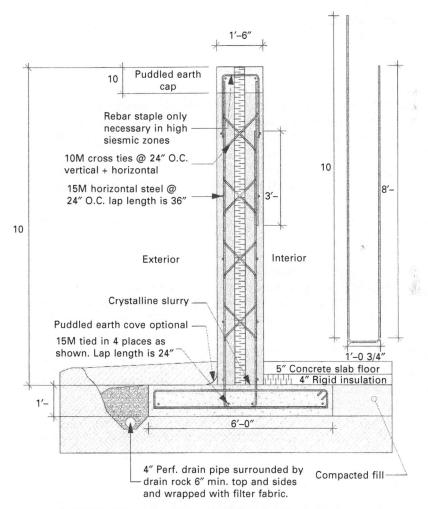

3.4 SIREWALL construction detail for stabilised insulated rammed earth walls incorporating cross-braced steel reinforcement and vertical restraint, which is designed for applications in seismic regions or high wind loading (© SIREWALL Inc., 2011).

together, coupled with vertical restraint from the reinforced foundation slab and puddle earth cap.

3.3   Thermal bridging theory

In order to implement effective and competent design of the thermal insulation a major obstacle has to be overcome. Thermal bridges are the main source of thermal losses especially in new buildings. They can contribute as much as 50% of the total negative heat exchange (Schnieders, 2009). Therefore, thermal bridges have to be carefully investigated and checked when the architect/ engineer is detailing the outer thermal envelope. Thermal bridges are defined as points along a building's envelope where the surface temperatures and/or the heat flow rate changes. They occur most often at junctions or when there is a change in the material composition of a building element. Thermal bridges can cause not only significant heat losses but also undesired condensation and mould growth (Mao and Johannesson, 1997). Lowered temperatures on the internal surface of the building can lead to vapour condensation, which in turn creates a suitable environment for mould and fungal spores to develop, thus causing damage to the construction (Kornicki, 2011) and posing health risks to the occupants. Thermal bridging therefore gives rise to two and three-dimensional heat flow paths, which can be approximated accurately using numerical methods (the most common is iterative Finite Element Analysis). Since numerical methods require a significant number of iterations, it is almost impossible to create and solve a model without being assisted by a computational tool (Tilmans and Orshoven, 2009).

A common way to evaluate thermal bridges is by using numerical methods. Most dynamic simulation programmes are very simplified and assume one-dimensional (1D) heat flow through the thermal envelope (e.g. Energy+, TRNSYS, CODYBA 6.0) (Gao *et al.*, 2008). The 2D or linear thermal bridges (Figure 3.5) can be analysed by bespoke thermal bridging tools such as Heat3, TB3D/FMD for heat flow at local positions or for analysis of temperature fields. KOBRU86 software (Physibel, 2011) allows 2D steady-state thermal bridge analysis; and offers in combination with the EUROKOBRA database a number of catalogues with more than 3000 thermal bridge details. Several research studies have shown that this approach provides a user-friendly way to analyse thermal bridges (Strachnan *et al.*, 1995; Hopfe and Manolov, 2011; Ben-Nakhi, 2003).

An example of a 2D thermal bridge study on window performance, using THERM 5.2, is shown by Cappelletti *et al.* (2011). They particularly show the difference in a number of window–wall junctions for externally insulated walls and for walls with cavity insulation. The outcome shows the importance of correct frame positioning, and shows similar outputs are achievable for both the external insulated and cavity wall examples.

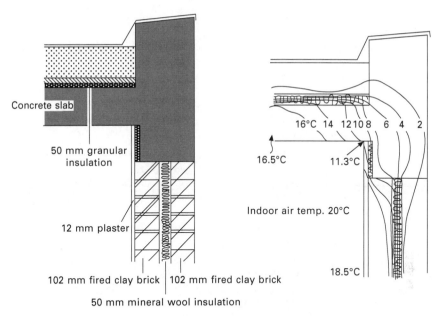

Concrete slab

50 mm granular insulation

12 mm plaster

102 mm fired clay brick | 102 mm fired clay brick

50 mm mineral wool insulation

16°C 14 12 10 8   6   4   2

16.5°C   11.3°C

Indoor air temp. 20°C

18.5°C

3.5 Example of thermal bridging and temperature gradient due to lack of continuity in fabric insulation at the eaves.

Another 2D calculation case study conducted by Theodosiou and Papdopoulous (2008) in Greece using TRNSYS, shows the impact of thermal bridges on the total energy demand in buildings by analysing four different wall configurations involving both external and cavity wall insulation. 3D or point thermal bridge studies appear in 3D corners or in the case of an insulated wall that is perforated by 'an element with high thermal conductivity' (Ben Larbi, 2005). In the context of a whole building simulation (1-year simulation with 1-h time step), a 3D calculation will place exceptionally high demands on computational power and time. An example of 3D thermal bridge studies can be found in Ben-Nakhi (2002), who developed a 3D gridding module that is integrated in the Building Performance Simulation BPS tool ESP-r for dynamic 'whole building' thermal bridging assessment. Results are tested and validated with VOLTRA from Physibel (Physibel, 2009). A more simplified model is proposed by Gao *et al.* (2008) who introduced a reduced heat transfer model in 3D that can be potentially implemented into standard BPS.

From a practical perspective it is appropriate to investigate the available software packages designed to calculate multidimensional heat flow in a building (Tilmans and Orshoven, 2009). An overview of some of the more commonly used software is shown in Table 3.2. One essential point is the conformance with the European Norm BN EN ISO 10211: 2007 'Thermal bridges in building constructions – Heat flows and surface temperatures –

Table 3.2 Numerical modelling simulation tools for assessing thermal bridging in building fabric

Tool	Key words	Website
AnTherm	Thermal bridge 2D, 3D, steady-state heat transmission (conduction), ISO standard conformant	www.kornicki.com/
Delphin	Heat, air and moisture transport, building envelope	http://bauklimatik-dresden.de
Flixo	2D heat transfer, frame U-value, thermal bridge	www.infomind.ch/bph/en/
Trisco Physibel	Thermal bridge 2D, 3D, free form, rectangular forms for steady-state and transient	www.physibel.be/
THERM	2D heat transfer, building products, fenestration	http://windows.lbl.gov/software/therm/therm.html
WUFI	Hygrothermal modelling, combined heat and moisture transport	www.wufi-pro.com/
Psi-Therm 2011	Thermal bridge, DIN 4108 incl. Beiblatt2, EN ISO 10211:042008, EN ISO 6946	www.psi-therm.de/

Detailed calculations'. The above standards set requirements and defined methods for accurate calculation of two- and three-dimensional heat flow. Tools that conform to this standard can be successfully used in the design and implementation of an accurate and standardised method for assessing the quality of the proposed thermal envelope (British Standards Institution, 2009).

Note that only two-dimensional flow will be shown in the following. It is assumed that two-dimensional heat flow modelling is adequate to assess thermal bridging to an acceptable level of accuracy in most cases (Tilmans and Orshoven, 2009). Heat flow through thermal bridges cannot be adequately assessed using one-dimensional models as stated above. Two-dimensional models can however predict heat flow behaviour to an accurate extent. In two-dimensional modelling cut-off planes are defined to isolate the full extent of any thermal bridge present (i.e. glazing system-to-wall connection, balcony-to-wall, roof ridge junctions, etc.). The cut-off plane represents a cross-section of the element to a point where the heat flow becomes once again one-dimensional. By inputting appropriate boundary conditions a model can thus be used to determine the sum of all two- or three-dimensional heat flows.

3.3.1   Two-dimensional calculations

Several factors have to be addressed when calculating a two-dimensional flow and, in particular, 2D thermal bridges. A major requirement is that the model conforms to the geometrical requirements stated in BS EN ISO 10211:2007. Once an accurate model of the desired problematic area has been created, materials have to be assigned in accordance with the architect's specifications. Thermal conductivity values of the material can be found in BN EN ISO 10456 (British Standards Institution, 2009). Where these are not listed, values that conform to CE testing procedures should be used in favour of manufacturers' stated values, which are in many cases misleading. Finally, appropriate boundary conditions in terms of temperatures and surface resistance of the corresponding environments are to be assigned in accordance with EN ISO 120011-1 (Ben Larbi, 2005). Once the model has been implemented, calculation of the thermal coupling coefficient and linear thermal transmittance can be executed according to the specified requirements. It is therefore appropriate to define the components contributing to the total (two-dimensional) heat exchange between both environments (Ward and Sanders, 2007).

Thermal coupling coefficient (L_{2D}) is defined as the heat flow rate per temperature difference between internal and external environments. According to EN ISO 10211:2007, 2D thermal coupling coefficients can be calculated as shown in Equation 3.2:

$$L_{2D} = \frac{Q}{\theta_i - \theta_e} \qquad [3.2]$$

Linear thermal transmittance, known as the Psi value (Ψ), is the residual heat flow when the one-dimensional heat flow (all elements inc.) is subtracted from the total heat flow per unit temperature difference, i.e. the 2D thermal coupling coefficient (Ward and Sanders, 2007) as shown Equation 3.3:

$$\Psi = L_{2D} - \Sigma U \cdot l \qquad [3.3]$$

Point thermal transmittance, known as the Chi value (χ), is the residual heat flow when the one-dimensional heat flow and the linear transmittance are subtracted from the three-dimensional coupling coefficient (Ward and Sanders, 2007).

3.3.2   Total heat flow

Total heat transfer through the building envelope can be defined by Equation 3.4.

$$H_D = \Sigma_i A_i \cdot U_i + \Sigma_k l_k \cdot \Psi_k + \Sigma_j \chi_j \qquad [3.4]$$

The first term $A_i \cdot U_i$ represents the thermal transmittance through the area of the plane element. If only this term is considered, then the problem becomes one-dimensional. It is commonly the most significant source of heat loss within Equation 3.4 (Ward and Sanders, 2007) and occurs at plane elements including roofs, external walls, floors, and doors/windows (Kuhnenne, 2008).

The second term $l_k \cdot \Psi_k$ represents the sum of all linear thermal bridges in the envelope, where Ψ is the linear thermal coefficient and l is the length of the thermal bridge. Ψ can be estimated by subtracting the thermal transmittance component $A \cdot U$ from the two-dimensional coupling coefficient defined previously as L_{2D} (Ward and Sanders, 2007). This term is relevant to heat transfer that occurs at thermal bridges including eaves and corners, i.e. a geometric thermal bridge (Kuhnenne, 2008) as shown by Figure 3.5.

The third term χ_j represents the point thermal bridges within the structure. They occur as a result of the interaction of the linear thermal bridges or where the residual heat flow at either boundary is three dimensional. Usually, they can be omitted from the total heat transfer calculation because their contribution is insignificant. However, in certain situations, they have to be included for a more accurate solution, e.g. at penetrations where the point thermal bridges are either frequent or very large, such as poorly insulated structural frame elements including steel stanchions. Therefore calculating accurate U and Ψ will be of major interest when comparing the value of computational packages (Ward and Sanders, 2007) and is of significant importance in the design of low energy buildings.

3.3.3   Humidity and condensation

One of the main problems associated with thermal bridges aside from the thermal losses are the cold regions of minimum temperatures on the internal surfaces of the building. Lowered temperatures can lead to condensation, which is an undesirable effect in contemporary design. Condensation leads to mould growth and premature deterioration of the construction materials (Kuhnenne, 2008). It is therefore vital to numerically determine the temperatures on the internal surface of the model. Once the minimum temperature on the internal surface is known, the temperature factor f_{Rsi} can be estimated. The temperature factor must be a minimum value of 0.7 in order to avoid condensation and mould growth (Kuhnenne, 2008). The temperature factor f_{Rsi} can be defined by Equation 3.5.

$$f_{Rsi} = \frac{\theta_{si,min} - \theta_e}{\theta_i - \theta_e} \geq 0.7 \qquad [3.5]$$

$\theta_{si,min}$ is the threshold internal surface temperature below which condensation can occur. Note that condensation depends upon the predefined internal relative humidity within the model (Kuhnenne, 2008).

3.4   Thermal bridging simulation tools

The Finite Element Method (FEM) is a calculation method for a structural element consisting of small elements, where the thermal quality of each element can be almost exactly defined. If a net is laid over the structural element, all junctions are defined. From this nodal grid the thermal properties of the whole element can be calculated. There are a number of well-established finite element computer programs for the calculation of two- or three-dimensional heat fluxes. A brief summary of a number of simulation software tools that can be used to predict thermal bridges in buildings is presented in Table 3.2.

Figures 3.6 and 3.7 show the modelling of an external wall-to-ground floor joint from a PassivHaus project in Ebbw Vale, South Wales. This particular section is one of the most difficult parts to design in terms of achieving the minimum threshold U-value of equal or less than 0.15 W/m^2 K that is

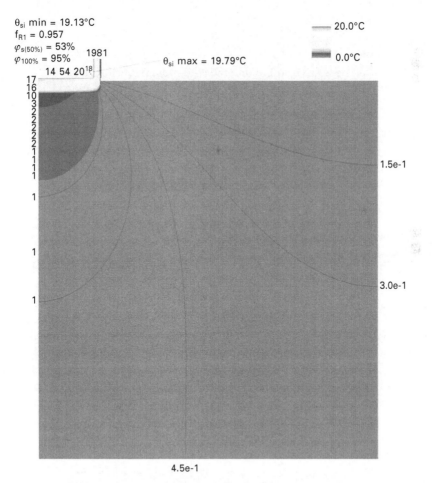

3.6 Temperature distribution of wall-to-ground junction, showing the min/max values, humidity and isotherms simulated with Flixo (Hopfe and Manolov, 2011).

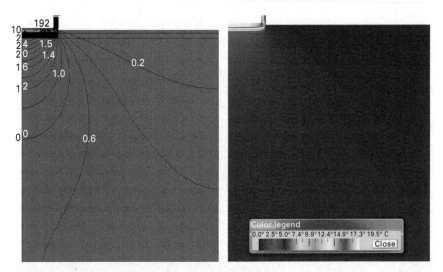

3.7 Temperature distribution of wall-to-ground junction, showing the isotherm lines and the heat flux diagram simulated with THERM (Hopfe and Manolov, 2011).

required in the PassivHaus standard. The pictures show the simulation of this problematic junction of the PassivHaus with two different simulation programs, Flixo and THERM.

3.5   Acoustic reverberation

In addition to hygrothermal behaviour and indoor air quality, the acoustic isolation of a building is very important in terms of providing occupant comfort. This requires, for instance, research into the acoustic properties of building components and materials (e.g. solid or hollow bricks, sealing and insulating materials, sound absorptive qualities of materials). Typically this research is carried out using in-situ experimental design to isolate the acoustic properties of materials. Subsequently computational and experimental optimisation is possible including the complex interactions in buildings. Numerous different computational methods exist and can be compared to real-world measurements when determining the acoustic properties of an enclosed space. This analysis leads to the practical possibilities of sound reduction and acoustic harmonisation of buildings and their components – the sound absorption of suspended ceilings for instance, or floor coverings.

The importance of acoustic comfort is increasing as noise pollution and annoyance in buildings is growing. In many countries guidelines and legal requirements exist that help building towards sound isolation. In the UK, Building Regulation Part E (resistance to the passage of sound) sets such restrictions in terms of protection against sound from other parts of the building, protection within dwellings, reverberation in common internal parts or acoustic conditions in schools (Building Regulation Part E, 2010). Acoustic performance is significantly affected by the building structure, which can be seen for example in terms of the performances of heavyweight buildings (e.g. rammed earth, concrete or brick constructions) compared to lightweight structures (e.g. insulated timber frame). Even though new structures are being improved with respect to their structural performance, sustainability and adaptability, this is often achieved at the expense of sound isolation (Schönwald, 2008). More precisely, if a building is designed to deliver good thermal performance that does not automatically confer the preferred acoustic qualities. Particularly when looking at lightweight structures, certain types of insulation can actually harm acoustic behaviour and have an increasing deleterious effect on noise reverberation and the sound absorption of the structure.

The thermal and acoustic properties of materials may exhibit contradictory behaviour. Building materials and wall constructions that provide good acoustic absorption usually have a low thermal inertia and vice versa, which will be shown in the following. It is therefore important to understand the physics involved in acoustic comfort, to know about the relevance of thermal inertia and the absorption coefficient, and to find the right balance between these factors in order to simultaneously deliver acoustic and thermal comfort in buildings.

3.5.1   Reverberation time

The most commonly used simplified model to calculate reverberation time, is that of the Wallace Sabine approximation. This assumes the speed of sound to be constant at 343.3 m/s, and at 20°C the term $2.76/t^{0.5} = 0.161$ s/m. If sound is reflected off a surface several times, the reverberation time is the time required for reflections to decay by 60 dB. The reverberation time is frequently stated as a single value, however, it can be assessed based on three versions of the diffuse-field theory (Citherlet and Macdonald, 2003). Typically, the reverberation time differs depending on the frequency band being measured. The frequency-dependent reverberation time can therefore be expressed by Equation 3.6.

$$t_r = \frac{2.76}{t^{0.5}} \cdot \frac{V}{A_f^{tot}}$$

[3.6]

The Wallace Sabine equation states that t_r is directly proportional to the volume, V, of the room, but inversely proportional to the room's effective surface area, A_f^{tot}. The effective surface area is the sum of the individual products of an area covered by a particular material multiplied by the material's absorption coefficient (α) as shown in Equation 3.7.

$$A_f^{tot} = \sum_{i=1}^{n} \alpha_i \cdot A_i = \alpha_1 \cdot A_1 + \alpha_2 \cdot A_2 + \ldots + \alpha_n \cdot A_n \qquad [3.7]$$

The reverberation time in a room is thus affected by the size and shape of the enclosure and the materials used in the construction of the room. Furthermore, every object that is in the room will impact on the reverberation time, including occupants and furniture. The theoretical model, from which these equations are derived, requires the decaying sound field to be perfectly diffuse. According to Citherlet and Macdonald (2003):

> 'This idealised condition is sufficiently fulfilled in practice when: (a) no room dimension is markedly different from the others; (b) room dimensions are large compared to the wavelength, which is the case in most building acoustics; (c) absorption is distributed almost uniformly over the enclosure boundaries.'

3.5.2 Acoustic absorption

Sound is absorbed because of its reflections at the enclosure boundaries during the propagation. At each successive reflection, the enclosure boundaries and all objects within the enclosure absorb a fraction of the sound energy. The fraction of absorbed sound is dependent upon the frequency of the emitted sound and the capacity of the reflecting material to absorb this frequency. For instance, concrete is a rather poor sound absorber compared to fibreboard, as shown in Fig. 3.8. The total absorption is the product of the absorption coefficient of all materials multiplied by the area. In practice this calculation has to be conducted at a number of different frequencies due to the fact that the absorption coefficient is not constant across frequencies. The absorption coefficient of a material is a number between 0 and 1 that indicates the proportion of sound absorbed by the surface compared to the proportion which is reflected back into the room. For example, in the situation of an open window without any sound reflections whereby sound would pass straight outside, this would be a perfect absorber and have an absorption coefficient of 1. The opposite of this might be a thick concrete ceiling, which is smooth and painted, and is from an acoustic point of view the equivalent of a mirror and has an absorption coefficient close to 0. The absorption coefficient varies with the frequency and so the reverberation time is a function of frequency. Some contractors have experimented by manipulating the surface texture of rammed earth walls, by adjusting the soil particle grading, in an attempt to alter the acoustic absorption coefficient SIREWALL (2011). This is an interesting avenue for further research as no detailed experimental data are currently available and yet could have significant potential for application in performing arts buildings, recording studios and other similar building typologies.

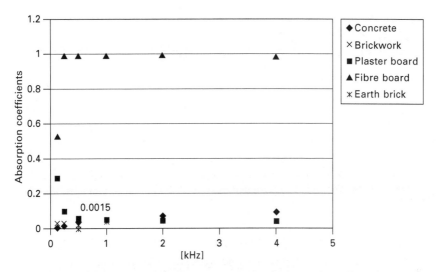

3.8 Comparison between the acoustic absorption coefficients of conventional and earth building materials.

3.5.3 Correlation with thermal inertia

Thermal inertia describes the combined effect of heat storage capacity within the wall and the thermal resistance to movement of heat through the wall. The following equation refers solely to thermal mass and does not take into account the thermal resistance of the wall, however it is an often-used formula to describe the thermal inertia of a material. Table 3.2 shows the impact of different materials and frequencies of the absorption coefficient. The thermal inertia, I, of a material is defined as the square root of the product of the material's thermal conductivity, density and specific heat capacity. It therefore quantifies the capacity of a material to store surrounding heat and to release it when surrounding temperature drops, as defined in Equation 3.8:

$$I = \sqrt{\lambda \rho C} \hspace{6cm} [3.8]$$

Note also that this is a simplified equation and does not, for example, describe solar absorption coefficients as either part of the boundary equation or the limits of 24-hour cycling.

In contrast to its acoustic performance, concrete has a higher capacity to store heat (thermal inertia) than fibreboard, because of its higher density and thermal conductivity. This higher thermal inertia generally increases absorption of solar gains, and reduction in internal heat gains may improve occupants' thermal comfort, especially for spaces with large glazed areas and no cooling plant. To provide a well-designed building, a balance must be found between acoustic absorption and thermal inertia. Earth brick walls make a significant contribution towards achieving adequate acoustic

Table 3.3 Different materials and their thermal inertia and absorption coefficient in relation to different frequencies

	Thermal inertia (J m^2/K $s^{-1/2}$)(Wh/m^3K)	250 Hz	500 Hz	1 kHz	4 kHz
Concrete	1936	0.02	0.04	0.06	0.1
Brickwork	918	0.03	0.03	0.04	0.07
Plasterboard	357	0.1	0.06	0.05	0.04
Fibreboard	134	0.99	0.99	0.99	0.99
Earth brick[1]	1069 (calculated from CIBSE Guide A data)		30-60 dB for 400 mm wall		

[1] Adobe brick 380 Wh/m^3K and Rammed Earth 590 Wh/m^3K (Houben and Guillard, 2003)

performance (Morton *et al.*, 2005). The performance of earth brick walls is very good in terms of thermal and acoustic performance and could be enhanced by avoiding flanking transmission. It is also good practice to repair initial shrinkage cracks, which may affect acoustic performance. The reason why soil functions particularly well in terms of thermal inertia is due to the combination of relatively high volumetric heat capacity and thermal conductivity. With its low thermal diffusivity, earth walls possess the ability to dampen and delay internal thermal variations in response to external inflows.

3.5.4   Acoustics and building performance simulation

It is important to realise that with regard to building acoustics it is all about getting the shell of the building right in the first instance. It is almost impossible to retrofit proper acoustic performance (May, 2009). Elsewhere as in Europe traditionally, buildings typically consisted of heavy monolithic constructions such as masonry walls with earthen or stone floors. With the advent of concrete blocks and monolithic construction, the first models for the prediction of sound isolation were based on a simple mass law where only transmission through the partition between two rooms was considered (Schoenwald, 2008). A large number of building simulation tools that provide computational models to predict sound isolation and acoustics in buildings are now widely available. The weighted Sound Reduction Index (R_w) of a solid wall is strongly dependent on the dry density of the material. SRE is a monolithic wall fabric where the dry density (ρ_d) is typically high, ranging between 1900 and 2200 kg/m^3. According to the 'acoustic mass rule' for solid walls, as defined by BS 8233 (British Standards Institution, 1999), R_w can be calculated using:

$$R_w = 21.65\log_{10} m' - 2.3 \qquad [3.9]$$

Table 3.4 Overview of simulation tools available for assessing the acoustic performance of buildings

Tool	Key words	Website
ACOUSALLE	Acoustics, codes and standards	http://lesowww.epfl.ch/anglais/ Leso_a_frame_sof.html
Acoustics Programme	HVAC acoustics, sound level prediction, noise level	www.trane.com/commercial/ software
CYPE-Building Services[1]	Analysis of acoustic behaviour, building services, energy simulation, sizing, HVAC, electricity, solar	http://instalaciones.cype.es/
ECOTECT	Geometric and statistical acoustic analysis, environmental design and analysis, conceptual design, validation, solar control, overshadowing, thermal design and analysis, heating and cooling loads, prevailing winds, natural and artificial lighting, life cycle assessment, life cycle costing	www.squ1.com
ESP-r	Integrated energy modelling tool for the simulation of the thermal, visual and acoustic performance of buildings	www.esru.strath.ac.uk/ Programs/ESP-r.htm
DONKEY	Self-generated noise, room sound pressure level, duct sizing, static regain, balanced pressure drop, duct acoustics	www.ozemail.com. au/~acadsbsg

[1] adapted from US Department of Energy (DOE, 2011)

According to the Building Regulations (2000) Approved Document E: 'Resistance to the passage of sound', the laboratory values for new internal walls within houses, flats and rooms for residential purposes must have a minimum R_w of 40 dB. A typical 300 mm thick solid SRE wall (without insulation), would have a weighted Sound Reduction Index (R_w) of 58.3 dB, assuming dry density to be 2100 kg/m^3. A brief summary of a number of the simulation software tools that could be used to better predict the acoustic performance of buildings are listed in Table 3.4.

3.6   Sources of further information

The following section provides brief details and links to sources of further information that will enable readers to tackle specific and advanced problems relating to thermal bridging and heat transfer in insulated and non-insulated building fabric. Numerical simulation/methods and tools can be generally divided into the following four categories:

1   FEM (finite-element method) such as Ansys, LSDyna, Nastran, Abaqus

2   Multibody system such as ADAMS, Dymola, Partial Flow Code PFC
3   CAS (computer Algebra system) such as Maple, Matlab/Simulink
4   CFD (computational fluid dynamics).

Ansys/CFX

Ansys/CFX is a fluid dynamics program with integrated solver that captures any type of phenomena related to fluid flow.
More information on www.ansys.com/.
Tutorial on http://seam.ustb.edu.cn/UploadFile/20080512040742250.pdf.

Ansys Fluent

The software package Fluent is a commercial CFD package with ancillary software called Gambit that is used to create mesh representations of the object that is being studied (typically purchased with Fluent).
More information on www.ansys.com/and www.ansys.com/Products/Simulation+Technology/Fluid+Dynamics/ANSYS+FLUENT/Features.
Tutorial on www1.ansys.com/customer/content/documentation/121/fluent/flwbtg.pdf.

PHOENICS

The CFD package Phoenics from CHAM has a range of applications models that can be implemented including architectural airflows, safety, fire spread, environmental pollutant dispersal, simultaneous stress/flow analysis, amongst others.
More information on www.cham.co.uk/.

Fire Dynamics Simulator (FDS)

The free software FDS was developed by the National Institute of Standards and Technology (NIST) of the United States Department of Commerce, in cooperation with VTT Technical Research Centre of Finland. It is a CFD model of fire-driven fluid flow. The software solves numerically a large simulation form of the Navier–Stokes equations appropriate for low-speed, thermally driven flow, with an emphasis on fire smoke and heat transport.
More information on http://fire.nist.gov/fds/.

Others

More information with respect to freeware preprocessing (meshing), and solver programs available on http://wiki.ubuntuusers.de/CFD_-_Str%C3%B6mungssimulation.

3.7   References

Ben Larbi, A., 2005. Statistical modelling of heat transfer for thermal bridges of buildings, *Energy and Buildings*, 37(9), pp. 945–951

Ben-Nakhi, A.E., 2003. Development of an integrated dynamic thermal bridging assessment environment, *Energy and Buildings*, 35(4), pp. 375–382

British Standards Institution, 2009. BS EN ISO 10211:2007. Thermal bridges in building constructions – Heat flows and surface temperatures – Detailed Calculations. London: BSI

British Standards Institution, 1999. BS 8233 Sound insulation and noise reduction for buildings: Code of practice, London: BSI

Building Regulation part E, 2010. resistance to the passage of sound, 2003 edition incorporating 2004 and 2010 amendments, HM Government, available from www.planningportal.gov.uk/uploads/br/BR_PDF_ADE_2003.pdf

Building Regulation part L, 2010 Part L. Conservation of fuel and power, HM Government, available from www.planningportal.gov.uk/uploads/br/BR_PDF_ADL1B_2010.pdf

Cappelletti, F., Gasparella, A., Romagnono, P., Baggio, P., 2011. Analysis of the influence of installation thermal bridges on windows performance: The case of clay block walls, *Energy and Buildings*, 43(6), pp. 1435–1442

CIBSE A, 2006. *Guide A: Environmental design*; Category: Heating, Air Conditioning and Refrigeration. London: CIBSE

Citherlet, S. and Macdonald, I. 2003. Integrated assessment of thermal performance and room acoustics. *Energy and Buildings*, 35 (3) pp. 249–255

Department of Energy, tools directory, http://apps1.eere.energy.gov/buildings/tools_directory/, last accessed March 2011

Department of Trade and Industry, 2001. *Energy consumption in the United Kingdom*. London: Energy Publications, Department of Trade and Industry

Gao, Y., Roux, J.J., Zhao, L.H., Jiang, Y., 2008. Dynamical building simulation: A low order model for thermal bridges losses, *Energy and Buildings*, 40(12), pp. 2236–2243

GreenSpec, 2010. Introduction to Passivhaus.[online]. Available at www.greenspec.co.uk/passivhaus-introduction.php (accessed 11 February 2011)

Hall, M. and Djerbib, Y., 2004. Moisture ingress in rammed earth: Part 3 – The sorptivity and the surface inflow velocity, *Construction and Building Materials*, 20 (6) pp. 384–395

Hens, H., Janssens, A., Depraetere, W., Carmeliet, J., Lecompte, J., 2007. 'Brick cavity walls: A performance analysis based on measurements and simulations', *Journal of Building Physics*, 31, p. 95

Hipworth, M., 2011. Thermostat settings in English houses: No evidence of change between 1984 and 2007. *Building and Environment*. 46 (3), pp. 635–642

Hopfe, C.J. and Manolov, S., 2011. Overview of simulation tools for 2D thermal bridge calculation, internal report, Cardiff: Cardiff University

Houben, H. and Guillard, H., 2003. *Earth Construction: a comprehensive guide*. London: ITG Publishing

Howell, J., 2008. *The rising damp myth*. Woodbridge: Nosecone Publications

Institut für preisoptimierte energetische Gebäudemodernisierung GmbH, 2011. www.ipeg-institut.de/index.php?article_id=1 (last accessed June 2011)

International Organisation for Standardisation (ISO), 1993. ISO 9613/1 Acoustics – Attenuation of sound during propagation outdoors – Part 1 – Calculation of the absorption by the atmosphere, London: ISO

Jansen, H., 2010. Übersicht verschiedener Kerndämmstoffe zur Hohlwandsanierung und die wichtigsten Eigenschaften im Vergleich. Available at www.daemmen-sanieren.de (last assessed 2010)

Kuhnhenne, M., 2008. Thermal bridges – Sandwich panel constructions, In: *Thermal bridge junctions*, Feldmann, M. ed., Edinburgh: European Quality Assurance Association for Panels and Profiles, pp. 1–33

Kornicki, T., 2011. AnTherm. 6.2. Computer program. Vienna, Austria: T. Kornicki Ltd.

Krus, M., Sedlbauer, K., Künzel, H., 2005. 'Innendämmung aus bauphysikalischer sicht', presentation at the event Innerdämmung – eine Bauphysikalische Herausforderung, 21 April, Münster, Germany

Langdon, D., 2010. Study on hard to fill cavity walls in domestic dwellings in Great Britain, DECC ref: CESA EE0211, Undertaken by Inbuilt Ltd & Davis Langdon, Inbuilt ref: 2579-1-1

Mao, G. and Johannesson, G., 1997. Dynamic calculation of thermal bridges. *Energy and Buildings*, 26 (3), pp. 233–240

May N., 2009. Why build better? Available at www.nbtconsult.co.uk/resources.html (last accessed 2009)

Morton, T., Stevension, F., Taylor, B., Charlton Smith, N., 2005. Low cost earth brick construction. Available at www.arc-architects.com/downloads/Low-Cost-Earth-Masonry-Monitoring-Evaluation-Report-2005.pdf

Physibel, TRISCO and KOBRU86, 2002. User Manual, last updated on 8 February 2011

Physibel, Voltra and SECTRA, 2002. User manual, last updated on 08.02.2011, available at www.physibel.be/v0n2vo.htm

Schnieders, J., 2009. A quantitative investigation of some passive and active space conditioning techniques for highly energy efficient dwellings in the South West European region, 2nd corrected edition, Darmstadt: Passivhaus Institute

Schoenwald, S., 2008. Flanking sound transmission through lightweight framed double leaf walls – Prediction using statistical energy analysis, PhD thesis, TV Eindhoven, the Netherlands

SIREWALL - Stabilized Insulated Rammed Earth; available at: http://www.sirewall.com/portfolio/residential-projects/lakeside-grand-piano/(Accessed 23 September 2011)

Stirling, C., 2000. *Good Building Guide (GBG) 44: 2000, Part 2: Insulating masonry cavity walls: principal risks and guidance*. Watford: IHS BRE Press

Strachnan, P., Ben-Nakhi, A.E., Sanders, C., 1995. Thermal bridge assessment, *Proceedings of Building Simulation* 95, pp. 563–570

Theodosiou, T.G. and Papadopoulos, A.M., 2008. The impact of thermal bridges on the energy demand of buildings with double brick wall constructions, *Energy and Buildings*, 40 (11), p. 2083–2089

Tilmans, A. and Orshoven, D.V., 2009. Software tools and thermal bridge atlases (4 March), Belgian Building Research Institute (webinar). Available at: www.bbri.be/homepage/index.cfm?cat=publications (Accessed 25 November 2010)

Walker, P. and Stace, T., 1996. Properties of some cement stabilised compressed earth blocks and mortars, *Materials and Structures*, 30, pp. 545–551

Ward, T. and Sanders, C., 2007. BR497 – Conventions for calculating linear thermal transmittance and temperature factors – part of the research programme of the Sustainable Buildings Division of the Department for Communities and Local Government. Watford: HIS BRE Press

3.8   Appendix: nomenclature

f frequency (Hz)
I thermal inertia, also known as thermal effusivity, β (W s$^{1/2}$/m^2 K)
l length (m)
m' wall surface mass (kg/m^2)
V volume (m^3)
T absolute temperature, e.g. of air (K)
A_f^{tot} total equivalent area of an enclosure for frequency f (m^2), that includes absorption due to boundaries, furniture, occupants and air.
Q total heat flow (W/m)
U Thermal transmittance (W/m^2 K)
θ_i Internal temperature (°C)
θ_e External temperature (°C)
$\theta_{si,min}$ Threshold internal surface temperature (°C)
Ψ linear thermal transmittance
χ point thermal transmittance

Modern earth building codes, standards and normative development

H. SCHROEDER, Bauhaus University Weimar, Germany

Abstract: This chapter provides an overview of standards and normative documents in the field of building with earth developed in the last 30 years. Building with earth is usually regarded as a 'non-engineered' construction technique with roots in a rich tradition of building heritage that needs to be maintained. As a consequence, building standards in the field of earth building have been drawn up in only a few countries. In the last decade, however, the use of earth in construction has become increasingly widespread in many countries. The 33 different standards examined in this chapter come from 19 different countries and provide varying degrees of technical information. With regard to their scope of application, the documents can be classified into three types, each dealing with a particular aspect: soil classification, earth building materials and earth construction systems. Our analysis shows that standard internationally accepted terminology is still lacking. This, however, is an essential general prerequisite for developing standards and normative documents.

Key words: building with earth, standards, soil classification, earth building materials, earth building techniques.

4.1 Introduction: a short history of building codes for using earth as a building material

Earth is one of the oldest building materials known to mankind. According to archaeological excavations, the first use of earth as a building material dates back to the Neolithic period in approximately 10 000 BC in the warm and dry climate of the eastern Mediterranean and Mesopotamia in what is today Anatolia (Turkey), Syria, Jordan, Lebanon, Israel (Palestine) and Iraq.

For thousands of years earth was the prevailing building material for house construction in many regions of the world with appropriate soils and climatic conditions. As a result, it became necessary to develop rules for using this material for building purposes. Written or painted documents describing earth building served as early 'rules' for using earth as a building material. Fig. 4.1[34] depicts the process of building with mud blocks in ancient Egypt in about 1500 BC and shows quite some detail including the block sizes, the block bonds in the wall construction and the use of a plumb as an expression

72

4.1 Technical standard of mud block building in ancient Egypt 1,500 BC[34].

of the technical standards and quality control mechanisms of masonry work at the time.

In Central Europe the history of technical rules in the field of earth building is closely related to the development of cities in the 14th and 15th centuries. As a result of their rapid growth, the availability of timber as primary building material became scarce. Moreover, timber structures were susceptible to fire damage, and fires were responsible for wiping out entire portions of cities, either by accident or as a result of war.

Both problems led to the drawing up of the first mandatory building codes across several German states. The *Ernestine Building Code*, introduced in 1556 in the state of Thuringia, did not permit houses to be solely constructed of timber but only as timber frame structures in combination with fired bricks, adobe or natural stone, or alternatively as Weller-structures, the German variant of cob. The *Saxony Forester Code* drawn up in 1575 permitted the use of timber for new house constructions only when the first floor could not be built of stone or cob. Two hundred years later, other building codes for Saxony (1786) (Fig. 4.2[35]), Prussia (1764) and Austria (1753) (called 'Egyptian stones') for wall construction. For hundreds of years building codes enforced the use of earth for building purposes by restricting the use of timber for house constructions.

The use of earth was characterised by a high degree of manual work. After around 1850, the Industrial Revolution brought about a fundamental change in the way building materials were produced in many industrialised countries: the mechanisation of production processes meant that building materials such as fired bricks could now be produced in large factories more economically and at better quality than before. Nevertheless, from the end of the 19th century new building materials such as steel, cement, concrete and reinforced concrete began to displace the use of earth until it was only rarely used. In Germany, earth experienced a brief revival after each of the

Deſſen Generale,

das Bauen mit Wellerwänden betreffend, vom 8. Auguſt 1786.

Ao. 1786. Friedrich Auguſt, Herzog zu Sachſen ꝛc. Churfürſt ꝛc.

Generale vom 2. Aug. 1763. Jn dem an ſämmtliche Creyß-Hauptleute und die Forſt-Aemter erlaſſenen Generali vom 2. Aug. 1763. *) iſt bereits unter andern vorgeſchrieben, daß die Aufführung neuer Gebäude mit hölzernen Schroten weiter nicht zu geſtatten, vielmehr darauf zu ſehen ſey, damit das untere Stockwerk derer Wohngebäude, ingleichen die Scheunen und Ställe, wo möglich, von Steinen aufgeführet, oder doch mit Ziegeln ausgeſetzt oder allenfalls geklebt, oder auch, nach Gelegenheit des Orts, ſogenannte Wellerwände von Lehm gefertiget werden ſollen.

Die Bauart von Wellerwänden ſoll Materiale an Lehm zu erlangen iſt, und hingegen da, wo der erforderliche Steine zum Bauen anders als mit vielen Koſten Lehm zu er- nicht zu haben ſind, die Bauart von Wellerwänden langen iſt, allgemeiner eingeführt wiſſen, begehren daher an- und es an *) C. A. cont. Abth. I. S. 1531.

Wir laſſen es hierbey noch ferner bewenden, und wollen beſonders an denjenigen Orten, wo das

durch gnädigſt, und befehlen, ihr wollet in vorkommenden Fällen die Unterthanen auf obige Vorſchrift bringen, auch diejenigen, welche Bauen na- aufmerkſam machen, auch diejenigen, welche Bauen vor- holz aus Unſern Waldungen bekommen, dahin, daß ſie die neu aufzuführenden Gebäude wenigſtens mit leimernen Wellerwänden zu errichten, auch die Dachung von Lehmſchindeln zu veranſtalten und hierzu ſich verbindlich zu machen haben, anweiſen, demnächſt in ſolchen Fällen, wo wegen unentgeldlicher Abgabe von Bauhölzern Bericht zu erſtatten iſt, wenn dieſe Bauart in einem oder andern einzelnen Falle nicht anwendbar wäre, ſolches ſammt den Urſachen beſtimmt mit anmerken. An dem geſchiehet Unſer Wille und Meinung. Gegeben zu Dresden, am 8. Aug. 1786.

Aus dem geheimen Finanz-Collegio.

An die Forſt-Aemter.

4.2 Decree of the Saxony's emperor Friedrich August from 1786 concerning the use of cob for ground floors[35].

world wars. As a consequence of industrialisation, building standards or regulations for building with earth were developed in just a few countries.

The German Earth Building Code (the *Lehmbauordnung*), drawn up in 1944, was the first contemporary technical standard in Europe dedicated to earth as a building material. The code summarised the entire technical knowledge of building with earth available at the time. In 1944, a year before the end of the Second World War, the industrial basis had been destroyed along with vast numbers of houses: millions of people were homeless and urgently needed new shelter. Very often earth was the only locally available building material. The German State Building Authority's intention was to regulate the use of earth as a building material for the period of reconstruction after the war. However, as a result of post-war reorganisation, the code could only be put into effect seven years later in 1951 as DIN 18951. In the five years that followed its issue, a number of further DIN-codes were drafted, dedicated to different fields of building with earth. Of these only the first, DIN 18951, came into force – all the others did not progress beyond draft status.[1] By the 1970s, the use of earth as a building material had all but disappeared as a result of industrialisation, and in 1971 the DIN was withdrawn and not replaced.

At the beginning of the 1980s, after the experience of the oil crisis in 1973 and health scandals caused by so-called modern building materials, a new way of thinking arose in some industrialised European countries: alongside 'traditional' economic and technical aspects such as durability, strength and so on, consumers started to give greater consideration to ecological aspects such as health, recyclability and low embodied energy as well as aesthetic and 'soft' design factors such as indoor climate, colour and surface qualities.

These 'new' ecological aspects are now anchored in the European Union framework of building regulations. The principle of 'life cycle assessment LCA' as a method for evaluating the sustainability of building materials will lead to a new generation of building standards. In the context of these developments, earth as a building material can be seen in a new light.[2,42]

4.2   Types of 'standards' for earth buildings

The International Standards Organisation (ISO)[3] distinguishes between the terms 'standards' and 'normative documents' as follows:

A *standard* is a 'document, established by consensus and approved by a recognised body, that provides, for common and repeated use, rules, guidelines or characteristics of activities or their results, aimed at the achievement of the optimum degree of order in a given context' and 'should be based on the consolidated results of science, technology and experience, and aimed at the promotion of optimum community benefits'. In this context 'consensus' is a general agreement characterised by 'the absence of sustained opposition to substantial issues'.

The 'promotion of optimum community benefits' means that standards primarily serve economic and social aims, facilitating the exchange of goods and services, protecting the consumer (safety, product quality, etc.) and ensuring a good quality of life (health, hygiene, environment, etc.).

The results of science, technology and experience (the 'state of the art') are not static and change continually as new developments arise, which in turn must be reflected in the standards. Standards therefore need to be periodically renewed, revised and updated. This process is the responsibility of the standards writing body.

Technical standards include building standards. The latter are usually established norms or requirements approved and recognised by state building authorities. Building standards can be classified into *material* standards, which describe the 'means' to achieve a 'result' (construction) and *construction* standards that describe the manufacture of materials necessary to achieve a construction with particular performance characteristics. Special topics (building with earth) can also be included as separate chapters of general building standards.

Producers of building materials can apply for a licence from the state building authority for the use of a specific building product. This 'technical permit' is a special type of building standard for 'repeated use' by the producer. It is not published publicly and is valid only for that manufacturer's specific building product and for a specified time period. It defines the qualities for production and terms for the ongoing control of the production process for a single building product.

A *normative document* is a 'document that provides rules, guidelines and characteristics for activities or their results', and as a result has neither the scope nor the endorsement of a standard, although it can become a 'standard' after adoption by a governmental body. Normative documents are developed by a group of specialists or organisations with proven competence in the respective field and are published for general use.

Technical standards, including those for building, can also be classified into different groups according to the geographic area and corresponding issuing organisation.

4.2.1   International standards

These are issued by the International Organisation for Standardization (ISO) in Geneva with about 150 members represented by the national organisations for standardisation.

4.2.2   Regional standards

These are issued in Europe by the European Committee for Standardization (CEN) in Geneva as European Standards EN. Members of the CEN are all member states of the EU as well as Switzerland and Norway. In addition, there are also European Building Codes ('Eurocodes EC') that describe uniform standards for the design, measurement and construction of buildings in the EU, e.g. the 'measurement and construction of brickwork' (EC 6). Earth blocks do not feature in this building code.

4.2.3   National standards

These are issued by the National Standards Bodies (NSB), in Germany the Deutsches Institut für Normung e.V. (DIN), which in turn are members of the CEN or ISO. National standards can also be issued by special (building) organisations or associations acting on a national level if they follow a predefined process of approval to acquire 'legal status'. In order to be recognised as a building standard by the national state building authorities, a draft code has to pass a predefined process of approval:

1   The draft has to be developed by a group of specialists or organisations with proven competence in the respective field (e.g. building with earth). This group is also known as a 'Technical Committee'.
2   The draft has to be submitted for public discussion to a broad audience of specialists with approved competence (in building with earth) with the aim of reaching a 'consensus among specialists'.

3   The 'consensus draft' is then presented to the national state building authorities for public consultation for a limited time. Comments and suggestions resulting from public enquiries are evaluated and if necessary considered. A revised 'final draft' is drawn up and submitted for ratification as a national/regional standard to the relevant authorities.

4   After ratification by the national state building authorities the draft is accorded the status of a building standard and is recommended for implementation.

5   European national standards must be registered with the department of standards at the European Commission in Brussels in order to disseminate the required information to all appropriate departments of the member states.

6   The building standard is published in a state decree and comes into force.

4.2.4   Local standards

In the USA several regional standards bodies exist that are responsible for issuing local building codes for a specific federal state, county or even a city. Each federal state, county or city may join depending on their geographic situation. Building standards issued by State Building Authorities are mandatory for building contracts. Normative documents issued by special national organisations or associations have not passed an approval process and therefore do not have any 'legal status', serving instead as a recommendation. Both types of documents must represent the 'state of the art' of a process or a technical development.

4.3   Normative documents for earth building

In recent years several studies have undertaken international surveys of standards and normative documents in the field of earth building. These were usually produced as part of financed projects by national technical groups in order to establish a knowledge basis for the development of own (national) standards or normative documents. Examples of such studies include Jiménez and Delgado and Cañas Guerrero, 2005,[4] which details various techniques using unstabilised earth, Maniatidis and Walker, 2003[5] which examines rammed earth codes and Cid et al., 2011[41] which gives a worldwide, overview about earth-building normative documents. Some handbooks on earth building also include chapters on the state of the art in this field, for example Houben and Guillaud, 1994[6]. McHenry, 1989[7] provides an overview of regional earth building standards in the USA.

4.3.1   Types of documents

Table 4.1 provides an overview of existing earth building standards and normative documents according to document type. Thirty-three different documents have been identified from 19 countries published by regional, national or local standards bodies over the last 30 years. The technical information they contain varies considerably.

The 'Australian earth building handbook HB 195-2002' was published by the national organization Standards Australia, but 'the Handbook has not been published under the auspices of the Standards Australia Committee BD-083, and therefore it should not be taken as representative of the views of the committee members'.[39] This document has not reached a 'consensus among specialists', which is an essential requirement of the approval process. Therefore, it was not included in Table 4.1.

In addition, numerous technical notes have been issued by state building research organisations (Overseas Building Notes and Overseas Information Papers/UK, Commonwealth Experimental Building Station/Australia, etc.) and non-governmental organisations (UNCHS/HABITAT, ILO, CYTED, Gate-BASIN, etc.). These have not been considered in this overview, but could nevertheless contain information that may be valuable for the analysis of existing standards or the development of new earth building standards.

Other national standards have been cited as reference documents for the resolution of specific aspects of earth building. These include standards that address soil classification, soil mechanics, testing procedures, the design of load-bearing earth walls and general codes for the structural design of buildings. Likewise, non-governmental bodies have also drawn up numerous technical notes in the field of conservation of historical earth buildings (cob, earthen infill, etc.). While these are not standards, they could be valuable for the drafting of new standards or normative documents on earth building because the conservation of historic earthen architecture constitutes a significant proportion of earth building activities. In the last few years, standards for the conservation of historic earthen buildings have also been developed by regional and national governmental organisations (Italy 2006,[43] Chile draft 2010).

The documents analyzed are almost always national standards (S) or normative documents (ND). Sometimes these take the form of individual chapters on earth building within broader national building codes (BC). In the USA, building standards are issued by the regional standards bodies, federal states, counties or even cities (L). One of the standards listed has been developed at a continental level (African Regional Standard ARS). The documents can be classified into three types with regard to their scope of application (see page 82):

Table 4.1 An overview of the types of earth building standards and normative documents

No.	Country	Document	Scope of document				Geogr. level	Ref.
		Name	Type	Soil	Building material	Construction system		
01	01	02	03	04	05	06	07	08
01	Africa	ARS 671-683 (1996)	S		EB	EBM	R	[16]
02	Australia	CSIRO Bulletin 5, 4th ed. (1995)	ND	E	EB, CSEB, EMM	RE, EBM	N	[22]
03	Australia	EBAA (2004)	ND	E	EB, EMM	EBM, RE	N	[23]
04	Brazil	NBR 8491-2, 10832-6, 12023-5, 13554-5 (1984-96)	S		CSEB		N	[44]
05	Brazil	NBR 13553 (1996)	S			CSRE	N	[44]
06	Columbia	NTC 5324 (2004)	S		CSEB		N	[45]
07	France[b]	AFNOR XP.P13-901 (2001)	S		EB		N	[27]
08	Germany	Lehmbau Regeln (2009)	S	E	C, LC, EB, EM, CP	RE, C, EBM, EP, EI, WL	N	[8]
09	Germany	RL 0803 (2004)	ND		EP		N	[11]
10	Germany	TM 01 (2008)	ND		EP		N	[9]
11	Germany	TM 02 (2011)	D[d]		EB		N	[10]
12	Germany	TM 03 (2011)	D[d]		EMM		N	[32]
13	Germany	TM 04 (2011)	D[d]		EP		N	[40]
14	Germany	TM 05 (2011)	ND	E			N	[36]
15	India	IS: 2110 (1998)	S	E, ES		RE	N	[19]
16	India	IS: 13827 (1998)	S		EB	EBM, RE[a]	N	[20]
17	India	IS 1725 (2011)	D[d]		CSEB		N	[37]
18	Kenya	KS02-1070 (1999)	S		CSEB		N	[47]
19	Kyrgyzstan	PCH-2-87 (1988)	S	E, ES		RE[a]	N	[25]

Table 4.1 Continued

No.	Document Name	Scope of document			Construction system	Geogr. level	Ref.
		Type	Soil	Building material			
01	02	03	04	05	06	07	08
20	New Zealand NZS 4297-9 (1998)	S		E, EB	RE, EBM, EP[a]	N	[18]
21	Nigeria NIS 369 (1997)	S		CSEB		N	[48]
22	Nigeria NBC 10.23 (2006)	BC	E		EBM, RE	N	[26]
23	Peru NTE E.080 (2000)	S		EB	EBM[a]	N	[17]
24	Spain MOPT Tapial (1992)	ND	E		RE	N	[28]
25	Spain UNE 41410 (2008)	S		CEB		N	[46]
26	Sri Lanka Specification for CSEB, SLS 1382 part 1-3 (2009)	S		CSEB	EBM	N	[21]
27	Switzerland Regeln zum Bauen mit Lehm (1994)	ND	E	EB, LE, EM	EBM, RE, EI, WL	N	[12]
28	Tunisia NT 21.33, 21.35 (1998)	S		CEB		N	[49]
29	Turkey TS 537, 2514, 2515 (1985-97)	S		CSEB		N	[50]
30	USA UBC, Sec. 2405 (1982)	BC			EBM[a]	L	[15]
31	USA 14.7.4 NMAC (2006)[c]	BC		EB, EMM	EBM, RE[a]	L	[14]
32	USA ASTM E2392/E2392M (2010)	S	E	EB, EM,	C, EBM, RE, EM, WL[a]	N	[13]
33	Zimbabwe SAZS 724 (2001)	S	E		RE	N	[24]

[a] Concerns the earthquake resistance of earth construction systems.

[b] In 2010 the French national earth building organisation AsTerre began drafting new French earth building standards covering the main earth construction systems prevalent in France, including EB, rammed earth (pisé) (RE) cob (C), light earth (LE) and earth plaster (EP).

[c] A number of local adobe building codes (L) exist that bear similarity to the NMAC (San Diego/CA, Tucson/AR, Marana/Pima/AR, Boulder/CO).

[d] Draft of NSB

S Standards issued by national standards bodies (NSBs) or specialised organisations that have passed a predefined procedure of approval and are recognised by a state building authority.

BC Building codes issued by NSBs with one or more chapters on building with earth.

ND Normative document issued by specialised organizations that have passed a predefined procedure of approval but are *not* recognised by a state building authority.

L Local
N National
R Regional
C Cob
CP Clay panel
E, ES Earth, earth stabilised with cement
EB Earth block; Adobe; compressed earth block, CEB; compressed stabilised earth block, CSEB; poured earth blocks, PEB
EBM Earth block masonry
EM Earth mortar; earth plaster, EP; earth masonry mortar, EMM; earth spray mortar, ESM
EI Earth infill; wattle and daub, WD; poured earth infill, PEI
LC light clay
RE Rammed earth; cement stabilized rammed earth, CSRE
WL Wall lining

1   soil
2   earth building material
3   earth construction system.

This classification relates to the life cycle assessment (LCA) of earth as a building material (Fig. 4.3[2]). Each section of this cycle has its own specific requirements that guarantee the usability of the material or product in that section. Usability requirements are defined in standards by corresponding parameters. Typical parameters include:

Soil: grading, plasticity, natural constituents content, linear shrinkage
Building material: strength/deformation characteristics
Construction system: strength/deformation characteristics, aspect ratio, sound and fire performance, thermal characteristics, earthquake resistance.

4.3.2   Classification of earthen materials and construction systems

In the standards listed in Table 4.1, the terminology used to describe earthen building materials and earth construction systems reflects the differing levels of technological development in different parts of the world. The definitions

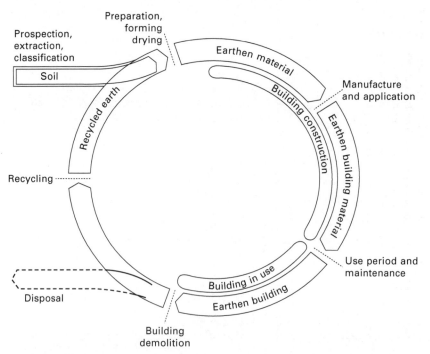

4.3 Building with earth presented as self-sustaining life cycle[2].

used are often derived from local building traditions and are not always appropriate for use as contemporary technical terminology. Nevertheless, the existing building stock of traditional earth buildings must also be considered in standards concerning conservation work. On the other hand, 12 of the 33 standards and normative documents are less than 5 years old, demonstrating an increasing acceptance of earth as a viable contemporary building material. At present, there is no common internationally accepted terminology for earth building. This, however, is an essential prerequisite for establishing earthen building materials alongside other conventional building products for contemporary building. In this chapter we recommend the following terminology based primarily on recent standards documents and guidelines.[8,9,10,13,18,29] The recommended terminology encompasses all earthen building materials and earth construction systems commonly used today and detailed in (earth) building standards. In some cases it was necessary to modify regional definitions in order to avoid misunderstandings.

Soil

A granular material derived from weathered parent rock, often transported and sedimented by natural processes. It can contain natural constituents (organic content/humus, soluble salts). A soil that is already suitable for building purposes is classified as 'earth'.

Earth (E)

A soil suitable for building purposes, which contains an appropriate mixture of silty, sandy and/or gravelly particles, together with clay minerals as a natural binder and water. Earth does not set chemically but hardens in the air. The specific properties of clay minerals mean that hardened earth can be softened again to a plastic mass through the addition of water (re-plasticisation). As a result, it can be recycled as often as required, which is one of the ecological qualities of the material. Other natural constituents of soils (e.g. lime, iron oxides) can also function as a natural binder but cannot be re-plasticised. Earth for building purposes must be prepared to a homogeneous mass, shaped and dried in order to produce an earth-building material or product.

Earth building materials

These can be produced in situ for individual constructions or manufactured industrially according to standardised reproducible procedures as defined in standards or norms. *Unstabilised* earth-building materials harden in the air with clay minerals acting as a natural binder. *Stabilised* earth-building materials contain artificial tempering admixtures (e.g. granular or fibrous particles), which are added during the preparation process. Stabilising admixtures can also constitute artificial binders (lime, cement) that change the clay minerals chemically (i.e. chemical stabilisation) in order to improve the mechanical properties of the earth building materials. A chemically stabilised earth material cannot generally be re-plasticised by adding water, and when possible only slightly. *Unshaped* earthen building materials are dry or moist ready mixtures for specific uses such as rammed earth (RE), cob walling (C), light clay (LC), earthen infill (EI), wall linings (WL), earthen mortars (EM), etc. *Shaped* earthen building materials are manufactured from moist unshaped earthen building materials by moulding in conjunction with different types of compaction (i.e. mechanical stabilisation), for example, earth blocks (EBs) or clay panels (CPs). In a dry state they are used for a specific purpose such as earth block masonry (EBM), EI, wall linings (WL), etc.

Earth mortars (EM)

These are earthen building materials containing appropriate tempering admixtures. They are delivered as unshaped dry or moist ready mixtures and are mixed to serve a specific type of application, which must be declared:

• Earth *plaster* mortar (EPM) is applied to the surfaces of (earth) building constructions and usually contains sand and/or natural fibrous tempering materials. A large number of different types of EPMs exist with coarse/ fine aggregates, particular colours or stabilised with artificial binders or tempering admixtures for special granularity or colour effects, etc.
• Earth *masonry* mortar (EMM) is used for constructing EBM and usually contains sand as a natural tempering material
• Earth *spray* mortar is used for manufacturing EBs or parts of a building by spraying into or against formwork. They contain appropriate tempering admixtures and in many cases artificial binders.

Earth blocks (EB)

A shaped earth building product manufactured from moist unshaped earth building materials with appropriate admixtures in a moulding process. The mechanical properties of earth blocks are influenced significantly by the type of compaction (manual/mechanized) and the amount of mixing water. Unlike conventional bricks (ceramic bricks), EBs are air-dried and not fired.

• *Adobe*: air-dried masonry block without any specific type of moulding or compaction. Adobe blocks made of a wet earth mixture are sometimes called 'mud' blocks

- *Compressed*: EBs formed in a block mould with the addition of (static) mechanical compression (CEB) or (dynamic) compaction. The use of compressed stabilised earth blocks (CSEB)/cement stabilised compressed earth blocks (CSCEB) is widespread in many countries
- *Extruded*: EBs formed by a process of extrusion
- *Cast*: EBs manufactured by pouring or spraying a slurry of (chemically) stabilised earthen materials into forms.

Clay panels (CP)

These differ from EBs in size and thickness as well as in their method of manufacture. They are typically used for non-load-bearing partitioning interior walls. *Thin CPs* are usually up to 30 mm thick and of a similar size to conventional plasterboard panels. They are mounted on a supporting construction of wood or metal studs. *Thick CPs* are large-format EBs that are self-supporting and can be laid with mortar or glued.

Earth construction systems

A variety of construction systems exist that employ earthen building materials either entirely or in part for load-bearing and non-load-bearing elements in different parts of a building. Common earth construction systems include earth block masonry (EBM), rammed earth (RE), cob (C) and earthen infill (EI).

Earth block masonry (EBM)

A construction system using earth blocks (EB)/compressed earth blocks (CEB)/cement stabilised compressed earth blocks (CSCEB) and earth mortars (EM) for building structures as well as for the earthen infill (EI) of framed or half-timbered constructions.

Cob (C)

A construction system in which a moist unshaped cob mixture (straw clay) is lightly tamped into place without formwork to form monolithic walls.

Rammed earth (RE)

A construction system that uses a slightly moist unshaped rammed earth mixture. The earth mixture is filled into prepared formwork in layers that are individually compacted (rammed). The series of compacted layers produces the desired shape of the final structure. The formwork can be removed before the compacted material has fully dried. Special decorative effects can be achieved by using earth mixtures of different granularity or colour.

Poured earth (PE)

A construction system in which a slurry containing earth mortar and artificial binders is sprayed against or poured into formwork similar to those used for in situ concrete.

Earth plaster (EP)

Earth plasters are made of earth plaster mortars (EPM). They are applied manually or mechanically in one or more layers to the interior surfaces of buildings or to exterior walls protected against exposure to water. The overall thickness of all earth plaster layers is usually no more than 20 mm for a new application. EPs help to improve the indoor climate by regulating humidity.

Earthen infill (EI)

A construction method used in conjunction with framed or half-timbered constructions that serve as the load-bearing system. The panels between the timber studs, rails and braces are filled with a non-structural 'infill' or 'nogging' in a variety of different techniques:

- *Wattle and daub* (WD): There are numerous different regional variants. The wattle and daub technique typical in Europe employs wooden struts jammed between the timber members that are interwoven with a wickerwork of split willow branches. This supporting 'wattle' is then daubed from both sides with a straw clay mixture
- Earth block masonry (EBM) infill (nogging)
- Light clay (LC), Poured Earth (PE) infill using formwork.

Wall linings (WL)

This technique, sometimes called an 'inner leaf', is often used in renovation work to improve the thermal insulation, wind-proofing and noise insulation of existing thin (historical) exterior walls, in particular where external insulation is not an option because the half-timbered elevations must remain visible. WLs can be executed as masonry wall linings using earth blocks (EB), as light clay wall linings using moist unshaped light earth or earth spray mortar/poured earth (PE) or as a lining of dry stacked earth blocks to create a suitably massive heat retentive thermal mass. Clay panels (CP) can also be used.

4.4   Selecting the parameters for earth building standards

Historically, earth construction techniques were 'non-engineered' building systems. Earth building materials were produced in-situ, quality tests were carried out according to quick and simple *field tests* in the form of macroscopic visual observations and manual handling. The equipment required was inexpensive and easily accessible, but the test conditions were not exactly reproducible. As such, the results could be classified as 'estimated values' and depended on the subjective observation of the tester. For non-engineered building systems field tests were entirely sufficient.

Since the late 20th century, the increased interest in building with earth has resulted in a change in the character of production methods from predominantly 'hand-made' to large-scale industrial production. The corresponding building techniques have now become 'engineered' and standardised construction systems. As a result, the character of test procedures also changed from field tests to *laboratory tests*. 'Engineered' construction systems require industrially produced building materials with defined parameters that are arrived at using standardised and reproducible testing procedures and quality controlled by authorised laboratories with qualified personnel using standardised test equipment.

An analysis of the standards and normative documents listed in Table 4.1 shows that uniform test procedures for determining the mechanical parameters of earth building materials as well as common design criteria for earthen construction systems do not currently exist. Instead, test procedures from other disciplines are often used, e.g. from soil mechanics for soil classification or from concrete testing for determining compressive strength. The applicability of these 'adopted' procedures for earth building materials has as yet not been proven.

It is, therefore, necessary to bring the test procedures for earth building products up to the same level as for other 'standardised' building materials in order to improve their range of use and competitiveness in the building sector. While field tests can support the results of laboratory tests, they cannot replace them. As such, they can be included as non-mandatory information in the appendices of standards and normative documents (e.g. ASTM, 2010[13]). A central reason for establishing proper standardised testing and quality control procedures is to ensure the quality of industrially manufactured earth building materials so that producers have a legal status for their products.

The differences between the analysed standards can be demonstrated using three examples of laboratory testing procedures and design criteria covering key areas of earth building:

1   soil classification and selection: granular composition and plasticity

2   compressive strength and optimum compaction for rammed earth
3   slenderness of wall structures.

4.4.1   Soil classification and selection

The primary parameters for selecting soils suitable for earth building purposes are the granular composition, which serves as the 'load-bearing skeleton', and the plasticity or cohesion caused by the type and amount of clay particles ($d \leq 0.002$ mm). These particles act as a natural binder holding together the coarse grains. The plasticity can also be influenced by natural admixtures such as humus and soluble salts. The nature of the binding properties of clay minerals mean that soil can be prepared, including mixing with admixtures, in a wet state but also retains its shape and stability, e.g. as a block or panel, during desiccation and when dry. Some natural soils exist that exhibit a nearly 'ideal' combination of plastic properties and grain composition for specific uses, e.g. loess-type soils for use as earth plaster. Secondary soil parameters, such as shrinkage, can be derived from the primary parameters. The interaction between these physical and chemical – mineralogical parameters is very complex and cannot be analysed independently of one another.

Industrially produced earth building materials consist of different ingredients which are exactly dosed and prepared according to defined recipes. Soil as a natural binder characterised by its plastic properties is only one part of this mixture. It is often replaced by industrially produced clay powder or other artificial binders. The importance of natural soil as a binding agent in the mixture therefore depends on the specific composition of a particular earth building material. Discussions concerning the definition of general or specified limits of plastic soil properties and soil grading in earth building standards does not lead to satisfactory and reliable results.

The new German Technical Recommendation DVL TM 05 'Quality control of soil as an ingredient of industrially produced earth building materials'[36] defines controls for the soil 'quality' in the form of a series of appropriate test procedures for determining typical soil parameters including plasticity, grain composition and soluble salt content level. A producer of such soils can declare these parameters to aid earth building product manufacturers in sourcing appropriate soils. The control of mechanical properties (compressive strength, shrinkage, etc.) that influence the usability of the resulting earth building product, e.g. earth mortar EM and earth blocks EB, including the definition of permissible limits, will be defined in 'product standards'.[32,10,40]

Grain composition and plasticity

Two approaches are used in earth building standards to quantify the parameters 'grain composition' and 'plasticity' for soils suitable for building:

The first approach defines numerical ranges for each of the soil parameters 'grading' and 'plasticity'. This approach was developed by Houben and Guillaud, 1989, 1994[6] on the basis of standard geotechnical testing procedures used by the Unified Soil Classification System (USCS). The American ASTM standards use other maximum grain diameters: silt 0.002–0.075mm; sand 0.075–4.75mm; gravel > 4.75mm. They propose two separate diagrams for the ranges of granular composition and plasticity with respect to their different uses: rammed earth (RE), adobe, compressed earth blocks (CEB) and cement stabilised earth blocks CSEB. 'The limits of the zones recommended are approximate, the permitted tolerances vary considerably. Present knowledge does not justify the application of narrow limits. The zones are intended to provide guidance and are not intended to be applied as a rigid specification.'[6]

Nevertheless, these diagrams were adopted by different earth building standards and normative documents as given in Table 4.1: ARS 680-681 (1996)[16] for CEB and earth masonry mortar EMM, AFNOR (2001)[27] for CEB, MOPT (1992)[28] for RE and the Swiss earth building guidelines (1994)[12] for RE and adobe (plasticity diagrams only). Figure 4.5 on page 92 shows the respective diagrams according to a study by Jiménez Delgado and Cañas Guerrero, 2005.[4]

Table 4.2 shows the following recommendations for unstabilised and cement stabilised rammed earth and earth blocks respectively with regard to soil gradation and plasticity.

The second approach used by the German Lehmbau Regeln (2009)[8] issued by the Dachverband Lehm e.V. also employs the geotechnical parameter 'soil plasticity' in conjunction with a second recommended parameter, 'binding force', which summarises the physical-chemical effects of 'grading' and 'plasticity' including the influence of possible natural admixtures in a single parameter (DIN 18952-2[1]). The respective testing procedure is known as the '8-shaped' test in which the ultimate tensile strength (i.e. binding force or cohesion) of an 8-shaped specimen made of a 'normative' test consistency is determined (Fig. 4.4). The tested soils are classified numerically according to their binding force on a scale ranging from 'lean/poor' to 'rich/clayey'.

The 'binding force' classes correlate numerically to 'linear shrinkage' classes. A soil suitable for a specific use, e.g. rammed earth, earth blocks, etc., should exhibit a defined binding force classification, which is given in the general requirements of soil selection for a specific earth building product[8].

Table 4.3 shows the numerical correlation between binding force and linear shrinkage according to DIN 18952-2[1] and their limits of suitability for a specific use.[8] Soils with a binding force < 0.005 N/mm^2 cannot be analysed exactly using the 8-shaped test but these are unsuitable for many

Table 4.2 Numerical ranges of recommended soil gradation and plasticity for stabilised rammed earth and earth blocks

No.	Country	Document (Table 4.1)	Earth building material	Soil gradation	Soil plasticity PI (%)
01	Brazil	NBR 8491-2, 10832-6, 12023-5, 13554-5 (1984-96) [44]	CSEB		$\leq$ 18, LL $\leq$ 45
02	India	IS: 2110 (1998) [19]	CS earth	Content of sand fraction $\leq$ 35%; d = 0.075–4.75 mm[a]	8.5-10.5, LL $\leq$ 27
03	India	IS 1725 (2010) [37]	CSCEB	10–15 % clay (d < 0.002 mm) > 65 % sand (d = 0.075–4.75 mm)	$\leq$ 12
04	Kyrgyzstan	PCH-2-87 (1988) [25]	CSRE	content of soil particles d < 0.005 mm: < 10–30%[b]) content of soil particles d > 0.005mm: < 70-90%[b])	2 – 9
05	Peru	NTE E.080 (2000) [17]	EB, un-stabilised	55–70% sand 15–25% silt 10–20% clay	
06	Spain	UNE 41410 [46]	CEB	$\geq$ 10% clay	
07	Sri Lanka	SLS 1382-1 (2009) [21]	CSCEB	10–15% clay (d < 0.002 mm) 5–20% silt (d = 0.002–0.06 mm) > 65% sand + gravel (max. particle size $\leq$ 12 mm)	$\leq$ 12
08	Zimbabwe	SAZS 724 (2001) [24]	RE, un-stabilized	50–70% fine gravel and sand 15–30% silt 5–15% clay	

[a] grain fractions according to US standard systems.
[b] according to Russian standard system СНиП (SNiP) the maximal particle diameter of clay is 0.005 mm.

earth building purposes. A numerical correlation between the parameters 'soil plasticity' (first approach) and 'binding force' is recommended in Schroeder, 2010[2] as a first approach. Further research is needed in this area. The permissible maximum grain diameter and minimum percentage of clay and recommended plasticity parameters for a specific use can be derived from Fig. 4.5.[4]

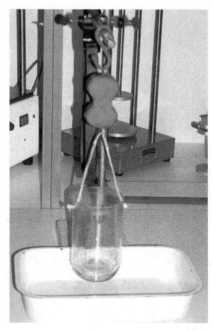

Binding force [g/cm²]	Classification
50–80	very poor
> 80–110	poor
> 110–200	almost clayey
> 200–280	clayey
> 280–360	very clayey

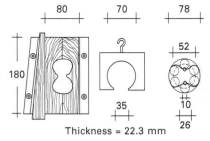

Thickness = 22.3 mm

4.4 The 'binding force' test[2,6,8].

Table 4.3 Correlation between binding force, linear shrinkage and plasticity[1]

No.	Property	Binding force (N/mm²)	Plasticity	Linear shrinkage (%)	Suitable for use as according to [8]
01	Very lean	0.005–0.008	None – low	0.9–2.3	EI
02	Lean/poor	>0.008–0.011	Low	0.9–2.3	EI, RE, C, EB, EM
03	Semi-rich	>0.011–0.02	Medium	1.8–3.2	RE, C, LC,EB, EM
04	Rich/clayey	>0.02–0.028	Medium	2.7–4.5	RE, C, LC, EB, EM
05	Very rich	>0.028–0.036	High	3.6–9.1	LC
06	Clay	>0.036	High	>9.1	LC

The ASTM E2329 2010[13] also contains an appendix with 'non-mandatory information'. This part recommends two field tests ('ribbon' and ball test) in order to estimate the plasticity and the cohesion of a test soil.

Permissible salt content levels

Naturally occurring soils exhibit a 'three-phased' system consisting of soil solids, water and air. The solid phase is formed by the 'mineral skeleton' as well as other natural ingredients such as lime, soluble salts and organic matter. The soluble salt content in natural soils is considerably higher in dry

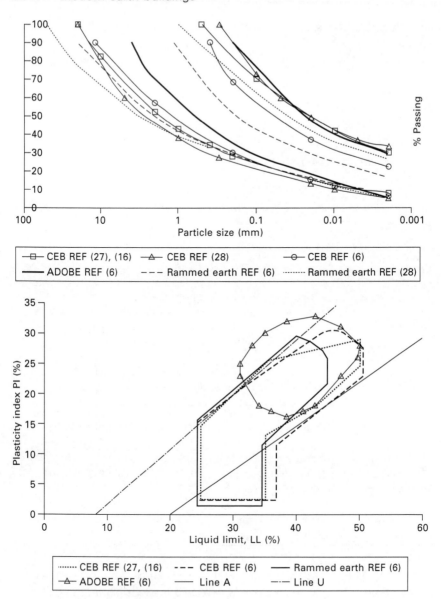

4.5 Recommended limits of grain composition and soil plasticity in earth building standards[4].

regions with low precipitation and where the ground water is between 2 and 3 m beneath the soil surface. The soluble salts usually consist of sulfates, chlorides and nitrates of calcium, sodium, potassium and magnesium.

If water is present in the structure, soluble salts can lead to typical damage patterns in buildings structures. Possible water sources include rain ingress or vapour or moisture penetration (rising damp) from the ground where a damp proof course is lacking. Moisture from the surrounding ground rises through capillary action transporting additional salts into the wall. The moisture eventually evaporates from the wall surfaces leaving behind salts in concentrated quantities, which appear as crystalline efflorescence on the wall surface. The salts additionally absorb water hygroscopically, reducing the strength of the earth building materials.

Only very few earth building standards define permissible levels of soluble salts in soils for use as a raw material for the production of earth building products as shown in Table 4.4. The permissible levels also differ considerably. Sulfates will damage cement and can, therefore, be a special problem for cement stabilised earth constructions. In this case, the use of sulfate-resistant Portland cement is recommended.

4.4.2   Compressive strength and optimal compaction

The mechanical strength of a soil or earth building material is one of the most fundamental items of information for earth construction design, especially for load-bearing wall constructions. The prepared soil or earth mixture must be compacted into a dense state of optimal compaction to minimise the void content. In this state the strength increases as it dries to a maximum while volumetric changes are minimal. The mechanical strength of a soil or earth building material is normally determined by the unconfined compressive strength (UCS) of test specimens as a result of uni-axial compression.

The maximum compaction of a given soil sample (maximum dry density MDD) is usually determined by heavy manual compaction, internationally known as the PROCTOR test and standardised in national norms in the field

Table 4.4 Permissible salt content levels in earth building standards

No.	Country	Document (Table 4.1)	Building material	Entire content of soluble salts (% by mass)
01	Australia	EBAA (2004) [23]	EB, EMM	To an extent which will not 'impair the strength or durability of a wall'
02	Germany	TM 02, 03 (2011) [10] [32]	EB, EM	< 0.12
03	India	IS: 2110 (1998) [19]	CS earth	≤ 1; sodium salts ≤ 0.1
04	Kyrgyzstan	PCH-2-87 (1988) [25]	CSRE	≤ 3
05	New Zealand	NZS 4298 (1998) [18]	E, EB	To an extent which will not 'impair the strength or durability of a wall'
06	Peru	NTE E.080 (2000) [17]	EB	Free of 'alien materials'
07	USA/New Mexico	14.7.4 NMAC (2006) [14]	EB, RE	≤ 2
08	Zimbabwe	SAZS 724 (2001) [24]	RE	Free

of soil mechanics (e.g. BS 1377-4, 1990). This general approach is also used in several earth building standards and normative documents for determining the UCS of soil specimens. For most practical earth building purposes an MDD $\geq 0.90\%$ will be accepted.

The soil preparation process usually involves the addition of water to cover the particle surfaces with a thin film of water. During compaction these films allow the soil particles to slide over each other more easily. Finer particles fill voids created by the coarser ones. This process works well for coarse-grained soils because the water film is extremely thin in comparison to the grain diameter. Up to a certain point the addition of water replaces air in the voids in the soil. After a relatively high degree of water content has been reached, the amount of entrapped air remains essentially constant. This 'optimum moisture content' (OMC) for a given soil and a defined compaction energy allows one to derive a maximum weight of soil per volumetric unit.

This process only works to a limited degree with fine-grained soils. The smaller particles are likewise covered with thin coatings of clay minerals, which also adsorb films of water. Due to the higher overall proportion of clay minerals and water in fine-grained clayey soils, the films of water are correspondingly 'thicker' compared with coarse-grained soils and resist significant compaction. The MDD is therefore lowest for fine-grained soils using the same compaction energy. The density can be increased for all soil types by using a greater compaction energy. In the case of fine-grained soils, a water content level higher than the OMC according to PROCTOR during compaction results in a higher compressive strength as a result of systematic laboratory tests.[2]

When considering strength properties, one should be aware that laboratory test conditions (size of test specimen, compaction energy, defined loading rate per time unit) cannot easily be transferred to real construction site conditions. 'Engineered' earth building structures require comparable design parameters determined by reproducible uniform test procedures. One laboratory test sample includes a series of a minimum of three and more individual specimens from which the arithmetic *average* value is determined for the UCS. The *characteristic* value denotes the standard deviation of the test sample. The *design* value (permissible working stress) also incorporates a safety factor to take into account likely worst-case ambient conditions that may affect the earth construction during use, in particular with regard to moisture.

Table 4.5 provides an overview of the recommended design values and specimen details/characteristic UCS values for rammed earth (RE) in the earth building standards and normative documents listed in Table 4.1.

Table 4.5 shows that uniform test procedures for determining the UCS of soils or earth building materials do not exist. The test specimens alone have different shapes and dimensions (cube, prism, cylinder). The UCS of

Table 4.5 Recommended design values and specimen details/ characteristic UCS values for rammed earth (RE)

No.	Country	Documents (Table 4.1)	Type	Design UCS (N/mm²)	Specimen details/characteristic USC β_D (N/mm²)
01	Australia	CSIRO Bull. 5, 4th ed. (1995) [22]	S	0.7[a]	Height to thickness or diameter ratio (0.4–5.0 or more) with aspect ratio factors/5.2²)
		EBAA (2004) [23]	ND	1.0[a]	
02	Germany	Lehmbau Regeln (2009) [8]	S	0.3–0.5	Cube, 20 cm³/2–4; 1 sample of 3 or more individuals, the lowest is standard
03	India	IS: 2110 (1998) [19]	S	1.4[a,b]	Cylinder, height/diameter (mm): Fine-grained soils: 100/50 Medium-grained soils: 200/100
04	Kyrgyzstan	PCH-2-87 (1988) [25]	S	0.63–3.6[b]	Cube 15 cm³/0.95–4.7; 1 sample of 3 or more individuals after 3, 7 and 28 days
05	New Zealand	NZS 4297-9 (1998) [18]	S	0.5	Height to thickness or diameter ratio (0.4–5.0 or more) with aspect ratio factors/1 sample of 5 or more individuals, the lowest must be > 1.3 N/mm² if aspect ratio factor = 1
06	Spain	MOPT Tapial (1992) [28]	ND	0.2[c] 0.1[c] wet environment	Cube, 30 cm³; 1 sample of 10 individuals/compress. crushing strengths at 5 cm³ cubes cut from a wall: Low strength 0.6 Medium strength 1.2 High strength 1.8
07	Switzerland	Regeln zum Bauen mit Lehm (1994) [12]	ND	0.3–0.5	Cube, 20 cm³/2–4
08	USA	14.7.4 NMAC (2006) [14]	SC		Cube, 10.2 cm³/2.07; 1 sample of 5 or more individuals, the lowest may be < 2.07 N/mm² but ≥ 1.725 N/mm²
09	Zimbabwe	SAZS 724 (2001) [24]	S		1.5 for one-storey walls up to 400 mm thick 2.0 for two-storey walls

[a] 'Design' and 'characteristic' values not distinguished.
[b] Cement-stabilised.
[c] E/I: Exterior/interior walls.

unstabilised soils is in the range 0.7–4.0 N/mm^2 and for (cement) stabilised soils it is significantly higher. In this case the UCS is determined after 28 days (additionally 3 and 7 days) along with the 'wet' UCS.

Only very few documents distinguish between 'design' and 'characteristic' values of the UCS. The German standards[8] and Swiss normative documents[12] recommend design values for UCS which include safety factors of 6.7– 8.0. The smallest 'design' value is 0.3 N/mm^2, the largest permissible 0.5 N/mm^2.

The Spanish normative document[28] recommends 'approximate but safe values of permissible stresses' (design values) for 'low risk projects',[31] for interior walls not exposed to wet environments and for exterior walls. For 'important projects' the characteristic UCS should be determined in a laboratory using 30-cm^3 cubes and then divided by a recommended safety factor of 6.0 for exterior walls exposed to the weather, or 3.0 for interior walls in order to obtain the design strength. For 'most cases' a 'low/medium/ high crushing strength' can be chosen for the respective actual situation and than compared with the real situation from in-situ testing of 5-cm^3 samples cut from test units by simple means.[31]

Walker et al., 2005[30] recommend partial safety factors ranging between 3.0 and 6.0 to account for variations in materials and the quality of work. The selection of the partial safety factor is a matter discretion for the designer, who needs to consider the possible consequences of failure and likelihood of accidental damage.

The SLS 1382: Part 3[21] recommends the determination of characteristic compressive strength for CSEB masonry on the base of BS 5628: Part 1.

4.4.3   Wall thickness

Limits for wall thickness and slenderness are key parameters required in building standards in order to limit the likelihood of excessive cracking and compression buckling under service load. The slenderness of a wall h/t is the ratio of wall height to thickness. In traditional building this ratio is usually 10 for freestanding walls and 18 for walls laterally restrained at the top and bottom. In seismic regions[13,14,17,18] these values are significantly lower and special reinforcement of the walls is required. Most standards limit earth constructions to buildings with a maximum of two-storeys. The SLS 1382[21] is applicable for load-bearing CSEB masonry construction up to three stories. Table 4.6 provides an overview of the limits for wall thickness and slenderness of laterally restrained load-bearing earth walls in different construction techniques.

Table 4.6 Recommended values for wall thickness and slenderness of earth walls

No.	Country	Document (Table 4.1)	Construction system (Table 4.1)	Wall type[a]	Min. thickness t of wall (mm)	Max. height h (mm)	Max. slenderness h/t (–)	Max. distance of laterally supported walls (mm)
01	Australia	EBAA (2004) [23]	EBM	E/I	200/125	4000/2700 3000/4000	20/21.6	3500
			RE	E/I	200/200	3500	15/20	3500
02	Germany	Lehmbau Regeln (2009) [8]	RE	E/I	325/240	3250	10	4500
			EBM	E/I	365/240	3250	13.5	4500
			C	E/I	400/400	3250	8.1	4500
03	India	IS: 2110 (1998) [19]	CSCEBM, CSRE	L/NL	300/200	3200	10.7/16	
04	India	IS: 13827 (1998) [20][f]	EBM, RE	E/I		$8 \times T$		$10 \times t$ or $64 \times t^2/h$
05	New Zealand	NZS 4297-9 (1998) [18]	RE		250	3300	6[c]	
			CEBM		130	3300	16[c]	
06	Peru	NTE E.080 (2000) [17]	EBM		400–500	2.400–3.000	≤ 6	[d]
07	Sri Lanka	SLS 1382-1 (2009) [21]	CSCEBM				acc. to BS 5628:Pt.1, clause 25	
08	Switzerland	Regeln zum Bauen mit Lehm (1994) [12]	RE	E/I	300/500[b]	3500	11.6/7	5000
			EBM	E/I	200/300[b]	3500	17.5/11.7	5000
09	USA	14.7.4 NMAC (2006) [14]	RE	E/I	457/305[b]	2438–3048		7315 (24 ft) max. length
			EBM		254/356[b]			
10	USA	ASTM E2392/E2392M (2010) [13]					[e]	
11	Zimbabwe	SAZS 724 (2001) [24]	RE		300		12 unstab. 16 cement stabilised	9000 max. length

[a] E/I: Exterior/interior wall; L/NL: load-bearing/non-load-bearing.
[b] Single storey/bottom wall of a 2-storey house.
[c] Earthquake zone factor Z > 0.6 (10 and 24 at Z ≤ 0.6 resp.).
[d] Maximum length of the wall: $12 \times t$
[e] Empirical design recommendation: $h \leq 8 \times t$ for medium seismic risk; $h \leq 6 \times t$ for high seismic risk
[f] Limits of heights for adobe buildings: seismic zones (SZ) V + IV one storey; SZ III two-storey buildings. Important buildings should not be constructed in SZ IV + V, in SZ III only one storey

4.5   New developments in earth building standards

4.5.1   Ecological aspects

Global climate change, excessive consumption of energy and resources and sustainable development are aspects of considerable concern to modern society. Sustainability means that our generation shall make careful and economic use of the available resources in order to guarantee appropriate living conditions for future generations. The building industry can make a major contribution to this aim. Construction systems with a smaller environmental impact over their life cycle and building materials with low primary energy consumption are becoming increasingly significant. In this respect, earthen building materials and construction systems are particularly favourable. This new development is also reflected by an increasing number of earth building standards in recent years that focus on industrially produced earthen building materials.

The assessment of the environmental impact of building materials, buildings and other necessary services throughout their lifetimes is known as Life Cycle Assessment (LCA). The assessment analyses the entire life cycle of a building product from raw material excavation (soil), through the processing of raw materials (soil and additives), manufacturing of the (earthen) building materials, the construction, utilisation and maintenance (of earth buildings), as well as recycling and final disposal. A model LCA for earthen building materials is described in Fig. 4.3.[2]

Energy impact

A simplification of the LCA is the life cycle energy analysis (LCEA), which focuses only on the measure of energy as environmental impact. The energy consumption employed by a building can be classified in:

- energy for production of (earthen) building materials incl. transport to the site (Primary Energy Impact, PEI or 'embodied energy', Tables 4.7 and 4.8[2]
- energy consumed during the building process
- energy used for maintenance and repair of the building
- energy used in demolition of the building at the end of its useful lifetime.

The PEI is a criterion commonly used to ascertain a rough estimation of the ecological balance of a building material, allowing one to compare the energy efficiency of production processes in terms of 'embodied' energy. In the case of earth building materials, the PEI is a favourable parameter because it is extremely low: comparing the PEI of earth with that of typical building materials such as concrete, reinforced concrete or steel shows just

Table 4.7 Primary energy impact values (PEI) of typical building materials comparing with earth building materials

Building material	PEI (kWh/m³)
Earth	0–30
Straw insulation panels	5
Timber, home	300
Timber materials	800–1,500
Fired bricks	500–900
Cement	1700
Concrete, 'normal'	450–500
Lime sandstone	350
Glass panes	15,000
Steel	63,000
Aluminium	195,000
Polyethylene (PE)	7600–13,100
Polyvinylchloride (PVC)	13,000

Table 4.8 Primary energy impact values (PEI) for different means of transport

Means of transport	PEI (kWh/t·km)
Railway	0.43
Car, Western Europe	1.43
Truck 40t	0.72
Truck 28t	1.00
Truck 16t	1.45
Truck < 3.5t	3.10
Boat transport, overseas	0.04
Boat transport, rivers	0.27

how much difference this makes. This is a further reason why the use of earth for construction purposes has become increasingly significant in many countries, and this new development is also reflected by a rising number of earth building standards.

The PEI of building materials is just one aspect of energy consumption over the entire life cycle of a building. In most countries with cold and temperate climates, buildings that are in regular use have to be heated during the winter period. Building standards in these countries define limits for the heat energy demand (e.g. in Germany, DIN 4108, EnEV 2009), particularly for residential buildings, and require appropriate heat insulation. The reduction of energy consumption achieved by thermal insulation over the lifetime of a construction (e.g. ~100 years for a residential house) outweighs the 'embodied' energy of the building materials significantly.

Earth building materials are not good thermal insulators (λ – values, Table 4.9). They generally do not fulfil the thermal insulation requirements for exterior walls in new buildings as stipulated in the respective national building codes in regions with cold and temperate climates. Earth building

Table 4.9 Thermal conductivity values for earth building materials[2,8]

Dry density ρ_d [kg/m^3]	Thermal conductivity λ [W/mK]	Earth building materials
2,200	1.40	RE
2,000	1.10	RE
1,800	0.91	RE, EM, CP
1,600	0.73	C, EI, EM, EB, CP
1,400	0.59	C, EI, EM, EB, CP
1,200	0.47	LC, EI, EM, EB, CP
1,000	0.35	LC, EI, EM, EB, CP
900	0.30	LC, EI, EM, EB, CP
800	0.25	LC, EI, EM, EB, CP
700	0.21	LC, EI, EM, EB, CP
600	0.17	LC, EI, EM, EB, CP
500	0.14	LC, EI, CP
400	0.12	LC, EI, CP
300	0.10	LC, EI, CP

Earth building materials in comparison to: polyurethane, 0.02 W/mK, aluminium, 200 W/mK.

materials can comply with building code requirements for thermal insulation in two general ways.

In *new buildings* earth materials can be used in exterior walls in combination with additional thermal insulation materials e.g. straw insulation panels. In addition, the capacity of constructions with a heavy mass (e.g. rammed earth) to absorb and re-radiate heat can be used to even out the degree of air temperature fluctuation in rooms in buildings made of lightweight constructions. Earth plasters also have good sorption properties: airborne moisture is absorbed through the pores of the earth building material, and as the interior room climate becomes drier the earth plaster releases moisture back into the air. As such, earth materials help regulate and filter the interior room climate.

Today, the improvement of the Indoor Environmental Quality (IEQ) is becoming an increasingly important criteria for the selection of building materials. Growing consumer awareness of hazardous emissions from building products has given rise to a new class of discerning consumers seeking building materials with few to no negative emissions (formaldehyde, volatile organic compounds, pesticides, radiation, etc.). Clay and earth building products are able to meet such criteria for healthy living. Producers of earth building materials declare the product constituents on their product labels, and ecocertificates[11] also underline their zero-emissions credentials.

Historical buildings, including those that use earth building materials in external walls, are not required to fulfil the same level of thermal insulation as new buildings. In Germany, light clay (LC) is often used for lightweight earth wall linings applied to the inside of the external walls of historical buildings, particularly half-timbered constructions with earth infill panels. In this way, earth building materials can contribute significantly to improving the thermal performance of historical constructions. Not all 'modern' thermal insulation materials are appropriate for use with earth constructions and there have been many cases where damage resulting from such 'repairs' has itself needed repairing, particularly of half-timbered constructions.

Environmental impact

A comprehensive evaluation of a building material's life cycle (LCA) analyses the energy consumption of all processes, including maintenance and repair, over its entire 'lifetime' (Fig. 4.3). It also considers the environmental effect of related energy production and consumption. Descriptive parameters are used to quantify the environmental impact (e.g. global warming potential GWP = 'CO_2-equivalent').

In the last ten to twenty years, assessment methodologies for evaluating the environmental impact ('ecological footprint') of technological (building) processes have been developed. The LCA methodology is generally defined in ISO 14044: 2006 'Environmental management, life cycle assessment, requirements and guidelines' and ISO 14040: 2009 'Environmental management, life cycle assessment, principles and frameworks'.[38] It consists of four stages:

- balancing (definition of goal and scope, inventory analyses, impact assessment, improvement analyses)
- feedback and iteration
- interpretation
- output in terms of functional unit(s).

Based on the parameters calculated, the final evaluation can provide a variety of different information:

- an analysis of construction alternatives with a view to identifying preferred variants (e.g. earth building materials)
- an estimation of ecological effects
- identification of effects relating to existing environmental impacts.

The quality of a LCA depends to a large degree on the quality of the inventory analysis data. LCA studies are directly related to the availability and quality of these data. These can consist of direct measurements, industrial reports, laboratory measurements and other documents and reports. There are numerous public LCA databases developed by academic, industrial, institutional, commercial and governmental organisations.[33]

This new 'way of thinking' will gradually be reflected in all (building) standards. At present, however, only very few earth building standards consider ecological criteria for the planning of building processes today. On the other hand, in comparison to other building materials, earth building products exhibit considerable advantages with regard to PEI and Indoor Environmental Quality (IEQ). This is an important factor for the promotion of earth building products, although it is currently still not reflected in the respective standards.

The ASTM 2392[13] and SLS 1382[21] include a verbal description of aspects of sustainable development, such as the PEI, the IEQ and the energy efficiency of earth building systems. But there is no reference to ISO 14040 and LCA procedures.

The DVL TM 02 and TM 03[10,32] detail procedures for determining the global warming potential (GWP) on the basis of (DIN EN) ISO 14040 and in TM 03[32] for the vapour adsorption of earth plaster (EP) on interior wall and ceiling surfaces in order to evaluate the IEQ.

4.5.2   Production process

In future, all European national standards will be replaced by European EN norms in response to the process of convergence in regulations. In the context of building products this will be achieved by the Construction Products Directive (CPD) of the European Economic Commission from 1989 (89/106/EEC). In March 2011, the European Commission replaced this document with a new 'Regulation of the European Parliament and the Council laying down harmonised conditions for the marketing of construction products'.[42]

The aim of this document is to foster the free movement and use of construction products in the internal European market. The emphasis of the revised proposal includes the use of a common technical language and clear terms for applying the CE mark. The CE mark guarantees defined qualities and reproducible testing procedures for the quality control of the production process of building products.

The Dachverband Lehm e.V. (DVL) has developed three standard drafts for approval by the German NSB DIN which will be issued in advance of the approval process as Technical Recommendations: DVL TM 02 'Earth blocks' (EB),[10] DVL TM 03 'Earth masonry mortars' (EMM)[32] and DVL TM 04 'Earth plaster mortars' (EPM).[40] A fourth document – DVL TM 05 'Quality control of soil as an ingredient of industrially produced earth building materials'[36] – will also be issued but is not currently part of the DIN approval process. The future DIN standards will have the following designations: DIN 18945 'Earth blocks'; DIN 18946 'Earth masonry mortar'; DIN 18947 'Earth plaster mortar'.

The DIN standard drafts comply with the requirements of the Building Products Act, the German national version of the European CPD. Earth building materials defined by DIN earth building standards will be classified as 'regulated building products'. They will receive the 'Ü mark' (conformity mark), the national version of the European CE mark which guarantees requirements for the quality control of the production process. The transition from this to a CE mark is subject to a special procedure and will be a future project for the DVL.

Today, earth building materials can be produced in two ways: manual 'hand-made' production and 'industrialised' factory fabrication using appropriate technical equipment. The manual production of earth building materials characterises the traditional way of building with earth in many parts of the world. The industrialised production of earth building materials changes the traditional, non-engineered into an engineered construction system. This shift means that the material is subject to the same quality standards for testing material parameters and for quality control of the production process as other industrially produced building materials such as lime, cement, concrete, etc. It also requires uniform standards for the design, measurement and construction of buildings with earth building materials.

4.6   Conclusions

In the last decade the use of earth for construction purposes has become increasingly significant in many countries because of its favourable ecological properties and its contribution to sustainable development. This new 'way of thinking' is also reflected by an increasing number of earth building standards that have arisen in the same period. Nevertheless, in absolute terms, the number is still very small compared with other building materials and systems. The earth building standards analysed were published by regional, national and local standards bodies over the last 30 years.

There is no common and internationally accepted terminology for earth building materials and systems. This, however, is an essential prerequisite for developing standards and normative documents and for the establishment of earthen building materials in contemporary building alongside other conventional building products and systems. This chapter describes recommended terminology for relevant earthen building materials and earth construction systems commonly used today and which are covered in (earth) building standards.

The 33 standards from 19 different countries provide widely differing technical information. With regard to their scope of application, the documents can be classified into three types dealing with typical parameters:

Soil: grading, plasticity, natural constituents content, linear shrinkage
Earth building material: strength/deformation characteristics
Earth construction system: strength/deformation characteristics, aspect ratio, earthquake resistance.

There are no uniform and internationally accepted laboratory testing procedures for determining material properties and design values for earth building construction. The test procedures currently in use were often originally developed for testing soils (soil mechanics) or concrete and did not take into account the specific properties of earthen materials. The suitability of these 'adopted' procedures for earthen building materials and systems is still to be proven. It is necessary to bring testing procedures for earth building products up to the same level as other 'standardised' building materials in order to improve their scope and competitiveness in the building sector.

Three examples of laboratory testing procedures and design criteria covering important aspects of earth building were discussed that demonstrate some of the differences in the standards documents:

1 *Soil classification and selection*: granular composition and plasticity, content of soluble salts
2 *Earth building material*: compressive strength and optimum compaction for rammed earth
3 *Earth construction system*: slenderness of wall structures.

In terms of the *granular composition* and *plasticity* of soils suitable for building purposes, it is impossible to justify the application of narrow limits on the basis of existing earth building standards. The new German earth building standards include a recommendation for the quality control of soil as an ingredient of industrially produced earth building products.

With regard to the *strength characteristics* of earth building materials (unconfined compressive strength UCS) there are no uniform and internationally accepted testing procedures. As a consequence, this makes comparability very difficult if not impossible. Very few of the standards documents examined distinguished between 'design' and 'characteristic' UCS values. There is practically no information on partial safety factors. The new German building standard for earth blocks defines 'strength' classes for different kinds of application.

The *slenderness* or aspect ratio of wall structures is a factor depending on the degree of regional seismic activity. There are very few countries subject to seismic activity that permit the use of earth building materials in construction systems and regulate their use through building standards. In many such countries, the use of earth as a building material for new building is excluded.

A new generation of building standards will be developed that take *sustainable development* into account by assessing the entire life cycle of a building product. Traditional evaluation systems that currently address material properties, construction systems, durability and economic features will be augmented by parameters that provide an indication of the 'ecological footprint' of a building material or system. The LCA is an appropriate methodology for analysing and evaluating the environmental impact of building materials, buildings and other necessary services throughout their lifetimes. In this respect, earth building materials compete very favourably. At present, however, such methodologies are rarely part of statutory standards.

Building with earth can contribute to sustainable development by reducing environmental impact when compared with other building materials and systems. This aim can only be achieved by developing earth building into an 'engineered' building system on the basis of building standards issued by national standards organisations.

4.7   References

1   DIN 1169   Earth mortars for masonry and plaster (06/47)
    DIN 18951   Earth buildings (01/51)
                    01: Construction
                    02: Comments
    DIN VN 18952 Earth
                    01: Designations, Types (05/56)
                    02: Tests (10/56)
    DIN VN 18953 Earth, earth construction elements
                    01: Use of earth (05/56)
                    02: Adobe walls (05/56)
                    03: Rammed earth walls (05/56)
                    04: Cob walls (05/56)
                    05: Light-clay walls in timber-framed structures (05/56)
                    06: Earth floors (05/56)
    DIN VN 18954 Construction of earth buildings, design (05/56)
    DIN VN 18955 Earth, earth construction elements, waterproofing (08/56)
    DIN VN 18956 Building with earth, plaster on earth constructions (08/56)
    DIN VN 18957 Building with earth, earth roofing tiles (08/56).
    (VN = draft)
2   Schroeder, H (2010), *Lehmbau – Mit Lehm Ökologisch Planen und Bauen*, Wiesbaden Germany, Vieweg + Teubner.
3   ISO (2004), *Standardisation and Related Activities – General Vocabulary: ISO Guide 2*, Geneva, ISO.
4   Jiménez Delgado, M C and Cañas Guerrero, I (2007), 'The selection of soils for unstabilised earth building: A normative review', *Construction and Building Materials*, 21, 237–251.
5   Maniatidis, V and Walker, P (2003), *A Review of Rammed Earth Construction, Innovation Project 'Developing Rammed Earth for UK Housing'*, Bath, University of Bath, UK; Bath, Dept. of Architecture & Civil Engineering, Natural Building Technology Group.
6   Houben, H and Guillaud, H (1994), *Earth Construction – A Comprehensive Guide*, London/Villefontaine, Intermediate Technology Publications/CRATerre-EAG.
7   McHenry, P G and May, G W (1989), *Adobe and Rammed Earth Buildings – Design and Construction*, Tucson, AR, The University of Arizona Press.
8   Dachverband Lehm e.V. (2009), *Lehmbau Regeln – Begriffe, Baustoffe, Bauteile*, Wiesbaden, Germany, Vieweg + Teubner, 3rd ed.

9   Dachverband Lehm e.V. (2008), *Anforderungen an Lehmputze*, Weimar, Germany, Technische Merkblätter Lehmbau – Blatt 01.
10   Dachverband Lehm e.V. (2011), *Lehmsteine – Begriffe, Baustoffe, Anforderungen, Prüfverfahren*, Weimar, Germany, Technische Merkblätter Lehmbau – Blatt 02.
11   Natureplus e.V. (2004), *Richtlinien zur Vergabe des Qualitätszeichens 'natureplus' – RL 0803 Lehmputzmörtel*, Neckargemünd, Germany.
12   Schweizerischer Ingenieur- und Architekten-Verein SIA (1994), *Regeln zum Bauen mit Lehm/Lehmbau-Atlas, Dokumentationen*, D 0111 + D 0112, Zurich, Switzerland.
13   ASTM International (2010), Standard Guide for Design of Earthen Wall Building Systems: ASTM E2392/E2392 – 10, West Conshohocken, PA, ASTM International.
14   Construction Industries Division of the Regulation and Licensing (2006), *New Mexico Earthen Building Materials Code*, CID-GCB-NMBC-14.7.4, Santa Fe, NM.
15   International Conference of Building Officials (1982), Uniform Building Code Standards, Section 2405, Unburned Clay Masonry, Whittier, CA.
16   Centre for Development of Industry (CDI), African Regional Organization for Standardization ARSO (1996), *Compressed Earth Blocks (CEB)*
     ARS 670 – Standard for terminology
     ARS 671 – Standard for definition, classification and designation of CEB
     ARS 672 – Standard for definition, classification and designation of earth mortars
     ARS 673 – Standard for definition, classification and designation of CEB masonry
     ARS 674 – Technical specifications for ordinary CEB
     ARS 675 – Technical specifications for facing CEB
     ARS 676 – Technical specifications for ordinary mortars
     ARS 677 – Technical specifications for facing mortars
     ARS 678 – Technical specifications for ordinary CEB masonry
     ARS 679 – Technical specifications for facing CEB masonry
     ARS 680 – Code of practice for the production of CEB
     ARS 681 – Code of practice for the preparation of earth mortars
     ARS 682 – Code of practice for the assembly of CEB masonry
     ARS 683 – Standard for classification of material identification tests and mechanical tests
     Brussels/Nairobi, African Regional Standards (ARS).
17   National Building Standards (2000), *Technical Building standard NTE E. 080: Adobe*, Lima, Peru, SENCICO.
18   Standards New Zealand (NZS) (1998)
     NZS 4297: *Engineering Design of Earth Buildings*
     NZS 4298: *Materials and Workmanship For Earth Buildings*, incorp. Amend. No. 1
     NZS 4299: *Earth Buildings Not Requiring Specific Design*, incorp. Amend. No. 1, Wellington, New Zealand.
19   Bureau of Indian Standards (BIS) (1980, reaff. 1998), *Code of Practice for In Situ Construction of Walls in Buildings with Soil-cement*, New Delhi, Indian Standards IS 2110.
20   Bureau of Indian Standards (BIS) (1993, reaff. 1998), *Improving Earthquake Resistance of Earthen Buildings – Guidelines*, New Delhi, Indian Standards IS 13827.
21   Sri Lanka Standards Institution (2009), *Specification for Compressed Stabilized Earth Blocks*, Colombo, Sri Lanka, Sri Lanka Standard (SLS) 1382

Part 1: Requirements

Part 2: Test methods

Part 3: Guidelines on production, design and construction.

22 Middleton, G F (revised by Schneider, L M, 1987) (1995), *Earth Wall Construction*, Bull. 5, North Ryde, NSW Australia, CSIRO Division of Building, Construction and Engineering, 4th ed.

23 Earth Building Association of Australia (EBAA) (2004), *Building with Earth Bricks and Rammed Earth in Australia*, Wangaratta, Australia, EBAA.

24 Standards Association of Zimbabwe (2001), *Rammed Earth Structures*, Harare, Zimbabwe, Zimbabwe Standard Code of Practice SAZS 724.

25 State Building Committee of the Republic of Kyrgyzstan/Gosstroi of Kyrgyzstan (1988), *Возведение малоэтжных зданий и сооружений из грунтоцементобетона PCH-2-87 (Building of low-storied houses with stabilized rammed earth)*, Frunse (Bischkek) Republic of Kyrgyzstan, Republic Building Norms RBN-2-87.

26 Shittu, T A (2008), *Earth building norms and regulation: A review of Nigerian building codes*, in: LEHM 2008, Proceedings of the 5th International Conference on Earth Building, 40–47, Weimar Germany, Dachverband Lehm e.V.

27 AFNOR (2001), *Compressed Earth Blocks for Walls and Partitions: Definitions – Specifications – Test Methods – Delivery Acceptance Conditions*, AFNOR XP.P13-901, St Denis de la Plaine, France AFNOR.

28 Ministerio de Obras Públicas y Transportes (MOPT) (1992), *Bases para el diseño y construcción con tapial*, Madrid, Spain, Secretaría General Técnica.

29 Dachverband Lehm e.V. (2004), *Building with Earth – Consumer Information*, Weimar, Germany Dachverband Lehm e.V.

30 Walker, P, Keable, R, Martin, J and Maniatidis, V (2005), *Rammed Earth – Design and Construction Guidelines*, Watford, UK, BRE Bookshop.

31 Jiménez Delgado, M C and Cañas Guerrero, I (2006), Earth building in Spain', *Construction and Building Materials*, 20, 679–690.

32 Dachverband Lehm e.V. (2011), *Lehm-Mauermörtel – Begriffe, Baustoffe, Anforderungen, Prüfverfahren*, Weimar, Germany, Technische Merkblätter Lehmbau – Blatt 03.

33 Menzies, G F, Turan, S and Banfill, P F G (2007), 'LCA methodologies, inventories, and embodied energy: A review', *Proceedings – Institution of Civil Engineers, Construction Materials*, 160, 135–143.

34 Fathy, H (1973), *Architecture for the Poor – An Experiment in Rural Egypt*, Chicago and London, The University of Chicago Press.

35 Güntzel, J (1986), *Zur Geschichte des Lehmbaus in Deutschland*, Kassel, Gesamthochschule, Dissertation Band 1.

36 Dachverband Lehm e.V. (2011), *Qualitätsüberwachung von Baulehm als Ausgangsstoff für industriell hergestellte Lehm-Bauprodukte – Richtlinie*, Weimar, Germany, Technische Merkblätter Lehmbau – Blatt 05.

37 Bureau of Indian Standards (BIS) (2011), *Indian Standard Code of Practice for Manufacture and Use of Stabilised Soil Blocks for Masonry*, Indian Standards IS 1725, Part I: Specifications for stabilized soil blocks for masonry; Part II: Code of practice for manufacture and construction using stabilized soil blocks, New Delhi, BIS.

38 International Standards Organisation (ISO) (2006–2009), '*Environmental Management, Life Cycle Assessment*,

ISO 14040: Principles and frameworks (2009)

ISO 14044: Requirements and guidelines (2006)
Brussels, Belgium.

39 Walker, P, Standards Australia (2002), *The Australian Earth Building Handbook*, HB 195-2002, Sydney, Australia, Standards Australia International.

40 Dachverband Lehm e.V. (2011), *Lehm-Putzmörtel – Begriffe, Baustoffe, Anforderungen, Prüfverfahren*, Weimar, Germany, Technische Merkblätter Lehmbau – Blatt 04.

41 Cid, J, Mazarrón, F R and Cañas, I (2011), *Las normativas de construcción con tierra en el mundo*, Informes de la Construcción, Vol. 63, 523, 159–169.

42 Regulation no. 305/2011 of the European Parliament and the Council for the Marketing of Construction Products, 9 March 2011, Legal deposit of the EU L88/5 (4 April 2011).

43 Italia. Regione Piemonte L.R. (2006) 2/06, *Norme per la valorizziazione delle costruzioni in terra cruda*, B.U.R. Piemonte, n° 13.

44 Associação Brasileira de Normas Técnicas ABNT, Rio de Janeiro (1984–1996)
NBR 8491 EB1481, *Tijolo maciço de solo-cimento* (1984).
NBR 8492 MB1960, *Tijolo maciço de solo-cimento – Determinação da resistência à compressão e de absorção d'água* (1984).
NBR 10832 NB1221, *Fabricação de tijolo maciço de solo-cimento com a utilização de prensa manual* (1989).
NBR 10833 NB1222, *Fabricação de tijolo maciço e bloco vazado de solo-cimento com a utilização de prensa hidráulica* (1989).
NBR 10834 EB 1969, *Bloco vazado de solo-cimento sem função strutural* (1994).
NBR 10835 PB 1391, *Bloco vazado de solo-cimento sem função estrutural* (1994).
NBR 10836 MB3072, *Bloco vazado de solo-cimento sem função estrutural – Determinação da resistência à compressão e de absorção d'água* (1994).
NBR 12023 MB 3359, *Solo-cimento – Ensaio de compactação* (1992).
NBR 12024 MB3360, *Solo-cimento – Moldagem e cura de corpos-de-prova cilíndricos* (1992).
NBR 12025 MB3361, *Solo-cimento – Ensaio de compressção simples de corpos-de-prova cilíndricos* (1990).
NBR 13554, *Solo-cimento – Ensaio de duribilidade por moldagem e secagem* (1996).
NBR 13555, *Solo-cimento – Determinação da absorção d'água* (1996).
NBR 13553, *Materiais para emprego em parede monolítica de solo-cimento sem função estrutural* (1996).

45 Instituto Colombiano de Normas Técnicas y Certificación (ICONTEC) (2004), *Bloques de suelo cemento para muros y divisiones. Definiciones. Especificaciones. Métodos de ensayo. Condiciones de entrega*. NTC 5324. Bogotá.

46 Asociación Española de Normalisación y Certificación (AENOR) (2008), *Bloques de tierra comprimada para muros y tabiques. Definiciones, especificaciones y métodos de ensayo*. UNE 41410, Madrid.

47 Kenya Bureau of Standards (KEBS) (1999), *Specifications for Stabilised Soil Blocks*. KS02-1070:1993 (1999), Nairobi.

48 Standards Organisation of Nigeria (SON) (1997), *Standard for Stabilized Earth Bricks*. NIS 369:1997, Lagos.

49 Institut National de la Normalisation et de la Propriété d'Industrielle (INNOPRI) (1998), *Blocs de terre comprimeé – Spécifications techniques*. NT 21.33 (1996),

Blocs de terre comprimeé – Définition, classification et désignation. NT 21.35 (1996), Tunis.
50 Turkish Standard Institution TSE (1995–1997)
    Cement Treated Adobe Bricks. TS 537 (1985),
    Adobe Blocks and Production Methods TM 2514 (1997),
    Adobe Buildings and Construction Methods TM 2515 (1985), Ankara.

5

Passive house design: a benchmark for thermal mass fabric integration

L. RONGEN, University of Applied Sciences, Erfurt and
Rongen Architects, Germany

Abstract: We live in a time when energy is becoming more precious.
Worldwide energy resources are finite and they seem to be running out
faster than expected. At this time when energy consumption is rising
dramatically in the emerging markets, the opposition to nuclear energy,
which in fact cannot be any solution to dwindling energy resources while
open security issues are not resolved, is growing. On 30 June 2011, the
federal government of Germany decided by law Germany's final exit from
nuclear energy no later than 31 December 2022. There is also the problem of
climate change. We will have to reduce CO_2 emissions into the atmosphere
dramatically if we want to preserve tolerable conditions of life on planet
earth 'for which there are no spare parts' (Hans Joachim Schellnhuber,
director of Potsdam Climate Institute). The biggest challenge of our time is
to minimise worldwide energy consumption. The 'Passive House Standard'
is the worldwide acknowledged leading standard in energy-saving building.
Building to the Passive House Standard is one answer to the most important
questions of our time. The chapter supplies answers to the questions: 'Why
build to the Passive House Standard?', 'What is a Passive House?' and 'How
do Passive Houses work?', and outlines planning approaches.

Key words: Passive House Standard, energy saving building, compactness,
airtightness, thermal bridges.

5.1   Introduction

'The house should be open to the exterior, and permeated by light, air and
sunshine. Letting the sun in is a new and imperative task of the architect.'
Le Corbusier, 1943.

Today, any true green architecture project has to use the word 'sustainable'.
Sustainability has become one of the most over-used concepts not only in
the field of architecture but also in the economy generally – especially in
the advertising industry – and is becoming little more than a catchword. The
origins of 'sustainability' are rooted in forestry. In forestry, sustainability
means not cutting down more trees than are replanted. Sustainable development
is economic and social development that meet the demands of the present
generation without compromising the demands of the coming generations.

High quality in architecture is not only a question of beauty, it is also a
question of function, a question of construction, a question of economy, a

110

question of ecology and, last but not least, it is a question of the well-being of its users. One of the greatest challenges for architects is to drastically reduce energy consumption. This has been a major focus since at least the 1970s. The Organization of the Petroleum Exporting Countries (OPEC) imposed an oil embargo in 1973 after the Yom Kippur War. This prompted the deepest international recession since the great depression of 1929, as a result of increasing oil prices and the throttling back of petroleum production. It also prompted major interest in saving energy.

In November and December 1973, for example, the German Federal Government tried to save energy by decreeing speed limitations and prohibiting driving private cars on Sundays. From 1973 to 1974, the price of petroleum in Germany climbed up from 82.20 Deutsch Marks (DM) to 223.87, despite depreciation of the dollar, which is an increase of around 172.2%. Altogether the Federal Republic of Germany spent nearly 23 billion DM on petroleum in 1974, an increase of 152.7% compared to 1973. From then on, energy costs have increased continuously up to a peak in July 2008, when the price was US$146 per barrel of crude oil. Experts believe that the demand for and price of crude oil will continue to increase.

Even more important than saving energy is the need to protect the environment and reduce carbon dioxide (CO_2) emissions into the atmosphere. The phenomenon of global warming is now widely accepted by most climate scientists. The ozone layer area over Antarctica has increased to more than 27 million km^2 (News Network Internetservice GmbH, 2006). The potential results include flooding, more extreme weather and extensive loss of biological diversity. Today, CO_2 concentration in the atmosphere, an estimated 380 parts per million (ppm), is the highest it has been for 650,000 years.

The International Energy Agency (IEA) has suggested how these CO_2 reductions can be achieved with the largest reduction (of about 53%) through improving energy efficiency. There are significant potential savings to be made in the building sector. It has been estimated that 50% of global energy is used for heating buildings, 25% for transportation, 25% for industry. Today, developing countries are growing which means energy consumption is likely to increase worldwide. One chance to save energy is to reduce the energy consumption used for heating and cooling buildings and to build energy-efficient buildings. Drawing on past experience, the Federal Government in Germany has, since 1976, constantly tightened legal measures to save energy in the construction sector (see Table 5.1).

As Table 5.1 shows, the requirements for energy efficiency in German buildings have increased steadily and we can already foresee the outcome. The Passive House Standard – the highest energy efficiency standard for buildings worldwide – will soon become the normal building standard in Germany, as well as in all EU countries. In 2006 the EU Commissioner responsible for development, Andris Piebalg, said: 'The Commission will

Table 5.1 A history of increasing energy efficiency in buildings in Germany

Year	Event
1952	For the first time, a minimal thermal insulation is demanded by law (DIN 4108) to avoid building damage caused by water condensation.
1976	As a consequence of the oil boycott by OPEC countries, the 'Energieeinspargesetz' (Energy Conservation Act) was passed, laying down principles for thermal insulation and heating systems, etc.
1977	The first 'Wärmeschutzverordnung' (*Heat Protection Ordinance*) was passed, regulating the increased thermal insulation of buildings. For the first time, maximum permitted heat energy consumption of buildings was limited by law – 270 kWh/m^{2}a.
1984	The second 'Wärmeschutzverordnung' (*Heat Protection Ordinance*) followed. The maximum heat energy consumption of buildings was limited to 180 kWh/m^{2}a.
1995	By the third 'Wärmeschutzverordnung' (*Heat Protection Ordinance*), the heat energy consumption of buildings was limited to 100 kWh/m^{2}a.
2002	The 'Energieeinsparverordnung (EnEV)' (Energy Conservation Code) 2002 was passed. The heat energy consumption of buildings was limited to 75 kWh/m^{2}a. The so called 'Low Energy House' was obligatory for new buildings.
2007	The 'Energieeinsparverordnung (EnEV)' (Energy Conservation Code) 2007 limited permitted primary energy consumption of buildings to about 30% (compared to 2002).
2009	The 'Energieeinsparverordnung (EnEV)' (Energy Conservation Code) 2009 was passed by law in October 2009. The permitted primary energy consumption of buildings was reduced about a further 30% (compared to 2007).
Next?	The 'Energieeinsparverordnung (EnEV)' (Energy Conservation Code) has already been proposed and is due in 2012.

propose action to ensure that in the longer term our buildings become "near zero emitting" – also called "Passive houses".'

5.2   Description of Passive House

A 'Passive House' is a house that is energy efficient, comfortable, economic and environmentally friendly, all at the same time. 'Passive House' is not a brand name but rather an idea which has proved itself in practice, and the 'Passive House Standard' is the global leading standard in energy saving building.

'Passive Houses' are defined as buildings with a negligible heating energy requirement and therefore they do not need an active heating system. Such houses can be kept warm 'passively' by using existing internal heat sources, solar energy and by heating with fresh air supply. A 'Passive House' is therefore a building in which comfortable temperatures can be reached without any separate heating or cooling systems both in winter and summer.

Testimony for the feasibility of 'Passive Houses' has been discussed in the thesis, 'Passive houses in central Europe' (Feist, 1993) using computerized simulations of the energy balance of buildings. In this paper, construction elements that determine the energy consumption of buildings were systematically varied and optimized based on energy efficiency, installation expense and living value. But the first really operable and full Passive House was not a house but rather a polar ship; Fridtjof Nansen's 'Fram'. Nansen wrote:

> The walls are covered with tarred felt, followed by a cork filling, then panelling made from fire wood, then another thick felt layer, then airtight linoleum and, finally, wood panelling again. The ship walls have got a thickness of approximately 40 centimetres. The windows are protected by triple panes to avoid incoming drafts. Here a warm, cozy atmosphere exists. Whether the thermometer shows 5 degrees or 30 degrees below zero, we have no fire in the stove The ventilation is excellent ... because fresh winter air floats down the ventilator. I am mulling over taking away the stove, since it is only in the way.
>
> Nansen, 1897

Of course, the polar ship 'Fram' was not designed as a Passive House as we know it today, but Fritjof Nansen's description corresponds very well with a modern Passive House.

After the 'low energy house' (see Table 5.1) was successfully proposed the 'Passive House' concept was developed in May 1988 by the German physicist Wolfgang Feist and his host, Professor Bo Adamson, during a research tenure (in the field of building construction) at the University of Lund, Sweden. The first Passive House pilot project was built in 1991, in Darmstadt Kranichstein (Germany), under the direction of Wolfgang Feist. At that time, Feist did not want to call it a 'House without heating system' because it still required a heat generator to secure its very low leftover heat demand, though no radiators were necessary for heat distribution. Since then, more than 25,300 Passive Houses have been built in Germany and Austria (IG Passivhaus AT, IG Passivhaus DE), both residential buildings and non-residential (such as office buildings, commercial buildings, sports halls, schools, kindergartens, etc.). There are no boundaries in the use or design of Passive Houses. In architectural competitions, the Passive House Standard is increasingly required.

The energy consumption for heating and cooling in a Passive House amounts to a maximum of 15 kWh/(m²a) equivalent to 1.5 litres (l) of oil or 1.5 cubic metres (m³) gas per square metre (m²) of 'heated area' per year. The energy requirement for heating and cooling a Passive House is 80% less in comparison to a new building in Germany in accordance with statutory requirements. In comparison with an average German building standard, a

Passive House consumes 90% less energy. These enormous energy savings are achieved mainly as a result of the basic principles of avoiding heat losses, gaining sun energy, thermal bridge-free building and a controlled mechanical ventilation system with heat recovery. The heat load is limited to 10 W/m^2. For example, in a Passive House, just two healthy adults as the only energy source are enough to bring a 20 m^2 room up to (and keep it at) a comfortable temperature, since one healthy adult has a heat output of about 100 W. An active heating system is therefore not required.

The additional investment costs of a Passive House compared to a standard house are about 5% for small buildings (e.g. single family houses). The larger a Passive House is, the lower the relative costs of more investment. The Passive House Standard is not only a solution for new buildings; it is also possible to refurbish old buildings by using Passive House building components.

Energy optimization for buildings should not be limited to heating energy: all household energy consumption must be minimized. It does not make sense for example, to reduce the heating energy requirement to 'zero' by using inefficient electrical devices that create high internal gains (such as incandescent light bulbs, for example). In a Passive House, the primary energy consumption (total energy consumption including energy consumption for warm water supply, household machines, energy transformation and energy provision) is limited to 120 kWh/m^2a.

Apart from a very low heating or cooling demand, the Passive House provides its users with several advantages in comparison to a conventional building. There is a much higher comfort factor because surface temperatures of the inner building façades (walls, windows) are almost uniformly warm. Air draughts as a result of temperature fluctuations do not exist in Passive Houses. 'Comfort ventilation' provides continuously fresh and more hygienic air and there is a very low CO_2 concentration. Experience has demonstrated that pupils do not get tired as fast in 'Passive House Schools' as they do in others. Figure 5.1 shows measured CO_2 concentrations in a German school, typical of most schools. During class breaks, when windows are opened, CO_2 concentration decreases. CO_2 concentration increases significantly as lessons continue. Figure 5.2 shows measured CO_2 concentrations in a German Passive House School. The level of CO_2 concentration is permanently under 1500 ppm (even during lessons), which is the permitted level for schools in Germany.

5.2.1   How a Passive House works

Fresh air is drawn into the house from outside at a height of about 2.5 m above ground level because 'waste' air (such as CO_2, contained in noxious car fumes) is heavier than fresh air and gathers on the ground. After being

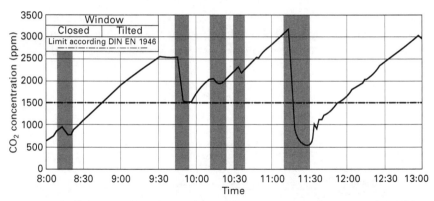

5.1 Shows the measured CO_2 concentration in a German school. The measurement result is typical for schools. During class breaks, when windows are opened, the concentration of CO_2 decreases. The CO_2 concentration increases significantly during continual lessons.

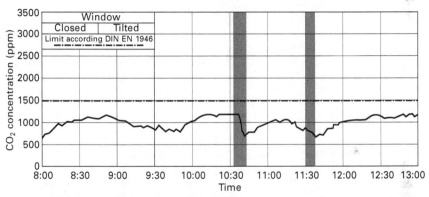

5.2 Shows the measured result of CO_2 concentration in a German Passive House School. The level of CO_2 concentration is under 1,500 ppm, which is the permitted level for schools in German, even during lessons.

sucked in, the fresh air is precleaned by a succession of filters, each finer than the last; pollen and allergen filters can also be added. It is clear already that with such a permanent fresh air supply, the internal air supply of Passive Houses is cleaner than that of standard buildings.

After passing through the filters, the fresh air flows into a ventilation device with an air–air heat exchanger. Any waste air from within the house (sucked out of the bathrooms, kitchens and toilets) also enters the air–air heat exchanger. The heat exchanger extracts a minimum of 75% (and in modern heat exchangers, almost 90%) of the waste air's heat, feeding the heat to the fresh air (but without mixing foul and fresh air together). Clearly, only a small residual amount of heat energy is required to keep the rooms at comfortable temperatures.

Additionally, the fresh air often flows through a 'ground heat exchanger' over a longer distance, after passing the filter boxes. Such ground heat exchangers mainly consist of endless lengths of pipes, laid to a depth of about 1.5 m to 2.0 m. By going through this ground heat exchanger, the fresh air takes heat from the terrestrial heat, and is preheated before flowing into the heat exchanger with heat recovery (Fig. 5.3).

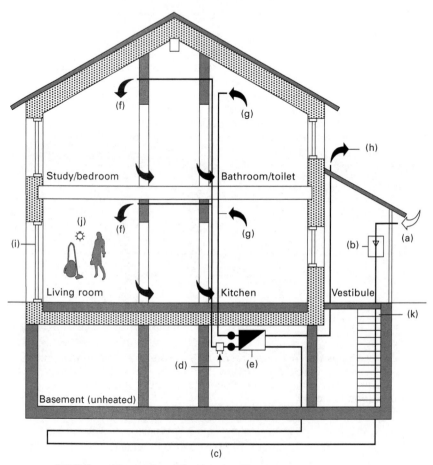

5.3 Schematic section of a Passive House
(a) Fresh air
(b) Fresh air/air filter
(c) Earth heat exchanger
(d) Incoming air heater battery
(e) Air/air heat exchanger
(f) Incoming air
(g) Outgoing air
(h) Exhaust air
(i) Passive house suitable
(j) Internal heat sources
(k) Basement access.

The preheated fresh air then flows with a very low inflow velocity into the rooms, so that there are no uncomfortable drafts or undesirable noises (the ventilation ducts are usually also equipped with silencers). Because of the low humidity ratio of the outside air, lower relative air humidity can arise inside the building during the heating period. If the relative humidity drops below 30% on a permanent basis, this could disturb the comfort zone of residents, since the air would then be perceived to be too dry. The air change rate of the ventilation system should therefore not be set too high. As required, it might also be advisable to reduce the air change rate in winter (ventilation plants with humidity recovery are now on the market to prevent this problem).

5.3   Functional principles of Passive House

A 'Passive House' is not high tech, but rather an absolutely uncomplicated, very well thermally insulated and, above all, airtight building. Because of their airtightness, Passive Houses are particularly energy efficient, keeping almost all the heat produced by living creatures and technical appliances (e.g. computers, lamps, etc.) inside the building. To achieve this, careful workmanship and construction are essential. The planning and construction of a Passive House is based on the optimization of all energy-relevant construction components, including those of the opaque building envelope, windows and doors, ventilation, heating, warm water supply and electricity. Therefore, an interdisciplinary collaboration amongst the experts of the different specialist fields is very important.

A passive house-planning-package (PHPP) is a spreadsheet-based design tool to help architects to design to the Passive House Standard. It is the central document for the assessment of the specific values of the construction project, and assists with the choice and sizing of building components. Passive House designers use a PHPP from the very first preliminary design through to execution planning and building construction. In addition, the PHPP is very useful for checking up on the actual consumption values of a realized Passive House. The PHPP clearly shows the necessity of interdisciplinary collaboration.

As a rule, the same planning principles apply for larger Passive Houses as for smaller (e.g. single family and semi-detached) houses. The main principles are:

- Energy consumption should be a maximum of 15 kWh/m²a for heating and cooling, with maximum primary energy consumption of 120 kWh/m²a for heating, cooling, warm water supply and electrical current (primary energy consumption is the energy consumption of a source of energy plus the amount required for energy regeneration, energy converting and energy sharing).

- Thermal bridges (materials that conduct heat and therefore automatically result in energy losses) should be avoided. Passive Houses have to be built without thermal bridges to avoid unnecessary heat loss through the building envelope. The best-known thermal bridge is metal, e.g. a metal spoon, which conducts the heat from hot water. Passive House windows are triple glazed and heat losses by transmission through the windows are therefore considerably reduced. Solar energy enters the rooms inside through the windows, but hardly any heat is lost through them.

- The building should have an extremely thermal insulated outside envelope. The energy saving emphasis in a Passive House is the reduction of energy losses by transmission and ventilation. This can be achieved by a very good thermal insulation of all external envelope surfaces (roof, external walls, foundations, windows). *The thermal transmittance value (U)* of exterior walls, roof and floor should have a maximum value such that $U \leq 0.15$ W/m$^2 \times$ K.

- The building should be compact. A compact building helps to achieve an optimal relationship between building volume (V) and its exterior surface area (A), its compactness being determined by A/V. The essential difference between smaller passive houses and larger passive houses is their compactness. This is explained further below.

- The building envelope should be as airtight as possible (which applies not only to Passive Houses). Structural damage caused by condensation can be avoided only if the building envelope is airtight. The quality of airtightness can be checked by a 'blower door test', a measuring device with a fan which is incorporated into a window or an exterior door. The test creates a pressure differential of 50 Pascals (P) between the building interior and the air outside, which corresponds to a wind force of 4 to 5 on the Beaufort scale (comparable to a wind speed of 20–38 km/h). During the blower door test, all other windows and exterior doors are closed, and all interior doors in the section being measured are opened. The quantity of air that influxes through leaks into the building envelope under the increased pressure differential is then measured. To achieve sufficient airtightness, relevant details have to be developed carefully in the early planning phase. Successful implementation requires neat and tidy work.

- There should be a ventilation system with heat recovery. Passive House ventilation systems with heat recovery guarantee an economic mode of action.

- Windows should be high quality. Passive House windows have thermally *separated* frames and are triple-glazed with an inert gas filling between the panes (argon is now commonly used, since the better insulating gas krypton is no longer available in sufficient quantities). The U-value of Passive House windows is ≤ 0.8 W/ m$^2 \times$ K, whilst the g-value (level of energy transfer) is > 50–60% (see Fig. 5.4). In the past, it was often said (incorrectly) that it was not possible to open the windows in a Passive House; in fact, windows can be opened any time, but it is not necessary to do so to secure a fresh air supply. In the early days, thermal isolation requirements resulted in relatively wide window frames and this was a disadvantage from an architectural perspective; however, very thin frames have now long been available on the market (Fig. 5.5).

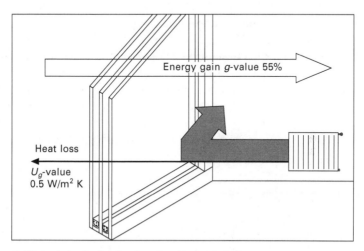

5.4 Principle draft of a Passive House window.

5.5 Passive House window 'ENERsign' with especially thin window frame (Pazen windows).

- In the case of timber constructions the interior airtight surface is the vapour barrier. Such a vapour barrier (in most cases a plastic foil) has to be adhered carefully with special adhesive tape to all connecting structural elements (Fig. 5.6). However, a further special insulation level should generally also be provided to protect against penetration of the *airtight level* (Fig. 5.6).
- In the case of solid stone buildings, the interior plaster works *as the* airtight surface. It is therefore important to ensure that the interior plaster continues down to the concrete floor (Fig. 5.7).
- *Site and orientation*. When realizing an optimized Passive House it is also important to pay attention to the site of the building and its orientation. The optimal orientation of Passive Houses is to the south, except in hot climates. This is an essential condition to gain sufficient solar (passive) energy gains in winter. The house should not be overshadowed by neighbouring buildings or trees because it will receive less sunlight, which is useful for heating the building. An optimal site would be on a

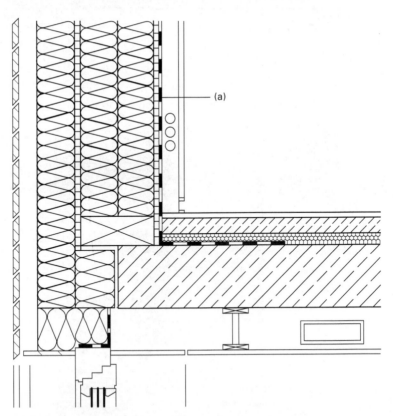

5.6 Passive House external wall as a timer construction.
(a) Airtight foil (vapour barrier).

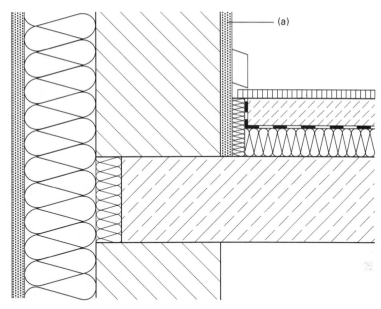

5.7 Passive House external wall of a solid building
(a) Inner plaster = airtight foil.

south-west facing incline, near to the sea, which serves as heat protection in summer, and near trees (but not overshadowed) to be protected against wind.

5.3.1   Compactness of the building

Compact buildings lose less heat than less compact buildings through the envelope (walls, roofs, windows, building sole or surface against unheated basement) during heating periods (as 'transmission losses'). The smaller the envelope in relation to the surrounding volume ('the *A/V* ratio'), the lower the volume's specific heat demand. Accordingly, Passive Houses should be as compact as possible. The most compact form of a building would be a sphere, since the volume of a sphere can be wrapped with the smallest envelope area, in relation to all other geometric shapes (a cube being the next most compact geometric shape). However, this does not mean that all Passive Houses should *always* be sphere- or cube-shaped (to consider this idea further, think about the envelope and *A/V* relationship of three examples (Fig. 5.8), each with the same volume: the first, arranged as a single cube; the second, as a row eight cubes long; and the third as eight single cubes). Figure 5.9 presents ideal *A/V* ratios, depending on the building size under consideration. In Fig. 5.10, three shapes (a sphere, a cube and a rectangle), all with exactly the same volume of 1000 m^3, are considered. In this case, the

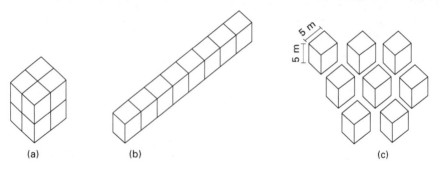

5.8 A/V relations in comparison, V respective 1000 m³
(a) A = 600 m² A/V = 0.60
(b) A = 500 m² A/V = 0.85
(c) A = 1200 m² A/V = 1.20.

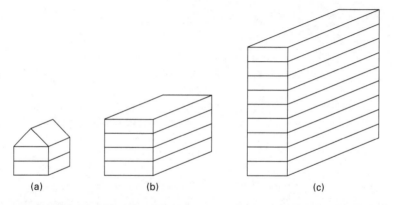

5.9 Recommended A/V relations (source: Passive House Institute, Darmstadt, Germany)
(a) Single family house A/V < 0.8
(b) Building of flats A/V < 0.4
(c) Office buildings/tall buildings < 0.2.

most compact building shape is the sphere. However, on the basis of costs alone (spheres requiring a great deal of construction), few owners would select a sphere as their favoured building shape. But it is important to note the general principle that the building shape of a Passive House should be as compact as possible, and therefore should not be more complicated than necessary. Of course, this does not mean that high-quality architecture is impossible in cases of compact building shapes: Ludwig Mies van der Rohe (the famous German architect) famously said: 'less is more'. And a wise Chinese man said: 'In simpleness and refinement there is richness'.

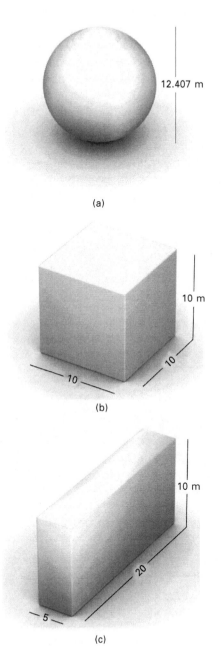

(a)

(b)

(c)

5.10 Different forms = different A/V relations
(a) Circle diameter = 12.407 m, V = 1000 m^3, A = 483.6 m^2, A/V = 0.48
(b) Cube V = 1000 m^3, A = 900 m^2, A/V = 0.6
(c) Rectangle V = 1000 m^3, A = 900 m^2, A/V = 0.9.

5.3.2   Further design principles

Thermal insulation

Figure 5.11 shows the external wall structure of a Passive House as a solid construction. In this case, the building base is thermally insulated below the floor slab. By doing this, heat bridges in the transitional 'structural floor/exterior wall' area can be avoided. The thermal insulation used must be approved as 'load-bearing thermal insulation'. In Fig. 5.12 the thermal insulation of the floor slab is placed above the bottom plate. If the building base is thermally insulated above the floor slab in the case of a building without a cellar, then the first masonry layer of the exterior walls should be produced from special offset blocks with a particularly high thermal transmission resistance to avoid further thermal bridges. Figures 5.11 and 5.12 show examples of building envelope structures with accordingly good U values. In the case of Passive

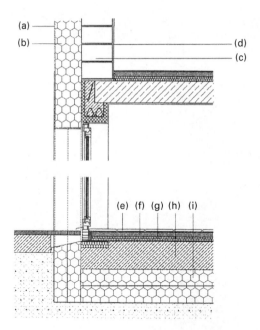

5.11 Monolithic exterior wall built to Passive House Standard (without cellar) – wall make up and soil structure
(a) Exterior plaster, 10 mm
(b) Thermal insulation (mineral wool) 200 mm
(c) Aerated concrete, 240 mm
(d) Interior plaster, 10 mm
(e) Flooring, 15 mm
(f) Floor screed, 40 mm
(g) Impact sound insulation, 45 mm
(h) Reinforced concrete base, 250 mm
(i) Polystyrene rigid foam, 300 mm (2 × 150 mm).

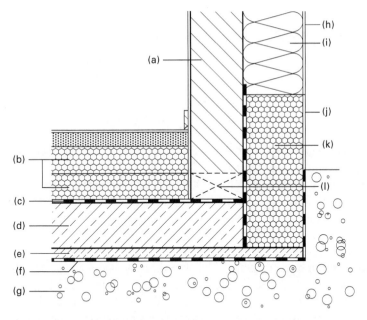

5.12 Bottom plate, insulated from the top
(a) Aerated concrete, 240 mm
(b) Thermal insulation
(c) Horizontal damp proof course
(d) Steel reinforced concrete floor plate
(e) Blinding layer (poor concrete)
(f) Polyethylene foil, 0.5 mm with overlap
(g) Gravel for capillary barrier, 200 mm
(h) Plaster coat, 5 mm
(i) Composite thermal insulation system
(j) Base plaster
(k) Perimeter insulation
(l) Offset block to minimize thermal bridging.

Houses with a basement, the entrance into the basement should be from the outside of the thermally controlled volume, for example via an unheated porch or exterior stairs. If this is the case, complicated penetrations of the airtight 'Passive House envelope' can be avoided; alternatively, an additional 'Passive House suitable' exterior door could be fitted.

The earth as a feasible heat storage system

Using the earth as a heat storage device is increasingly popular in Passive Houses. Excess heat from a solar thermal plant is stored in the soil under the building with the help of an inexpensive brine register, installed in the soil. When heat is needed, it can be removed using a heat pump. The bottom of the building should be very well thermally insulated so that, in winter, large transmission heat losses through the base into the ground cannot occur.

Passive cooling

Just as it is possible to heat a building 'passively', a building can also be cooled 'passively'. This is done most effectively by sun protection, geothermal heat-exchange, 'night' or 'free cooling' using outside air, solar cooling and cooling by heat pump (air conditioning or heat pump in combination, for example, with an earth probe). 'Night' or 'free cooling' works as a result of the cooling effect due to lower external temperature compared to internal room temperature. Whenever the outside temperature is below a point set for incoming air or room temperature (which frequently happens on summer nights), room air temperature can be lowered by controlled air ventilation. This can be useful not only on summer nights but also in the transition periods (though this assumes that the directly operated heat exchanger is equipped with so-called 'bypass valves' and a second outside air inlet; thus the cooler outside air can be transported into the interior of the building whilst heat recovery is avoided, otherwise the air would be 'heated' again by the heat recovery unit) (Fig. 5.13). However, the performance of 'free' or 'night' cooling is relatively low compared to a water-based cooling system, because the flow rates are lower and (in any case) air is not as effective a heat transport medium as water.

The thermal mass of a building (including components with a high weight, such as concrete ceilings and solid walls) cools down the stored surface

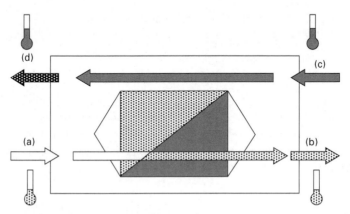

5.13 'Free cooling' principle sketch
Source: www.kampmann.de
(a) Fresh air
(b) Incoming air
(c) Outgoing air
(d) Exhaust air.

temperature and discharges it into the rooms again. The thermal storage mass can be increased by using phase changing materials (PCMs) in walls and ceilings. If, in addition, the windows are opened during the coldest 12 hours a day, then there is an increased exchange of air and thus a passive lowering of the internal temperature. No wind is necessary to allow increased air exchange compared to a controlled air ventilation system. In this case, the internal temperature does not decrease below a user-defined minimum (e.g. 22°C). In the case of high humidity of the outside air natural cooling by fresh air should not be used because of the danger of condensation.

Passive Houses can also be cooled by so-called 'adiabatic cooling' whereby water is injected into the ventilation system. For environmental health/ sanitary reasons, the water is injected into the exhaust pipe in front of the heat recovery unit. The result is 'evaporative cooling', which cools air before it enters the heat exchanger. The fresh air cooled by the heat recovery unit flows through the air intakes into the rooms and cools them down. At first, it was possible to cool down fresh air this way by around 3–4 K; today, heat exchangers with adiabatic cooling are on the market that make it possible to cool down fresh air by 11 K. The very small residual heat energy left can be easily used by a small refrigeration supply system, electrically powered by a photovoltaic system ('green power').

If the Passive House being considered has a panel heating system, e.g. underfloor heating, wall heating, an overhead radiator system or activated building components beside the controlled air ventilation system, then rooms can be cooled down by these surfaces without much additional effort. However condensation has to be avoided (Fig. 5.14).

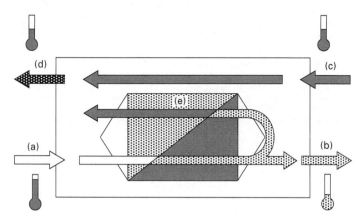

5.14 'Adiabatic cooling' principle sketch (source: www.kampmann.de)
(a) Fresh air
(b) Incoming air
(c) Outgoing air
(d) Exhaust air
(e) Process air.

5.4   Case studies of Passive Houses in different climates

At first, the Passive House concept was developed for the Central European climate. Increasingly, however, it is being requested not only for very cold but also for very hot climates. Achieving the Passive House Standard is feasible in all climates.

Under instruction from German Federal Environmental Foundation (DBU), Rongen Architects (Wassenberg, Germany) undertook a research project 'Passive Houses for different climates' together with the Passive House Institute and Prof. Dr Feist, Darmstadt, Germany. The project was funded by DBU and by the French group Saint Gobain. The locations studied were Dubai (as an example of a hot and humid climate), Las Vegas (for a hot and dry desert climate), Yekaterinburg in North Russia (as an example of a very cold climate), Shanghai (hot and humid in summer, cold in winter) and, last but not least, Tokyo (with a similar climate to Shanghai but with other requirements for residential buildings). The design for each location was also required to evaluate the respective regionally typical construction methods, design languages and cultural needs of their inhabitants. The buildings described below have not yet been built, but there is great interest in their realization in the countries concerned.

5.4.1   Dubai

In Dubai, the winters are so warm that well-insulated buildings do not require heating, so the design interpretation for this Passive House has been optimized for summer. Bright surfaces and sun protection glazing are also

Table 5.2 Dubai Passive House requirements

Dubai Passive House requirements	
• 25 mm insulation on the roof • Airtightness, energy recovery for cooling • Triple solar protective glazing • Fixed shading • No night ventilation • Small windows • Supply air cooling if possible	
Calculated energy consumptions	
Heating	0 kWh/m²a
Cooling	40 kWh/m²a
Dehumidification	10 kWh/m²a
If uninsulated (for comparison): Calculated energy consumptions	> 300 kWh/m²a

required here, as are triple glazed windows and immovable shading. The Dubai Passive House presents itself outwardly as a one-storey house and is largely introverted (Fig. 5.15). The bedrooms are completely located at basement level. The windows are mainly oriented towards the inner courtyard, which is covered by a translucent membrane. The water areas inside the inner courtyard help to cool it down by evaporation (Fig. 5.16). The water areas are fed partly by grey water.

5.4.2   Ekaterinburg (North Russia)

A multi-storey, mixed-use building was designed for Yekaterinburg. In the typical very long and cold winter periods in the city, excellent thermal insulation is required, which can be realized exactly by today's available Passive House components. The supply-air heating system provides comfort in all rooms, which are supplied with fresh air by the controlled mechanical ventilation system. In the bathroom, it is occasionally necessary to turn on a small radiator, which can also be used for warming up and drying towels. Because of the very long and severe winters, building periods are very short in Yekaterinburg and we therefore recommended a prefabricated modular building system. Because of the strict fire prevention requirements, a timber structure was not a possibility and so we developed steel modular constructions to the Passive House Standard, with a composite thermal insulation system. The stair enclosure and the passenger lift are located outside the thermally controlled volume in a separate block in front of the actual residential building (Figs 5.17 and 5.18). By doing this, only one Passive House door for each floor is necessary and airtightness, especially in view of the lift, is easier to implement.

Table 5.3 Ekaterinburg Passive House requirements

Ekaterinburg Passive House requirements	
• 40–80 cm insulation	
• Very good heat recovery	
• Very good airtightness	
• Good triple glazing (or better?)	
• South orientation	
• Night ventilation	
Calculated energy consumptions	
Heating	22 kWh/m^{2}a
Cooling	0 kWh/m^{2}a
Dehumidification	0 kWh/m^{2}a

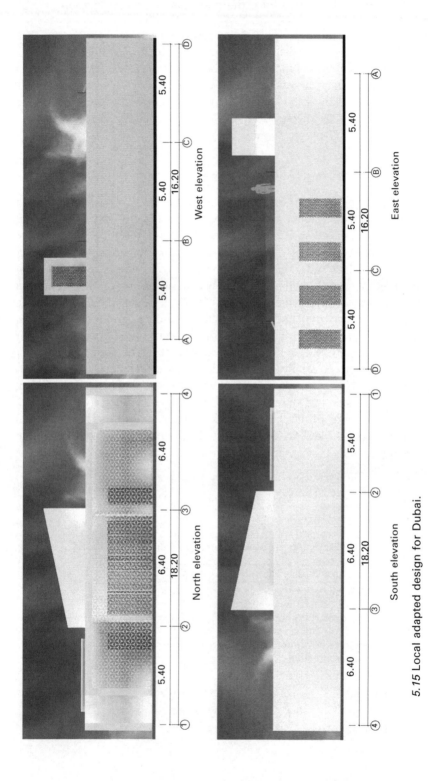

5.15 Local adapted design for Dubai.

5.16 Inner courtyard with water.

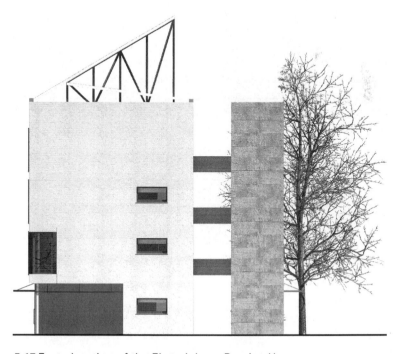

5.17 East elevation of the Ekaterinburg Passive House.

5.18 North perspective of the Ekaterinburg Passive House.

5.4.3   Las Vegas

In the desert climate of Las Vegas, with moderately cold, sunny winters, impact sound insulation in the basement ceiling already fulfils the duties of thermal insulation. The winter in Las Vegas is not so harsh, so the energy balance can handle even double glazed solar glazing. For the façades and the roof, light colours ('cool colours'), which do not absorb sunlight should be used (see Fig. 5.19).

Table 5.4 Las Vegas Passive House requirements

Las Vegas Passive House requirements	
• c.20 cm thermal insulation	
• Cool colours	
• Energy recovery (winter)	
• Night ventilation	
• Double solar protective glazing	
• Fixed shading	
Calculated energy consumptions	
Heating	15 kwh/m²a
Cooling	15 kwh/m²a
Dehumidification	0 kwh/m²a

5.19 South-east perspective of Passive House in Las Vegas.

Table 5.5 Shanghai Passive House requirements

Shanghai Passive House requirements	
• c.20 cm thermal insulation	
• Energy recovery, bypass with humidity control (summer)	
• No night ventilation	
• Double glazing	
• Movable shading	
• Separate cooling and dehumidification	
Calculated energy consumptions	
Heating	15 kWh/m^{2}a
Cooling	1 kWh/m^{2}a
Dehumidification	6 kWh/m^{2}a

5.4.4   Shanghai

For Shanghai, Rongen Architects designed a high-rise building with three underground parking floors, a three-storey mixed-use base and 23 residential floors. In this building, the overhanging and thermal bridge-free balconies shadow the floors below (see Fig. 5.20). The depth of the balconies varies from 0.84–1.92 m, depending on the solar altitude on the hottest summer days (Fig. 5.21).

5.4.5   Tokyo

Tokyo has a very similar climate to that of Shanghai and, here, a moderate thermal insulation and double glazed windows are sufficient. However, because of the confined construction site (the distance to a neighbouring house is often only 1.20 m and houses often shade each other), single family houses

5.20 Perspective of Passive House in Shanghai.

Table 5.6 Tokyo Passive House requirements

Tokyo Passive House requirements	
• c.20 Cm thermal insulation insulation	
• Energy recovery, bypass with humidity control (summer)	
• No night ventilation	
• Double/triple clear glazing	
• Movable shading	
• Separate cooling and dehumidification	
Calculated energy consumptions	
Heating	15 kWh/m²a
Cooling	1 kWh/m²a
Dehumidification	6 kWh/m²a

in Tokyo rarely gain direct solar radiation. Thus, triple-glazed windows are required. These relatively small houses have a bad A/V ratio and the habit of parking a car on the individual plots of land does not immediately facilitate the realization of a Passive House. The roof surface is used for the family 'garden' (Fig. 5.22).

5.5   Examples of Passive House architecture in Germany

Passive House technology is a successful method for realizing a highly energy efficient building standard, which is also sustainable. Some examples of Passive Houses that have already been built can be found below.

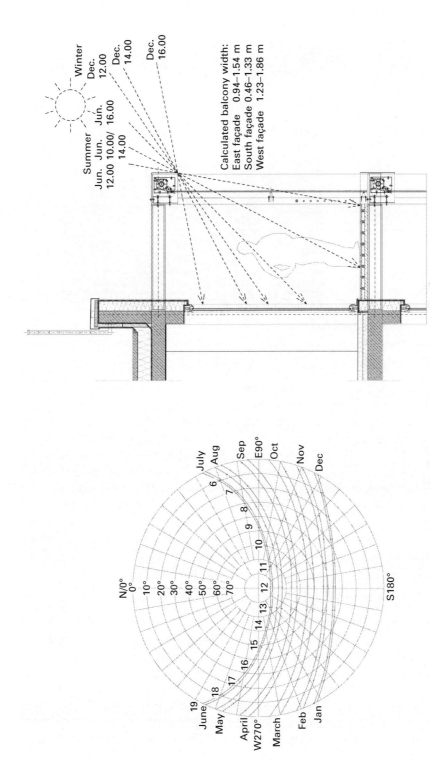

5.21 (a) Sunset Shanghai, China, (b) calculated balcony width. The depth of the balconies varies from 0.84–1.92 m, depending on the solar altitude on the hottest summer days.

Calculated balcony width:
East façade   0.94–1.54 m
South façade 0.46–1.33 m
West façade   1.23–1.86 m

5.22 West perspective of Passive House in Tokyo.

5.5.1   Houben-Engels House, Hückelhoven

Architects: Rongen Architects GmbH

This single family house opens to the south and to the west with a wonderful view of the landscape. The photovoltaic panels shade the bedrooms in the upper floor in the summer when the sun is very high and, in this way, overheating of the bedrooms is avoided without using movable sun protection elements (see Fig. 5.23). On the ground floor, the façades are covered with varnished larch timber battens to anticipate the silver grey appearance of natural weathering. Larch wood is water resistant and therefore it will not be necessary to paint the timber battens again in the future. On the upper floor, the façades are plastered and painted white, in contrast to the wooden façades of the ground floor.

5.5.2   Hanssen-Höppener House, Selfkant

Architects: Rongen Architects GmbH

The 'Hanssen-Höppener' House, in Selfkant, is a sustainable CO_2-neutral Passive House. The large window areas on the east, south and west sides supply the interior with bright daylight. However, despite the large windows, movable sun protection screens (which would require electric energy) are not needed, since the building does not overheat in summer. The high southern sun does not penetrate into the open living area and, on the upper floor, the framed overhanging balcony prevents the lower floor from overheating;

5.23 House Houben-Engels. On the ground floor the façade is covered with timber battens, on the first floor it is plastered. Photo courtesy of Rongen Architects GmbH.

the overhang shadows the windows below (Fig. 5.24a and 5.24b). A pond in the garden is oriented to the southwest, from where the warm summer wind blows. By placing it here, the water evaporates and creates evaporation cooling to cool the floor-to-ceiling glazing. As a result of warm water supply from a solar-water heater, and heat supply from a wood pellet oven and a heat pump, and also electric power supply from photovoltaic panels on the roof, the 'Hanssen-Höppener' Passive House is CO_2 neutral. The façades of the Hanssen-Höppener' House are partially covered with sustainable wood. The larch timber battens will weather naturally and over time will become more and more resistant to the influence of the weather.

5.5.3   Dr and Dr Sander House

Architects: Rongen Architects GmbH

The 'Dr and Dr Sander' House is a 410 m^2 luxury mansion built to the Passive House Standard as a prefabricated timber construction (see Figure 5.25). Figures 5.26a and 5.26b show the building under construction. In the first two years, the energy costs for heating and cooling the house were less than €250 per year. Because all the windows can be opened, the summer heat can exit via a chimney effect, dispensing with the need for movable sun protection screens. The sun's energy warms up the pool.

(a)

(b)

5.24 Passive House 'Hansen-Höppener', (a) rear view, (b) terrace.
Photo: Bullik Photography.

5.25 House 'Drs Sander', Wassenberg (Germany). A luxury mansion of more than 410 m². Photo courtesy of Thomas Drexel.

5.5.4   The Caritas House, Neuwerk

Architects: Rongen Architects GmbH

The new Caritas elderly care centre in Mönchengladbach-Neuwerk is the first elderly care centre in Europe built to the Passive House Standard, and combines a modern group-living concept with advanced environmental technology (Fig. 5.27). Not only does the Passive House require extremely little active heating energy, it also provides a much higher than average quality of room air and thus an improved quality of living. Large areas of green space lie in the immediate vicinity. The direct vicinity of the church and other public facilities will assist in promoting community relations. The 'Caritas House Neuwerk', as it has been named, is particularly intended to provide those suffering from dementia with dignity and living accommodation appropriate to their illness. The intention was to create a building identified by key concepts such as security, esteem, self-determination, freedom, closeness and a homely atmosphere.

The new elderly care centre can accommodate a total of 80 residents and is divided into eight apartments, each with 10 residents. This living structure allows each group of 10 to establish a family-type way of life. All fundamental activities of a structured daily nature are carried out within the protected bounds of this stable family group. In each case there is one direct care and nursing area for every two or three groups of residents. This area includes a care room, a duty room, a care bathroom and other complementary rooms.

(a)

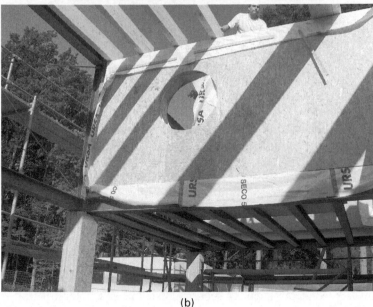

(b)

5.26 (a) and (b) show the building under construction. Photo courtesy of Rongen Architects GmbH.

Due to the high standard of sanitary facilities and equipment – all rooms are single rooms with their own shower – it can be assumed that the ward bathroom with its lift-assisted bath tub will not often be used during the normal daily routine. Its main function – in accordance with the care concept – is to be used as a pleasant care and bathing experience. The offer

5.27 View to the entrance and dining room of The Caritas House Neuwerk. Photo courtesy of Rongen Architects GmbH.

of such opportunities leads to a clear gain in quality of life, especially for the increasing number of geronto-psychiatric people, whilst the building helps to ensure that basic care and treatment have priority over individual showers and baths.

The manageable size of the groups assists the therapeutic concept. Residents and nursing personnel know one another, housekeeping tasks (i.e. using the facilities) help to retain orientation. The readiness of orientated and, in some cases, less orientated residents with dementia to take on a certain joint responsibility has increased. The positioning of the entrances to each of the group apartments in the general living room/rest room area is also intended to 'bring this zone to life'. Due to the restricted number of residents in each group, the frequency of contacts and movement remains manageable for the residents.

The design/plan as realized provides for an administrative area of the house on the ground floor. A room for worship and meditation provides an opportunity for mutual prayer as well as communication, but it is also a room in which to withdraw and be alone with oneself. The dining room, which is in the immediate neighbourhood of the main entrance and the meditation room, opens up towards the entrance to the building and to the Engelbecker Straße, which means there is a lot to see outside (Fig. 5.28).

The single rooms have been built in such a way that two single rooms can be adapted to make a double room (for married couples), which can

5.28 Unmovable outside sun protection for the dining room. Photo courtesy of Rongen Architects GmbH.

easily be converted back into single rooms with no rebuilding work required. Thus, should one of the partners die, the remaining resident can stay in the old familiar group and does not need to move into another new group.

The supporting structure of the Caritas House Neuwerk is a partly cellared, reinforced concrete, skeleton construction. The exterior walls are mainly composed of prefabricated sandwiched panel elements made of wooden materials and, on the ground floor, have a front-shell of bricks (clinker). Although built to the Passive House Standard, the overall building is highly structured, which makes it easy to comprehend and understand. It is not an anonymous accommodation machine. The partly protruding lean-to roofs give the building a certain lightness of appearance which, up until now, Passive Houses have often lacked. The shaping and layout of outdoor areas invites residents to spend time there and carry out activities in the fresh air (Fig. 5.29). In addition, these areas offer places of peace and quiet outside of the house.

A problem brought repeatedly to our attention – long before our very first draft – which occurs regularly in care homes and which had not yet been solved was the smell of urine, which is typical for such establishments. It was for this reason that we came up with the idea of proposing a solution with 'controlled permanent ventilation'. Controlled permanent ventilation reduces the problem of 'urine smell' considerably, and will perhaps even dispose of it entirely. Additionally, the permanent and comfortable quality

5.29 Inner courtyard. Photo courtesy of Rongen Architects GmbH.

of room air is ensured because, whenever humidity gets too high, the excess moisture is withdrawn.

The materials selected for the exterior façade – red clinker, varnished wooden boards, coloured window frames (wooden window frames to Passive House Standard) together with the window sunshades – act together to form a lively building but without creating an unsettled impression. Originally it seemed that varnished wooden boards would not be allowed as an element of the façade. Following long negotiations with the fire brigade and the building authorities responsible, we were able to convince them to refrain from applying requirement B1 (slightly inflammable) to the façade elements, thus avoiding a costly and tedious technical building licensing procedure (i.e. through the Institute for Building Technology in Berlin). After all, with regards to the question of fire protection, the prefabricated sandwiched panel elements that form the basis of the construction are better than the simple, chemically treated wooden board elements classified as B1 in the fire protection classifications. Following negotiations with the Ministry for Building and Accommodation in North Rhine–Westphalia, the double-layered, load-bearing, thermal insulation beneath the ground slab (the perimeter insulation) was approved 'for this individual case', which led to a large saving in costs.

For the care home company and for the residents too who, in the end, will have to bear the costs, the very low heating costs were one decisive argument to build an establishment to the Passive House Standard. The particularly high level of quality of life offered weighs even more in favour of such a decision. For elderly people who still want to take part in life 'outside' whilst sitting inside (and who are therefore often found close to the windows), it is of great benefit to no longer suffer from the draughts. The permanently high quality of air, containing an extremely low concentration of CO_2 is also beneficial to the residents. The only possible problem encountered by residents, was getting used to 'properly airing' the Passive House, i.e. realizing that it is not necessary always to open windows; however, this is a problem equally encountered by younger generations.

The insulated building shell of the Passive House enables the reduction of heating levels. It is not necessary to install radiators in the residents' rooms, for example. The glazing from floor to ceiling leads to an increase in usable space of approximately 1 m^2 per resident room, which would otherwise be 'wasted' living space. Small meters set into suspended ceilings over the baths measure the air temperature and allow individual residual heating requirements to be achieved.

Fresh air at all times coupled with permanent vacuum removal of WC/bathroom odours ensures permanent and comfortable air quality, whilst at the same time high concentrations of moisture are removed. Outside air is sucked in through a ground thermal converter, pre-warmed in winter and cooled down in summer. Filtered outside air holds back dust and pollen and to a certain degree bacteria as well.

To avoid the usual necessity of fire protection flaps between fire areas, exterior air is directed up the façade from outside to decentralized ventilators at each floor level. Five resident rooms are orientated on the points of the compass, each forming one airing unit. Therefore, if cooling is required, nursing personnel are able to switch off the heat recovery system and let in cool air through the ground thermal converter. In this way, for example, in autumn, the southern side could be cooled whilst at the same time the heat recovery system or the heating system could heat the northern side.

Had the house building been financed by a loan from the bank, the cost to borrow the extra capital required to build to Passive House Standard, rather than the normal WSchVO 1995 standards would have been approximately €3835 (at 5.5% interest and with 1.0% repayment). This should be compared with the savings on heating costs made by building to Passive House Standard, which have been calculated to be approximately €14,087 per year. As a result, it can be assumed that had the house been financed in this way, there would have been a saving in operating costs of €10,252 per year.

5.5.5   The Caritas House, St Josef, Mönchengladbach-Giesenkirchen

Architects: Rongen Architects GmbH

The Caritas House St Josef (Mönchengladbach-Giesenkirchen) is the first elderly care centre constructed in a prefabricated steel modular system to the Passive House Standard. The existing elderly care centre in St Joseph Mönchengladbach-Giesenkirchen was originally going to be renovated but, after carefully considering the financial costs and social consequences (e.g. the mobile care required for current residents during the construction period), the client decided against an extensive refurbishment. In the end, the decision was made to build a new building and to partially demolish the existing care centre, which dates from the 1960s. The remaining portion of the existing building and the new building together constitute a new elderly care nursing centre.

At first, part of the building was vacated. One part was then rebuilt while the rest remained fully occupied. The new building has 84 single rooms, workplaces and care areas, etc., realized in a prefabricated modular steel system (see Fig. 5.30). Relocation of residents into the new rooms did not begin until the new building was completely built, after which the rest of the demolition followed. The main building was demolished down to the ground floor, which was then renovated, rebuilt and integrated into the new building (the entrance area, kitchen and dining room being located in it). The entire new building, including a chapel on the ground floor of the previous building, was built to the Passive House Standard. As a result of the decision to use a prefabricated modular system, the construction time was reduced from 20 to 5 months (module size = 3.89 × 15.25 × 3.20 m (height)).

The decision to realize the entire new building to Passive House Standard

5.30 The modules under construction. Photo courtesy of Kleusberg GmbH & Co KG, Wissen, Germany.

and to renew the remainder of the predecessor building using Passive House components, was made easy by the client (Caritas Association Mönchengladbach), who had already had very good experiences of the Passive House Standard in other care centres in Europe. The Caritas House St Josef was the first of their homes, indeed the first anywhere, not to smell of urine and, additionally, it cost less than originally calculated (Fig. 5.31).

5.5.6   Grammar School, Baesweiler

Architects: Rongen Architects GmbH

The Baesweiler city government commissioned an analytical study of 21 municipal buildings. After finishing the study, it was to be decided which building should be refurbished first and which energy standard should be achieved. The result was a decision to start with the grammar school, consisting of four buildings and a gymnasium. It was very important for the client to improve not only energy efficiency but also the artistic design (Fig. 5.32). The ambitious goal was to follow the 'Passive House Standard', making the project unique because it is the biggest project to have been refurbished to this standard, and is also the first redevelopment to Passive House Standard with a curtain wall.

The calculated savings in heating energy are about 90% and the CO_2 reduction will be more than 530 tons per year. In the summer, the school is 'passively' cooled. Sun protection prevents overheating of the classrooms

5.31 The new building will not look like a modular building, as shown in this image. Photo courtesy of Kleusberg GmbH & Co KG, Wissen, Germany.

5.32 Construction phases 3 and 4 of the school buildings before refurbishment. Photo courtesy of Rongen Architects GmbH.

5.33 Sections of the school buildings after refurbishment. Photo courtesy of Rongen Architects GmbH.

and, using automatically controlled window opening at night, the building mass is cooled down by cool night air. During the following day, the building mass stores heat (Fig. 5.33).

On 18 November 2010, the first stage was honoured as one of the winning projects in a Germany-wide 'Local Climate Protection' competition, winning prize money of €40 000.

5.5.7   Stiftsplatz 6 memorial, Wassenberg

Architects: Rongen Architects GmbH

This single family house is a listed memorial (Fig. 5.34).The originally dilapidated 'Stiftsplatz 6' memorial was restored with ecological materials and in sympathetic architectural language. The earlier renovation using pumice building blocks was removed completely. All exterior walls were thermally insulated inside using rush matting (Fig. 5.35), in front of which a new interior wall of clay bricks (plastered with clay plaster) was built. This wall structure reached a thermal quality equivalent to a very good 'Low Energy House'. The frame walls inside were also filled in with clay bricks, or renewed with clay plaster (Fig. 5.36). Clay has very positive physical properties; one of which is the ability to regulate moisture.

5.6   Future trends

The high demand for Passive Houses in Europe has caused enormous development of appropriate structural components, which has resulted not only in an improvement of the quality but also in lower costs. And there appears to be no end in sight for further development of innovative structural components, especially for Passive Houses. One of the most remarkable

5.34 Memorial Stiftsplatz 6, Wassenberg, Germany. Photo courtesy of Rongen Architects GmbH.

5.35 Inside thermal insulation with rush mat. Photo courtesy of Rongen Architects GmbH.

5.36 Frame walls with clay bricks. Photo courtesy of Rongen Architects GmbH.

innovations is vacuum glazing, which easily reaches a *U*-value of 0.8 W/m^2K for the glass plates with only double glazing. The total thickness of such windows is only 10 mm. Such windows are not only clearly thinner but also lighter than conventional double glazing. It is assumed that vacuum glazing will become a mass product in the near future and will thus become an economical alternative to triple glazed windows (Fig. 5.37).

Very good thermal insulation of the building envelope of Passive Houses can now be realized in an attractive, aesthetically pleasing way. A number of innovative products are available such as high-performance insulation in CFC-free rigid foam *Resole*, a building material with a thermal conductivity of 0.22 W/m²K.

While very thick insulation can be planned comparatively easily in the case of new buildings, the situation in the case of old buildings is very different. All too often, thick insulation packages are unfeasible and also not recommended. Vacuum insulation panels (VIP) with average λ-values of 0.004 W/mK require much thinner thermal insulation for the same insulating properties as conventional insulation, opening up new possibilities; but even vacuum insulation panels are still relatively expensive.

The transparent high-performance insulation material Nanogel, which has a thermal conductivity of 0.0018 W/mK, is also increasingly used. Nanogel is an aero gel based on silicic acid. Because it consists of 97% air, it is very light, weighing between 60 and 80 g/m³. Despite the low weight, the sound insulation properties of Nanogel are very good. Aero gels are among the easiest and most effective insulation materials available today, and are hydrophobic and thus moisture- and mould-resistant.

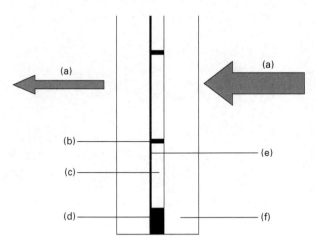

5.37 Schematic sketch of vacuum glazing
U_g – value < 0.5 W/m₂K
(a) Heat flow
(b) Spacing stay
(c) Evacuated interface between the glass plates c. 0.7 mm
(d) Vacuum tight glass sealing
(e) Functional layer
(f) Glass, 4 mm.

So far, Passive House design has very much been focused on energy efficiency. In building Passive Houses, architects will begin to use more and more ecological building materials in the future. In this respect, clay is an ideal example. Clay is a healthy and environmentally compatible material, today enjoying a return to popularity. The energy demand for its production as a building material is very low and its use as a building material therefore significantly reduces primary energy demand. Clay is also a good thermal conductor. Because of its high specific heat capacity, clay walls are able to compensate for temperature differences. In hot climates, therefore, thick adobe walls provide a pleasant indoor climate. And because of its rapid moisture absorption, clay influences the indoor climate positively and thus promotes healthy living. Due to the equable heat radiation of massive clay walls, air drafts do not happen and this too increases living comfort. Clay emits no pollutants, has no known effects on allergies and has a lot of positive ecological properties. Moreover, clay has very good acoustic properties and is able to shield high-frequency electromagnetic radiation. Adobe walls are 100% recyclable and can be returned for disposal directly back into the environment.

5.7   Sources of further information

Books

Hestermann, Ulf and Rongen, Ludwig (2010) *Frick/Knöll Baukonstruktionslehre 1, 35th Edition*. Vieweg+Teubner Verlag, Wiesbaden, Germany
Neumann, Dietrich, Hestermann, Ulf and Rongen, Ludwig (2008) *Frick/Knöll Baukonstruktionslehre 2. 33rd Edition*. Vieweg+Teubner Verlag, Wiesbaden, Germany

Websites

www.fiz-karlsruhe.de (BINE Information service)
www.kampmann.de
www.kleusberg.de
www.bullik-kunkler.de
International Passive House conference 2011, available at: www. passivhaustagung.de/sechzehnte/Englisch/index_eng.php

5.8   References

Feist, W. (1993) Passive Houses in Central Europe, Dissertation, University of Kassel, Germany
Nansen, F. (1897) *In Night and Ice, the Norwegian Polar Expedition, 1893–1896*, F A Brockhaus, Leipzig

News Network Internetservice GmbH (2006) 'Rekordozonloch über Südpol: 27,45 Mio. km^2 – Fläche von USA und Russland zusammen', available from: www.news.at/articles/0642/35/154507/rekordozonloch-suedpol-27-45-mio-flaeche-usa-russland

Part II

Earth materials engineering and earth construction

6

Soil materials for earth construction: properties, classification and suitability testing

L. N. REDDI, Florida International University, USA,
A. K. JAIN and H-B. YUN, University of
Central Florida, USA

Abstract: Each soil has a different particle size and/or mineral structure, based on the process of formation. This leads to different properties, which distinguish the constructional abilities of a soil. A brief discussion of soil types and classification is provided. Chief mineral constituents of clays along with their structures are explained. Different soil properties which make soil a suitable construction material are also discussed. Some soil property tests to identify soil properties, such as consistency and compaction are briefly explained. Different compaction procedures suitable for different soils are discussed. Water content in a soil plays an important role in providing strength to a soil during compaction.

Key words: clay, minerals, consistency, loam, soil classification, compaction.

6.1   Introduction

The term *soil* has different meanings, carrying different senses to different groups. Soil for use as a building material, is considered to include all naturally occurring loose or soft deposit overlying the solid bedrock crust, which is produced by the physical and chemical disintegration of rocks (weathering), and which may or may not contain organic matter. The use of soil as a building material is old as mankind itself. Since, the time the very first dwelling was built, humankind has been confronted with soil problems – most important being its highly varied nature. Depending on the methods of soil formation, a soil has distinct properties, which leads to its different uses.

6.2   Soil formation

The variation in soil properties can be explained by variations in the type of weathering, transportation, deposition and upheaval, again followed by weathering, and so on. Weathering is accomplished through physical and chemical agencies. The physical agencies causing weathering of rocks are

155

periodical temperature changes, impact and abrasive actions of flowing water, ice and wind, and splitting actions of ice, plants and animals. Cohesionless soils are formed due to the physical disintegration of rocks. The chemical weathering of rocks, or decomposition as it is called, is caused mainly by the following principal reactions: oxidation, hydration, carbonation and leaching by organic acids and water. Clay minerals are produced by chemical weathering.

The products of rock weathering up to the size of gravel are termed soil, and beyond that size are considered rock fragments. Soils so obtained may be residual or transported. Residual soils are those which remain in place directly over the rocks from which they were formed. If they have been removed from the original bedrock and re-deposited somewhere else, they are known as transported soils. These deposits may be considerable in depth and their homogeneity or heterogeneity depends upon the manner of their transportation and deposition.

The various agencies of transporting and re-depositing soils from their original place of formation to other locations are water, ice, wind and gravity. Alluvium soils, marine deposits and lacustrine soils result from water being the agency of transport. These are usually stratified because of fluctuations in the stream velocity, and the average particle size decreases with increasing distance from the source of the stream. Glaciers cause the formation of drift, till, moraine or varve. These are a heterogeneous mixture of rock fragments and soils of varying sizes and proportions and, except the stratified drift deposited by glacial streams, are without any normal stratification. Dune sand and loess are the wind-blown (aeolin) deposits. Sand dunes consist of wind-drifted uniformly sized sand. Loess is wind-blown silt or silty clay with little or no stratification. Soils transported by gravitational forces are termed colluvial soils, such as talus, which consist of soil fragments and soil material collected at the foot of cliffs or steep slopes. The accumulation of decaying and chemically decomposed vegetable matter under conditions of excessive moisture results in the formation of cumulose soils such as peat and muck. Peat is a fibrous spongy mass of organic matter under various stages of decomposition in which plant forms can still be identified. Muck is essentially a thoroughly decomposed peat in which plant forms cannot be recognized.

Tremendous earth upheavals have taken place in the geological past, resulting in an enormous lifting, tilting and folding of soil deposits and rock formations. As soon as upland is formed, the cycle of weathering, transportation and deposition begins again, resulting in the further formation of soil.

6.3   Soil types

A soil is normally classified by the size of its particles. The various modes of weathering determine particle size, and then the form transportation causes a mixture of different soils. A group of particle sizes may have different behaviors and are classified accordingly. The individual soil particles can range from gravel size to clay size. The size of the soil particles is determined using sieve analysis and a particular size group refers to a particular soil. A number of classification systems exist to identify the general characteristics of soils (Das, 2006). All of the systems take into consideration the particle size and consistency of soils. AASHTO (American Association of State Highway and Transportation Officials) classification system, Unified Soil Classification System, Massachusetts Institute of Technology (MIT) Classification System, British Standards Classification System, Canadian System of Soil Classification, Indian Standards of Soil Classification, etc. are among many of the prevalent classifications that are followed in different regions. A typical soil classification by MIT (Scott and Schoustra, 1968) classifies various soil groups according to the individual soil grain size (mm):

Gravel: > 2.0
Sand: 0.06–2.0
    fine sand: 0.06–0.2
        medium sand: 0.2–0.6
        coarse sand: 0.6–2.0
Silt: 0.002–0.06
    fine silt: 0.002–0.006
    medium silt: 0.006–0.02
    coarse silt: 0.02–0.06
Clay: < 0.002

Sands and gravels are termed as coarse-grained soils and the silts and clays as fine-grained soils. A known weight of soil is made to pass through a stack of sieves, with the sieve sizes decreasing towards the bottom and based on the sizes of the classified soil groups. The coarse-grained soils are cohesionless soils exhibiting zero plasticity. It is the fine-grained soils, especially the clays, which are plastic, that contribute more to building materials for earthen construction. For the fine-grained soils, further classification is carried out using a hydrometer or sedimentation analysis, which uses Stokes's law for the fall of suspended individual small-sized soil particles (silts and clays).

Figure 6.1 shows some shapes of particle size distribution curves. A steep slope (curve *a*) indicates a uniform soil. Humps in the curve indicate a soil to be a mixture of two or more uniform soils (curve *b*). A steep curve in the sand sizes that gradually flattens into a long, flat curve in the fine sizes would characterize a soil that was formed by weathering (curve *c*). A flat

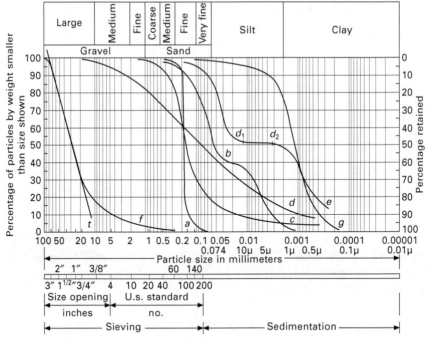

6.1 Gradation curves for different soils (Jumikis, 1967).

curve shows a wide range in particle sizes, which indicates a well-graded soil (curve *d*). A flat portion in the curve indicates a deficiency of particle sizes at the flat interval, $d_1 - d_2$, of curve *e*. Soils in the early stages of their development are mainly the results of physical weathering, characterized by the presence of large amounts of coarse material (curve *f*). Soil *g* varies widely in its coarser range, with the finer fractions being more uniform (Jumikis, 1967).

Laser diffraction analysis is also a widely used technique for particle size analysis. Instruments employing this technique are considered easy to use and particularly attractive for their capability analyzing particle sizes in the range 0.02 to 2000 μm. Rapid data acquisition and ease of verification are two important factors that make it a better technique than conventional sedimentation analysis.

6.3.1 Clays

Acids dissolved in pore water and soil solids combine to form a solution of grains. These minerals combine and crystallize under differing conditions of temperature and pressure, giving rise to new minerals, which may have entirely different arrangements of molecules to the original minerals. The new arrangement gives rise to smaller sized mineral particles, which may be needle- or plate shaped, with lengths or diameters tens to hundreds of times their thickness. The new minerals thus formed are known as the clay minerals, which vary in size from hundredths of a micron to tens of a micron in diameter. The surface area of these particles is very large leading to high intermolecular forces on the surface and edges, which attracts bipolar water molecules. This property that binds water to the soil grain is greatly exploited in its use as a construction material. The amount of water present in clayey soils plays an important role in defining the clay characteristics, but equally important is the role of type of mineral(s) present in the clay.

Structure of clay minerals and their surface activity

As mentioned earlier, the characteristic properties of clay minerals are high specific surface and high specific activity. The behavior of conglomerations of such minerals is governed largely by surface phenomena. It is therefore necessary to study the atomic structure of these minerals. X-ray diffraction analysis of the clays (Mitchell, 1993) has shown the clay minerals to have a crystalline structure. There are two fundamental building blocks for the clay minerals, which are complex silicates of aluminum, magnesium and iron. One is silica tetrahedral unit (Fig. 6.3a), in which four oxygen atoms or hydroxyl molecules with the configuration of a tetrahedron enclose a silicon atom. The tetrahedra are combined in a sheet structure so that the oxygen of the bases of all the tetrahedra are in a common plane, and each is shared by two tetrahedra (Fig. 6.3b). The silica tetrahedral sheet alone may be reviewed as a layer of silicon atom between a layer of oxygen atoms and a layer of hydroxyl molecules (tips of the tetrahedra). The silicon sheet is represented by the symbol shown in Fig. 6.2, representing the oxygen basal layer and the hydroxyl apex layer.

6.2 Silicon sheet.

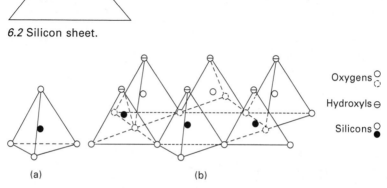

(a)   (b)

6.3 Structural units in the silica sheet (Grim, 1968).

The second building block is an octahedral unit in which an aluminum, iron or magnesium atom is enclosed by six hydroxyls with the configuration of an octahedron (Fig. 6.5a). The octahedral units are put together into a sheet structure (Fig. 6.5b), which may be viewed as two layers of densely packed hydroxyls with a cation between the sheets in octahedral co-ordination (Grim, 1968). When the cation is aluminum, this unit is known as *gibbsite*; when it is magnesium this unit is called *brucite*. This unit is shown in Fig. 6.4.

About 15 minerals are ordinarily classified as clay minerals, and these belong to four main groups: kaolin, illite, montmorillonite and palygorskite (Punmia *et al.*, 2005). The structure of the first three minerals, which are the most common, is discussed here.

Kaolinite is the most common mineral of the kaolin group. Its structural unit consists of gibbsite sheets (with aluminum atoms at their centers) joined to silica sheets through the unbalanced oxygen atoms at the apexes of the silicas (i.e. the apexes of the silica layer and one of the gibbsite molecules form a combined layer). This structural unit (Fig. 6.6) is about 7 Å thick. The kaolinite mineral or crystal is a stack of 7-Å-thick sheets, which can be symbolized as shown in Fig. 6.7a. The structure is like a book with each leaf of the book 7-Å-thick. Such successive layers are held together with *hydrogen bonds* (Fig. 6.7b). The kaolinite crystal occurs in clay as platelets from 1000–20,000 Å wide by 100–1000 Å thick. The hydrogen bond is

6.4 Gibbsite.

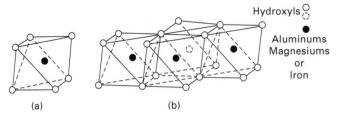

Hydroxyls

Aluminums
Magnesiums
or
Iron

(a)   (b)

6.5 Structural units in the octahedral sheet (Grim, 1968).

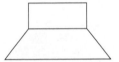

6.6 The basic unit of kaolinite mineral.

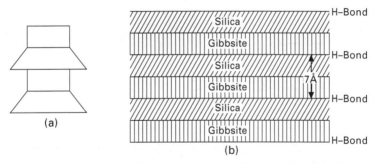

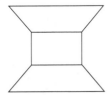

6.7 (a) Symbolization of kaolinite mineral, (b) structure of kaolinite.

6.8 The basic unit of montmorillonite mineral.

fairly strong and crucially the intra-lamellar spacing is very small, hence it is extremely difficult to separate the layers, and, as a result, kaolinite is relatively stable and water is less able to penetrate between the layers, thus it exhibits relatively little *swell* on wetting. The platelets carry negative electromagnetic charges on their flat surface, which attract thick layers of adsorbed water thereby producing plasticity when kaolinite is mixed with water.

Montmorillonite is the most common of all the clay minerals in expansive clay soils. It is made up of sheet-like units. Each unit is made up of a gibbsite sheet sandwiched between two silica sheets, shown in Fig. 6.8. The thickness of each unit or sheet is about 10 Å and the dimensions in the other two directions are indefinite. The gibbsite layer may include atoms of aluminum, iron, magnesium or a combination of these. In addition, the silicon atoms of the tetrahedrals may interchange with aluminum atoms. These structural changes are called *amorphous changes* and result in a net negative charge on the clay mineral. Cations which are in soil water (i.e. Na^+, Ca^{++}, K^+, etc.) are attracted to the negatively charged clay plates, and exist in a continuous state of interchange. The base 10 Å thick units are stacked one above the other in layers and are symbolized as shown in Fig. 6.9a. There is a very weak bonding between the successive sheets and water may enter between the sheets causing the minerals to swell (Fig. 6.9b). The spacing between the elemental silica–gibbsite–silica sheets depends on the amount of available water to occupy the space. For this reason, montmorillonite has an expanding lattice. Each thin platelet has the power to attract to each flat surface a layer

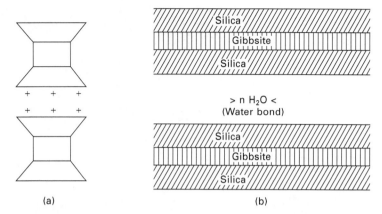

6.9 (a) Symbolization of montmorillonite mineral, (b) structure of montmorillonite.

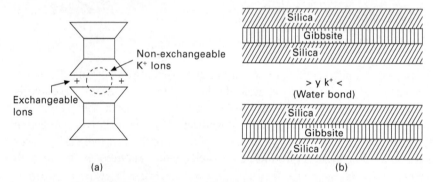

6.10 (a) Symbolization of illite, (b) structure of illite.

of adsorbed water approximately 200 Å thick. Soils containing these minerals exhibit high shrinking and swelling characteristics, depending on the nature of exchangeable cations present.

Illite has a structure similar to montmorillonite except that there is substantial (± 20%) replacement of silicon atoms by aluminum in the tetrahedral layers and potassium atoms are between the layers serving to balance the charge resulting from the replacement and to tie the sheets together. The basic unit is represented as shown in Fig. 6.10a. The cation bond of illite is weaker than the hydrogen bond of kaolinite, but stronger than the water bond of montmorillonite. Due to this, the illite crystal (Fig. 6.10b) has a greater tendency to split into ultimate platelets consisting of a gibbsite layer between two silica layers, than that in kaolinite. However, the illite structure is less expansive because of movement of water between the sheets, as in the case of montmorillonite. Illite particles may be 50–500 Å thick and 1000–5000 Å in lateral dimensions.

Clay minerals have the distinctive property of adsorbing certain ions at the surface. This suggests that the mineral surfaces are polar, like water. The surface thus attracts ions, polar molecules or polar radicals, which in turn attracts ions and polar molecules from the environment. The electrostatic attraction between the opposite charges thus leads the process of adsorption. When the distribution of charges is equaled, i.e. negative and positive, a similar distribution of positive and negative charges is seen in the adsorbed layer. If the distribution of the charges on the surface of minerals is not uniform, the adsorption of ions will be of a different degree. As a result, an *electric double layer* will be formed at the surface, where a remarkable difference is seen in the electric potential of the mineral surface and of the surrounding solution.

Moisture entering an adobe brick causes clays to swell and release their bonds so that it disintegrates. To prevent disintegration, soil stabilizers are added to the basic soil mix to waterproof or increase the weathering resistance of the adobe brick. The most commonly used stabilizers are sand, straw, Portland cement, lime and bituminous and asphalt emulsions (Clifton, 1977), although as many as 20 different materials have been found in use (Wolfskill *et al.*, 1970). Further details of soil stabilization are given in Chapter 9.

6.4   Soil consistency

The term consistency states the relative ease with which a soil can be deformed. It denotes the degree of plasticity of a soil and is indicated by terms: *soft*, *firm*, *stiff* or *hard*. In practice, only the fine-grained soils, particularly clayey soils, for which the consistency is related to a large extent to water content, are described using these terms. Depending on the water content the following four *states of consistency* are used for the behavior of a soil: (i) the liquid state, (ii) the plastic state, (iii) the semi-solid state and (iv) the solid state. The water contents at which the soil passes from one state to the next are known as *consistency limits*. The significance of these limits was first demonstrated by Atterberg, a Swedish soil scientist, and they are also known as *Atterberg limits* (Das, 2006).

In a very wet state, a fine-grained soil acts as a viscous liquid and is said to be in the liquid state. As the soil dries, it starts acquiring plastic properties and is said to be in the plastic state. The soil no longer flows like a liquid, but continues to be deformable, or plastic, without cracking. With further reduction in water content the soil ceases to be plastic and becomes brittle. It is then said to be in the semi-solid state. Up to the semi-solid state, reduction in the volume of soil is nearly equal to the volume of water lost. When the volume of the soil stops reducing, even with further reduction of the water content, the soil is said to be in the solid state. The limiting water contents expressed as a percentage of the dry mass, when the soil passes from liquid to plastic, plastic to semi-solid and semi-solid to solid states of consistency are respectively termed *liquid limit*, *plastic limit* and *shrinkage limit*. The transition between each state is gradual and there is no abrupt change in the physical properties of a soil. Arbitrary tests are adopted to precisely define the limits.

The liquid limit can be defined using the Casagrande technique, as the minimum water content at which a pat of soil cut by a groove of standard dimensions will flow together for a distance of 13 mm under the impact of 25 blows in a standard liquid limit apparatus. It is the water content at which the soil shows a definite shearing resistance, though small, as the water content reduces. In the cone penetrometer test, the liquid limit of the soil is the water content at which an 80 g, 30° cone sinks exactly 20 mm into a cup of remolded soil in a 5 s period.

The plastic limit is the minimum water content at which the soil can be rolled into a thread of approximately 3 mm in diameter without breaking.

The shrinkage limit is the minimum water content at which the reduction in water content will not cause a decrease in volume of the soil. It is the minimum water content at which a soil just becomes completely saturated.

The range of consistency within which a soil exhibits plastic properties is called the plastic range and it is indicated by the term *plasticity index*. The plasticity index is defined as the numerical difference between the liquid limit and the plastic limit:

$$I_P = w_L - w_P \qquad\qquad [6.1]$$

Cohesionless coarse-grained soils have no plastic state of consistency, and the liquid and plastic limits may be said to coincide, i.e., I_P is zero. When the liquid limit or the plastic limit cannot be determined (in the case of sandy soils, the plastic limit should be determined first), the plasticity index is reported as *NP* (non-plastic). When the plastic limit is equal to or greater than the liquid limit, I_P is reported as 'zero'.

The behavior of saturated fine-grained soils with their natural water contents may be predicted by their *consistency index*. The importance of the consistency index for cohesive soils is comparative to that of the density index for cohesionless soils. The consistency index is defined as the ratio of liquid limit minus the natural water content to the plasticity index:

$$I_C = \frac{w_L - w}{I_P} \qquad\qquad [6.2]$$

A soil with I_C as zero is at its liquid limit and with I_C as unity is at its plastic limit. If $I_C < 0$, the natural water content of the soil is greater than the liquid limit and the soil behaves like a liquid. If $I_C > 0$, the soil is in a semi-solid state and is stiff.

The liquidity index is defined as the ratio of the difference between the natural water content and the plastic limit to the plasticity index:

$$I_L = \frac{w - w_P}{I_P}$$ [6.3]

The liquidity index determines in what part of its plastic range a given soil mass lies.

Until the soil reaches its shrinkage limit by reduction in the water content, the total volume change is equal to the water removed. The ratio of a given volume change, expressed as a percentage of the *dry* volume, to the corresponding change in water content above the shrinkage limit is called the *shrinkage ratio*:

$$SR = \frac{\frac{V_1 - V_2}{V_d} \times 100}{w_1 - w_2}$$ [6.4]

where V_1 = volume of soil mass at water content w_1
  V_2 = volume of soil mass at water content w_2
  V_d = volume of dry soil mass.

It can be shown that shrinkage ratio is equal to the mass specific gravity of the soil in its dry state, i.e.

$$SR = \frac{\gamma_d}{\gamma_w}$$ [6.5]

Based on knowledge of shrinkage ratio, we have two important terms, which are frequently used in the construction of building with earthen materials. The *volumetric shrinkage* is defined as the decrease in the volume of soil mass, expressed as a percentage of the dry volume of soil mass, when the water content is reduced from a given percentage w_1 to the shrinkage limit:

$$VS = (w_1 - w_s)SR$$ [6.6]

Linear shrinkage is defined as the decrease in one dimension of a soil mass expressed as a percentage of the original dimension, when the water content is reduced from a given percentage to the shrinkage limit:

$$L_S = 100\left(1 - \left\{\frac{100}{VS + 100}\right\}^{\frac{1}{3}}\right)$$ [6.7]

The above mentioned plasticity characteristics vary depending on the mineral present in the soil, which is responsible for the swelling or shrinkage characteristics. The plasticity index is typically 5–10 for low plasticity and 20–40 for high plasticity soils. For montmorillonite clays, it may be greater than 200. The shrinkage limit varies typically from 8.5–30 (Mitchell, 1993), with a lower value indicating expansive clays. Linear shrinkage values are typically one-half of the plasticity index value (Bell, 2000).

The proportions of expandable/non-expandable clay minerals and the amount of the clay-size fraction in adobe soils help to control the quality of adobe bricks, pressed-earth blocks and rammed-earth walls. Clay-size fractions high in expandable clay minerals relative to non-expandable clay minerals increase the compressive strengths of adobe materials. That is, a soil with a small clay-size fraction high in expandable clay minerals will make an adobe product high in compressive strength. Generally, inclusion of expandable clay minerals in an adobe soil results in greater compressive strength than inclusion of non-expandable minerals; however, too much of expansive clay in the soil may make a poor adobe product. In such soils, depending on the presence or absence of water, expansion and contraction of the clay may cause excessive cracks in the adobe product. More silt, sand or straw can be added to the soil to dilute the effect of the expandable clay minerals (Smith and Austin, 1989).

6.4.1   Loam

Earth used as a building material is often given different names. Referred to as loam, it is a mixture of clay, silt, sand and sometimes larger aggregates like gravel or stones (Minke, 2000). Loam does not stand for a standard building material, as the percentage amounts of its different constituents may differ according to the site. Binding strength and workability is achieved by adding water to the mixture, i.e. loam. The evaporation of water causes shrinkage cracks. This disadvantage of loams can be reduced by decreasing the clay content, desired for a particular mode of construction such as mortar and mud bricks, rammed earth or compressed soil blocks.

Loam has been found to have many advantages compared to common building materials. Its composition may make it weak, swelling with water contact; however, under the influence of vapor, it remains solid by absorbing the humidity and retaining rigidity, and hence it balances the air humidity to a higher extent. It has been found in many such houses built in Germany to have a relative humidity of about 50% throughout the year, fluctuating less than 10%, with lower levels in summer than in winter. Loam stores solar heat, balancing the indoor climate. Further, loam is energy efficient; on-site production accounts for only about 1% of energy required for production and transportation of reinforced concrete or burnt bricks. The most impressive aspect is that it is reusable, thus saving material and transportation costs. Not all types of soils can be used for earth construction, especially soils with high organic content and with shrinking/swelling properties. The US National Bureau of Standards even has a formula: 17% clay, 25% silt, 19% coarse angular sand and 42% fine sand. More than 30% clay should be avoided to prevent shrinkage cracks.

6.5   Compaction of soil

Compaction of soil is defined as the process of packing soil particles closely together by mechanical manipulation, thus increasing the dry density or dry unit weight of the soil. Practically speaking, this process refers to a reduction *mainly* in the air voids under a loading of short duration, such as the blow of a hammer or passing of a roller, or due to vibration. Study of soil compaction is important as it enables us to have earthen materials for building construction of the desired strength, especially in stabilized rammed earth and cob constructions.

Compaction is measured quantitatively in terms of the dry density to which the soil sample can be compacted. A number of laboratory tests have been developed for studying the compaction of soil. These methods are based on any one of the following methods or types of compaction: dynamic or impact, kneading, static and vibratory. In dynamic or impact compaction, soil is compacted under the blows of a rammer dropped from a specified height. In kneading compaction, a tamping foot, relatively small in cross-sectional area, is used to compact the soil. During compaction the penetration of the tamping foot has a kneading action on soil, which induces a relatively greater degree of remolding or change in structure. The compaction of soil in a mold is termed static compaction. In the vibratory method of compaction, employed for sandy soils, soil is compacted by vibrations. The main aim of these tests is to arrive at a standard that may serve as a guide and a basis of comparison for field compaction. A typical compaction curve for four types of soils is shown in Fig. 6.11. The impact compaction method has been used. The curves show that the grain size distribution, shape of soil grains, specific gravity of soil grains, and the amount and type of clay minerals greatly influences the maximum dry unit weight and optimum moisture content.

The moisture (water) content principally affects the resistance to relative movement of soil particles, particularly the fines of a soil. The resistance to particle movement is provided by the friction between soil particles and the attractive and repulsive forces of the adsorbed water layers. When only a relatively small amount of water is present in soil, it is firmly held by electrical forces at the surface of the soil particles, which leads to a low interparticle repulsion and the particles do not move over one another easily. Thus at a low water content, less compaction or low dry density with high percentage of air voids is obtained. The increase in water content results in reduction in the net attractive forces between the soil particles, which permits the particles to slide past each other more easily into a more oriented and denser state of packing together. The increase in dry density with reduction

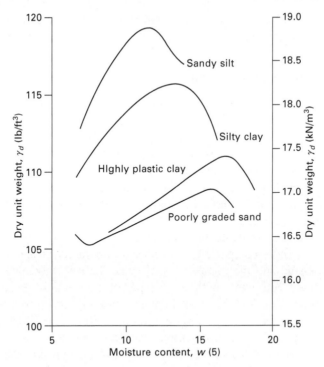

6.11 A typical compaction curve for four different soils (Das, 2006).

in air voids continues till the *optimum moisture content* is reached. Each soil type has its own optimum moisture content for a given compactive effort.

After the optimum moisture content is reached, the air voids approach approximately a constant value as further increase in water content does not cause any appreciable decrease in them, even though a more orderly arrangement of particles may exist at higher moisture contents. The total voids due to water and air in combination continue to increase with increase of moisture content beyond optimum, and hence, the dry density of the soil fails.

For all types of soil and with all methods of compaction, the effect of increasing the compactive energy is to increase the maximum dry density and to decrease the optimum moisture content. A typical pattern of moisture-density curves that results in increasing compactive effort is shown in Fig. 6.12. In general, increasing compactive effort shifts the position of the entire moisture-density curve upwards and to the left. The line joining the peaks of the curves, termed the *line of optimums*, follows the general shape of the zero air voids line or the saturation line.

The maximum dry density that can be obtained by compaction depends upon the type of soil. Well graded coarse-grained soils attain a much higher density than the fine-grained soils. Heavy clays attain relatively low densities.

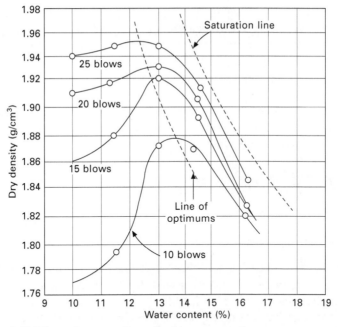

6.12 Effect of compactive effort on compaction.

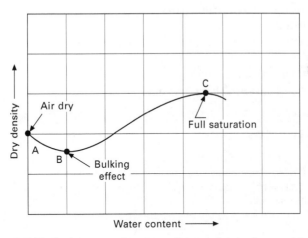

6.13 Typical compaction curve for a cohesionless sand.

Because of the greater surface area of fine particles, fine-grained soils require more water for their lubrication and thus have higher optimum moisture contents.

In the case of cohesionless soils, which are devoid of fines, the dry density decreases with an increase in moisture content, in the initial stage of the curve, as shown in Fig. 6.13. This is due to the 'bulking of sands' wherein the capillary tension resists the soil particles to achieve a denser state. On further addition of water, the meniscus is destroyed, letting the soil particles pack closer together, resulting in an increase in dry density.

Theoretically, there exists a particular gradation that, for a given maximum aggregate size, will produce the maximum density. This gradation would involve a particle arrangement where successively smaller particles are packed within the voids between larger particles, resulting in a minimum void space between particles and producing a maximum density. In 1907, Fuller and Thompson developed a widely used equation to describe a maximum density gradation for a given maximum aggregate size:

$$P = \left(\frac{d}{D}\right)^n \times 100 \qquad\qquad [6.8]$$

6.6   Conclusion

There are many varieties of soil because of the different processes through which they were formed. Each soil has a different structure and composition, and hence the same kind of treatment for each is not appropriate. For clays, consistency is an important property and contributes to the binding power of the soil when combined with cohesionless soils. Compactive effort, along with optimum water content is an effective tool to strengthen the soil used in building construction.

Our current expertise in soil sciences gives us an opportunity to dig deeper in this age-old mud technology and make it up-to-date, by augmenting it with different materials and adopting superior methodologies while keeping the primitive base concept, which was energy efficient.

6.7   References

Bell, F. G., 2000. *Engineering Properties of Soils and Rocks.* Wiley-Blackwell.

Clifton, J. R., 1977. Preservation of historic adobe structures—a status report, Institute for Applied Technology, National Bureau of Standards, Washington, D.C., NBS Technical Note 934, 30 pp.

Das, Braja M., 2006. *Principles of Geotechnical Engineering*, 6th ed., Cengage Learning.

Grim, Ralph E., 1968. *Clay Mineralogy*, McGraw Hill.

Jumikis, A. R., 1967. *Introduction to Soil Mechanics*, D Van Nostrand Company, Inc.

Minke, Gernot, 2000. *Earth Construction Handbook*, WIT Press.

Mitchell, James K., 1993. *Fundamentals of Soil Behavior*, John Wiley & Sons, Inc. 2nd ed.

Punmia, B. C., Jain, Ashok Kumar and Jain, Arun K., 2005. *Soil Mechanics and Foundations.* Laxmi Publications.

Scott, Ronald F. and Schoustra, Jack J., 1968. *Soil Mechanics and Engineering*, McGraw Hill.

Smith, Edward W. and Austin, George S., 1989. Adobe, pressed-earth, and rammed-earth industries in New Mexico. *Bulletin 127, New Mexico Bureau of Mines & Mineral Resources.*

Wolfskill, L. A., Dunlap, W. A. and Gallaway, B. M., 1970. Hand-book for building homes of earth, Department of Housing and Urban Development, Office of International Affairs, Washington, DC, 160 pp.

6.8   Appendix

w_L : liquid limit (%)

w_P : plastic limit (%)

w_S : shrinkage limit (%)

I_P : plasticity index (%)

w : natural water content (%)

I_C : consistency index

I_L : liquidity index

SR : shrinkage ratio

w_1, w_2 : water content, expressed as percentage

V_1 : volume of soil mass at water content w_1

V_2 : volume of soil mass at water content w_2

V_d : volume of dry soil mass

γ_d : dry unit weight of soil

γ_w : unit weight of water

VS : volumetric shrinkage

L_S : linear shrinkage

P : percent finer than an aggregate size

d : aggregate size being considered

D : maximum aggregate size

n : parameter which adjusts curve for fineness or coarseness (for maximum particle density $n \approx 0.5$ according to Fuller and Thompson).

7

Alternative and recycled materials for earth construction

A. DAWSON, University of Nottingham, UK

Abstract: Waste and by-product materials have great potential for use as additives to conventional materials that are used to make modern earth buildings, or even as a partial or total replacement for natural soils. The chapter gives an overview of potential materials, their variety, potential, benefits, disadvantages and characteristics. Special attention is paid to the issues of leaching, age-dependent strength and the use of fine-grained by-products as pozzolanic or self-cementing binders in place of conventional cement. An introduction to long-term evaluation and ultimate recycling of these materials is also provided.

Key words: waste and by-product, pozzolan, recycling, leaching, life cycle assessment.

7.1 Introduction

In a world that increasingly values its resources and the need to conserve them, more and more construction materials are being obtained by reusing materials from life-expired or redundant constructions and by recycling waste and by-product outputs from other human activities into building materials. This chapter attempts to give a broad overview of the types of materials that are available, some of the issues that must be considered and assessments that may have to be made.

In the context of 'earth' building, many of the available bulk recyclates from construction or from other industries can be used as compacted materials that have some strength enabling them to act in a similar manner to rammed earth. Many have binding properties so can act as binders in rammed earth or can act as a complete replacement for stabilised compacted earth, comprising both mass and binder in one material. Either alone, or in combination with soil-derived materials, some have the potential to provide special thermal or air-conditioning effects, whilst others might provide similar properties but at lighter mass. Most will have a minimal intrinsic environmental footprint compared to conventional construction materials, although this may not be meaningful if the environmental 'costs' of transport are included and the result compared with soil sourced from the construction site.

New materials are developing rapidly in order to make best use of available

172

alternative materials and to reduce reliance on primary materials. They also often provide an economic means of obtaining particular properties not found in conventional materials. Yet, in many ways, alternative materials are no different from conventional materials. As this chapter shows, the same scientific and technical thinking that has been developed in the context of traditional materials can be applied to alternative and recycled materials to get a clear picture of their behaviour and potential.

7.1.1   Wastes and by-products

Modern societies produce very large quantities of bulk material that are generated incidental to the primary activity with which any particular part of society is engaged. Figure 7.1 gives some data for the UK as an illustration of the volume and types available. Household, commercial units and industrial activity all generate waste as a consequence of the consumption of the inputs to their respective activities. Thus households produce municipal solid waste largely composed of paper, packaging, food preparation off-cuts and surpluses, life-expired clothing and household objects, cleaning debris and garden waste. Offices dispose of large amounts of used paper in addition to smaller quantities of the types of waste just listed. Industrial enterprises can produce very large quantities of wastes specific to their activity – chemical sludges, trapped dusts, clinker, etc. – but their waste streams are, typically, homogeneous and continuously produced when compared to the sporadic and heterogeneous nature of household wastes.

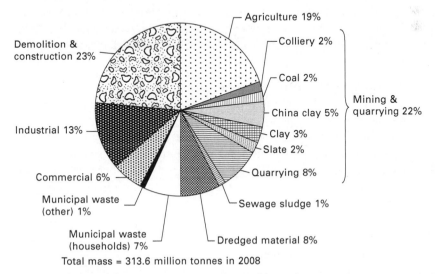

Total mass = 313.6 million tonnes in 2008

7.1 Annual waste arisings in the UK by source (approximate proportions).

In industrial societies, household waste is of the order of 0.5 to 0.7 tonnes per capita per annum and total waste 1–7 tonnes per capita per annum, the latter figure varying a lot from country to country depending on the size of that country's mining industry and its definition of waste (OECD, 2010).

But, as the saying goes, 'where there's muck, there's brass' – one person's waste is another person's opportunity. When is a waste a by-product of the primary activity? These issues make it difficult to define a waste. Depending on the jurisdiction, there may be legal definitions of waste implemented so as to regulate disposal and to provide a measure for waste reduction. For example, the European Union defines a waste as any material or object that is produced by an activity that was undertaken for a purpose other than the production of that material or object. Thus, nations burn a lot of coal to generate electricity and, incidental to the electricity generation, produce a lot of *waste* ash. That ash has, however, many beneficial uses so is rightly seen by its producers as a *by-product* of electricity generation. In order that waste-disposal regulations are not circumvented by rogue waste producers who might simply declare their waste to be a by-product, it is normal that the waste must pass through some kind of processing and quality control/quality assurance procedure.

7.1.2   Benefits and drawbacks

In this way those materials that would otherwise have been disposed of as wastes can become useful alternative materials for some purpose or another. As many of these processed wastes, now by-products, arise because of some industrial or bulk post-consumer treatment process, they are usually available in large quantities (although maybe only in a single locality if the waste supply or processing is localised). Such materials have, in principle, several advantages:

1   aside from transportation costs, they ought to be economic as they have been largely funded by the primary activity of which they are the by-product
2   they incur little or no environmental impact in their sourcing – although the primary activity may have had a large impact
3   their use saves the environmental impact that would otherwise have been incurred in the sourcing of conventional construction materials (e.g. in quarrying clay)
4   their use saves the environmental impact of the processing method of the conventional construction materials that they replace (e.g. the energy consumed in making cement).

Furthermore, aside from these direct benefits, the source of these by-products tends to be near centres of industry and population. Thus transportation costs and associated environmental impacts are usually kept low. Individual by-product materials may have other benefits as described later.

However, these benefits have to be offset against the disadvantages. The common disadvantage is perceptual – a waste is often, or even usually, considered an undesirable material and thus, in some ill-defined way, less than perfect for the application. Of course no material is perfect, but there can be many benefits to using alternative materials that are not obtained from conventional materials (these are mentioned in Sections 7.2, 7.3 and 7.4). Thus many alternatives can be better than conventional materials. Nevertheless, some materials may genuinely introduce problems that need to be specially addressed. Those problems of substance are listed below. Although few, if any, alternative materials will suffer from all these disadvantages, they form a checklist against which candidate materials should be assessed.

- *Particle strength.* Naturally sourced materials derive from geologic processes involving magma cooling, metamorphism, lithification and long periods of time to form hard, strong particles. On the other hand, many alternative materials have been through relatively light strength-forming processes so durability and resistance to damage due to compaction are issues on which to keep watch
- *Radionuclides in hot-processed materials.* All natural materials carry radioactive atoms because they occur naturally in the environment. Concentrations are often greater in ore-bearing rocks. Radioactivity is often associated with denser minerals and these will report preferentially to slag and bottom ash streams in smelters and incinerators. Thus some alternative materials may be a little more radioactive than conventional materials. Section 7.4.4 considers this aspect a little more
- *Leachability of pollutants.* Almost all materials (perhaps glass is an exception), conventional or alternative, will leach chemicals from particle surfaces or from the interior of particles when in contact with water. Sections 7.4.4 and 7.6 discuss these issues in more detail.
- *Chemical reactivity.* Some alternative materials, because they come direct from an industrial process without exposure to the environment, are reactive. Often this is beneficial as they, thus, have self-cementing abilities. However, the chemical reactions that they undergo when put into use can lead to problems of cracking and heave (due to chemical reaction products occupying more space than the reactants) and to premature drying (due to exothermic reactions). Some of these chemical reactions are long term and can lead to reductions in strength over months or years. Usually these issues are small or non-existent and, if not, can be minimised by limiting access to water, which is usually a necessary reagent.

7.1.3   Maximising use of alternative materials

Like any other material, some alternative materials are more suited to certain applications than others. In the past, having been seen as wastes of little value, the use of alternative materials has often been rather indiscriminate on the basis of their low cost. However, with increasing use and experience the technical value of many of the alternative materials has become much better recognised. This has had a number of significant effects:

- Value, and hence price, has increased
- The increased price has increased the possibility of processing in order to deliver reliable, quality controlled materials
- The increased cost has driven, and the possibility of processing has allowed, a desire to fully exploit each material's beneficial properties, i.e. to use each material in its highest value application
- Composite materials, combining several materials together, are thus frequently used in order that the optimum material properties, no more and no less, are available for each application
- In this way, environmental, cost, short-term and longer-term mechanical properties, insulating properties, etc. may be balanced in the desired manner
- Many alternative materials are only locally available. Like all bulk materials, transportation economics militates against their applicability a long way from their point of production. Exceptions to this rule apply to those alternative materials that are used as binders or activators at small quantities in mixtures as the improvement in mechanical behaviour that they deliver can sustain the additional transportation costs.

But this maximising of value is only part of the issue for recycling and more consideration must wait until Section 7.8.

7.2   Classification

7.2.1   Origins

One way to classify materials is by considering the temperature of their production. This is potentially useful because the mineral basis of the material is likely to be highly dependent on the temperature at which it was produced. Most of the mining and quarrying wastes named in Fig. 7.1 will be of direct geologic origin without any recent high temperature processing, whereas most of the commercial, industrial and municipal wastes in the figure will have no reusable value in construction without such processing. Some wastes may merely need sorting and selection (e.g. construction and demolition wastes).

Very high temperature smelting

Slag-derived alternative materials are made from residues that form during the smelting of metals. Dependent on the metal involved, the temperature of production is, typically, well over 1000°C. At high temperatures such as these, non-crystalline, perhaps glassy, forms of metal oxides are commonly produced. The exact types produced are dependent on the raw material input, on the fluxing agents that are added to the furnace and on the method of smelting and reduction that is practised. In many cases the minerals formed are similar to those formed in the production of cement. The quality and characteristics of materials produced in this way depends to a large extent on the changes that have been wrought by the high temperature treatment.

High temperature incineration

Although the flue gases of incinerators are elevated to a very high temperature for a few seconds in order to destroy dangerous components in the gases, the temperature at the actual incinerator grate is normally somewhat lower, probably less than 1000°C. At these temperatures the incoming material is not, for the most part, melted, but all volatiles and combustible components should be destroyed. The quality and characteristic of materials produced in this way depends mostly on the non-combustible elements present in the feed. Temperature has acted as a sorting mechanism to remove the materials that are (for the most part) weak. In practice, some combustible but unburnt material may find its way through such incineration by a 'Baked Alaska' process, whereby an outer charred layer acts to insulate inner, unburnt material.

Wastes from geological sources

Mining and quarry wastes produce rock-derived materials. Clearly, these have been through a variety of geologic processes in the past including, perhaps, high temperatures. However their condition now depends not only on this rock-forming history, but also on the cold weathering in history and the recent industrial processing that generated them. Typically this will mean that they are composed of fragments that may be weaker, smaller or more flaky than, say, an ideal aggregate with which to make concrete.

Cold processing of wastes

Many alternative materials are produced mechanically without high temperature treatment. The quality and characteristics of these materials depends entirely upon the mechanical processing applied and the characteristics of the input stream. Mechanical processing can both separate out undesirable elements (e.g. steel or plastics in demolition waste) and can impart some degree of homogenisation – e.g. by shredding and blending a single input stream or by mixing different stockpiles of the raw input stream to create a more consistent supply.

When the source is consistent (for example over- or under-size fractions of rock from a quarry) then the consistency of supply can be very good over long time periods. In such circumstances, users can apply the material in confidence knowing its properties and taking appropriate action to get the performance they want. Where the material comes from a geographically or temporally varying set of sources (e.g. trench arisings, demolished buildings, seasonally changing waste streams), then far more care is needed to assess the source at the time of use to determine its contents and properties so that any post-processing is selected on the basis of what is actually available and not merely on the basis of routine.

7.2.2   Particle size and activity

Another way of considering alternative materials is according to their particle size. Materials with particles the same sizes as conventional aggregates can often be considered as a replacement aggregate. Certainly, properties will be different due to difference in particle angularity, particle strength, grading, etc., but the same basic considerations are needed. Several alternative materials are very fine grained, either because they are formed as a dust, or because they have been ground to a very fine particle size. Most of these fine materials have an 'active' capability, that is they are pozzolans – materials having a silicate mineral content which, when combined with calcium hydroxide (lime) in the presence of water, react to form calcium silicate hydrates exhibiting cementitious properties – or will produce self-cementing binders themselves that are activated when water is added to them. The resulting minerals are the same as some of those chemicals present in hydrated Portland cement.

The pozzolanic or self-cementing abilities of these fine materials allows us to produce hydraulic bound mixtures (HBMs). HBMs typically show slower strength development than do conventional Portland cement-bound mixtures (i.e. Portland cement concretes). This allows for greater workability periods – while 2 hours is a typical limit for a Portland cement-based mixture, 24 hours or even several days is easily possible for many HBMs. This allows off-site mixing, transportation to site and at-will use. Figure 7.2 gives some data for a range of mixtures showing the greatly extended workability of the majority.

With appropriate mix design the ultimate strength can be limited, preventing the HBM from becoming a brittle mix that will only deform by cracking (as is the case with most Portland cement concretes), but allowing some flexibility as seen in lime mortared brickwork, for example.

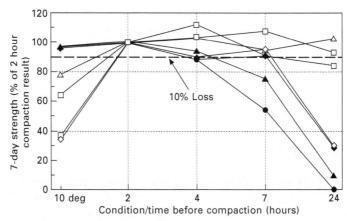

Conventional concrete (lowest line on the right hand side of the figure) has zero 7-day strength if compacted 24 hours after mixing. Two of the mixtures evaluated showed almost no loss of strength if compacted 24 hours after mixing. Five of the seven mixtures were fully workable 7 hours after initial mixing. Seven-day strengths are compared to the 7-day strength of the same mixture compacted at 2 hours after mixing. Effect of curing at 10° instead of 20°C is also shown on the same basis (data from Atkinson *et al.*, 1999).

7.2 Extended workability of various hydraulic bound mixtures.

7.3   Types of alternative material

There are, of course, an almost infinite number of different alternative materials that might be used in compacted (and, perhaps, treated) construction material. With increasing pressure from the public, politicians and regulators to recycle and reuse materials that would otherwise have been discarded, new candidate materials are appearing all the time, so the following list cannot be considered complete. Nevertheless, bearing in mind the broad classification just mentioned, the following lists the more common, generic, types of alternative material that might find a use in compacted construction materials.

7.3.1   Slags from mineral processing

Metal production, from ore, requires very high temperatures to reduce the compound in which the metal exists into its pure, metallic form. Typically this is done in the presence of a fluxing agent (e.g. limestone) and an energy source. Sometimes energy is provided in the form of an electric arc, whereas at other times it might be provided as a solid fuel mixed with the ore. The metal, being dense when liquid, sinks and the non-combustible residues float to the top from where they are removed as a liquid slag. This slag is solidified in a variety of ways, perhaps by spraying it into a sheet of falling water, or simply by air cooling in a hollow in the ground, or by some other means. This usually generates a material with high alkalinity and pozzolanic properties (see Section 7.2.2).

Coarse slags are usually crushed to form a granular aggregate that can then be used in concretes, fills or, in the present context, in compacted fills for construction processes. As the fines are generally cementitiously active (see below), the resulting graded aggregate can slowly self-develop a weakly bound structure (a little like a weak concrete) with some tensile strength if given some hydration.

Due to their rapid cooling and chemical make-up, coarse slags may not be stable. The slag formed in converting iron into steel can, depending on the process used, suffer from instability. Linz-Donawitz process steel slag is a case in point. The hydration products of lime or magnesia may occupy more space than the chemicals from which they formed causing localised swelling of coarse ash particles and, hence mass swelling of compacted slag. An alternative explanation is to do with the reactions between lime, phosphorus or other chemicals (which are in the slag as a result of its origins) and the silicate minerals that make up the mass of the slag. Yet another reason could be transitions from high temperature-to low temperature-forms of dicalcium/ tricalcium silicates (or the transition from the tricalcium to dicalcium forms), the high temperature forms being 'locked in' due to the rapid cooling of the slag into a glass form and which must now revert to a low temperature form, which requires expansion (and hence cracking) of coarse particles. Probably, all these reasons are true for some slags in some circumstances and the exact reason will seldom concern the constructor planning to use slags. But that constructor must be aware of the possibility and make due allowance. Readers with a desire to educate themselves further on steel slag swelling can refer to Verhasselt and Choquet, 1989 and Champion et al., 1981.

'Due allowance' will certainly necessitate use of a coarse slag that has been crushed to the desired grading and then left to weather, probably for at least six months. Crushing should not be performed just prior to use, as this would release newly exposed lime, or magnesia, that had not weathered. This might then permit the very swelling in the construction that the weathering was designed to allow to complete. 'Due allowance' may also mean allowing individual coarse particles to expand and crack causing local blemishes (which may even be architecturally attractive) and to allow for some mass expansion (by no more than around 3–5% by volume).

When fine slag particles are under consideration, it is no longer expansion that is the issue of relevance, but their binding ability. The activity of such slags depends both on their chemical make-up and on their fineness. Fine material has a large surface area per unit of mass. As the slag is a solid, the chemical processes of hydration and cementation have to take place at the surface of the solids. Therefore, the finer the slag, the greater is the surface area for the same mass and the more active the slag becomes.

Many slags have residual quicklime (CaO) or its hydrated form, calcium hydroxide, in them. Such materials, when fine grained, can be self-cementing in the presence of water alone. Other fine-grained slags need lime to be added so as to initiate the reactions that develop cementation. Given the wide range of possible slags available, it is not possible to list all possible candidates that have these binding properties, but a few are well known.

Ground granulated blast furnace slag (*ggbs*) is formed by the grinding to a fine powder of quench-cooled slag formed in an iron-making blast-furnace. Although its binding action may be slower than conventional cement (as are most slag cementation reactions), it can be as effective, or even more effective, in the long term. 'Long term' might be more than one year. *ggbs* is often used as a partial cement replacement to reduce flash curing of pure cement, to extend the durability of concrete, to lower the permeability of the concrete to water and to reduce the opportunity for sulfate attack on the cemented phases.

Silica fume is a by-product of the production of silicon and ferro-silicon alloys. Extremely fine-grained (the particles have a size approximately 100 times smaller than those in cement), it has a very large specific surface area. Not only does this make it somewhat pozzolanic when activated by lime and water, the fineness means that it can effectively block the pores in a granular material, thereby reducing water permeability and the diffusion of chemicals in water.

7.3.2   Ashes

Ashes arise from incineration either of fuels (a fairly controlled source) or from waste (a fairly uncontrolled source), particularly municipal solid waste (MSW), which is burnt so as to reduce its volume prior to landfilling, and to retrieve the energy of combustion. The two will be considered in turn.

Combustion by-products (fba, pfa)

When coal is burnt to produce electricity, three solid residues may result. The clinker falls through the grate, is cooled (often by water quenching) and is then crushed to form furnace bottom ash (*fba*). Compared to conventional crushed rock aggregates, this material has a lower density, more uniform grain size (it is light on fines and, typically doesn't have the largest particles) and weaker particles. Because of its open grading, it doesn't readily compact into an adhering mass, so it is normally used in a cemented mix. Lightweight concrete blocks typically have *fba* as a major component, for example. Given the ease with which it can be used to make such blocks, it may not be freely available for purchase, all or most of the production having been committed for such uses.

The second product of burning coal is the so-called fly ash, or pulverised fuel ash (*pfa*). This is the smoke component; ash that is carried up the chimney in hot flue gases. To prevent atmospheric pollution this ash is trapped by electrostatic precipitators and/or filters and sent to an external heap or lagoon where it is allowed to weather. The ash is fine grained so, rather like *ggbs*, it has pozzolanic properties though it may contain insufficient lime to be self-cementing. It is an important component of most commercially available cements as it acts to prevent rapid hardening yet provides cementing properties. Thus it becomes a benign dilutant of conventional Portland cements. *pfa* has great potential as a binding agent when activated by lime and is a prime candidate for amending compacted materials that have low fines and/or are in need of some cementation action. By itself, this cementation action will be rather slow, but it can easily deliver strengths as great or greater than can conventional cement binders given sufficient time (e.g. 1 year).

pfa is readily available in countries that generate power from coal combustion, although this is usually limited to areas close to the power stations. However, the type of *pfa* does vary quite a lot from one power station to another depending on the type of coal being burnt and the particular furnace arrangements that are in place there. Two classes of *pfa* are commonly defined – Class F and Class C. Class F, coming from the combustion of older coals, has less than 20% CaO and is pozzolanic but not self-cementing. Class C has more than 20% CaO and is pozzolanic and, often, self-cementing.

The third product from coal combustion is obtained in those power stations that practise flue-gas desulfurisation. This is the subject of Section 7.3.3.

Incinerated waste residues (incinerator bottom ash, incinerated sewage sludge ash and paper sludge ash)

Municipal solid waste is often incinerated in preference to landfilling. Like the combustion of coal, it results in a bottom and a fly ash. Unlike coal, the fly ash is not usable due to high levels of dioxin (a carcinogen), so this must be landfilled. However, the incinerator bottom ash (IBA) is not affected and may be used. It is cooled (typically by quenching with water) and then left to weather. Remaining contamination concentrations typically decline significantly due to leaching, biological action and chemical activity during this period. Economically valuable components are separated (mostly iron-based components that can be magnetically separated) and the remaining material is crushed to provide a broadly graded aggregate. This material is not usually pozzolanic, but can be readily cemented to form a weak (or strong) concrete. In broken or compacted faces it often has a certain attractiveness due to colourful components such as broken pottery, pieces of brick and glass, melted aluminium and crushed metal fragments. IBA has the advantage over many materials that, due to its origin, it is available near centres of population where construction needs are greatest.

Lower volumes of other incinerated materials are also available. Incinerated sewage sludge ash (ISSA) is also available near centres of population though environmental restrictions may hinder its beneficial reuse as heavy metal concentrations can report to this ash stream. This should not be an issue in dry applications. ISSA is of low density and high water absorbency so has the potential to be used to reduce the mass of construction and help to moderate ambient moisture levels inside buildings.

Paper sludge is a by-product of the recycling of post-consumer paper and card. It contains cellulose fibres, kaolin and calcium carbonate (both are used as fillers in paper) and residual inks and sundry chemicals. Due to its high organic content, energy can be extracted by burning it leaving behind an ash (paper sludge ash, PSA) with reactive silica and alumina and lime (CaO). Therefore the ash is usually pozzolanic and may have the ability to be self-cementing. It can therefore be used as a cement replacement in a similar manner to *ggbs* (see Section 7.3.1 above). Further details may be found in Dunster (2007).

7.3.3   Flue gas desulfurisation (FGD) gypsum

The flue gases from coal burning contain sulfur dioxide, a gas that reacts with rainwater to form acid rain. To prevent this, a number of power stations across the world have installed flue gas desulfurisation equipment. In the vast majority of cases this involves spraying either lime or limestone (usually in a wet slurry form), producing calcium sulfite. Once oxidised and hydrated it will form calcium sulfate ($Ca_2SO_4 . H_2O$), otherwise known as gypsum. Commercially, such material is used to produce building plaster (calcined gypsum) and plasterboard (dry walling).

Calcined gypsum reverts to regular gypsum on the addition of water; hardening and binding in the process. Thus, it may be added to compacted granular material as a binding agent. Commonly, it has been used with fly ash and lime to form quite strong materials, albeit with a slower rate of strength gain than is usually seen in Portland cement-bound mixtures. Swelling of such combination materials by a few per cent can occur and constructors need either to check and adjust the mix design (usually by controlling the amount of lime) or use it in such a way that the swelling can be accommodated.

7.3.4   Cement kiln dust (*ckd*)

High temperature kilns are used to convert the ingredients of Portland cement into the cement clinker that is then taken away and crushed to make powdered cement. During the operation of the furnace, solid particles are carried away in the flue gases, trapped and collected. Similar to cement (and already in powder form as it can be carried by the flue gases), this cement kiln dust (*ckd*) differs from cement in a number of ways.

Firstly, full combustion/alteration of the solids to cement has not yet been achieved – meaning that the free lime content will be higher than with conventional cements. For this reason it makes a good clayey soil-binding agent, acting somewhat in an intermediate way between Portland cement and lime. Secondly, it tends to contain much higher proportions of chloride and sulfur than does cement. Reactions with this sulfur can lead to swelling problems if mixed with some soils and kept moist as ettringite formation then becomes possible. (Ettringite is a calcium aluminium sulfate mineral that occupies considerably more space than the minerals from which it forms, due to the needle-like shape of its crystals, which may force apart the material surrounding them if it forms after the material has initially stabilised. This action can cause bulk expansion of a compacted and stabilised assembly.)

7.3.5   Recycled construction and demolition debris

In metropolitan areas, reconstruction and redevelopment generates large volumes of demolition debris. New construction also generates its own waste (e.g. from over-ordering, off-cuts, damaged pieces). Four main types of demolition waste are now described.

Recycled concrete aggregate (RCA)

Recycled concrete aggregate (RCA) comes from demolition of Portland cement concrete. Given that the original concrete might have been strong or weak, dense or open graded, fresh or weathered, then the aggregate's pieces can be expected to vary similarly. If the RCA comes from a central recycling plant the consistency will have been addressed, to some extent, by blending of materials from different sources. If the material is coming from an on-site crushing plant then it will reflect more directly, and more immediately, the type of concrete being crushed.

The crushing process produces agglomerations of the original concrete's aggregates with adhered mortar. These agglomerations are, typically, more angular than conventional aggregates. Also the crushed concrete will produce fines from the mortar element, the amount being controlled to a large extent by the strength of the original concrete. Thus high-strength concrete will typically crush to produce very sharp, even lance-like, blade aggregates with low proportions of fines, whereas the weakest concrete may crush to produce almost the original coarse aggregates plus a large proportion of fines made of the old mortar. In the crushed mortar component, unhydrated cement will be newly exposed. The effect of this will be a slow strength gain as this cement starts hydrating either with water that has been deliberately added, or with water attracted hygroscopically from the surrounding environment. Thus RCA is, to some degree, a self-cementing material with RCA from strong concretes (those with high cement contents in the original mix) often exhibiting a higher self-cementing ability.

Recycled brick

Like recyclate from Portland cement concrete, the quality and characteristics of recyclate from old bricks depends a lot on the strength of the bricks. High-strength ('engineering') bricks can produce reasonable aggregates, but softer bricks will generally produce a higher proportion of fines in the mix with less flaky coarse pieces. Lower quality bricks usually have water absorbing capabilities, especially when broken to reveal more open faces. They may, therefore, be unsuitable for use in external walls in wet or humid situations. On the other hand, derived materials may be used beneficially in internal walls to help passive air conditioning. Stabilisation of some kind will probably be needed as bricks do not have pozzolanic or self-cementing properties when crushed. The colour of the old brick may be architecturally attractive.

Some suppliers blend mixtures of RCA and crushed brick. This can have many advantages. The self-cementing activity of the RCA fines are available to improve more aggregate; the low fines content of some RCAs is rectified by fines from the brick, making a more stable and integral blend; the poor mechanical performance of the crushed brick is addressed; and the volume of good quality material is increased. The resultant material may also have a better appearance due to the mixing of particles with two colours.

If the brick (or concrete) is coming from a plastered building, it is important to control the gypsum plaster components. Sulfate reactions with pozzolans and cement will produce weaker and/or delayed bond strengths and may lead to swelling problems (see Sections 7.3.3 and 7.7.2) although benefit can, with care, be extracted from such mixes (see Section on 'Gypsum from waste plasterboard'). Figure 7.3 shows the use of coarse demolition waste, largely brick, into a compaction, gabion-enclosed wall. Note the careful positioning of some painted elements from the demolished source.

Recycled aggregate

Conventional aggregates can be recycled quite readily, either directly or by crushing of old, large aggregate to form new, smaller aggregate (e.g. size reduction of rail track ballast). The quality of such materials depends a lot on the source and on the processing. Most specifications allow up to 1% by

7.3 Wall made of coarse compacted demolition waste in gabions (photo reproduced by kind permission of www.gabionbaskets.net).

mass of extraneous material (plastic, wood, etc.) to allow for contaminants that have been picked up during previous use or the reclamation process. Although 1% by mass may sound small, plastic typically has a solid density 3–4 times smaller than that of the rock-forming minerals, so a 1% ratio by mass may indicate a 3–4% ratio by volume.

Because of the processing needed to collect and, probably, clean old aggregates, recycled aggregate can be low in fines unless these are specifically added. Therefore effective compaction may require manipulation of the grading and/or the addition of a binder.

Reclaimed asphalt planings (RAP)

Reclaimed asphalt planings (RAP) are obtained when an asphalt pavement is cut prior to resurfacing or other rehabilitation. The material will, typically, be granular, formed of agglomerations of old aggregate and asphaltic mortar, but with almost no fine particles. When compacted the bituminous content allows the agglomerations to deform and engage with one another, resulting in an overall material that has some of the properties of an aggregate and some of an asphalt concrete. Thus the compacted RAP will exhibit some viscous behaviour (deformation under long-term loading, increasingly so in warm conditions) and some tensile strength capacity. This strength requires the material to be well compacted (to promote good contact between stones and old binder) and may need time and a little deformation (creep) for full development because the bitumen's viscosity resists the rapid application of load by, e.g., vibratory compaction. Warmth will aid this process (again, an effect of the material's viscosity).

One implementation of RAP in buildings is the 'Bitublock' concept (Forth *et al.*, 2004, 2006 and 2010), in which RAP is compacted into brick-sized blocks and then hardened to resist creep by heating to a high temperature ($\geq 200°C$) and, if possible, a raising to a high air pressure. Temperature and pressure accelerate oxidative hardening, which limits subsequent creep behaviour at normal temperatures and pressures. Forth *et al.* (2006) experimented with the inclusion of *pfa*, *fba*, MSW-IBA, glass cullet, steel slag and ISSA in 'Bitublocks' reporting that successful mixes were possible for all of these additives. Interested readers are referred to those papers for further details.

Often recycled planings are combined with recycled aggregate due to a full-depth removal of an old highway pavement. Re-compaction of such material leads to a granular material with some colour texture given that the RAP element is almost certainly much blacker than the rock aggregate evidenced in the other large particles.

Gypsum from waste plasterboard

Large amounts of waste plasterboard (drywall) are obtainable from demolition and construction off-cuts. This gypsum from plasterboard/drywall can be mixed with *ggbs* and *ckd* to make an effective sulfate-activated pozzolan that is self-cementing (Claisse *et al.*, 2007 and 2008; Sadeghi Pouya *et al.*, 2007). Such use is relatively new and, like all binders involving several components, some experimentation is needed to determine which proportion of each component is optimal. As with other pozzolans, strength gain of this HBM is slower than with Portland cement.

7.3.6   Processed, post-consumer solids: tyres and glass

There is a wide range of post-consumer residues, but only two are described here in any detail, being those that might be found in many locales.

Rubber

Rubber from car and truck tyres is a major material because landfilling of old tyres is either forbidden or strongly deprecated in most jurisdictions. Thus uses must be found for old tyres. One process is to separate the rubber from the carcass of the tyre (mostly fibre), such that the rubber is in the form of crumbs. These crumbs can then be treated as a sand-sized material and added to aggregate or soil to form mixes with modified thermal properties (more insulating) and reduced densities. The structural success of such composite materials depends a lot on the degree of integration that can be achieved. In Portland cement-based mixtures it is difficult to achieve a good bond between highly compressible rubber and relatively inflexible mortars. In soil-based mixtures the adherence may be better.

Rubber is also available as shreds that are formed by chopping the tyre into many small pieces. These are more difficult to integrate into mixtures but can be bagged or used gabion-style. The ultimate is to bundle whole tyres into bales, held and compressed together with steel bands (Fig. 7.4). These bundles have a very high insulation ability once covered and are of low density, but covering them needs special consideration as the relative movement between one bundle and the next tends to cause localised cracking in any cover. A particular use is as an insulating and lightweight sub-floor fill used to bring ground level up to a desired height without imposing much additional ground stress (that might promote settlement of the ground). In such use a reinforced cover slab is needed so that small rocking and lateral movements of the bales relative to each other can be accommodated (see Fig. 7.5).

A working standard is available for tyre bales (PAS 108).

Glass and glass cullet

Glass cullet from the crushing of post-consumer glass is readily available in urban areas. It can be added in small quantities (probably < 20%) into many mixes, but its smooth surface means that it is difficult to achieve a good bond with the surface. Glass dust is a hazardous irritant, so appropriate health and safety precautions must be taken (see Section 7.4.3). Residual sugars on glass may also delay hydration of Portland cement and hydraulic binders. Use of intact glass bottles is a relatively easy and effective means of putting light-passages into compacted walls and has attracted several architects although usually in Portland cement concrete rather than in compacted materials (e.g. Hundertwasser – see Fig. 7.6).

Other discarded materials

Coir (the waste fibre from the hulls of coconuts) and rice husk ash are locally available in some countries and could be applied as reinforcement and a pozzolan, respectively. In all probability there are particular sources near most centres of construction that could be exploited given information and, perhaps, incentives.

(a)

(b)

7.4 (a) Tyre bale being made, compressed but not tied together, (b) tyre bales ready for use.

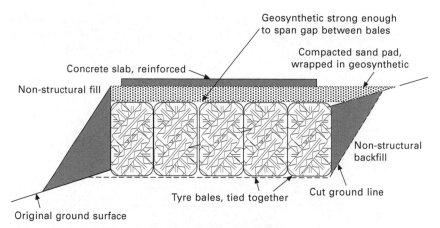

Geosynthetic strong enough
to span gap between bales

Compacted sand pad,
wrapped in geosynthetic

Concrete slab, reinforced

Non-structural fill

Non-structural
backfill

Tyre bales, tied together   Cut ground line

Original ground surface

7.5 Schematic of floor arrangements on tyre bales.

7.6 Wall with bottles in Hundertwasser's public conveniences at Kawakawa, New Zealand.

7.4   Characteristics of alternative and recycled materials

Of course, the characteristics of each alternative material is heavily influenced by its origin, make-up, form, chemistry and content. However there are a number of properties that should be assessed, either by specific testing or by review so as to limit risks and to ensure that as complete a view of a candidate material is achieved as is reasonably possible.

7.4.1   Mechanical

The strength of an alternative material rests on the same items as does the strength of a conventional compacted material – strength of particles, compacted skeleton, pore suction characteristics and strength of any binder. Many of the alternative binders available typically in the short term produce materials with low tensile strength, although with time (which may be several months or even years) the strength may increase to rather high values, e.g. HBMs, see Section 7.5.2. Strength developed in this way typically involves a less brittle bonding action so the resultant mass is more ductile than a conventional compacted mass that is bound with Portland cement. Compressive strength is the result of all the factors of the mix acting in composite, so the binder's properties are only one aspect to consider.

Conventional compacted materials may have been made with a feed that had a particular particle strength requirement. With alternative materials this is not possible – the feed is whatever it is! So it is entirely possible that weakness in a compacted material made of an alternative material is a result of low particle strength. The only way to deal with this is to amend the grading to a denser one in which stress is carried through the compacted mass by more particles and more particle contacts. Then individual particles are not so highly stressed under the same external loading and particle survival is greater.

7.4.2   Thermal

Most alternative materials will not exhibit thermal properties that are distinct from those of conventional materials. However, the inclusion of significant amounts of rubber could reduce thermal conductivity and thermal mass, leading to better insulation properties but reduced thermal mass (and, hence, building temperature stability). Steel and other metalliferous slags are somewhat more conductive but, in particular, tend to have a much higher specific gravity than conventional aggregates. For this reason they can improve the thermal mass considerably whilst occupying the same volume as would a conventional material.

7.4.3   Health and safety

All bulk compacted materials, whether soil, aggregate or an alternative material have handling and dust issues that must be considered in health and safety plans. For some alternative materials, the dust generated during handling and compaction could have specific respiratory or even carcinogenic risks. For example, crushed demolition waste could include asbestos contamination if correct procedures have not been observed in the demolition process. To overcome this an approved and documented demolition plan should have been followed. Glass cullet and ISSA can produce irritant dusts that should not be breathed in. Damping of glass cullet will helpfully reduce dust generation to a great extent as an additional strategy to the deployment of personal protection equipment (PPE). Crushed high-strength concrete and glass cullet are two materials that can easily cut the skin, so appropriate PPE to protect against this would be required for such materials.

The high alkalinity of many alternative materials can cause skin burns. Even more concern is associated with the highly active, high pH, fine-grained materials. These often contain, or behave similarly to, quicklime (CaO). This material is highly hygroscopic, undergoing exothermic reaction as it hydrates. Therefore breathing in its dust will burn the lining of the lungs, eyes and mouth as it draws water out of, and heats, the surfaces of the body that it comes into contact with.

A few sentences on health and safety matters cannot be complete given the vast number of alternative materials that are available. Instead, this section is included to highlight the need to make a proper risk assessment of any material that is considered and, especially, to make a thorough assessment where there is no previous history of this material being used. Not only does this advice extend to entirely new materials, but also to new mixtures of known materials. Sometimes the unexpected occurs when materials are used in a new combination. Only a clear risk assessment will reveal this.

As a counter to this somewhat negative aspect, most alternative materials are no worse than, and can even be preferable to, conventional candidate materials. Understandably, but largely without warrant, natural materials are often considered 'good' simply because they are natural and, when compared to alternative materials, they receive relatively low levels of attention as regards health and safety and environmental impact (see the next section).

7.4.4　Environmental

Not only are humans at risk during construction in the ways just mentioned, but there may be continuing risks to people during the service life of the object made with the compacted material. Strategies may need adopting to prevent dust or sharp particles from impinging harmfully on the users of the construction into which the material has been incorporated. There may also be continuing risks to the environment that need to be checked. The chemistry of a compacted material may be an issue (and not only for an alternative material) if it is possible for the compacted material to become wet, and for water to wash contaminating species from its surface or if water will seep through it and wash them out. This is discussed further in Section 7.6.8.

Two other environmental issues of concern are combustibility and radiological activity. Few compacted materials will be combustible. Indeed they are mostly very inert and resistant both to flames and to heat. In most applications the thickness necessary to provide thermal and structural capacity means that fire action will only affect a thin surface. There may, however, be a few alternative materials – e.g. those containing rubber – that may need special fire-resistance consideration. Radiological hazard is of concern with some incinerator ashes. The burning process tends to mean that the heavy atoms of radionuclides report to the bottom ash. Thus a bottom ash may only represent 5% of the mass of MSW introduced to the incinerator, but may contain almost 100% of the radioactive fraction introduced. In this way, the incineration process acts as a radionuclide concentrator. Construction materials made of large volumes of such a residue would then have a radioactivity somewhat higher than is found in conventional materials although still very low. However, it might then be necessary to restrict use of the resulting construction material to places where people will not be in close, long-term contact. So its use in a warehouse would be acceptable whereas its use in a domestic residence would not.

7.5 Form of recycled and alternative materials: bulk or binder

The materials being considered in this chapter may be used in two ways – as a bulk material, replacing the earth that would have been compacted to form the construction, or as a low-volume additive to the earth to aid with binding. Thus, although the bulk of this chapter considers the bulk usage, alternative binders may have a greater routine application. As already mentioned in Section 7.1.2, binders can travel in a way that bulk materials cannot. As most soils contain some active silica, they can be pozzolanically stabilised by the addition of lime or other hydroxide-rich material – like *pfa* or *ggbs*. Increased activity can be encouraged by the fineness of the additive and the addition of other pozzolanic elements in the additive. Only testing can determine how strong, how rapid and how reliable such blends of unconventional binders with conventional earth or aggregate may be, but the example of a conventional aggregate with fly ash, lime and gypsum (see Fig. 7.7) illustrates the immense potential of such binding agents.

7.5.1 Treatment

From the foregoing it should be apparent that many alternative materials can be handled and compacted very much in the same manner as conventional

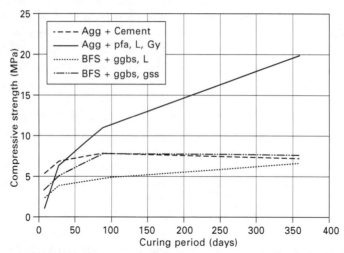

Compressive (cube) strength determinations on samples of four aggregate –
binder mixtures. Codes are as follows:
Agg = conventional aggregate
BFS = blast furnace slag aggregate
pfa = pulverised fuel ash
L = slaked lime
Gy = gypsum
ggbs = ground granulated blast furnace slag
gss = ground steel slag
(Data from Atkinson *et al.*, 1999)

7.7 Strength development over time of four aggregate and binder
mixtures.

materials. Where they need added strength, then the same binders can be
added as for conventional cases. One common difference is that many
alternative materials have a pozzolanic or even a self-cementing capability.
So, for these materials, it is often sensible to see if blending of conventional
and alternative source materials will produce a better result than either by
themselves.

Blending can deliver significant improvements, but it can also cause
problems. Inappropriate combinations can cause materials to swell as they
hydrate or place chemicals together that will react in the medium term to
cause a material to degrade. Other combinations might hinder the binder
action of the pozzolan or cement. Alkali–aggregate reaction is well known
in the Portland cement concrete industry. Ettringite formation, as mentioned
in Section 7.3.4 (causing volumetric swelling and disruption of the material's
fabric) can be a problem when lime-based bonding is employed. Most of
these degradation processes need water to allow the reactions to proceed,
so are not an issue in dry usages. Blending may also hinder maximisation
of the benefit that can be obtained from the material when the construction
is demolished and it must, in its turn, be recycled (see Section 7.10).

7.5.2 Age-related aspects

The development of early versus long-term strength (see Sections 7.3.1 and 7.4.1) is a key aspect of many materials with alternative binding action. Figure 7.7 shows some illustrative data for some laboratory mixtures. Note that the cement-bound material has the highest strength at 7 days and just maintains this advantage at 28 days. However, beyond this age it hardly gains any further strength, whereas the mix with the lowest early strength has almost the same strength at one year and looks as though it will continue to gain strength thereafter. The strongest mix (probably too strong for the applications considered here) is almost three times stronger than the cement-base mixture and seems likely to gain even more strength.

For those materials with a slowly developing strength, design needs to allow for this. Conservatively, one could design for all loadings based on the early strength, but this might result in very uneconomic and/or bulky solutions. Therefore a two-stage design may be more appropriate – with self-weight and construction loads being assessed against a short-term strength assessment, while full, in-use, live loadings (including extreme wind and snow) are evaluated against longer-term strengths.

In the long term, carbonation of calcium hydroxide in Portland cement paste and in many of the alternative binders can be anticipated, especially those low in fines where large voids will allow the movement of carbon dioxide through the compacted mass. Carbonation will probably cause a small increase in strength and stiffness though will also make the material a little more brittle. It shouldn't cause deterioration of the material unless it contains corrodible steel.

Because strength gain can be slow (see Section 7.2.2 and Fig. 7.2), water might be required over a longer period than for materials bound with conventional Portland cement so that full hydration can take place. This, combined with the slow strength gain, may require temporary supports and moist conditions to be provided for a longer period than in other cases. This may have implications for construction sequencing and, in some cases, may mean that an alternative material has to be dropped from consideration for pragmatic, rather than technical, reasons.

7.6   Leaching

7.6.1 Mechanisms

Almost all chemicals, if at sufficiently high concentrations and in an environment intolerant to them, can be considered to be pollutants. Thus a limestone-based material may leach calcium and carbonates at relatively high concentrations that could be considered pollutants where the local water (in the ground or in a nearby drainage ditch) is naturally acid (e.g. in a heathland).

In the present context, the leachate chemistry of a compacted material can only be of issue if the material could become thoroughly wet. Mere wash-off from a surface (e.g. from rain impact) cannot generate significant leachate (though it might be of concern from a solids erosion point of view).

With most alternative materials, the concentration of contaminants that can be liberated, even when saturated and water is made to flow through the compacted bulk, is small. In the case of materials that have been through a high-temperature process with rapid cooling this can be because contaminants are locked up in glassy mineral phases and are not, therefore, readily leachable. Furthermore, they may not be very soluble at the pH level engendered by the mixture's composition or at the pH of the leaching water.

7.6.2   Species

The leachate quality is, of course, dependent on the contaminant in the source material. So slags tend to have high metallic contents – the exact species of chemicals being dependent on the source – although they may not be very available to water in the vicinity. Aluminium, arsenic, barium, boron, cadmium, calcium, cobalt, copper, iron, lead, lithium, magnesium, molybdenum, potassium, selenium, sodium, strontium, tin, vanadium and zinc are more or less commonly studied with nitrate, nitrite, sulfate, sulfite, chlorine, fluorine and ammonia anions also being commonly measured.

Organic species should be considered when organic materials such as ashes (e.g. *fba*, *pfa*, MSW-ISA, ISSA), coal mine waste or tyres are being considered. Total organic carbon and total inorganic carbon are usually measured where the source has a carbon element. More specific organic molecules may sometimes be investigated – usually because those particular compounds are known to be of concern with a particular source material.

However, the species in leachate do not only depend on the source chemicals in the materials through which water seeps. Speciation is also affected by pH, contact time, solubility of the species and many other factors.

7.6.3   Pathways and receptors

If the compacted construction material is a source of contaminants this is of no concern until those contaminants move and arrive at a place, person, plant or animal where they are unwelcome at the concentration upon arrival. Thus we need to know the *pathway* of any movement of the contaminant – and any changes in contamination as it travels that pathway, and we need to know the tolerable levels at the *receptor*. For most purposes the receptor is defined as a body of water (stream, river, lake or groundwater) and the tolerable concentration is defined either by a value set in some regulation or by a site-specific value set after a specific analysis of the special features and condition of the receptor. In practice, we expect most contaminants not to experience an increase in contamination concentration as they move down the pathway but, probably, a reduction due to various attenuation processes in the pathway (such as sorption and bio-degradation). This being the case, an assessment of these processes is usually able to demonstrate that the concentration on arrival at the receptor will be satisfactory.

7.6.4   Assessment

Levels of chemical species in leachate resulting from chemical leaving the compacted solids and moving into water that is adjacent or seeping through the solids can be assessed by simple batch or tank leaching test (e.g. the US TCLP test (EPA, 1992) or the EU shaker (CEN, 2002b) or monolith (CEN, 2002a) tests). These tests give an index of leachability that is usually well in excess of that which can be expected in situ because of their intensity (i.e. the degree of interaction between solids and water in these tests is unlikely to be experienced in situ). Taking into account the lack of saturation that most construction sources will experience, the lack of agitation with water as applied in the tests as well as the pathway and receptor issues described in Section 7.6.3, then even if the concentration leaving the compacted material is indicated by the laboratory test as being excessive, then processes of (e.g.) sorption in the pathway may be sufficient to reduce contamination to satisfactory levels. Then, as part of a risk assessment exercise, it may be a relatively simple matter to demonstrate regulatory compliance. It is worth noting that application of these tests to conventional materials could often produce an excessive reading, so it is worth making comparison between the behaviour of a conventional and an alternative material if any test readings from the alternative material initially suggest a problem.

Most assessments need to be site-specific – both the sensitivity of the receptors at the site and the conditions under which the construction will be called to operate have an influence on defining acceptability. Thus, if used in an industrial or commercial setting where the water will have atypical pH or chemical make-up before encountering the compacted material, assessment would need to take this into account as well. Similarly, the concentration of a chemical species in a leachate might be acceptable where the ground or surface waters are already contaminated, whereas the same concentration would be unacceptable near a river used for fly-fishing.

7.7   Physical and mechanical properties of alternative and recycled materials

For the most part the same physical and mechanical properties are of relevance as for conventional compacted materials. However there are some specific properties that may be of relevance to alternative materials that seldom have to be considered for conventional version, so it is these that are considered in this section.

7.7.1   Properties to be considered

Swelling is a particular feature of some alternative materials due to the complex pozzolanic and sulfate interaction reactions that may continue, slowly, for a long period after compaction. Depending on the application, it may be easy or impossible to make allowances for future swelling. Creep is experienced in materials with a visco-elastic component – principally mixes with a significant RAP content. Creep will be greater in hot climates (or high temperature indoor conditions) for these materials. Limited creep can, again, be allowed for in many designs, but like swelling, if the amount is large modification of the mixture's make-up should be made to reduce the magnitude of the problem. Forth and co-workers (e.g. Forth *et al.*, 2010) proposed specific curing regimes to remove creep in blocks of RAP (see section on 'Reclaimed asphalt planings (RAP)'). Delayed curing, and the conditions to encourage its completion have already been mentioned in Section 7.5.2.

7.7.2   Testing specific to alternative materials

Standard materials tests are available to assess both swell and creep. Swell (ASTM, 2008) may be assessed either by measuring the amount of surface heave achieved when a specimen is constrained in a cylindrical mould with an open top, or the stress can be measured that is necessary to keep the top of a cylinder in place when a specimen tries to expand longitudinally. The former is more suitable for unconfined situations and the latter better for estimating loads that might be imposed on adjacent, immovable construction elements. Creep (Neville *et al.*, 1983) may be assessed by applying an axial load to an unconfined cylinder of the material in question. As the operating temperature of viscous materials is so important to assess creep, the test should be carried out at an elevated temperature at a variety of representative stress levels to reflect the worst possible conditions.

7.8   The use and reuse life cycle

In order to assess the benefits and impacts of any proposed use of recycled or alternative materials, a life cycle assessment (LCA) could be performed to assess the environmental impacts associated with all the stages of the construction material's sourcing, production, use and disposal or recycling (the so-called 'cradle-to-grave' evaluation). In principle, a LCA can be extended to cover other aspects of value such as economic and social benefits and impacts so as to ensure that decisions yield 'bearable, viable and equitable' solutions .... or can be seen to be more bearable, more viable and more equitable than others. Regrettably, some LCAs are becoming little more than evaluation of atmospheric carbon generation (or carbon savings) as governments impose various carbon targets on constructors.

Figure 7.8 presents an illustration of the concept of LCA. First a system boundary has to be drawn – what is going to be included in the evaluation and what isn't. This decision can have a major influence on the ease of performing the analysis, on the usefulness of the study and on the conclusions reached. Set the boundaries too wide ('Do I include a relevant fraction of the transport 'costs' of the company that supplies the wheelbarrows to the shop from which I got them?') and the process becomes unmanageable, set them too narrow and so little is included that the results have little worth. However, within the range of sensible choices it is easy to reach a totally different conclusion by selecting different, yet still reasonable, sub-sets of the possible boundaries.

Once the boundaries have been selected, the benefits and impacts of each activity must be evaluated – often using data that are of uncertain pedigree or applicability, at least as far as concerns the particular project or activity being evaluated. Figure 7.8 imagines that four major process areas are evaluated (inside the central box) from project start until recycling of the construction with some 'cost' being associated with the inputs being consumed (on the left) and more 'costs' being associated with the emissions on the right. Repeating this calculation for different recyclates allows the preferred option to be selected and its impact evaluated. The detailed process for evaluating greenhouse gas emissions is set out in the UK specification, PAS2050 and in ISO 14064. More generally the LCA procedure is set out in ISO standard 14040 (and associated documents).

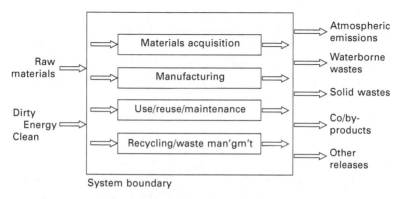

7.8 The reuse cycle.

7.9   Future trends and conclusions

Recycling is not just about the use of what would otherwise be waste and by-product materials, nor about the best environmental use of such materials. It must also consider the potential for re-use of the present construction once that ceases to serve a useful purpose – the basis of the so-called 'cradle-to-cradle' – see below. When the next generation comes to build, they ought to be able to take our generation's building materials and use them to make their own. Most of the materials mentioned in this chapter (as well as the primary materials mentioned elsewhere in this book) should be amenable to such recycling/reuse, but persistence will be needed to ensure that they are, indeed, reused at the same level.

As mentioned in Section 7.1.3 of this chapter, reuse of materials has often been into lower value uses in the past. Clearly it is beneficial to maximise the use, but the 'cradle-to-cradle' concept (C2C) goes beyond this and says that reuse must be to at least the same level of use. C2C proposes that the eco-efficiency idea ('doing more with less') is ultimately a negative activity (albeit not as negative as traditional resource consumption) because we ought to be doing everything with what we already have. Instead, C2C goes further and includes three key principles, only the first of which has so far been discussed in this chapter:

1   waste from one activity becomes the resource for something else
2   solar derived energy needs to be used to meet energy needs
3   biological, cultural and conceptual diversity have to be promoted and combined.

The second of these principles is included to ensure that the full cycle of the construction, use, demolition and reconstruction can be performed not only without any new raw material consumption, but also without degrading the planet's energy resources in order to save its construction resources. The third is included to ensure that the construction materials aren't sourced at a net human and ecological cost. Thus C2C doesn't permit benefit in one sphere to be achieved at the expense of another sphere. Clearly such concepts go well beyond what can be adequately covered in this chapter so readers are pointed towards the C2C source book (Braungart and McDonough, 2002).

Several of the materials described in this chapter envisage a blending of materials, but this could make the process of recycling more problematic – for example a glass cullet fraction might limit the recycling processes that could be applied due to health problems with generation of dust. Use of rubber in a mix might limit the possibility of a recycling process involving heat. In practice, there seem to be few problematic issues of this kind as there are usually common, practical recycling processes that can be applied without causing such problems. However, it is necessary to consider each candidate materials in detail so as to ensure that its ultimate reuse at an equal (recycled) or higher (upcycled) performance level is not compromised by decisions taken at the present.

In this chapter, only the materials have been considered. However, no practical use can limit consideration to the materials. The materials of which the building is constructed have immense impacts on the energy consumption, human scale, environmental fit, etc., of the building during its life. In many cases the scale of these factors is orders of magnitude larger than those of the issues involved in the materials themselves, even when considered over the whole life of the building from cradle-to-cradle. It is no benefit to use a recycled material that saves money, energy, raw material resources and human effort and then produces an expensive-to-operate, unpleasant, energy-inefficient, high-maintenance building. But such discussion is well beyond the scope of this chapter.

7.10   Sources of further information

There are few (maybe no) sources that address the topic of this chapter for compacted construction materials. The book by Jagadish *et al.* (2007) contains a little on the topic and much coverage of closely related material. The majority of relevant work in this area is in the area of pavement engineering (where moderate strength materials, compared to structural concrete, are common) and in alternative brick manufacture. Many of the materials dealt with in this chapter are also covered, from a pavement perspective, by Sherwood (2001) though this is a little dated. The UK Waste and Resources Action Programme (WRAP – see www.wrap.org.uk) has an enormous website with reports, case histories, advice and background data on almost every material that might be suitable to make elements of modern 'alternative-earth' buildings.

The triennial WASCON conference series (originally 'WASte materials in CONstruction', but now 'International Conference on the Environmental and Technical Implications of Construction with Alternative Materials') has many relevant papers in it. Web links for these conferences and other relevant publications can be found at the parent society's website – www.iscowa.org. Another relevant, though newer conference is the 'Sustainable Construction Materials and Technologies Conference'. These do not have a common website but they can be found via an internet search engine.

7.11   References

ASTM, 2008, Standard test methods for settlement and swell potential of cohesive soils, Am. Soc. Testing Materials, Standard D4546-08.

Atkinson V.M., Chaddock B.C. and Dawson A.R., 1999, Enabling the use of secondary aggregates and binders in pavement foundations, TRL report 408, TRL Crowthorne, Workingham.

Braungart, M. and McDonough, W., 2002, *Cradle to Cradle: Remaking the Way We Make Things*, North Point Press, New York.

CEN, 2002a, Preparation of eluates by leaching of aggregates, EN 1744-3:2002. Tests for the Chemical Properties of Aggregates Comité Européen de Normalisation, Brussels.

CEN, 2002b, Compliance test for the leaching of granular waste materials and sludges, EN 12457-3:2002, Characterisation of Waste. Leaching. Comité Européen de Normalisation, Brussels.

Champion, P., Guillet, L. and Poupeau, P., 1981, *Diagrammes de Phases des Matériaux Cristallins*, 2nd edition, 244 pp, Masson, Paris.

Claisse, P.A., Ganjian, E. and Sadeghi Pouya, H., 2007, Use of recycled gypsum in road foundation construction, Waste and Resources Action Programme, Report PBD5-002, November. Available for download at www.wrap.org.uk/document.rm?id=4896.

Claisse, P.A., Ganjian, M. and Tyrer, M., 2008, The use of secondary gypsum to make a controlled low strength material, *Open Construction and Building Technology Journal*, **12**, pp. 294–305. Available for download at www.bentham.org/open/tobctj/openaccess2.htm

Dunster, A.M., 2007, Case Study: Paper sludge and paper sludge ash in Portland cement manufacture, A report of the WRT 177/WR0115 project on Characterisation of Mineral Wastes, Resources and Processing Technologies – Integrated waste management for the production of construction material, 8pp. Mineral Industry Research Organisation/Building Research Establishment/University of Leeds/Akristos Ltd/UK National Industrial Symbiosis Programme.

EPA, 1992, Toxicity characteristic leaching procedure, Method 1311, in SW-846, Test Methods for Evaluating Solid Waste, Physical/Chemical Methods, Ch 8.4, 35pp., Environmental Protection Agency, Washington DC.

Forth, J.P., Zoorob, S.E. and Dao, D.V., 2004, The development of a masonry unit composed entirely of recycled and waste aggregates, *RILEM Conference on the Use of Recycled Materials in Buildings and Structures*, Barcelona, pp. 341–350.

Forth, J.P., Zoorob, S.E. and Dao, D.V., 2010, Investigating the effects of curing methods on the compressive strength of Bitublock, ASCE, *J. Materials in Civil Engineering*, **22** (3), March, pp. 207–213.

Forth, J.P., Zoorob, S.E. and Thanaya, I.N.A., 2006, Development of bitumen-bound waste aggregate building blocks, *Proceedings of the Institution of Civil Engineers*, *Construction Materials*, **159** (CM1), pp. 23–32.

ISO 14040, 2006 Environmental management – Life cycle assessment – Principles and framework, International Standards Organisation, Geneva.

ISO 14044, 2006, Environmental management – Life cycle assessment – Requirements and guidelines, International Standards Organisation, Geneva.

ISO 14064-1, 2006, Greenhouse gases – Part 1: Specification with guidance at the organization level for quantification and reporting of greenhouse gas emissions and removals, International Standards Organisation, Geneva.

ISO 14064-2, 2006, Greenhouse gases – Part 2: Specification with guidance at the project level for quantification, monitoring and reporting of greenhouse gas emission reductions or removal enhancements, International Standards Organisation, Geneva.

ISO 14064-3, 2006, Greenhouse gases – Part 3: Specification with guidance for the validation and verification of greenhouse gas assertions, International Standards Organisation, Geneva.

Jagadish, K.S., Venkatarama Reddy, B.V. and Nanjunda Rao, K.S., 2007, *Alternative Building Materials and Technologies*, 218pp. New Age International, New Delhi.

Motz, H. and Geiseler, J., 2000, Products of steel slags, an opportunity to save natural resources, in proc. *Waste Materials in Construction*: WASCON 2000, eds Goumans, J.J.J.M., Woolley, G.R. and Wainwright, P.J., Elsevier, pp. 207–220.

Neville, A.M., Dilger, W.H. and Brooks, J.J., 1983, *Creep of Plain and Structural Concrete*, Construction Press, London.

OECD, 2010, *Factbook, Economic, Environmental and Social Statistics*, 300pp. Organisation for Economic Co-operation and Development, Paris.

PAS 108, Outlines specifications for the density, porosity and dimensions of tyre bales, Waste and Resources Action Programme. Available for download at www.wrap.org. uk/document.rm?id=3779

PAS 2050, Assessing the life cycle greenhouse gas emissions of goods and services, British Standards Institution. Available for download at http://shop.bsigroup.com/en/ Browse-by-Sector/Energy-Utilities/PAS-2050/

Sadeghi Pouya, H., Ganjian, E., Claisse, P.A. and Karami, S., 2007, Strength optimization of novel binder containing plasterboard gypsum waste, *ACI Materials Journal*, **104** (6), pp. 653–659.

Sherwood, P., 2001, *Alternative Materials in Road Construction*, 2nd edition, 176pp. Thomas Telford, London.

Verhasselt, A. and Choquet, F., 1989, Steel slags as unbound aggregate in road construction: problems and recommendations, in *Unbound Aggregates in Roads*, ed. Jones, R.H. and Dawson, A.R., pp. 204–209, Butterworths, London.

7.12   Appendix

ckd   cement kiln dust
fba   furnace bottom ash
FGD   flue gas desulfurisation
ggbs   ground granulated blast furnace slag
HBM   hydraulic bound materials
IBA   incinerator bottom ash
ISSA   incinerated sewage sludge ash
MSW   municipal solid waste
pfa   pulverised fuel ash, 'fly ash'
PPE   personal protection equipment
PSA   paper sludge ash
RAP   recycled asphalt pavement
RCA   recycled concrete aggregate
TCLP   toxicity characteristic leaching procedure

8

Soil mechanics and earthen construction: strength and mechanical behaviour

C. E. AUGARDE, Durham University, UK

Abstract: This chapter covers selected topics in mechanics and soil mechanics, which are important for an appreciation and understanding of the mechanical and hydraulic behaviour of unstabilised earthen construction materials. Stabilised materials are not covered in detail in this chapter, but aspects of their behaviour are clearly linked to those of unstabilised materials. The role of water is crucial as is the acceptance that many of the materials with which we are concerned are what is termed 'unsaturated'.

Key words: soil mechanics, friction, unsaturated soils.

8.1 Introduction

Understanding how earthen construction materials deform and carry loads is important in the development of design rules. For unstabilised earthen construction materials, the inherent properties of the soil mixture are of major significance. 'Soil mechanics' is the study of the fundamental principles governing the behaviour of all subsoil, and is a branch of civil engineering (subsoil being the 'earth' we are interested in, as opposed to topsoil, which we do not use for building). Soil mechanics principles are used by engineers designing foundations and retaining walls and for assessing the stability of slopes.

In this chapter some of the background necessary to understand the sources of strength in earthen construction materials will be covered. We will be focusing on unstabilised materials where strength derives from the properties of the earth mixture rather than from cementation supplied by a stabiliser (such as cement in modern stabilised rammed earth), however many aspects of the behaviour of stabilised materials are linked to these properties. To do this we will use the engineering science of 'soil mechanics', a discipline that emerged in the 1930s, which has since proved highly successful as a tool for civil and *geotechnical* engineers involved in designing foundations, retaining walls, tunnels, etc., in fact anything built in, on or with soil. Soil mechanics is also a vibrant area of research in many universities. It is important to state at the start that the soil in soil mechanics is subsoil, does not contain organic matter (unlike topsoil) and is the 'earth' used in earthen construction.

Basic mechanics is the foundation from which soil mechanics has developed and many complex soil behaviours can be understood using relatively simple mechanical concepts as will be demonstrated. The fundamental difference between soil and other civil engineering materials is that it is multi-phase, comprising soil grains and water. The changing balance between these components is the key to explaining its mechanical behaviour, and the simple concept of 'effective stress', which follows from recognition of this multi-phase nature, and which will be explained below, has proved of great power in development of methods for predicting movements and failure of geotechnical structures. While it is increasingly possible to measure the strength of soils in situ the majority of strength parameters used in calculations by engineers are obtained from laboratory tests on small samples of soils. In these tests the conditions in the ground, due to the presence of surrounding soil and changing water conditions, are mimicked as closely as possible.

Since the 1990s a subset of soil mechanics has grown up, termed *unsaturated* soil mechanics, in which it is recognised that many soils are three-phase, having air present in voids as well as water. This seemingly innocuous difference leads to wildly different behavioural features. It will be shown that many, if not all, modern unstabilised earthen construction materials can be regarded in part at least as manufactured unsaturated soils. For many stabilised materials, free water may also be present and therefore these could be regarded as *cemented* unsaturated soils. And therefore if we wish to build confidently with these materials we must also understand unsaturated soil mechanics.

8.2   Basic mechanics

Some basic mechanics is required to be restated so that the concepts to be discussed later are clear. Engineering mechanics can be divided into statics and dynamics. In the latter we consider problems in which there is acceleration, such as design of structures to withstand earthquakes, or the determination of the vibration modes of a bridge. In statics we consider that no acceleration is involved and the basic check is one of equilibrium, i.e. the principle that all the forces at a point which is not accelerating sum (in a vector sense) to zero. Statics and dynamics are not separate concepts, but together are Newton's second law; in statics we have no inertial forces since there is no acceleration. As well as making sure that equilibrium is satisfied we have also to check compatibility (the notion that objects subject to loads deform smoothly), the boundary conditions are met (the points of support of the object and the loads applied to it) and finally the constitutive relation is followed. In statics we tend to be either dealing with one of two different situations: (i) what is happening at a point in a piece of material, and (ii) what is happening to a structure built out of various components.

In this chapter we will be considering the first of these. Given a structure subject to loads, one of the most significant jobs of the engineer is to check conditions at material points within to assess likelihood of material failure.

8.2.1   The constitutive relation

The link between the effects of the applied loads and the deformations produced at a material point is known as the constitutive relation. Loads and deformations are scale-dependent and we instead write constitutive relations in terms of their scale-independent counterparts, stresses and strains. By 'effects' here we do not mean the resulting deformations but the stresses produced in the material by the loads. Stress is a quantity calculated as force divided by area (so stress increases if we either increase the force or decrease the area over which it acts). Stress is often measured in civil engineering in units of Pascals (Pa) where 1 Pa = 1 Newton per metre squared (m^2), and is usually given the Greek letter σ. We further classify stresses into normal and shear stresses. A normal stress is directed at right angles to the plane on which it acts, while a shear stress acts parallel to the plane. The difference between the two types of stress is illustrated in Fig. 8.1, which shows a rectangular block subjected to just shear stresses and just normal stresses. In the former the block is seen to change shape but not area. In the latter the situation is reversed. Note that Fig. 8.1a shows normal *compressive* stresses; stresses in the opposite directions, tending to stretch are termed *tensile*. For most materials it is shearing that leads to failure, and so we are most concerned with determining the maximum values of shear stress that can be experienced at a material point. The state of stress at a material point is visualised by thinking of an infinitesimally small cube placed at the material point whose sides are subject to normal and shear stresses. For each face of the cube there is one normal stress and two shear stresses, so for the whole cube there are

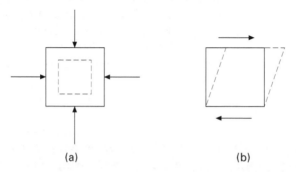

(a)   (b)

8.1 Definition of stresses: (a) normal; and (b) shearing.

18 different stresses. However due to equilibrium only six of these stresses are different. Strain, in its simplest form, is a dimensionless ratio of a change in a dimension over the original dimension measured and is given the Greek letter ε. In a one-dimensional problem, such as the extension of a uniform rod, there is a single strain (the axial strain) calculated as

$$\varepsilon = \frac{\text{change in length}}{\text{original length}} \tag{8.1}$$

The constitutive relation links stresses to strains via the material properties where for each stress on the infinitesimal cube mentioned above there is a strain. In most real calculations we have to consider the full combination of stresses and strains acting at a point in all three dimensions, rather than being able to use a single stress as in the example above, a complex task that will not be covered in detail here.

The simplest constitutive relation is linear elasticity in which there are two material properties, usually taken as Young's modulus, E, which is a measure of the stiffness of the material and Poisson's ratio, ν, which is a measure of how much something deforms in one direction when loaded in another direction. For instance, taking a small block of clay and squashing it leads to deformations in directions at right-angles to the direction of squash. This is what most materials do and is an indication of a positive Poisson's ratio. For the case of a material loaded in one direction only such as the rod in Fig. 8.2a Young's modulus can be measured as

$$E = \frac{\sigma}{\varepsilon} \tag{8.2}$$

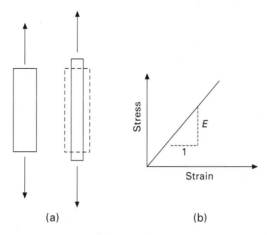

8.2 Young's modulus in one-dimension (a) one-dimensional test of a loaded rod; (b) elastic stress–strain curve.

where σ and ε are the axial stress and strain respectively. If the stress–strain relationship is plotted then Fig. 8.2b is produced and Young's modulus can be seen as the slope of the line. Table 8.1 gives Young's modulus values for a number of civil engineering materials (the units are gigapascals, i.e. 1 GPa = 10^9 Pa). Two out of four material properties can be used for linear elasticity; instead of E and v we sometimes replace one or both of these with bulk modulus K or shear modulus G where the conversions are:

$$G = \frac{E}{2(1+v)} \quad K = \frac{E}{3(1-2v)} \quad\quad [8.3]$$

Linear elasticity predicts (i) that the stress in the material is always proportional to the strain, (ii) that a material will continue to deform with increasing load forever and (iii) that on release of the load the deformation will entirely disappear. The second of these is clearly not realistic as materials fail, that is they either (i) reach a point where no more load can be applied and permanent deformations occur (which stay once the load is removed, known as ductile failure or yield) or (ii) the material fractures (known as brittle failure). Instead of the stress–strain behaviour looking like Fig. 8.2b plots look more like those in Fig. 8.3. Soils actually show very little elasticity, exhibiting permanent deformation (termed *plastic*) almost immediately upon loading. Therefore any useful constitutive relation must also include definition of the point of failure. E and v, are lumped together as *stiffness* properties (i.e. how much

Table 8.1 Young's moduli for a range of materials

Material	Young's Modulus (GPa)
Steel	205
Aluminium	70
Concrete	15–45
Brick	10–20
Soil	0.3–10
Rock	1–100

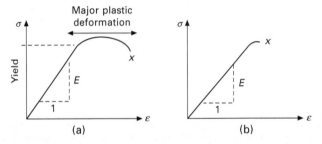

8.3 Stress–strain plots for: (a) ductile; and (b) brittle materials.

and what type of deformation occurs for a given load). The properties and rules that tell us about failure are associated with *strength*. Understanding this distinction is vital in mechanics of materials. For materials such as metals, we often talk of a yield strength, which is a single stress beyond which the material can be expected to experience permanent (or plastic) deformation. In fact it is only for a very limited series of loading situations that one can directly compare a single yield stress with a single value of stress one calculates as occurring in the material. In modelling continuum materials, as we do in soil mechanics, the strength rules (often called yield or failure criteria) are written in terms of all stresses acting at a point (on the infinitesimal cube) considering all three dimensions.

Brittle failure (Fig. 8.3b) is characterised as sudden fracturing with very little apparent plastic deformation; glass being a classic brittle material. Analysis of brittle failure is more complex than the ductile failure described above, largely because the process of brittle failure involves a sudden transfer of energy. Engineering fracture analysis tends to focus on the assumption of cracks of a certain length being present in a material and then answering the question: will those cracks get larger (propagate)? The material property most commonly used to assess this in metals is the fracture toughness K_{IC}; a value that is compared to a *stress intensity factor* calculated using the applied stresses. This is a complex procedure to consider applying to non-metallic materials and is not considered in this way for soils and rocks. Instead a simpler approach is sometimes taken so that cracking is assumed to occur when the soil is subjected to tensile stresses (soil often being assumed to have no tensile strength).

For earthen construction materials we are almost always focused on strength rather than stiffness. While structures built from earth will deform due to applied loads (mostly due to the weight of the earth itself) these movements are generally insignificant and we are mostly concerned with the likelihood of failure. Therefore in what follows we will focus on strength, its sources and its measurement in these materials.

8.3   Fundamental soil behaviour

What distinguishes soil behaviour from other civil engineering materials? Examples are quite easy to point to. A soil slope can fail both due to being overloaded and due to rainfall. What is happening here? Would we consider a concrete structure (where particles are cemented together throughout) to behave any differently when wet than dry? The key mechanical concept in soil strength is friction and both of the slope failures can be explained by considering this physical phenomenon. The other major distinguishing feature of soil is that it is a multi-phase material, i.e. it contains soil grains, water and sometimes in addition, air. When it contains only soil grains and water it is termed *saturated* (i.e. all the voids in the soil are filled with water) and this is assumed to be so in the majority of cases considered by engineers in design. Two commonly used measures of soil state are usefully stated at this point. Firstly, the *degree of saturation* of a soil sample is given by:

$$S_r = \frac{V_w}{V_v} \tag{8.4}$$

where V_w is the volume of water and V_v is the volume of voids (i.e. $S_r = 1$ means full saturation). Water content is defined as:

$$w = \frac{m_w}{m_s} \tag{8.5}$$

where m_w is the mass of water in a sample and m_s is the mass of solids in the sample.

It is now widely accepted that friction holds soils together, but before applying this idea to soils we should consider simpler models of friction of the type studied at school. Consider a block sitting on a flat plane (Fig. 8.4). The block is subjected to a horizontal force F but will move only when F exceeds a certain value determined as:

$$F = \mu N \tag{8.6}$$

where μ is the coefficient of friction (a material property of the interface between block and plane) and N is the normal force across the plane, perhaps resulting from the block's self-weight. The coefficient of friction can be replaced with the tangent of an angle ϕ as follows

$$F = N \tan \phi \tag{8.7}$$

where the angle is that made by the line of action of the reaction force from the plane to the vertical. As the normal force is increased so the frictional resistance to movement increases. As stated above, the majority of failures in materials are due to shear strength being exceeded, and we can easily see that the interface between the block and the plane is in shear, so that this frictional model is itself a model of shear strength. The first slope failure

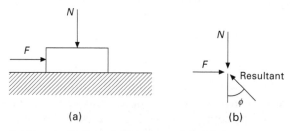

(a)   (b)

8.4 Simple friction: (a) block model; (b) forces.

mentioned above occurs due to overloading leading to shear *stresses* in the slope exceeding the shear *strength*. The rainfall-induced failure cannot however be explained in the same way as the applied load has not changed so the shear stresses in the slope must be the same. The explanation is a little more subtle and requires the concept of effective stress.

8.4   Effective stress

The most powerful idea in traditional soil mechanics is that of effective stress, first proposed by an Austrian Engineer, Karl Terzaghi in the 1920s, which has never proved to be wrong ever since. The concept of effective stress lies behind all the geotechnical construction that surrounds us. The easiest way to understand effective stress is by reference to an idealised view of what is happening at the grain level in a soil (Fig. 8.5). The applied loads are now termed the total stress σ. These are the stresses that we check against equilibrium and are applied to the soil. Within the water there will be a pressure which we call the pore water pressure u. The total stress will be pushing the soil grains together while the pore water pressure will be pushing them apart. Remembering that pressure is the same thing as a stress, the difference between the two is called the *effective stress* σ' where

$$\sigma' = \sigma - u \qquad\qquad [8.8]$$

Effective stress is the stress transmitted through the arrangement of grains and is therefore the stress that will control the frictional behaviour. Crucially, Equation 8.8 shows that the effective stress can be altered both by changing the loads applied or by changing the water conditions. This now explains how soils can fail both by an increase in load but also by changes in the water pressures, as in the rainfall-induced landslide example above. The first increases the applied shear stress beyond the limit of the material; the second *decreases* the strength of the material by weakening the frictional capacity between grains. So the strength rules we write for soils have to be written in terms of the effective stresses not the total stresses.

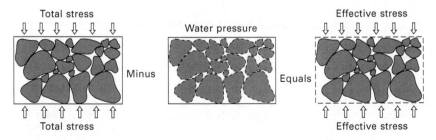

Total stress   Water pressure   Effective stress

Minus   Equals

Total stress   Effective stress

8.5 Saturated soil at the grain level explaining effective stress.

Since it is now clear that the water in between the grains in a soil plays a crucial role in its behaviour it is worth explaining a pair of terms widely used in soil mechanics, which are also widely misunderstood, 'drained' and 'undrained'. These terms refer to the conditions under which loads are applied to a soil sample, or soil at a location in the ground. Consider a sample of saturated soil loaded so that the pore water cannot escape. The applied load will be carried by an increase in the pore water pressure (the water is effectively incompressible as are the soil grains, so preventing volume change will increase the pressure). This is termed *undrained* loading. As we can see from Equation 8.8, this increase in pore water pressure will change the effective stress. Alternatively if the sample is free to lose water (but to still remain saturated throughout) then this is termed *drained* loading. Fine-grained soils, such as clays, have very low permeabilities (which means that water cannot travel very fast since voids are tiny and tortuous). Therefore drained loading in these soils must be carried out very slowly; otherwise the pore water pressure will build up as it fails to travel through the void space fast enough, and if pore pressure changes so does effective stress. We can consider these loading conditions in design too. If a load is applied quickly to a clay soil the initial behaviour will be undrained, for instance. In fact undrained and drained loading are two ends of a spectrum of loading. In drained loading, water is leaving the sample so the sample volume must be reducing, and this volume change is known as *consolidation*. This is an important feature of soil behaviour considered by engineers in design as it tells them that soil constructions may continue to move (by consolidation) long after initial construction is finished, and determining the magnitude and time frame of these movements is critical in many cases. Two further terms that may also confuse are 'normally consolidated' and 'overconsolidated' which actually refer to current and previous states of effective stress. A normally consolidated material is at a yield point it has never before experienced. An overconsolidated sample has been unloaded from a normally consolidated state. Why this is important is that deformation behaviour differs markedly between the two classes. A famous example of an overconsolidated soil is the clay under most of north London (called London clay).

8.5   Models of shear strength for soils

The most commonly used strength 'rule' or yield criterion for soils is known as the Mohr-Coulomb criterion. This is a mathematical representation of the simple model of friction expressed in Equation 8.6. The angle ϕ' is referred to as the effective angle of friction of the soil (since we are now concerned with effective stresses) and is a property that can be determined via a number of different soils tests. The superscript indicates we are dealing with effective stresses here (compare to Equation 8.7). Most commonly Mohr-Coulomb appears for the two-dimensional case as shown in Fig. 8.6 and can be expressed in terms of normal stress σ and shear stress τ as:

$$\tau = \sigma' \tan \phi' \qquad\qquad [8.9]$$

which is a statement of the stress state at failure, i.e. if the actual shear stress exceeds the value of the RHS of the equation then failure will occur. The Mohr-Coulomb criterion is related to drained loading. For undrained loading the shear strength is found to be independent of the normal stress and dependent only on the water content so that:

$$\tau = c_u \qquad\qquad [8.10]$$

where c_u is the *undrained shear strength*.

Many textbooks in the past, and even some today, include another term on the right-hand side of Equation 8.9 namely a cohesion c (or c'). This represents a strength inherent in the soil regardless of friction, i.e. some 'glue' holding the grains together. The notion that soil has an inherent strength available to it when the normal forces are zero is now recognised as incorrect for uncemented materials (it will however be present in stabilised earthen construction materials and some rocks). If there is no normal force at grain to grain contact, where is the frictional strength? In the past, supporters of cohesion have pointed to the electrostatic charges on clay particles as the source for this component, however analysis shows these forces to be an order of magnitude too low to be responsible. Others point to the visible evidence of standing clay faces in trenches as evidence of cohesion, however the reason for their stability (initially at least) lies in another mechanism entirely which will be considered in Section 8.8.

Beyond Mohr-Coulomb there is a wide range of advanced models for the shear strength of saturated soils. For the undrained case one can use models routinely used for metal plasticity (surprisingly) which have a single strength parameter, such as those due to Tresca and von Mises. For frictional behaviour by far the most popular models are based on the Critical State concept, a major advance in understanding the behaviour of soils developed at Cambridge University in the 1960s, which brings together shear strength

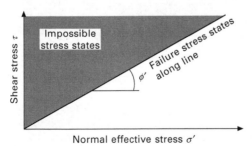

8.6 The Mohr-Coulomb failure criterion.

with features of observed deformations and is most successful when applied to normally or lightly over-consolidated clays. Models suitable for stabilised earthen construction materials must include an additional source of strength from the interparticle bonding through cementation. These models have been developed for soils but are of considerably greater complexity than the models described above. For a lightly stabilised material, a considerable proportion of the strength behaviour may also be governed by friction.

8.5.1   Experimental methods for determination of strength

We have established that the primary mode of failure in soils, as in most other materials is that of shearing and that the strength parameters we wish to find for a soil are ϕ' or c_u. We will now look at some of the most popular methods for determining strength parameters, whether these are required for the Mohr-Coulomb criterion described above or more complex models.

The shearbox test

Probably the simplest test is the direct shear (or shearbox) test which is explained in Fig. 8.7. The shear box test is mostly used for granular soils. The soil is placed or compacted into a small open chamber in the device and a vertical load is applied via a hanging weight applied through a grooved

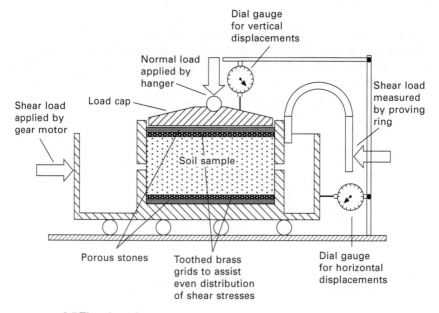

Dial gauge
for vertical
displacements

Normal load
applied by
hanger

Shear load
applied by
gear motor

Load cap

Shear load
measured
by proving
ring

Soil sample

Porous stones

Toothed brass
grids to assist
even distribution
of shear stresses

Dial gauge
for horizontal
displacements

8.7 The shear box test.

plate. The base of the device is then slowly displaced horizontally while the top half remains fixed thus shearing the sample. The horizontal load is measured (which gives the shear stress across the interface on division by the sample plan area) as well as the horizontal displacement. The peak shear stress divided by the normal stress gives the tangent of ϕ' from Equation 8.9. The test is repeated for a number of different normal stresses by emptying the box, refilling and using a different vertical load.

The vertical displacement of the hanging load is also measured and reveals another significant feature special to soils. Typical plots from shear box tests for materials in various states are shown in Fig. 8.8. where it can be seen that loose soils (i.e. those not compacted heavily into the chamber) show a downwards movement of the vertical load, while dense soils show an upwards movement of the vertical load and a peak strength that reduces to a residual level. So in shearing dense soils, the volume of the sample increases. This behaviour on yielding is unique to particulate materials and hints towards the idea that the soil is moving, in both cases, to a unique state where volume will not change, shearing will take place at a constant value of shear stress and, of course the normal stress will not change. This behaviour is one of the powerful features of Critical State Soil Mechanics and we can draw a parallel between loose samples in the shear box and normally consolidated clays, and between dense samples in the shear box and overconsolidated samples.

Triaxial tests

More complicated but of greater utility than the shearbox is the triaxial test where cylindrical samples surrounded by an impermeable membrane are placed in a chamber which is then filled with water, totally surrounding the sample. The sample is then squashed vertically at the same time as a constant pressure is applied via the water surrounding the sample. While it

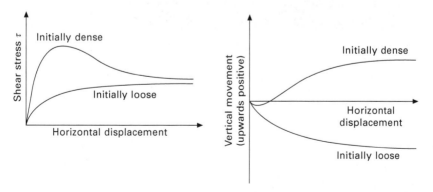

8.8 Typical results from shear box tests.

may not look as if this arrangement of loads is shearing in sense of the Fig. 8.1a it is still taking place since there is a difference between the stresses applied in the three different directions. Failure modes for these samples tend to fall into two classes, of shearing and barrelling as shown in Fig. 8.9. Triaxial testing provides much more information and control than the shear box test. It also does not force the sample to yield across a predefined plane, as happens in the shear box test. Triaxial tests are most easily carried out on clay materials since the cylindrical samples have to be self-supporting during preparation for the test, although there are techniques for carrying out these tests with sands. Drained and undrained tests are possible as the sample ends are supported on discs, which can be porous and linked to a drainage system. With the drainage tap turned off the water in the sample cannot escape and the test is therefore undrained. With the tap switched on a drained test can be carried out, although for the reasons above, this must usually be done very slowly. It is also relatively easy to include additional sensors in the triaxial test to monitor pore water pressures during shearing, and hence determine effective stresses. Complex triaxial testing equipment can be computer-controlled to subject samples to intricate *stress paths*.

Other tests

The shearbox and triaxial tests determine parameters for models of shear strength and also stiffness parameters if necessary. Determining the tensile strength of soils for the purposes of simple fracture prediction is very difficult, however this may be of major importance for stabilised materials and one simple test, borrowed from rock mechanics, can be used on soils which are either cemented slightly, or have tensile strength from a source to be discussed in Section 8.8. This is the Brazilian test. The testing arrangement

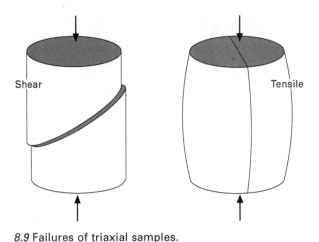

8.9 Failures of triaxial samples.

is shown in Fig. 8.10 and comprises the loading of a thick disc of material thickness t across its diameter D to failure load P. The Poisson's ratio effect means that the compressive loading across the diameter induces a tensile strain in the opposite direction, which causes the specimen to crack and via analysis one can link the failure load P to the approximate tensile stress acting horizontally at the centre of the disc σ_t as

$$\sigma_t = \frac{2P}{\pi Dt} \qquad\qquad [8.11]$$

The use of this test and development of equipment for earthen construction materials is further discussed in Beckett (2008).

8.6   Unsaturated soil behaviour

Terzaghi's theory of effective stress applies only to saturated soils, i.e. those where water fills the inter-grain pores entirely. In many applications, however, the assumption of full saturation is incorrect. Even in temperate climates such as the UK, the top half- to one-metre of natural soil may not be fully saturated at certain times of the year. The water table may lower leaving a zone (known as the vadose zone) above which some water will remain by capillary action but air will be present in the pores too. In hotter parts of the world the upper layers of soil, where most engineering construction takes place, will be unsaturated. So, the question is, is this significant for soil mechanics? Can Terzaghi's effective stress be adapted for use in unsaturated conditions? Before addressing these questions it is wise to consider the changes to mechanical behaviour on desaturation.

Unsaturated soils are known to be stronger than their saturated counterparts. This is easy to test yourself. Building sandcastles is an experiment in

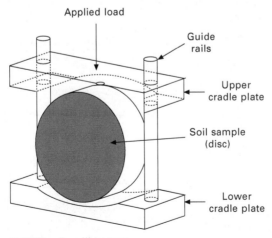

8.10 The Brazilian test.

unsaturated soil mechanics. Completely dry sand is useless as a construction material, as is very wet sand. The former blows away while the latter tends towards a fluid (and fluids in the main cannot withstand shear stress so are not candidates for building materials). There is, however, somewhere in between when sand is just wet enough to make a good building material, for sandcastles anyway, and that is a condition where there is water and air in the voids between particles. This must be an unsaturated case as, if it were saturated, one would not be able to add any more water to the mixture, and clearly you can. So what is providing the additional shear strength in this unsaturated case?

8.6.1   Suction

The explanation for this is 'suction'. Consider two sand grains in an unsaturated soil. Water will be held in menisci between grains, with curved surfaces as shown in Fig. 8.11. Some simple physics tells us that because of the curvature of the menisci surfaces the pressure of the water inside the meniscus must be tensile, i.e. the opposite case to that we have discussed above for saturated soils. So, instead of the water pushing the grains apart it pulls them together and this effect will be heightened if the air pressure in the voids u_a is not atmospheric. Therefore the normal frictional force at contacts is greater in the unsaturated case. A greater normal force leads to a greater frictional force and hence explains the greater shear strength. Suction is therefore defined as the difference between the air pressure and the pore water pressure, i.e. $s = u_a - u_w$. A tensile pore water pressure would appear as a negative quantity in this equation thus giving a positive value of suction.

Unsaturated soils have a number of other behavioural differences to saturated soils, but when considering earthen construction, suction is of most significance. It helps to explain why unstabilised earthen materials do not collapse, since while they may look dry they cannot be totally so. The small amount of water present provides suction to aid basic frictional shear

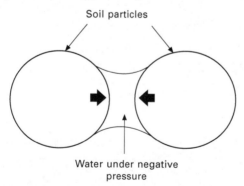

Soil particles

Water under negative
pressure

8.11 Particle model of an unsaturated soil.

strength. Constitutive modelling for unsaturated soils is an active area of research at present, but there is a lack of consensus as to what stress variables should be used to characterise strength. What is clear is that one cannot adopt Terzaghi's effective stress (Equation 8.8) for unsaturated soils. Instead many researchers use net stress $\overline{\sigma} = \sigma - u_a$ and suction. An example of a highly regarded model for unsaturated soils is the Barcelona Basic Model (BBM) developed by Alonso *et al.* (1990)

8.6.2   The soil water retention curve (SWRC)

One other feature of unsaturated soil behaviour is worth mentioning as it is linked to the suction expected in a sample at a given water content, namely the soil water retention curve (SWRC). This is the relationship between suction and either the water content or degree of saturation of a sample. Typical curves for both drying and wetting are shown in Fig. 8.12. At saturation, suction is zero since the pore water pressure will be zero and there will be no air in the voids to supply u_a. As a sample dries, water is lost and suction increases. This continues until the point where the remaining water is located in tiny quantities at grain contacts and around clay particles, and special techniques would be needed to entirely remove this water. This is the residual water content. If the sample is then wetted the path taken back to saturation is different, a hysteretic effect as shown in Fig. 8.12. This has consequences for the measurement of suctions as it is then a requirement to know if the sample was wetting or drying.

Mention has already been made of measurements of pore water pressures in the triaxial test and one question here is, can suctions be measured? The answer is sometimes, but not with great accuracy above a certain level of suction. Many devices have been developed, in universities, in the past 15 years to measure suction with varying degrees of success. High accuracy results up to suctions of 1.5 MPa can be made using small devices known as tensiometers (for examples see Lourenço *et al.*, 2008; Tarantino and Mongiovi, 2001; Ridley and Burland, 1993). At higher suctions the options are limited but include the filter paper test and psychrometers (Rahardjo and Leong, 2006).

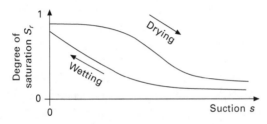

8.12 The soil water retention curve.

8.7   The use of soil mechanics in earthen construction

The preceding has covered the very basics of soil mechanics, which might be of relevance to those working with earthen construction materials. Other chapters in this book will cover stabilised materials in more detail and the means of design of various structural elements, such as foundations and retaining walls. For these types of construction the strength models outlined above will often still be of direct relevance, as part of the analysis will be to check the stability of the element itself, which will require assessment of the working stresses within and a comparison with the strength, which will in turn be based on the results of tests on small samples. For other structural members such as the use of highly stabilised earth to form beams and lintels, the emphasis on material point behaviour may be less immediately relevant although the coverage on fracture and tensile strength may be useful. Despite this, it seems obvious that those working with earth/soil should be aware of its fundamental properties and be able to explain behaviour, especially when things go wrong. The crucial role that water has to play in the strength of soils, with or without stabiliser, should be clear and hence the need for any designers in earth to be fully aware of this important aspect.

8.8   Future trends

The most vibrant area of soil mechanics research in universities is unsaturated soil mechanics and, worldwide, many researchers are working to develop new constitutive models, laboratory equipment and numerical modelling techniques. Centres of excellence in research are UPC (Barcelona), ENPC (Paris), and the Universities of Glasgow and Durham University in the UK. Considerable work is also ongoing attempting a more fundamental understanding of these materials at the microstructural level, using modern methods of non-intrusive investigation such as X-ray computed tomography (Hall *et al.*, 2010). Earthen construction materials have only recently been recognised as manufactured unsaturated soils in works such as Jaquin *et al.* (2009) and Gelard *et al.* (2007), and the application of soil mechanics principles to these materials is at an early stage. The same can be said for the development of models for bonded unsaturated soils, of use for analysis of stabilised materials. The conservation of historical earthen construction is likely to be a major beneficiary from our improving geotechnical understanding of these materials, as stabilisation is likely to be absent or at a low level. A clear gap in knowledge is our understanding of fracture processes in dry earthen construction materials, but this area remains open in geotechnical engineering in general.

8.9   Sources of further information

There are many textbooks covering soil mechanics aimed at undergraduates on civil engineering courses. Most contain sections on soil strength, hydraulic behaviour and testing, as well as sections on design in/on/with soil. Recommended textbooks are those by Powrie (2002) and Smith (2006). Online one can access full texts on soil mechanics at the website of Professor Arnold Verruijt (http://geo.verruijt.net) and at the Géotechnique information site www.geotechnique.info/. Schofield (2005) is an interesting account of major misunderstandings in soil mechanics that have not yet been entirely banished, and Likos and Lu (2004) is a readable introduction to unsaturated soil mechanics. Soil mechanics research is published in journals such as *Géotechnique, Computers and Geotechnics* and the *Institution of Civil Engineers Proceedings: Geotechnical Engineering*.

8.10   References

Alonso EE, Gens A, Josa A (1990) A constitutive model for partially saturated soils, *Géotechnique* 40, 405–430.

Beckett, CTS (2008) Cracking in rammed earth, MEng dissertation, Durham University, Durham.

Gelard D, Fontaine L, Maximilien S, Olagnon C, Laurent J, Houben H, Van Damme H (2007) When physics revisits earth construction: recent advances in the understanding of the cohesion mechanisms of earthen materials. *In: Proc. International Symposium on Earthen Structures, IIS Bangalore*, 22–24 August 2007, 294–299.

Hall SA, Bornert M, Desrues J, Pannier Y, Lenoir N, Viggiani G, Besuelle P (2010) Discrete and continuum analysis of localised deformation in sand using X-ray μCT and volumetric digital image correlation, *Géotechnique* 60, 315–322.

Jaquin PA, Augarde CE, Gallipoli D and Toll DG (2009) The strength of unstabilised rammed earth materials, *Géotechnique*, 59, 487–490.

Likos WJ and Lu L (2004) *Unsaturated Soil Mechanics*, John Wiley & Sons, London.

Lourenço SDN, Gallipoli D, Toll DG, Augarde CE, Evans FD, Medero GM (2008) Calibrations of a high suction tensiometer, *Géotechnique*, 58, 659–668.

Powrie W (2002) *Soil Mechanics: Concepts and Applications* (2nd edition), Taylor & Francis, Didcot.

Rahardjo H, Leong EC (2006) Suction measurements, *Geotechnical Special Publication (ASCE)*, 147, 81–104.

Ridley AM, Burland JB (1993) A new instrument for the measurement of soil moisture suction, *Géotechnique*, 43, 321–324.

Schofield AN (2005) *Disturbed Soil Properties and Geotechnical Design*, Thomas Telford, London.

Smith I (2006) *Smith's Elements of Soil Mechanics* (8th edition), Wiley-Blackwell, Oxford.

Tarantino A, Mongiovi L (2001) Experimental procedures and cavitation mechanisms in tensiometer measurements, *Geotechnical and Geological Engineering*, 19, 189–210.

9

Soil stabilisation and earth construction: materials, properties and techniques

M. R. HALL, K. B. NAJIM and P. KEIKHAEI
DEHDEZI, University of Nottingham, UK

Abstract: This chapter describes the advantages and disadvantages of soil stabilisation, within the context of both soil mechanics and construction materials, and with specific reference to stabilised compressed earth construction, i.e. rammed earth and stabilised compressed earth blocks. Each of the key types of inorganic binder stabilisers are discussed at length including specific sections on high-calcium and naturally hydraulic hydrated limes, Portland cement and composite cements including pozzolanic materials and bituminous emulsions. In addition, stabilisation technologies such as synthetic and natural binders, including polymers, resins and adhesives, as well as fibre reinforcement are included. The chapter gives advice on stabiliser type selection and dosage rates with regard to compatibility with key soil characteristics, complemented by details of a selection tool and decision chart.

Key words: cement, hydrated lime, hydraulic binder, non-hydraulic binder, bituminous emulsion, fibre reinforcement, polymeric binders, adhesives.

9.1   Introduction

The three forms of soil stabilisation collectively refer to various techniques of modification to enhance the physical and/or mechanical properties of a soil for a specific application. This can include altering the texture and plasticity of the soil by, for example, adding or subtracting sand, clay, etc. It can also include compacting the soil to increase its resistance to loading. Or it can include the addition of inorganic binders (e.g. cement) to either enhance the strength and durability, or to make an otherwise unsuitable soil useable. This chapter explains the various stabilisation techniques, how they work and how to select the right one depending on soil type.

Soil stabilisation can be defined as the controlled modification of soil texture, structure and/or physico-mechanical properties. Not only is it extremely rare for a natural soil in the 'as raised' condition to be perfectly suitable for earth construction materials without any soil grading modification, but also most earth building techniques involve some form of compaction to form a strong, stable material. The approaches to stabilisation can be broadly classified as physical, mechanical or binding (Burroughs, 2001). Therefore,

222

practically all earth construction materials are stabilised in some form, however the term is commonly applied only to the application of inorganic binder additives. Physical stabilisation is the modification of, for example, soil particle size distribution and plasticity by the addition/subtraction of different soil fractions in order to modify its physical properties. Mechanical stabilisation is the modification of soil porosity and inter-particle friction/interlock, for example by compaction or other means. An inorganic binder is an additive that improves the strength, durability or other properties of the earth walls. Since mechanical and physical stabilisation techniques are already implicit in modern earth construction techniques (e.g. rammed earth, compressed earth blocks) as part of their criteria for material selection and preparation, the focus of this chapter will be primarily on inorganic binder stabilisation. As with all bound aggregates (e.g. concrete) water content can strongly affect the physico-mechanical properties of stabilised soils, and so they have different properties when saturated compared to dry. It is well known that soil strength and bearing capacity is much higher in the dry state, compared with the saturated state, due to the higher plasticity of saturated soil. Obviously the most common state that exists is partial saturation, as discussed in Chapter 8, and so the exact amount of water content can vary widely as a result of ambient weather conditions or ground water level.

Many modern rammed earth construction companies routinely stabilise their soil with inorganic binders to ensure consistent quality in terms of mechanical performance, durability, resistance to moisture ingress and thermo-mechanical loading, i.e. thermal expansion–contraction. The most common forms of soil stabilisation are Portland cement, non-hydraulic lime, hydrophobic admixtures (see Chapter 10) and sometimes bituminous emulsion. One general approach to soil stabilisation is proposed by the *Australian Earth Building Handbook HB195* (Walker and Standards Australia, 2002) that non-hydraulic lime should be used to stabilise cohesive soils, and that hydraulic (e.g. Portland cement) and/or bituminous stabilisers should be used for granular soils. In line with this, the selection criteria for these stabilisation techniques can be summarised in the graph in Fig. 9.1 using the plasticity index and the relative amount of cohesive material (%wt soil mass below 80 µm particle diameter) as the key parameters.

However, there are several other options (in addition to lime and cement) for soil stabilisation including fibre reinforcement and polymers, for example. These will be discussed in the following sub-sections in terms of their composition, physico-mechanical properties and their adhesion/binding mechanisms. Other stabilisers include chemical admixtures that offer specific functional properties such as set acceleration/retardation of hydraulic binders, plasticisers and hydrophobic admixtures as discussed in detail in Chapter 10.

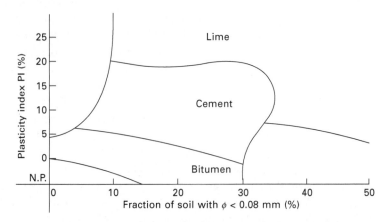

9.1 Selection criteria for common stabilisers with reference to soil characteristics (Houben and Guillaud, 1996).

9.1.1   Advantages and disadvantages of stabilisation

In soil stabilisation there is no inorganic binder that can be applied indiscriminately, as a means of stabilisation, to ensure good results (Houben and Guillaud, 1996). Stabilised earth walls, however, may be built more thinly and there may be no need for the application of expensive surface treatments to improve durability and water tightness. In terms of durability, cement-stabilised materials such as rammed earth rarely have problems meeting the requirements of even the most severe tests (Walker and Standards Australia, 2002). A 'modern' building often results from the use of stabilised soils, which can be distinguished from 'traditional' earth materials, and can have higher status in certain regions such as developing countries (Keable, 1996). The decision to stabilise the soil mix should only be made after consideration of the following advantages and disadvantages:

Advantages

- Speeds up the building process as the required wall thickness is generally much less and so less material and labour is required
- Significantly improves durability and strength, particularly where the locally available soil is poor
- May reduce or eliminate the need for expensive surface treatment or rendering.

Disadvantages

- Raw material costs are increased – soil is free/low cost and cement is comparatively expensive
- The stabilisation materials needed may not be readily available in some developing countries or may be expensive to transport

- The processes of mixing and building can become more complicated depending upon the type of stabiliser that is chosen. This can increase the chance of problems occurring thus affecting time/budget
- Potential environmental impact – e.g. the use of cement and lime can increase the embodied energy (and associated CO_2 emissions) of the wall materials
- Health and safety – cement and lime are both hazardous materials that can cause burns to the skin and eyes, some other chemical additives contain volatile organic compounds (VOCs).

Many guidelines are empirically derived and so are subject to change as the availability of experimental data and experience from industry practice constantly expands. A detailed study was conducted by Burroughs (2001) to produce an extensive database supported by statistical analysis to correlate the relationships between soil classification (Atterberg limits, particle grading and moisture content) and the 'optimum' stabilisation approach for lime and Portland cement using mechanical behaviour as the main assessment criterion. This has led to a more detailed selection framework and is discussed in Section 9.8.

9.2   Lime stabilisation

Lime is the common name given to the oxides and hydroxides of calcium and magnesium. High calcium limes are commercially produced through calcination of carbonate rock minerals (e.g. calcium carbonate; $CaCO_3$) in the form of crushed chalk or limestone. Limes can also be produced commercially as dolomitic lime ($Ca(OH)_2 + Mg(OH)_2$) consisting of calcium and magnesium oxides that are then, for example, pressure hydrated. Calcination of high calcium carbonate rock minerals occurs at ~900°C and atmospheric pressure yielding the highly reactive calcium oxide or 'quicklime' (CaO), which can then be hydrated to the hydroxide form $Ca(OH)_2$. This is achieved either using steam to produce a dry hydrate, or by slaking in water to produce a wet putty. This calcination–hydration–carbonation cycle is commonly known as the 'lime cycle' and is illustrated in Fig. 9.2.

Hydrated limes are commonly used for soil stabilisation in temporary road surfaces and for road sub-bases, especially where the sub-soil has a high percentage of cohesive fine aggregate and/or clay. This technique was first used by the Romans and other early civilisations for road construction and latterly was pioneered by civil engineers in the USA during the 1920s. Since this time it is approximated that millions of square metres of lime-stabilised roads have been constructed throughout the USA including large scale examples such as the ground infrastructure at Dallas Forth Worth Airport, which covers some 70 km^2 (Houben and Guillaud, 1996).

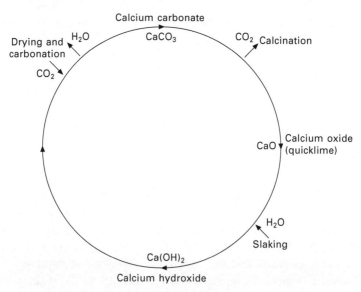

9.2 The calcium carbonate cycle for calcination–hydration–carbonation.

Naturally hydraulic limes (NHLs) are those that are formed by the calcination of clay-bearing limestone or chalk parent rock. This has the effect of creating small quantities of dicalcium silicate ($2CaO.SiO_2$), which exothermically hydrates to form calcium silicate hydrate ($3CaO.2SiO_2.3H_2O$), in addition to natural carbonation of the calcium hydroxide. For this reason NHLs are classed as 'partially hydraulic' and can be used as a higher strength, faster setting substitute for dry hydrate or lime putty. Roman cement was used historically to construct such buildings as the famous Pantheon in Rome. It consisted of hydrated lime mixed with very fine, reactive powder rich in amorphous silica. It was originally sourced as volcanic ash from the town of Pozzuoli in Italy, from which the present day term for these materials 'pozzolan' originates, but was also sourced in the form of fired clay brick dust and wood ash. There are several classifications of naturally hydraulic limes that are commercially available, as shown by Table 9.1. The names of each class are derived from the mechanical strength of the cured material, in which there is some flexibility due to the natural variation in composition of the clay-bearing limestone rocks from which it is manufactured.

The Horyuji temple site is protected by the United Nations (UNESCO) and was Japan's first World Cultural Heritage site. The extensive, lime-stabilised rammed earth wall that surrounds Horyuji Temple in Japan, as shown in Fig. 9.3, is thought to have been built between the dates 607 to 750 AD according to a local historian (Henman, 2002, pers. comm.) and is testament to the longevity of the material. Another example is the Taikoubei structure shown

Table 9.1 Classifications and properties of naturally hydraulic limes and allied materials

Lime binder type	Mean compressive strength range at 28 days (N/mm^2)	Initial setting time in water
Hydrated lime	0.3–1.3	N/A
Feebly hydraulic NHL 2.0	1.3–2.0	15–20 days
Moderately hydraulic NHL 3.5	2.0–5.0	6–8 days
Eminently hydraulic NHL 5.0	5.0–10.0	2–4 days
Roman cement	≥ 10.0	15 mins–1 day

9.3 The lime-stabilised rammed earth perimeter wall that surrounds Horyuji temple, Japan (photo: courtesy Darel Henman © 2002).

in Fig. 9.4, which is a lime-stabilised rammed earth wall connected to the main southern gate of the Sanjyusangen-Do Temple in Kyoto, Japan. It is thought to have been built in around 1610 by order of Toyotomi Hideyoshi. The Oonerihei wall structure was originally constructed somewhere between 1392 and 1466 AD for the Nishinomiya Shrine, in Hyogo-ken, Japan. In 1995, 24 of the 62 wall sections were damaged in the Kobe earthquake and were only repaired around ten years later. The wall panels are approximately 2.5 m tall, 4 m wide and 1.2 m thick, as shown in Fig. 9.5, and cover the entire perimeter of the shrine with a total combined length of approximately 247 metres. The soil that was available on-site is thought to have been mixed in equal amounts with a new soil, of low clay content, that was imported from elsewhere in order to improve the overall soil grading (Henman, 2002, pers. comm.). The soil mix was then stabilised with hydrated lime and 'nigari'; a magnesium chloride-rich mineral deposit formed during the process of salt extraction from sea water. Apparently, the use of nigari was

9.4 Taikoubei rammed earth wall – note the raised stone plinth, drainage channel and large eaves overhang on the roof (photo: courtesy Darel Henman © 2002).

9.5 Oonerihei, Japan – note the degree of weathering and surface erosion on some of the wall sections is only minor for a wall that is over 500 years old (photo: courtesy Darel Henman © 2002).

often recorded in ancient Japanese earth building although its use has now decreased. Today, magnesium- and calcium chloride-rich products are often used in the maintenance of earthen roads as a dust suppressant and so we may hypothesise that the use of nigari may have been for the same reasons.

Chemical reactions involving clay and lime can form cementitious reaction products in the form of calcium silicate aluminate hydrate minerals that can significantly contribute to the strength of lime- and/or Portland cement-stabilised earth (Akpokodje, 1985; Bell, 1996). The reaction(s) takes place between calcium hydroxide ($Ca(OH)_2$), free water and the silica and/or alumina minerals that comprise the clay particles. The $Ca(OH)_2$ is highly alkaline (pH ~12.5) and has the effect of dispersing the clay particles into solution, which aids the reaction (Bell, 1996). The reaction is slow but exothermic and so the rate can be increased with temperature. The $Ca(OH)_2$ is the main compound in high-calcium hydrated lime and is also a significant by-product during the curing of Portland cement. This reaction with the clay minerals is a form of the pozzolanic reaction referred to above, and is discussed in more detail in Section 9.4 below. Studies conducted by Reddy and Lokras (1998) have shown that for compressed earth blocks stabilised with both lime and the pozzolan 'pulverised fuel ash' (PFA), steam curing at ~80°C can produce high strength blocks with a compressive strength > 10 N/mm^2 that are ready for use in construction after only 3 days. The elevated temperature and availability of moisture accelerates the curing by pozzolanic reaction(s), whilst the alkalinity of the lime disperses the clay during mixing, aiding uniform distribution of the binder.

9.3   Cement and pozzolans

Portland cement was patented in England in 1824 by Joseph Aspdin, and was so called because it resembled the high-quality grey limestone from the Island of Portland (UK) that, for example, was used to construct Buckingham Palace and other high-profile buildings. Portland cement is a hydraulic binder which is defined as 'a finely ground inorganic material which, when mixed with water, forms a paste which sets and hardens by means of hydration reactions and processes and which, after hardening, retains its strength and stability even under water' (BSi, 2000). The raw ingredients are calcium carbonate-bearing rock (e.g. chalk, limestone), clays and iron ore (haematite). These are pulverised to a fine meal before staged heating in a rotating kiln starting at temperatures of ~700°C increasing up to ~1400°C. The cooled product is known as clinker, which contains four main reactive compounds, as shown in Table 9.2. Note that the C_3A compound is highly reactive and so a small quantity of gypsum ($CaSO_4.2H_2O$) is added in order to avoid flash setting, usually $\leq$ 5%wt.

The relative amount of each compound in the clinker depends upon the proportions of raw ingredients and also the duration of certain firing temperatures within the cement kiln. The clinker composition therefore determines the properties of the cement, e.g. higher C_3S content can increase setting speed and rate of strength gain, low C_3A content can yield sulphate-resistant cement, higher C_2S can yield lower heat of hydration cement, etc.

Table 9.2 Composition and reactivity of the four principal reactive compounds in clinker

Compound	Chemical equation	Shorthand notation	Time taken to 80% hydration	Heat of hydration (J/g)
Tricalcium silicate	$3CaO.SiO_2$	C_3S	10	502
Dicalcium silicate	$2CaO.SiO_2$	C_2S	100	251
Tricalcium aluminate	$3CaO.Al_2O_3$	C_3A	6	837
Tetracalciumaluminoferrite	$4CaO.Al_2O_3.Fe_2O_3$	C_4AF	50	419

It is possible to estimate the relative proportion of each compound that could be formed from a known quantity of the principal oxides contained within each raw ingredient. Since the raw ingredients are mainly limestone, clays and iron ore the principal oxides will be CaO, SiO_2, Al_2O_3, Fe_2O_3, plus several other minor quantity oxides. The %wt of each oxide can be entered into the empirically derived Bogue equation, as shown below.

$$\%wt\ C_3S = 4.0710 \cdot CaO - 7.6024 \cdot SiO_2 - 1.4297 \cdot Fe_2O_3 - 6.7187 \cdot Al_2O_3$$

$$\%wt\ C_2S = 8.6024 \cdot SiO_2 + 1.0785 \cdot Fe_2O_3 + 5.0683 \cdot Al_2O_3 - 3.0710 \cdot CaO$$

$$\%wt\ C_3A = 2.6504 \cdot Al_2O_3 - 1.6920 \cdot Fe_2O_3$$

$$\%wt\ C_4AF = 3.0432 \cdot Fe_2O_3 \tag{9.1}$$

One of the key assumptions is that all of the SiO_2 is consumed in the formation of C_2S, which is reasonable since this is formed at lower temperatures in the kiln. Any excess CaO then combines with C_2S at higher temperatures to form C_3S. The Bogue equation is therefore useful for predicting the total heat of hydration of a given clinker or estimating the rate of strength gain, resistance to chemical attack (e.g. sulphate) etc. Hardened cement has high Young's Modulus (Modulus of Elasticity) and compressive strength but is typically hard and brittle. Obviously it can impart these properties to granular materials when used as a binder. The properties originate in the microstructure of the hardened cement paste, mainly due to the formation of the reaction product CSH, which arises from hydration of the di- and tri-calcium silicate phases as follows:

$$2C_3S + 6H \rightarrow C_3S_2H_3 + 3CH$$

$$2C_2S + 4H \rightarrow C_3S_2H_3 + CH \tag{9.2}$$

Of particular interest, at this stage, is the pozzolanic reaction, which follows the basic form:

$$CH + S + H \rightarrow CSH + H \tag{9.3}$$

This is the basis of Roman cement, discussed in Section 9.3, where calcium hydroxide (CH) combines with silica in the presence of water to produce CSH. Since CH is a by-product of the calcium silicate reactions (see above), additional silica, in the form of a 'pozzolan', can be added to the cement prior to hydration. A good candidate for a pozzolan should be a finely divided powder that is (i) high in silica content (preferably amorphous) and, (ii) has a high specific surface area. Adding pozzolans to cement can result in significantly increasing the relative proportion of CSH in the hardened cement paste, thus increasing strength and stiffness. Examples of commonly used pozzolans include industrial by-products such as pulverised fly ash (PFA) from coal-fired power stations and condensed microsilica or silica fume (SF) from electric arc furnace silicon production. There are also manufactured pozzolans such as metakaolin (MK) produced from heat-treated kaolinite, and latently cementicious materials in the form of ground granulated blast furnace slag (GGBS) which contains both amorphous silica and C_2S. Some natural pozzolans are available in various parts of the world such as volcanic ash, wood ash, brick dust and rice husk ash. In addition to strength gain many of these materials can also impart additional physical properties to the cement paste such as improved interfacial aggregate bonding, resistance to chemical attack and improved workability. One of the characteristics for a pozzolan to be considered as a cement additive is the loss on ignition (LOI) value, which is the %wt of organic material it contains, and cannot be higher than 5%. Further information about cement chemistry, including composition, hydration and pozzolans, can be obtained from *Construction Materials: Their Nature and Behaviour* (Domone and Illston, 2010) and also the classic text *Properties of Concrete* (Neville, 1995).

Ordinary Portland Cement (OPC) is, in its purest form, > 95 powdered clinker plus gypsum powder, and in the form of a fine powder with a Specific Surface Area (SSA) of between 350 and 500 m^2/kg. The higher the SSA, the more quickly the cement gains strength when hydrating. It can also be blended or diluted with other materials including inert fillers and pozzolans. Under BS EN 197 (BSi, 2000) the different blends of cement composition are grouped into 'CEM' classes, in one of five categories:

- CEM I Portland cement
- CEM II Portland-composite cement
- CEM III blast furnace cement
- CEM IV pozzolanic cement
- CEM V composite cement.

The compositions of each of these categories are given in Table 9.3, and further details can be obtained from BS EN 197-1 (BSi, 2000).

For stabilised rammed earth white Portland cement is commonly preferred to grey ordinary Portland to preserve the natural colour of the soil. The

Table 9.3 Composition of the five different CEM classifications of Portland cement (adapted from BSi, 2000)

| Main types | Notation of the 27 products (types of common cement) | | Composition (percentage by mass[a]) | | | | | | | | | | |
| | | | Clinker K | Main constituents | | | | | | | | | Minor additional constituents |
				Balast furnace slag S	Silica fume D[b]	Pozzolana Natural P	Pozzolana Natural calcined Q	Fly ash Siliceous V	Fly ash Calcareous W	Burnt shale T	Limestone L	Limestone LL	
CEM I	Portland cement	CEM I	95–100	–	–	–	–	–	–	–	–	–	0–5
CEM II	Portland-slag cement	CEM II/A-S	80–94	6–20	–	–	–	–	–	–	–	–	0–5
		CEM II/B-S	65–79	21–35	–	–	–	–	–	–	–	–	0–5
	Portland-silica fume cement	CEM II/A-D	90–94	–	6–10	–	–	–	–	–	–	–	0–5
	Portland-pozzolana cement	CEM II/A-P	80–94	–	–	6–20	–	–	–	–	–	–	0–5
		CEM II/B-P	65–79	–	–	21–35	–	–	–	–	–	–	0–5
		CEM II/A-Q	80–94	–	–	–	6–20	–	–	–	–	–	0–5
		CEM II/B-Q	65–79	–	–	–	21–35	–	–	–	–	–	0–5
	Portland-fly ash cement	CEM II/A-V	80–94	–	–	–	–	6–20	–	–	–	–	0–5
		CEM II/B-V	65–79	–	–	–	–	21–35	–	–	–	–	0–5
		CEM II/A-W	80–94	–	–	–	–	–	6–20	–	–	–	0–5
		CEM II/B-W	65–79	–	–	–	–	–	21–35	–	–	–	0–5
	Portland-burnt shale cement	CEM II/A-T	80–94	–	–	–	–	–	–	6–20	–	–	0–5
		CEM II/B-T	65–79	–	–	–	–	–	–	21–35	–	–	0–5
	Portland-limestone cement	CEM II/A-L	80–94	–	–	–	–	–	–	–	6–20	–	0–5
		CEM II/B-L	65–79	–	–	–	–	–	–	–	21–35	–	0–5
		CEM II/A-LL	80–94	–	–	–	–	–	–	–	–	6–20	0–5
		CEM II/B-LL	65–79	–	–	–	–	–	–	–	–	21–35	0–5
	Portland-composite cement[c]	CEM II/A-M	80–94	–	–	–	–	6–20	–	–	–	–	0–5
		CEM II/B-M	65–79	–	–	–	–	21–35	–	–	–	–	0–5

CEM III Blast Surnace cement	CEM III/A	35–64	36–65	–	–	–	–	–	0–5
	CEM III/B	20–34	66–80	–	–	–	–	–	0–5
	CEM III/C	5–19	81–95	–	–	–	–	–	0–5
CEM IV Pozzolanic cement[c]	CEM IV/A	65–89	–	–	11–35	–	–	–	0–5
	CEM IV/B	45–64	–	–	36–55	–	–	–	0–5
CEM V Composite cement[c]	CEM V/A	40–64	18–30	–	18–30	–	–	–	0–5
	CEM V/B	21–38	31–50	–	31–50	–	–	–	0–5

[a] The values in the table refer to the sum of the main and minor additional constituents.
[b] The proportion of silica fume is limited to 10%.
[c] In Portland-composite cements CEMII/A-M and CEM II/B-M, in pozzolanic cements CEMIV//A and CEMIV/B and in composite cements CEM V/A and CEM V/B the main constituents other than clinker are declared by designation of the cement.

addition of cement can be as high as 20%wt of dry soil, but is commonly used at ≤ 10%wt. In predominantly granular materials inter-particle bonding and interlock is achieved by the hardened cement paste surrounding and adhering to the aggregates. Cement stabilisation is therefore most effective on low cohesion soils, partly because it is difficult to ensure good distribution of the anhydrous stabiliser amongst cohesive clays (Bryan, 1988a, 1988b) and also because the larger granular particles can be surrounded and coated by the cement paste. In cohesive soils, many particles are smaller than anhydrous cement grains and thus are more difficult to coat. In cement–clay mixtures three different forms of reaction can occur (Burroughs, 2001; IPRF, 2005):

1   The hydration reaction forms cement gels on the surface of clay aggregations, and the hydrated lime (calcium hydroxide) that is freed during hydration reacts with silica and alumina in the clay minerals
2   Clay agglomerations are disaggregated by the hydration products and are penetrated by the resultant cement gels
3   Cement gels and the clay aggregates become intimately bonded. This results in both an inert sand–cement matrix and a matrix of stabilised clay in the new structure.

When Portland cement and water is added to soils, at the mixing stage a proportion of the clay particles that were coating the coarse aggregates can become detached and are dispersed into the aqueous phase. This has the effect of adjusting the exothermic hydration of cement paste and hence affecting both the setting time and the strength gain (IPRF, 2005). The remaining clay particles stay attached to the aggregate surfaces, which adjusts the interfacial transition zone (ITZ) at the cement paste–aggregate surface interface. The relative proportion of clay dispersed into the aqueous solution, and therefore becoming part of the hardened cement paste microstructure, depends upon the mineralogy and particle diameter of the clays. The detachment and dispersion of kaolinite clay, for example, is increased from 50% to 79% when the pH is raised from its natural value of 2 up to 12, when an alkaline substance such as cement is added. Further research is needed to determine the precise relationships between these factors. Previous research has shown that highly expansive clay minerals (e.g. Na-montmorillonite) suffer macroscopic swelling, which decreases the rate of hydration and reduces the rate of strength gain in cement-stabilised soils. Conversely, volumetrically stable clays with minimal swelling (e.g. kaolinite) or those with only crystalline swelling (e.g. Ca-montmorillonite) can increase the rate of hydration and strength gain in stabilised soils (IPRF, 2005).

9.3.1   Mechanical and durability performance

The unconfined compressive strength of cement stabilised compacted earth materials generally shows a non-proportional increase with the %wt of added Portland cement, as shown by Fig. 9.6.

Interestingly the relative increase in strength, with %wt of added cement, is much greater for soils with a lower specific surface area (e.g. sandy soils as opposed to silt or clay-rich soil). This is most likely due to the increased interfacial bonding between the cement paste and the aggregate. This may be due to the non-homogeneity of the microstructure, which has been observed to comprise (i) a cement-bound granular matrix, (ii) a dispersed and partially-stabilised cohesive matrix and (iii) a matrix of unstabilised material (IPRF, 2005). The degree of non-homogeneity is less for concrete materials that use washed aggregates, and the proportional increase in strength with additional %wt cement declines significantly for stabilised soils as the cohesive proportion of the soil increases. This was also observed through practical work conducted by Keable (1996).

This presents a problem in interpretation of research data, since three soils can exhibit different proportional increases in compressive strength with additional cement content. The solution can be to improve the particle size distribution and soil plasticity to make it more compatible with cement stabilisation and thus reduce both the %wt of stabiliser needed and also the cost per square metre of the mix. Cement stabilisation can provide cohesion in a low cohesion soil (thus enhancing strength and durability) or reduce the effective clay content of a very high clay soil (Middleton, 1987), and can also be used to reduce linear shrinkage and water absorption (Keable, 1996). King (1996) noted that excessive salts can often impair strength and

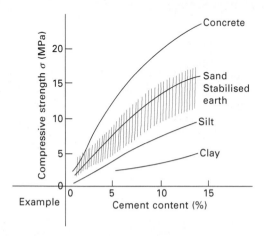

9.6 Relative increase in compressive strength with cement content for different groups of soils (Houben and Guillaud, 1996).

lead to excessive efflorescing in soils that are to be stabilised with cement. When using cement stabilisation in earth walls it must be mixed with dry soil immediately prior to mixing, and the wall must be shaded and kept moist for at least seven days to allow adequate cement hydration.

Previous research conducted by Hall and Djerbib (2004) compared the compressive strength of ten different soil types prior to stabilisation, which were found to exist between the range 0.7 and 1.5 N/mm^2 as shown by Fig. 9.7. The soils were produced using identical aggregate and clay mineralogy, but with closely controlled particle size distributions (i.e. soil grading). This process of soil blending and characterisation is described in more detail in Hall and Djerbib (2004). The minimum compressive strength value for structural earth walls is given as 1.3 N/mm^2 in the official document *Materials and Workmanship for Earth Buildings* by Standards New Zealand (1998). Middleton (1987) instead proposes a minimum compressive strength of 2 N/mm^2. In a separate study, the addition of 3, 6 and 9%wt of CEM IIa ordinary Portland cement stabilisation was performed for three of the soils used by Hall and Djerbib selected as high, intermediate and low classifications of moisture absorption. Clearly the relative increase in compressive strength varies depending on soil grading and does not always follow a linear trend, as shown by Fig. 9.8.

Heathcote (1995) proposed that the ratio between wet and dry compressive strength of earth wall materials could be used as an indicator of the likely

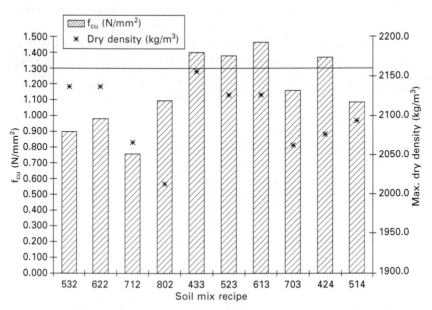

9.7 The relationship between dry density and compressive strength for ten different unstabilised soil types (Hall and Djerbib, 2004).

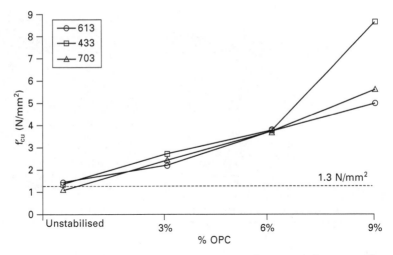

9.8 Relationship between cement content and compressive strength for three contrasting soil types (Hall, 2004).

durability of the wall components. Based on the experiments he conducted it was concluded that a ratio of between 0.33 and 0.50 may be regarded as an indicator of a suitable level of durability dependent upon the severity of the rainfall to which the material is to be exposed. Walker (1995) has suggested that, as a measure of durability, the wet/dry strength ratio is likely to be approximate at best and certainly no substitute for real testing owing to the inherent variability of the material. Both the dry and saturated strength of cement stabilised soil blocks was improved with increased cement content and reduced clay content.

Figure 9.9 shows the comparison between an unstabilised specimen and a 6%wt cement stabilised specimen, made using exactly the same soil, after exposure to a 60-minute initial surface absorption (ISA) test, as defined by BS 1881-208 (1996). The unstabilised material (left) has suffered from loss of structural integrity because the increase in moisture content caused a loss of inter-particle friction and cohesion at the micro-structural level, whilst the stabilised specimen (right) has remained intact.

Analysis of the morphology of cement stabilised earth can be revealed using Scanning Electron Microscopy (SEM). The electron micrographs shown in Figs 9.10, 9.11 and 9.12 were taken using a Philips XL30 Environmental Scanning Electron Microscope with Field Emission Gun (ESEM-FEG) in wet mode using a Gaseous Secondary Electron (GSE) Detector. The two larger particles are fine sand aggregates approximately 300 μm in diameter, inter-connected with a hardened cement paste with clay mineral inclusions. As the partial vapour pressure was increased, the sample temperature was reduced to below the dew point temperature using a Peltier cooled stage. This

9.9 A visual comparison between an unstabilised rammed earth cube sample (left) and a 6% cement-stabilised rammed earth cube sample (right) after 1 h exposure to water absorption in an ISA test (Hall and Djerbib, 2005).

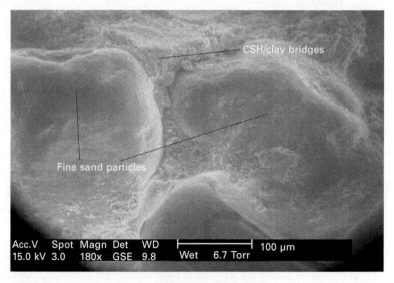

9.10 Electron micrograph of two fine sand particles interconnected with a clay–cement bridge (© 2006 Hall).

method allowed direct control over the relative humidity and spontaneous formation of water droplets by surface condensation. As the moisture content increases, a water meniscus forms between the aggregate particles, some of which may be absorbed by the cement–clay matrix. The presence of moisture reduces inter-particle friction and hence the strength is reduced, however the microstructure remains intact due to the relatively strong interfacial bonding with the hardened cement paste.

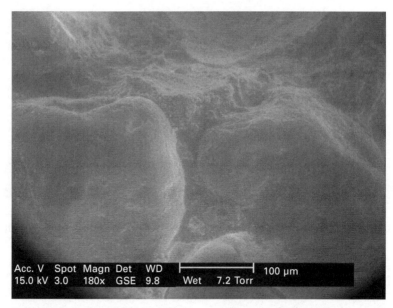

9.11 Early stages of condensation at a relative humidity of 95% showing some surface droplet formation and also capillary condensation within the clay–cement bridge (© 2006 Hall).

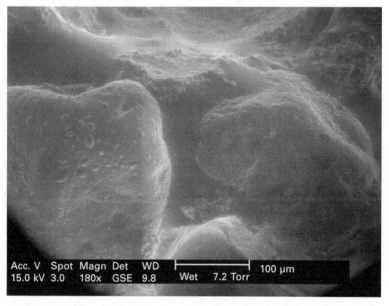

9.12 Late stages of condensation at a relative humidity of 95% showing multiple surface droplet formation and coalescence, and extensive capillary condensation within the clay–cement bridge (© 2006 Hall).

9.3.2   Moisture absorption and transport

Heathcote (1995) states that according to the Portland Cement Association stabilised soils are considered to have passed the ASTM D559 test for cyclic wetting and drying if the mass loss is less than 14% for well-graded sandy soils and 7% for clayey soils. Walker (1995) found that wetting and drying durability was improved with increased %wt cement stabilisation and by reduced clay content, the minerals of which appeared to disrupt the bonding action between the granular soil particles. For stabilisation with $\leq$ 10%wt cement, ASTM D559 is generally satisfied by soils having a plasticity index of 15–20%.

In relation to earth materials the height of a capillary wetting front in a clay soil, where the mean pore radius may be 0.5 μm, would be 3 m compared with the height of a wet front height of around 100 mm in a fine sand with a mean pore radius of around 20 μm (Allaby and Allaby, 1990). This clearly indicates the importance of particle size distribution for rammed earth materials in relation to the origins of moisture transport coefficients. The durability of many types of natural building stones has been related to pore size (BRE, 1983), where materials with high microporosity are often less durable than ones with a lower content of micropores. Other than the effect of wetting, the principal effect of rainfall is surface erosion due to the release of the kinetic energy associated with raindrop impact and dissolution (Heathcote, 1995). In previous research by Hall (2007), four stabilised rammed earth (SRE) test walls were successfully constructed and tested over four separate regimes in a climatic simulation chamber with realistic inner and outer wall conditions. The aim of the study was to evaluate the rate of moisture penetration and the likelihood of interstitial condensation when the walls were exposed to pressure-driven moisture ingress or temperature and humidity gradients. The study found that the SRE walls were able to far exceed the requirements for resistance to pressure-driven rainfall penetration based on BS 4315-2. Furthermore, after five days of exposure to pressure-driven rainfall ingress, there was no evidence of moisture penetration or significant surface erosion of the wall. It was also observed that no interstitial or internal wall surface condensation occurred despite no vapour barrier being installed and with exposure conditions of 20°C ± 1° and 40% relative humidity ± 5% (indoors) and 75% ± 5% outdoors varying at temperatures of +8°C, 0°C and –8°C. Measurements were taken using an embedded array of electronic sensors inside the walls and they detected no significant increase in the relative humidity or liquid moisture content of the test walls (Hall, 2007).

For freeze–thaw testing, the tests detailed in Houben and Guillaud (1996) and ASTM D 560 are apparently almost identical, and are tests originally designed and published for cement-stabilised soils in road design for the

North American climate. Some authors consider that this may be 'too severe' for assessing walling material, however the experiments performed by Bryan (1988a, 1988b) showed that with cement stabilisation, clayey soils disrupted at all levels of cement content and compaction pressure, whilst sandy soils remained intact. According to Laycock (1997) the presence of salts affects the freezing process, and in times of high humidity the salts act to hold water hygroscopically to the surface of the material, thus ensuring better ice crystal growth where uncontaminated samples dehydrate to some extent during freezing.

9.4   Bituminous binders and emulsions

Bituminous materials have been in known use for thousands of years, when bitumen mastic was used in Mesopotamia as water proofing for reservoirs (Read and Whiteoak, 2003). Many years before crude oil exploration and its industrial processing began, man had recognised the numerous advantages of bitumen application and had started the production of bituminous materials found in natural deposits. As reported in the *Shell Bitumen Handbook*:

> It is widely believed that the term bitumen originated in Sanskrit, where the word 'jatu' meaning pitch and 'jatu-kirt' meaning pitch creating referred to the pitch produced by some resinous trees. The Latin equivalent is claimed by some to be originally 'gwitu-men' (pertaining to pitch) and by others, pixtu-men (bubbling pitch), which was subsequently shortened to bitumen then passing via French to English.

> Read and Whiteoak, 2003

BS 3690-1:1989 (BSi, 1989) defines bitumen as:

> 'a viscous liquid, or solid, consisting essentially of hydrocarbons and their derivatives, which is soluble in trichloroethylene and is substantially non-volatile and softens gradually when heated. It is black or brown in colour and possesses water proofing and adhesive properties. It is obtained by refinery processes from petroleum, and is also found as a natural deposit or as a component of naturally occurring asphalt, in which it is associated with mineral matter'.

Bitumen is a product of oil refining and is a long chain complex hydrocarbon. It typically contains 82–88% carbon and 8–11% hydrogen, the rest being sulfur, oxygen and nitrogen. There are four fractional components making up the chemical composition of bitumen (Read and Whiteoak, 2003):

1   Asphaltenes – these are insoluble in n-heptane, black or brown amorphous solids, fairly high molecular weight (1000–100,000), polar and their particle size is 5–30 nm. Asphaltenes constitute 5–25% of the bitumen.

2   Resins – these are soluble and play a key role in bitumen structure, acting as dispersing agents (peptisers) for asphaltenes. They are solid or semi-solid, polar, and have a particle size of 1–5 nm and their molecular weight is 500–50,000.
3   Aromatics – these constitute 40–65% of the total bitumen; they are a dark brown viscous liquid with high dissolving ability. Aromatics have the lowest molecular weight of 300–2000.
4   Saturates – these are white in colour and constitute 5–20% of bitumen. They are usually found in the waxy bitumen.

One of the drawbacks of bitumen is that it will only adhere to aggregate particles if that aggregate is heated sufficiently to drive off all moisture, and this is a costly and energy-intensive procedure. In order to overcome the problem and make the bitumen workable at ambient temperatures bituminous emulsions can provide a suitable solution. A bituminous emulsion is a dispersion of bitumen in water plus emulsifying agents. Hot bitumen will break into very small droplets, typically 1–20 μm in size, by using a colloid mill. At the same time the emulsifying agent and water will be added to the hot bitumen (see Fig. 9.13). The emulsifier agent is a chemical with charge at one end, either positive or negative, and a long polymer tail with a strong affinity to bitumen, as illustrated by Fig. 9.14. When these emulsifier ions attach themselves to the bitumen droplets they are converted into charged particles. These charges are sufficient to prevent the droplets from coalescing, since bitumen and water have very similar specific gravities (1.00 and 1.03, respectively) and so the bitumen droplets float in the water.

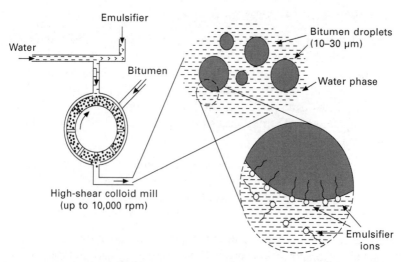

9.13 An illustration of the manufacture process for bituminous emulsion (Thom, 2008).

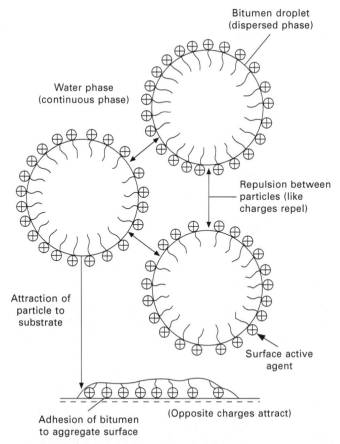

9.14 Schematic diagram of the surface charges on bitumen droplets (Read and Whiteoak, 2003).

Bituminous emulsions that are used in the UK are classified according to a three-part classification coding system. The first part of the code designates the emulsion as being either 'A' (for anionic) or 'K' (for cationic). The second part of the code is a number ranging from 1 to 4 that indicates the stability or breaking rate of the emulsion. Thus, the higher the number the more stable the emulsion. The third part of the emulsion classification code is a number that specifies the bitumen content of the emulsion. Thus, for example, a K1-70 emulsion means that it is a cationic emulsion, rapid-acting, with 70% by mass of residual bitumen (O'Flaherty, 2002).

The behaviour of bitumen stabilised materials (BSMs) is similar to that of unbound granular materials (e.g. earth materials), but with a significantly improved cohesive strength and reduced moisture sensitivity. The bitumen disperses only amongst the finest particles in BSMs, resulting in a bitumen-rich mortar between the coarse particles, and larger aggregate particles are not coated with bitumen. The benefits of bituminous emulsion for soil stabilisation can be summarised as follows (Asphalt Academy, 2009):

1   The BSMs could replace alternative high-quality materials in the upper layers of flooring, foundations or road construction due to the increase in strength associated with bitumen treatment
2   Lower quality aggregate or poorer soil grading can often be used
3   Improved durability, reduction in moisture sensitivity and the introduction of water-repellent properties
4   Reduction in temperature susceptibility compared to hot mix asphalts
5   Improved shear strength.

Materials that are usually treated with bitumen are crushed stone of all rock types, natural gravels such as andesite, basalt, chert, diabase, dolerite, dolomite, granite, limestone, norite, quartz, sandstone, and pedogenoc materials such as laterite/ferricrete, and also reclaimed asphalt materials. The factors that must be taken into account for bituminous emulsion stabilised soil mixtures are as follows (Ibrahim, 1998):

1   *Compatibility between the emulsion and the aggregate.* Interfacial bonding between the emulsion droplets and the aggregates is strongly dependent on aggregate electro-charge. Mineral aggregates bearing a negative net surface charge (depending on pH), such as sandstone, siliceous gravel and granite, are often highly compatible with cationic bituminous emulsions. On the other hand, mineral aggregates such as limestone with a positive surface charge are often more compatible with anionic emulsions
2   *Mixing consideration.* The emulsion–soil combination, the amount of premixing water, and the mixing time are important factors to be considered in order to ensure a sufficient uniform dispersal of emulsion throughout the soil mixture. The amount of water needed for good dispersion is not the same for all emulsions. In fact cationic emulsions require additional mixing water in order to achieve a satisfactory coating (Ibrahim, 1998). As can be seen from Fig. 9.15, in the case of medium setting emulsion (MS), the coating is not improved by the presence of excess water; however, slow setting emulsion (SS) is improved by the presence of large amounts of added water.
3   *Curing.* Curing of bitumen soil mixtures is the process where the mixed and compacted layer discharges water through various mechanisms including evaporation, particle-charge repulsion and pore pressure-induced flow paths. Curing time, environmental conditions such as temperature and humidity, and the air voids in the mixture are therefore factors affecting the rate of moisture loss. According to Finn *et al.* (1968) and Marais and Tait (1989), the curing period may be as much as 6 months in dry climatic regions and two years in wet climates.

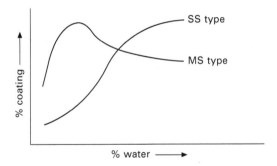

9.15 Comparison between the degree of aggregate coating vs. the amount of mixing water for medium and slow setting bituminous emulsions (Ibrahim, 1998).

The mechanisms involved in the stabilisation of a soil with a bituminous material (usually hot bitumen, cutback bitumen, or anionic or cationic bitumen emulsion) are very different from those involved with cement or lime. The main function of the bitumen is to add cohesive strength to the soil. The unconfined compressive strength (UCS) test is typically performed on soils that are stabilised with additives (e.g. bitumen, cement, lime, etc.), as opposed to the triaxial test. In the UK, stabilised soils are normally tested using 150 mm cubes for coarse-grained and medium-grained materials; for fine-grained (passing through a 5-mm sieve) soil specimens having a height to diameter ratio of 2:1 is recommended. Before testing, stabilised samples are compacted to a pre-determined dry density, which is usually the maximum value obtained during the Proctor moisture–density test. Compaction is either by static (e.g. for fine- and medium-grained soils) or dynamic (e.g. for medium- and coarse-grained soils) techniques (O'Flaherty, 2002). Huan *et al.* (2010) measured the UCS of compacted soil stabilised with bitumen. They used different combinations of crushed granite and crushed limestone along with different percentages of Class 170 virgin bitumen. Their result showed that the sample of 75% CRB and 25% CLS with 3% foamed bitumen showed the highest UCS value at 254.5 kPa (Fig. 9.16).

Marandi and Safapour (2009) investigated the use of combined stabilisers using Portland cement and bitumen emulsion with graded, granular material used for base courses in highways applications. The particle grading of these materials is typically the same as that for cement-rammed and is highly suitable for dynamic compaction techniques. They concluded that (i) the UCS increased with increase of %wt cement addition in the soil, and (ii) the UCS reached a maximum value when the bitumen emulsion was added at ~3%wt (see Fig. 9.17).

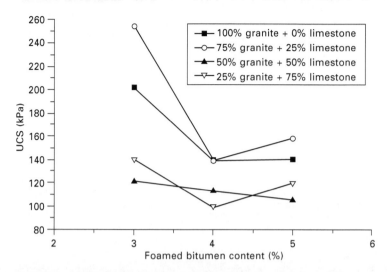

9.16 Plot of UCS versus foamed bitumen content for four different mixtures (Adapted from Huan *et al.*, 2010).

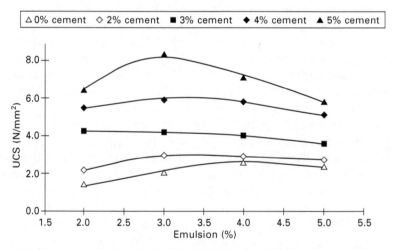

9.17 Twenty-eight-day UCS of combined cement/bitumen emulsion soil stabilisation (Marandi and Safapour, 2009).

9.5 Synthetic binders, polymers and adhesives

Several synthetic binders can be used for soil stabilisation instead of cement or hydrated limes including calcium sulfate dihydrate, commonly known as gypsum ($CaSO_4.H_2O$), tetrasodiumpyrophosphate ($Na_4.P_2O_7.10H_2O$), and calcium, sodium and potassium salts (Burroughs, 2001). In addition, several organic chemicals including fatty polyamides, lignin and casein can be used to enhance water repellence. Although polymers are relatively expensive compared with Portland cement and lime, for example, many different types are used as soil stabilisers. A large number of long-chain aliphatic and resin amines, amides and related cationic agents can be used for soil stabilisation (Burroughs, 2001). Polymers can be mixed with soil in the form of a liquid in order to fill the pores and harden the soil structure. In order for polymeric stabilisation to be used, the following requirements must be met:

- the polymer must have the ability to adhere to soil particles with the assistance of water (adhesive)
- it must be internally cohesive
- it must be capable of working sufficiently (polymerising) at high humidity and at low/non-elevated ambient temperatures
- it must be miscible with water to produce a low viscosity liquid.

Practically, polycondensational polymers are more suitable than polyadditional because (i) the method works even with large chains, (ii) when the polymerisation process stops, there are fewer possibilities for it to re-start and (iii) they are low cost and easy to prepare (Brandl, 1981; Coumoulos and Koryalos, 1983; Joshi *et al.*, 1981). There are many types of polymers used for soil stabilisation including resorcinol-formaldehyde resin, phenol formaldehyde resin, furan resins, polyacrylates and polyurethanes. However, urea formaldehyde resins (UFR) are the most commonly used due to their low price compared with the others (Lahalih and Ahmed, 1998). Additionally, there is a possibility of creating hydrogen bonds between free hydroxyl groups and the soil (clay) alumino-silicates with the CO groups in UFR. This substance is mainly used with granular (i.e. sandy) soils, and is suitable for reducing the permeability and/or enhancing strength. It is applied either by pressurised injection or spraying depending upon whether it is being applied for surface or deep stabilisation. The main drawback in this stabilisation method is the brittleness of the material and the reduction in ductility. Therefore, when high plasticity soil is required, such as in connecting regions between concrete structures with differential settlement, UFR stabilisation is not the preferred method (Levačić and Bravar, 1990).

9.6   Fibre reinforcement

Following the Second World War, soil stabilisation received increased attention to support overseas operations on rapid strengthening of weak soils to facilitate military traffic. Although cement and lime were still the common stabilising agent, innovation in non-traditional stabilising methods began, including fibre reinforcement, which can be used either alone or with conventional stabilising admixtures (Rafalko *et al.*, 2007). Earth reinforcement techniques with fibres are used to improve the shear strength and stability of soils, often at concentrations of up to 1%wt of the soil. These composite soil–fibre materials have relatively high tensile strength and can be used in many application ranging from road structures to embankments and retaining structures (Jamshidi *et al.*, 2010) as well as modern construction materials, as detailed in Chapter 20.

Recently, environmental concerns have led to increased interest in the utilisation of 'alternative' fibres, including solid waste and by-product materials, in soil reinforcement to improve the load-bearing capacity and durability of stabilised soil to use, for example, in retaining wall backfill and highway base layers. In addition to using raw materials as fibrous reinforcement armatures, alternative materials can be utilised as added-value materials in a variety of different shapes and textures, e.g. shredded, fibrous, sheets, smooth or rough, heavy- or lightweight, brittle or ductile, natural or synthetic, etc., whilst arising at lower cost and with less environmental impact (Jamshidi, 2010; Galán-Marin, 2010). To reduce the impact of natural disasters such as floods, landslides and earthquakes in susceptible areas, earth can be stabilised by fibre reinforcement to increase the shear strength of slopes as discussed in Chapter 19.

Soil fibre reinforcement is usually obtained by mixing the fibre to give random orientation within the soil structure. This can increase inter-particle cohesion and structural integrity by providing supporting armatures within the fabric creating a flexible structural mesh. The advantages of this technique are (i) different materials can be used as soil fibre reinforcement, (ii) there is minimal requirement for additional mixing machinery, (iii) recycled, waste and by-product fibres can be utilised, (iv) it is compatible with most soil types and (v) it can be applied over large soil volumes. The main limitations are that this technique can only be implemented at shallow depths, and also some fibres are biodegradable so can suffer from long-term performance degradation (Babu and Vasudevan, 2008). There are many different fibre types suitable for use in soil stabilisation (both natural and synthetic) including jute, sisal, bamboo, timber, cotton, glass, wool and shredded rubber fibre, as well as the following more common fibres types.

9.6.1   Polypropylene fibre (PP fibre)

Polypropylene (PP) fibres are manufactured in one of two main types: monofilament fibres that have a cylindrical shape and fibrillated fibres characterised by a flat, 'tape-like' shape which is susceptible to balling during the mixing process (Fletcher and Humphries, 1991). PP fibres are usually mixed with lime and/or cement-stabilised soils in order to decrease their brittleness and are added by up to 0.25%wt dry soil. The soil strength and toughness both increase with increasing %wt fibre content, whilst swelling and linear shrinkage is decreased (Cai *et al.*, 2006; Tang *et al.*, 2007).

9.6.2   Nylon fibres

The main use of nylon fibres is to enhance the ductility, durability and toughness of concrete, however a handful of investigations have also focused on applications in stabilising soils. Mixing nylon fibres with cement is expected to stabilise soil both mechanically and chemically due to its ability to absorb water and bond with the cement paste during curing (Rafalko *et al.*, 2007; Zellers *et al.*, 2002).

9.6.3   Poly(vinyl) alcohol (PVA) fibre

Poly(vinyl) alcohol (PVA) fibres are again commonly used in concrete to improve the ductile behaviour, but recently have started to be used in conjunction with cement-stabilised soils by taking advantage of the hydrogen bonds that can occur between cement particles and the hydroxyl groups. The main drawback to using this approach is that when the bond between PVA fibre and cement paste is extremely strong, rupture can occur rather than the preferred fibre pull-out mechanism, resulting in brittle failure of the composite. To overcome this disadvantage, PVA fibres can be submerged in an oiling agent before mixing to facilitate post-cracking fibre pull-out (Rafalko *et al.*, 2007; Kanda and Li, 1998).

9.6.4   Natural fibres

Recently, palm fibre was used in soil stabilisation and was found to have the ability to absorb significant quantities of water (up to 187%). After 24 hours of soaking the fibre dimensions increased by 2.5% in length and 11.11% in cross-sectional area. The main use for this fibre is to improve the soil-bearing capacity, which can has been found to increase by up to 26 times in a silty sand (Marandi *et al.*, 2008). Coir (coconut) fibre has a high tensile strength compared with other natural fibres, which exists even in wet conditions. It was found that up to 18% improvement in strength could be obtained when 2.5%wt (at 15 mm length) fibre is used for reinforcement of clays (Babu and Vasudevan, 2008).

9.7   Selection tool for modern stabilised earth construction

Burroughs (2001) developed a seven-stage flow chart of selection criteria for soil suitable for stabilisation for use in modern earth construction. It is the product of an extensive database of experimental data coupled with detailed analysis and validation. It enables the user to assess the suitability of a soil whilst minimising the need for independent (possibly expensive) experimental testing. Stage 1 involves identification of the provenance and availability of candidate materials (e.g. natural sub-soils, as-raised ballast or graded quarry waste). At this stage materials can be accepted or rejected based on initial screening for organic matter content (nominally $\leq 2\%$wt), which can be determined by loss on ignition, as well as high concentrations of contaminants such as sulfates, chlorides, etc. If the material is provisionally suitable then representative samples can be obtained for experimental testing. This begins at Stage 2 by characterisation of the linear shrinkage coefficient, the acceptance criteria for which are shown in Fig. 9.18.

Assuming that 'favourable' or 'satisfactory' linear shrinkage has been observed, Stage 3 involves the characterisation of the moisture contents at which the soil can be plastically deformed under load (i.e. non-elastic strain), known as the plastic limit (PL), and the moisture content at which the soil flows due to gravity known as the liquid limit (LL). The ratio between LL and PL is referred to as the plasticity index (PI), where the acceptance criteria for SRE materials are:

- favourable: PI < 16% and LL < 36%
- satisfactory: PI = 15–30% and LL = 36–45%
- unfavourable: PI > 30% and LL > 45%.

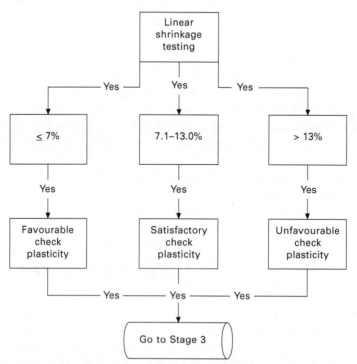

9.18 Stage 2 soil stabilisation selection criteria for linear shrinkage coefficient (adapted from Burroughs, 2001).

At Stage 4 the texture (or grading) of the soil is assessed by determining the particle size distribution. This can be achieved by a combination of wet sieve analysis for the granular fraction (i.e. particle diameters 0.063–20 mm+) and either sedimentation or laser particle analysis for cohesive fraction (i.e. sub-63 μm particle diameters), as explained in Chapter 6. The logic sequence and statistical likelihood of material suitability are shown in Fig. 9.19 where the acceptance criteria are as follows:

- favourable: cohesive fraction (silt + clay) 21–35%, or sand 30–65%, or gravel 3–5%
- satisfactory: silt + clay 36–45%, or sand 66–75%, or gravel 3–12%

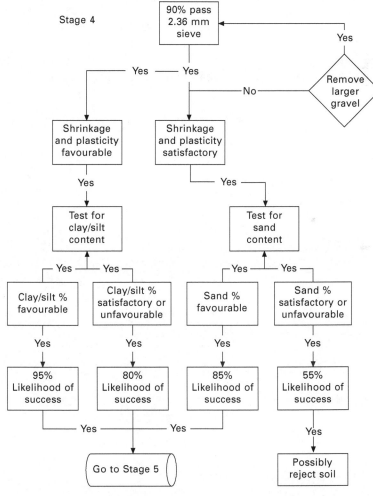

9.19 Stage 4 soil stabilisation selection criteria for particle size distribution (adapted from Burroughs, 2001).

- unfavourable: silt + clay < 21% or > 45%, or sand > 75%, or gravel < 3%.

The results of a previous study suggest that the physical characteristics of a soil are much more significant in determining the 'success' of stabilisation, for example in terms of the achievable mechanical strength and reduction in linear shrinkage, than is the exact %wt or type of added stabiliser. The Standards Australia HB195 *The Australian Earth Building Handbook* (Walker and Standards Australia, 2002), and also Houben and Guillaud (1996), suggest a generic approach whereby high-calcium lime stabilisation can be applied to soils with a high cohesive material content, whilst Portland cement is preferred for stabilisation of predominantly granular soils. The detailed study by Burroughs (2001) showed little experimental evidence to support these claims regarding grading and texture, and instead found statistically correlations between the type and %wt of stabilisation with both the shrinkage and the plasticity of the soil. This enabled the creation of the stabiliser selection criteria given in the chart in Fig. 9.20.

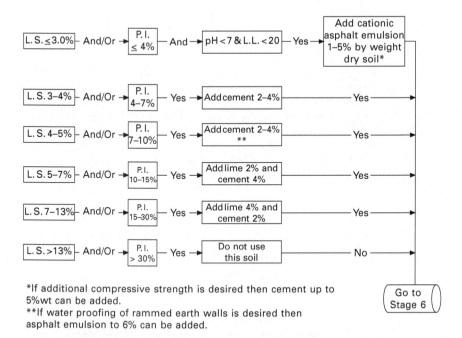

*If additional compressive strength is desired then cement up to 5%wt can be added.
**If water proofing of rammed earth walls is desired then asphalt emulsion to 6% can be added.

9.20 Decision chart for selection of soil stabiliser type and dosage rates (adapted from Burroughs, 2001).

9.8   References

Akpokodje EG, 1985, 'The stabilization of some arid zone soils with cement and lime', *Quarterly Journal of Engineering Geology and Hydrogeology*, 18 [2] pp. 173–180

Allaby A and Allaby M, 1990, *The Concise Oxford Dictionary of Earth Sciences*, Oxford University Press, Oxford

Asphalt Academy, 2009, *Technical Guideline: Bitumen Stabilised Materials – a Guideline for the Design and Construction of Bitumen Emulsion and Foamed Bitumen Stabilised Materials – Second Edition*, Asphalt Academy, Pretoria, South Africa

Babu GLS and Vasudevan AK, 2008, 'Strength and stiffness response of coir fiber-reinforced tropical soil', *Journal of Materials in Civil Engineering*, 20 [9] pp. 571–577

Bell FG, 1996, 'Lime stabilization of clay minerals and soils', *Engineering Geology*, 42 [4] pp. 223–237

Brandl H, 1981, 'Alteraction of soil parameters by stabilization with lime', *Proceedings of the 10th International Conference of Soil Mechanics and Foundations Engineering*, Stockholm, Vol 3, pp. 587–594

BRE, 1983, *BRE Digest 269: The Selection of Natural Building Stone*, HMSO, London

Bryan AJ, 1988a, 'Criteria for the suitability of soil for cement stabilisation', *Building and Environment*, 23 [4] pp. 309–319

Bryan AJ, 1988b, 'Soil/cement as a walling material – I: stress/strain properties', *Building and Environment*, 23 [4] pp. 321–330

BSi, 1989, BS 3690-1: 1989 +A2:2008 – 'Bitumens for building and civil engineering: specification for bitumens for roads and other paved areas', British Standards Institute, London

BSi, 1996, BS 1881-208: 1996, 'Testing concrete – Part 208: recommendations for the determination of the initial surface absorption of concrete', British Standards Institute, London

BSi, 2000, BS EN 197-1 2000, 'Cement–composition, specifications and conformity criteria for low heat common cements', British Standards Institute, London

Burroughs VS, 2001, 'Quantitative criteria for the selection and stabilisation of soils for rammed earth wall construction', PhD thesis, University of New South Wales, Australia

Cai Y, Shi B, Ng CWW and Tang CS, 2006, 'Effect of polypropylene fibre and lime admixture on engineering properties of clayey soil', *Engineering Geology*, 87 [3–4] pp. 230–240

Coumoulos DG and Koryalos IP, 1983, 'Grout mixtures for ground improvement – laboratory testing and quality control', in *Proceedings of the VIII European Conference of Soil Mechanics and Foundation Engineering*, 23–26 May, Helsinki

Domone PLJ and Illston JM, 2010, *Construction Materials: Their Nature and Behaviour*, Spon Press, Abingdon

Finn FN, Hicks RG, Kari WJ and Coyne LD, 1968, 'Design of emulsified asphalt treated bases', *Highway Research Record*, 239

Fletcher CS and Humphries WK, 1991, 'California bearing ratio improvement of remolded soils by the addition of polypropylene fiber reinforcement', *Journal of the Transportation Research Board*, 1295 pp. 80–86

Galán-Marin C, Rivera-Gómez C and Petric J, 2010, 'Clay-based composite stabilized with natural polymer and fibre', *Construction and Building Materials*, 24 [8] pp. 1462–1468

Hall MR 2004, 'The mechanisms of moisture ingress and migration in rammed earth walls', PhD thesis, Sheffield Hallam University, UK

Hall M, 2007, 'Assessing the environmental performance of stabilised rammed earth (SRE) walls using a climatic simulation chamber', *Building and Environment*, 42 [1] pp. 139–145

Hall M and Djerbib Y, 2004, 'Rammed earth sample production: context, recommendations and consistency', *Construction and Building Materials*, 18 [4] pp. 281–286

Hall M and Djerbib Y, 2005, 'Moisture ingress in rammed earth: part 2 – the effect of particle-size distribution on the rate of static pressure-driven moisture ingress', *Construction and Building Materials*, 20 [6] pp. 374–383

Heathcote KA, 1995, 'Durability of earth wall buildings', *Construction and Building Materials*, 9 [3] pp.185–189

Houben H and Guillaud H, 1996, *Earth Construction – A Comprehensive Guide: Second Edition*, Intermediate Technology Publications, London

Huan Y, Siripun K, Jitsagiam P and Nikraz H, 2010, 'A Preliminary study on foamed bitumen stabilisation for western Australian pavements', *Scientific Research and Essays* 5 [23], pp. 3687–3700

Ibrahim HEM, 1998, 'Assessment and design of emulsion-aggregate mixtures for use in pavements', PhD thesis, University of Nottingham, UK

IPRF, 2005, Effects of coarse aggregate clay coatings on concrete performance, *Technical report IPRF-01-G-002-01-4.2*, Innovative Pavement Research Foundation, Skokie, IL, USA

Jamshidi R, Towhata I, Ghiassian H and Tabarsa AR, 2010, 'Experimental evaluation of dynamic deformation characteristics of sheet pile retaining walls with fiber reinforced backfill', *Soil Dynamics and Earthquake Engineering*, 30 [6] pp. 438–446

Joshi RC, Natt GS and Wright PJ, 1981, 'Soil improvement by lime-fly ash slurry injection', in *Proceedings of the 10th International Conference of Soil Mechanics and Foundations Engineering*, 15–19 June, Stockholm

Kanda T and Li VC, 1998, 'Interface property and apparent strength of high-strength hydrophilic fiber in cement matrix', *Journal of Materials in Civil Engineering*, 10 [1] pp. 5–13

Keable J, 1996, *Rammed Earth Structures: A Code of Practice*, Intermediate Technology Publications, London

King B, 1996, *Buildings of Earth and Straw – Structural Design for Rammed Earth and Straw Bale Architecture*, Ecological Design Press, USA

Lahalih SM and Ahmed N, 1998, 'Effect of new stabilizers on the compressive strength of dune sand', *Construction and Building Materials*, 12 [6–7] pp. 321–328

Laycock EA, 1997, 'Frost degradation and weathering of the magnesian limestone building stone of the Yorkshire province', PhD thesis, University of Sheffield, UK

Levačić E and Bravar M, 1990, 'Soil stabilization by means of "LENDUR EH" ureaformaldehyde resin', *The Mining-Geological-Petroleum Engineering Bulletin*, 2 [1], pp. 137–143

Marandi SM and Safapour P, 2009, 'Base course modification through stabilization using cement and bitumen', *American Journal of Applied Sciences*, 6 [1], pp. 30–42

Marandi SM, Bagheripour MH, Rahgozar R and Marandi HZ, 2008, 'Strength and ductility of randomly distributed palm fibers reinforced silty-sand soils', *American Journal of Applied Sciences*, 5 [3] pp. 209–220

Marais CP and Tait MI, 1989, 'Pavements with bitumen emulsion treated bases: proposed material specifications, mix design criteria and structured design procedures for southern

African conditions', in *Proceedings of the 5th Conference on Asphalt Pavements for Southern Africa*, 5–9 June 1989, Royal Swazi Convention Centre, Swaziland

Middleton GF, 1987, *Bulletin 5 – Earth Wall Construction: Fourth Edition*, National Building Technology Centre, Chatswood, Australia

Neville AM, 1995, *Properties of Concrete – 4th Edition*, Pearson, Essex, UK

O'Flaherty CA, 2002, *Highways: the Location, Design, Construction and Maintenance of Road Pavements*, Butterworth-Heinemann, Oxford

Rafalko SD, Brandon TL, Filz GM and Mitchell JK, 2007, 'Fiber reinforcement for rapid stabilization of soft clay soils', *Transportation Research Record*, 2026, pp. 21–29

Read J and Whiteoak D, 2003, *The Shell Bitumen Handbook – 5th Edition*, Thomas Telford, London

Reddy BVV and Lokras SS, 1998, 'Steam-cured stabilised soil blocks for masonry construction', *Energy and Buildings*, 29 pp. 29–33

Standards New Zealand, 1998, *NZS 4298: 1998 Materials and Workmanship for Earth Buildings*, Standards New Zealand, Wellington, New Zealand

Tang C, Shi B, Gao W, Chen F and Cai Y, 2007, 'Strength and mechanical behavior of short polypropylene fiber reinforced and cement stabilized clayey soil', *Geotextiles and Geomembranes*, 25 [3] pp. 194–202

Thom NH, 2008, *Principles of Pavement Engineering*, Thomas Telford, London

Walker P and Standards Australia, 2002, *HB 195 The Australian Earth Building Handbook*, Standards Australia International, Sydney

Walker PJ, 1995, 'Strength, durability and shrinkage characteristics of cement stabilised soil blocks', *Cement and Concrete Composites* 17 [4], pp. 301–310

Zellers B and RCN Inc., 2002, 'Nycon nylon fibers add to hydration efficiency of cement', available from: www.nycon.com/techpapers/, accessed: 3 December 2010

10

Integral admixtures and surface treatments for modern earth buildings

R. KEBAO and D. KAGI, Tech-Dry Building Protection
Systems Pty Ltd, Australia

Abstract: This chapter discusses integral admixtures and surface treatments that impart important protection features to modern rammed earth buildings. The chapter first reviews the use of integral admixtures such as silicone water-repellents to provide protection against natural weathering. Other functional admixtures such as superplasticisers and set accelerators or retarders are then discussed. The chapter then considers surface treatments including vapour-permeable hydrophobic sealers such as silane/siloxane water-repellent impregnants, dust binders and surface coatings or renders.

Key words: silicone water-repellent admixtures, vapour-permeable hydrophobic sealers, surface binders coatings and renders, rammed earth buildings, silanes and siloxanes.

10.1   Introduction

Integral admixtures and post-surface treatments of modern rammed earth buildings not only provide protection against natural weathering, but also improve the quality of the rammed earth construction. This chapter discusses modern technologies for treating rammed earth building materials using integral functional admixtures and surface treatments after construction.

Due to the permeable and hydrophilic nature of rammed earth building materials, water can easily penetrate into rammed earth buildings. Therefore water penetration becomes an important factor affecting the durability of rammed earth construction. The inherent weakness of earth in water makes the earth structure an impractical choice due to the subsequent deterioration under natural climatic conditions. Modern rammed earth building materials often contain a small amount of cement as a stabiliser to improve the durability of the rammed earth structure. However, due to the hydrophilic nature of the earth/cement material, which attracts water penetration, cement cannot solve the problems associated with water penetration for rammed earth buildings under natural weathering conditions.

When cement is added to rammed earth material, water permeation often leads to another problem: efflorescence. Efflorescence can have a detrimental effect on the appearance of modern rammed earth buildings. Efflorescence

256

is caused by water movement within the capillaries of the masonry building material containing cement or lime. The moisture within the capillaries carries calcium hydroxide produced from cement hydration or existing lime within the substrate to the surface. The carbonation of calcium hydroxide by carbon dioxide in the atmosphere produces insoluble white calcium carbonate deposits. Efflorescence may also be caused by the transport of soluble salts either from the subsoil and aggregates or from ground water, which re-crystallise at the surface after water evaporation. This commonly shows as white streaks on vertical masonry wall surfaces. This deterioration is detrimental to the natural decorative appearance of cement-stabilised rammed earth buildings. Further, damp walls due to water penetration often lead to other water-based staining problems such as surface soiling or deposits of other staining materials.

Water penetration damages the thermal insulation properties of rammed earth buildings. Cold rammed earth walls caused by high moisture content within the wall require additional heating in winter. Cold walls caused by water penetration also often cause internal surface condensation resulting in problems such as mould growth and efflorescence on the internal wall surface.

Integral admixtures play an important role for rammed earth. A properly designed silicone water-repellent admixture effectively minimises water movement within the capillaries and hence not only reduces the water absorption, but solves other water-related problems as discussed above. The admixture does not affect the vapour permeability and the surface appearance of rammed earth buildings. Other water-repellent admixtures such as stearates or oleates are also effective for attaining water repellency-but these materials lack long term durability. Although it is uncommon to use other functional integral admixtures such as a plasticiser or a set accelerator or retarder for rammed earth materials, these admixtures, under special circumstances, can help to achieve certain desired properties for rammed earth materials.

Post-surface treatment is another way to protect rammed earth structures. Silicone water-repellents are useful sealers for rammed earth. The advantage of using silicone water-repellents is that this sealer penetrates into the capillaries and reacts with the substrate via strong siloxane bonding resulting in a long-term protection against natural weathering. Unlike other sealers, silicone impregnation does not affect the vapour permeability and the surface appearance of rammed earth buildings.

A water-resistant paint or coating such as a silicone masonry paint or a highly filled acrylic coating containing silicone water-repellent admixture can also provide a good protection against water penetration for rammed earth. Due to the permeable paint film structure, silicone masonry paint and highly filled acrylic coatings do not significantly affect the vapour permeability of rammed earth, but the coatings can alter the surface appearance of rammed earth construction. If a natural appearance is desired then water-repellent impregnation is a better choice.

Surface dust sealing is normally required for internal rammed earth walls Polyvinyl acetate (PVA), acrylic or styrene butadiene (SBR) latex emulsion are useful materials for sealing the internal surface of rammed earth walls as dust-binding sealers or dust suppressants. By selecting proper sealing material at a low concentration with a proper application method, internal dust sealing does not cause significant change in the surface appearance and vapour permeability of rammed earth walls.

To avoid reducing the vapour permeability of the substrate, it is not normal practice to undertake dust sealing for the exterior surface of a rammed earth wall. However, it is possible to carry out exterior dust sealing using a mixture of an acrylic or other polymer emulsion at low concentration and a silicone water-repellent admixture to achieve both dust binding and water repellency for rammed earth buildings if exterior dust sealing is required.

10.2   Integral admixtures for modern earth construction

Modern rammed earth construction is often stabilised by a small amount of cement. Cement remarkably improves the quality of rammed earth building. Other admixtures such as bitumen, acrylic or latex emulsions and other polymeric materials may also be useful as stabilisation agents.[1] However, cement, due to its low cost and wide availability, is the cheapest and the most efficient material to stabilise rammed earth material. It is not the main aim of this chapter to discuss various stabilisation admixtures, but the aim is to discuss integral functional admixtures whose purpose is to improve quality and to provide some important features to modern rammed earth building materials.

10.2.1 Hydrophobic admixtures

For many years, hydrophobic or water-repellent admixtures have been used in cementitious masonry building substrates including cement-stabilised rammed earth. At a small dosage such as 0.05% or less, water-repellent admixtures efficiently minimise moisture movement within the capillaries of substrates to reduce water absorption under natural weathering conditions. Such a water-repellent admixture does not block the capillaries of masonry substrates and therefore does not affect the vapour permeability of building materials and does not affect the surface appearance of the substrate. Figure 10.1 shows the effective water-repellent effect (or beading effect) on a pressed concrete block containing a silicone water-repellent admixture. Note: a pressed concrete substrate is made of sand and/or fine gravel, cement

10.1 Effective water-repellent effect (or beading effect) on a pressed concrete block containing a silicone water-repellent admixture.

and pigment as a colouring agent and has very similar properties to that of a cement-stabilised rammed earth substrate that is made of cement and soil, which mainly contains sand (and/or fine gravel) and clay which is like pigment in pressed concrete. In this chapter, unless otherwise stated, most test results were based on standardised pressed sand/cement test substrates for the consistency of testing, as rammed earth is a very variable substrate.

Oil- or fat-based water-repellent admixtures

Conventional hydrophobic admixtures are oil- or fat-based water-repellent admixtures such as stearates and oleates, which have been used for pressed concrete[2] and cement-stabilised rammed earth for many years. These admixtures may be in the form of raw oil such as vegetable oil or other fatty acids, or may be made from animal fat or vegetable oil, for example, ammonium or calcium stearates or oleates in the forms of either aqueous emulsions or solid powders. These materials can achieve a satisfactory water-repellent effect for cement-stabilised rammed earth at an addition rate of approximately 0.1% by weight based on the active solid content of admixture. The water-repellent property is achieved by the hydrophobic admixture molecules within the rammed earth matrix where the individual active ingredient such as the oil or fat molecule can react with calcium present in the rammed earth material or any other masonry substrate to form calcium stearate or oleate, which are insoluble salts. These insoluble calcium stearates or oleates within the rammed earth matrix convert the hydrophilic nature of rammed earth into a hydrophobic material. Figure 10.2 represents the hydrophobic salt deposits within a pressed concrete substrate.

The long-term hydrophobic performance of these oil- or fat-based admixtures is found to be unsatisfactory due to the fact that they are not permanently bonded to the substrate, but exist as individual hydrophobic salt deposits within the substrate. These individual hydrophobic deposits are subject to many deterioration factors under atmospheric conditions. UV and natural weathering, for example, can effectively destroy these materials over a period of time. These oil- and fat-based materials can be effectively re-hydrolysed under strong alkaline conditions where cement is used in the rammed earth building. These admixtures are also subject to biological deterioration particularly under humid or wet conditions. The durability of these oil- or fat-based admixtures compared to that of the new silicone water-repellent admixtures will be discussed further in the following section.

Silicone water-repellent admixtures

The most recent hydrophobic admixture technology developed for rammed earth is silicone-based water-repellent admixture. This silicone admixture was initially developed by Victoria University in Australia in 1996 for pressed concrete substrates. The technology was successfully commercialised by an Australian company for the pressed concrete industry. This silicone water-repellent admixture was soon found to be equally effective for cement-stabilised rammed earth substrates and other cementitious masonry materials. Since 1997, a new silicone water-repellent admixture was introduced to the rammed earth industries both in Australia and overseas.

This silicone water-repellent contains reactive silanes and silxoanes, which effectively form a nanomolecular polysiloxane lining within the capillaries of the rammed earth substrate rather than blocking the masonry

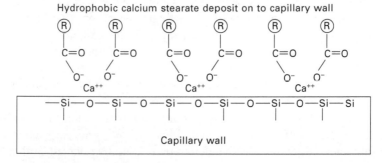

10.2 Molecular structures of stearate or oleate in capillary.

capillaries. Therefore, relatively small amounts of silicone are required to molecular-line the capillaries compared to the amount required to block the capillaries. However, molecular lining capillaries using small amounts of silicone can greatly reduce the surface tension of the substrate and convert the hydrophilic nature of masonry capillaries to a hydrophobic state. Figure 10.3 represents capillary rise due to a hydrophilic masonry capillary versus capillary depression due to a hydrophobic capillary treated with silicone water-repellent admixture.

The advantage of a silicone admixture is that the reactive silane and siloxane react with masonry ingredients via strong siloxane bonding and they also crosslink to form polysiloxane, which lines over the capillary wall of the substrate.[3, 4, 5, 6] Due to the chemical bonding, the polysiloxane lining becomes part of the rammed earth substrate resulting in long term durability compared to that of a traditional water-repellent admixture, which relies on a hydrophobic calcium stearate or oleate salt deposited within the capillaries. Figure 10.4 shows the silane reaction forming polysiloxane within a masonry capillary wall. The alkylalkoxysilane or siloxane first hydrolyses in the presence of water to form a silanol, which then reacts with the masonry substrate forming crosslinking nano-polysiloxane structures via siloxane linkages within the masonry matrix resulting in a permanent hydrophobic attachment to the treated masonry substrate.

The silicone admixture is effective in making cement-stabilised rammed earth water-repellent. At an addition rate of 0.05% based on the mass of the rammed earth, the silicone admixture achieves approximately 80% reduction in water absorption. Figure 10.5 shows a capillary water absorption result according to DIN 52617 using a pressed concrete test substrate containing a silicone admixture. A 24-hour water absorption result clearly indicates that the capillary water absorption of test substrates containing a silicone admixture is significantly reduced. The test result also indicates that the addition rate of 0.05% of silicone admixture was sufficient to make a water-resistant

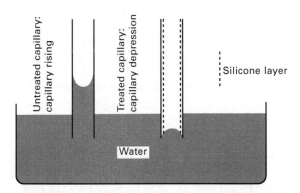

10.3 Capillary rising versus capillary depressing.

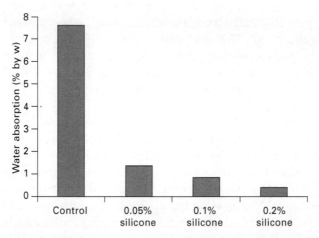

10.4 Molecular structures of polysiloxane in concrete masonry.

10.5 Capillary water absorption of substrate containing a silicone admixture.

rammed earth substrate. Feedback from the rammed earth industry from different countries also confirmed that addition at a rate of 0.05% silicone admixture is sufficient in practice for most cement-stabilised rammed earth buildings.

The efficacy of the silicone admixture is further confirmed by a test carried out to model the resistance to wind-driven rain (water penetration under pressure) for pressed concrete block walls conducted by the Australian Commonwealth Scientific and Industrial Research Organisation (CSIRO) according to ASTM E514-90. This test is equivalent to wind-driven rain at a wind speed of 120 km/h (or 500 pa) against a single skin pressed concrete block wall containing 0.05% silicone admixture. The test device is shown in Fig. 10.6. The result shows that the test wall successfully passed the

10.6 This device is used to test the equivalent to wind-driven rain at a wind speed of 120 km/h (or 500 pa) against a single skin pressed concrete block wall containing 0.05% silicone admixture.

4-day test period, which was far superior to the industrial standard (ASTM standard) of 4 hours. In comparison, a standard pressed concrete block wall without silicone admixture generally resisted water penetration for only 20 minutes under the same test conditions.

Please note that the above outstanding result of resistance to water penetration under pressure provided by the silicone admixture within the pressed concrete wall should not be applied under a tanking situation of a rammed earth building such as when a part of building is constructed below the ground or in underground structures, or under a slope or a retaining wall situation. An appropriate waterproofing treatment such as using bituminous, polyurethane or other polymeric membranes together with a fabric membrane and installing an appropriate drainage pipe or a similar waterproofing measure according to the common building practice to prevent groundwater ingress to the walls or footings of the building is highly recommended.

Rammed earth containing a silicone water-repellent admixture shows a good resistance to rising damp. Rising damp in a masonry building is generally caused by moisture rising through the building footings and is a difficult problem for remedial treatment in buildings. Figure 10.7 shows the result of a rising damp test that was conducted by placing a pressed concrete block in a water bath for 24 hours. Results shown in Fig. 10.7 reveal that

10.7 Result of a rising damp test conducted by placing a pressed concrete block in a water bath for 24 hours.

the block containing 0.05% silicone admixture was dry inside whilst the control was fully saturated with water due to rising damp (capillary water absorption). This test result clearly indicates that the silicone admixture in the treated substrate provided a good resistance to rising damp.

Despite a good resistance to rising damp offered by this silicone water-repellent admixture, an effective damp course of either chemical damp-coursing or appropriate fabric or plastic membrane damp-coursing material should be properly installed according to common building specifications when a rammed earth building is constructed.

The resistance to efflorescence endowed to rammed earth substrates by silicone water-repellent admixture is significant. Table 10.1 shows the results of an accelerated efflorescence test for a pressed concrete substrate. The test was conducted by laying the test substrate into a tray containing 10% sodium sulfate solution at a depth of 10 mm (less than 30% of the substrate total height) for 7 days. The top surface of the substrate above the solution was visually examined for evidence of efflorescence. The test results in Table 10.1 indicate that the substrate containing 0.05% silicone water-repellent admixture showed no evidence of efflorescence after 7 days of the test period, whilst the control was fully covered with efflorescence within 1 day. The efflorescence of the sample containing 0.025% of the silicone admixture was also significantly reduced compared to that of the control substrate.

An efflorescence test in outdoor conditions was conducted by using commercial pressed concrete blocks under natural weathering conditions in Tasmania, Australia for seven years (as shown in Fig. 10.8). The result indicates that the block containing 0.05% silicone admixture showed remarkable resistance to efflorescence against natural weathering. The treated samples remained in almost the original condition after seven years, whilst the control blocks showed a poor surface appearance mainly due to efflorescence (white salts) and other water-based staining problems associated with water penetration.

The durability of rammed earth is also significantly improved by incorporating silicone water-repellent admixture. A durability test was conducted by an accelerated salt erosion test. A pressed concrete substrate was dipped into a 10% sodium sulfate solution for 12 hours and subsequently dried at 80°C in an oven for 12 hours. This was repeated over many cycles. The ingress of salty water into the substrate and drying can result in serious damage to substrates due to repeated salt crystallisation, which expands in the capillaries of the substrate. In this test, treated samples containing 0.05% silicone admixture and a control of standard pressed concrete substrate were used. A substrate containing conventional stearate water-repellent admixture was also used as a comparison sample.

Table 10.1 Approximate surface coverage of efflorescence over seven days

	After 1 day	After 3 days	After 7 days
Control	100% covered	–	–
0.025% silicone	10% covered	20% covered	50% covered
0.05% silicone	Nil	Nil	Nil
0.10% silicone	Nil	Nil	Nil

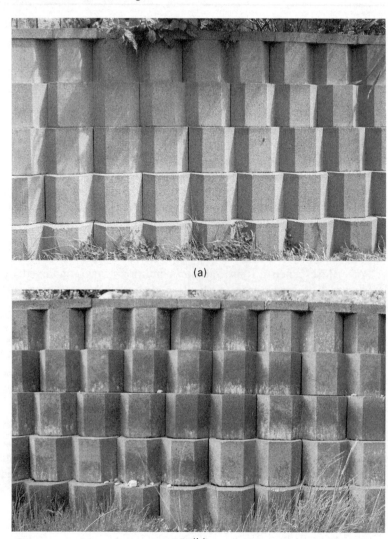

10.8 Examples of commercial pressed concrete blocks that underwent natural weathering conditions in Tasmania, Australia for seven years. (a) Pressed concrete blocks containing 0.05% silicone admixture, (b) Control blocks.

The salt erosion test demonstrated that after 15 cycles of dipping and drying, the disc with silicone admixture remained almost unchanged, whilst the control commenced to erode after five cycles. The sample with silicone admixture still showed a good water-repellent effect at the end of the erosion tests. The test also indicated that the sample with the stearate admixture showed some resistance to erosion during the early cycles, but the erosion accelerated after 10 cycles. Figure 10.9 shows the pressed concrete substrates before and after 15 cycles of the salt erosion test. This test confirms that the durability of the concrete with silicone admixture is significantly improved compared to that of the control and the test also indicates that the result with silicone admixture is superior to that of a traditional oily-based admixture.

Figure 10.10 shows the weight losses of the test substrates after 15 cycles. The substrate containing silicone admixture almost retained its original mass, whilst the control lost 12% and the sample with stearate lost 9% of its

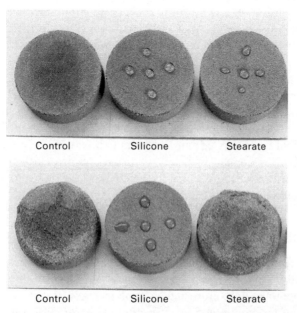

| Control | Silicone | Stearate |

10.9 Pressed concrete substrates before and after 15 cycles of the salt erosion test.

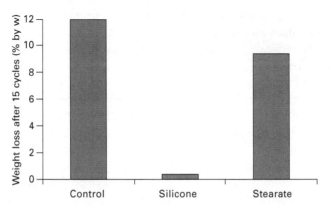

10.10 Mass loss of pressed concrete after salt erosion test.

total weight. This result again confirmed that the durability of the substrate containing silicone admixture was superior to that of the sample without admixture. The traditional stearate admixture was not sufficiently resistant for long term protection against erosion from natural weathering.

The method of use of silicone water-repellent admixture is simple and cost effective. The admixture is simply pre-mixed into the rammed earth soil mixture before being rammed into the form work. The water-repellent rammed earth substrate is already formed after the mould is removed. No further procedure or extra post-treatment is required for the construction. However, it is worthwhile to mention that, due to the immediate surface water repellency, further spraying of water onto the ready-formed rammed earth structure in order to help cement curing becomes difficult after the mould is removed. Therefore, it is important to retain the moisture level within the rammed earth substrate during the curing period in order for the cement to cure completely. It is recommended to keep the rammed earth in the mould for a slightly longer time in order to achieve better curing. It is also recommended to cover up the freshly made rammed earth wall before and after the mould is removed until the rammed earth is fully cured. In fact, preventing water contact with freshly made rammed earth walls also prevents initial efflorescence occurring, which is a common problem for cementitious materials during the early curing period.

10.2.2 Other integral admixtures

Apart from the polymeric binding materials (which are not discussed in this chapter) and the hydrophobic admixtures as discussed in the previous section, other integral admixtures, such as superplasticisers, set accelerators and retarders, are not commonly used in rammed earth. There have been hardly any case studies or practical applications of other integral admixtures for the stabilised rammed earth building materials or other earth building substrates. The integral admixtures discussed in this section will be limited to the cement-stabilised rammed earth materials with similar properties to that of dry-pressed concrete. For example, the rammed earth soil used in Australia often contains sand and gravels with limited clay content (generally at less than 8–10% clay) and approximately 8–10% cement is used as a stabiliser. The property of this rammed earth material is similar to that of dry-pressed concrete. Theoretically, many integral admixtures which are suitable for dry-pressed concrete should be useful for this type of rammed material if certain properties of the rammed earth material are required. However, as mentioned above, integral admixtures apart from hydrophobic admixture are not practically used in Australia and the authors have no feedback from the rammed earth industry that these admixtures are practically used in other parts of the world.

Workability and compaction enhancement

Superplasticisers are commonly used in dry-pressed concrete to improve the workability and to enhance the compaction of concrete for increasing density and to improve the surface finish of the concrete product. These admixtures should be equally effective for cement-stabilised rammed earth materials, which have similar properties to those of dry-pressed concrete. Lignosulfonate, sulfonated naphthalene condensate or sulfonated melamine or polycarboxylate ethers are commercially available superplasticisers that should be effective for rammed earth materials. Due to the variation of natural soil components from place to place, a test for the compatibility and effectiveness for local rammed earth soil should be individually conducted if an admixture is selected in practice.

A silicone water-repellent admixture has been found to be effective as a superplasticiser apart from being a water-repellent admixture for rammed earth materials. Silicone is used as a release agent and workability or compaction enhancement agent for many uses. It is not surprising that silicone is useful as a superplasticiser for rammed earth material. Feedback from rammed earth builders in Australia, who have used silicone water-repellent admixtures has confirmed that workability and compaction is enhanced by adding the silicone water-repellent admixture. A rammed earth wall containing silicone admixture tends to have higher density and better surface appearance. This superplasticiser effect offered by a silicone admixture may not be as effective as that of a commercially available purpose-designed superplasticiser, but is an added benefit for rammed earth building.

Set accelerators and retarders

Like concrete, rammed earth containing cement sets fast in summer and slowly in winter. Too fast or too slow setting adversely affects the quality of rammed earth construction. Therefore a set accelerator or retarder becomes useful in the above-mentioned cement-stabilised rammed earth under extreme climatic conditions to adjust the setting rate.

There are many commercially available cement set accelerators available on the market. A typical material used for cement acceleration is calcium chloride ($CaCl_2$). Sodium chloride ($NaCl$) may also be effective as a cement accelerator. Calcium chloride can effectively accelerate cement hydration to increase early strength and shorten the setting time for concrete. Theoretically, this set accelerator should be equally effective for the above-mentioned cement-stabilised rammed earth if fast setting is required. However, builders should conduct testing for the set accelerator selected for each project to alleviate negative effects due to the use of a set accelerator. It is also worthwhile to mention that chloride ions present in the set accelerator can potentially promote corrosion activity of steel reinforcement if used in the reinforced rammed earth construction, especially in moist environments.

Cement retarders are not commonly used in the concrete industry nor in the rammed earth industry. However, like accelerators, cement retarders used in the concrete industry should also be effective for cement-stabilised rammed earth. In practice, we have found that many chemical admixtures such as hydrophobic admixtures and superplasticisers have some effect on cement setting (generally a retarding effect). Such a retarding effect due to the chemical admixture may become significant when the temperature is low in winter. Evidence shows that a silicone admixture, for example, slows down cement setting for pressed concrete and cement-stabilised rammed earth, although such an effect may not be significant due to the low dosage rate. It is understandable that crosslinking between silcone molecules and cement and other rammed earth ingredients will adversely affect cement curing (cement crosslinking within rammed earth). In practice, we found that silicone admixtures slighly slow down the cement setting in rammed earth, but the effect is negligible under normal climatic conditions. However, it is worthwhile to maintain maximum curing conditions by keeping the rammed earth in the mould slightly longer in order to achieve sufficient curing, particularly when the atmospheric temperature is low. It is also important to prevent moisture loss from the rammed earth during the curing period, particularly when the atmospheric temperature is high.

10.3   Surface treatment for modern earth buildings

Surface treatment is commonly undertaken and can achieve many desired properties for rammed earth buildings. Surface treatment includes water-repellent sealing, surface rendering, waterproofing or decorative coating and surface binding or dust suppressant sealing.

10.3.1 Vapour-permeable water-repellent sealing

When choosing a sealer to repel water from exterior rammed earth walls, the vapour permeability of the treatment should be considered and the original surface finish of the rammed earth should also be maintained. Sealers made from polyvinyl acetate (PVA), acrylic or latex emulsions, and other polymeric materials are generally not ideal materials for water-resistant treatment of rammed earth. This is because these materials are film-forming materials that will block the vapour permeability and change the surface appearance of rammed earth. Although sealers such as acrylic or latex emulsions are water-resistant and durable materials, protection will not be achieved with these materials. This is because the water resistance relies on a continuous film formed on the rammed earth surface by the sealer. This film is generally a clear thin film that is easily damaged by either physical abrasion or natural deterioration such as UV radiation or harsh weathering resulting in poor durability of the treatment.

A silicone water-repellent is a highly effective material for sealing masonry building materials.[7] Commonly used silicone sealers include siliconate, alkylalkoxysilane and siloxane (see Fig. 10.11). A silicone sealer has a relatively small molecular size so can effectively impregnate into the capillaries of the substrate forming a deep water-repellent layer (or hydrophobic zone) within the substrate rather than forming a continuous thin film over the surface. A silicone is a highly reactive material, which can crosslink with masonry substrates forming a polysiloxane molecular lining within the capillary walls of the masonry substrate via chemical bonding. This polysiloxane molecular lining is highly hydrophobic, which can effectively reduce capillary water penetration into substrates. Figure 10.12 shows a practical hydrophobic zone formed by a silicone sealer in concrete. The treated zone becomes

$$HO\!-\!\underset{\underset{OK}{|}}{\overset{\overset{CH_3}{|}}{Si}}\!-\!OH \qquad RO\!-\!\underset{\underset{OR}{|}}{\overset{\overset{R}{|}}{Si}}\!-\!OR \qquad RO\!-\!\underset{\underset{OR}{|}}{\overset{\overset{R}{|}}{Si}}\!-\!O\!-\!\underset{\underset{OR}{|}}{\overset{\overset{R}{|}}{Si}}\!-\!O\!-\!\underset{\underset{OR}{|}}{\overset{\overset{R}{|}}{Si}}\!-\!OR$$

Potassium methyl siliconate   Alkylalkoxy silane   Alkylalkoxy siloxane or polysiloxane

10.11 Silicone water-repellent sealers for masonry substrate.

10.12 A practical hydrophobic zone formed by a silicone sealer in concrete.

hydrophobic due to the polysilloxane molecular lining from the silicone sealer. Figure 10.13 presents the water absorption data on rammed earth substrates impregnated with a silicone sealer. The figure indicates that the water absorption of the treated substrate is significantly reduced compared to that of the control.

As a hydrophobic molecular lining by a silicone impregnant does not block the capillaries, the water vapour permeability of treated substrates is not changed. Figure 10.14 demonstrates the unaffected vapour transmission

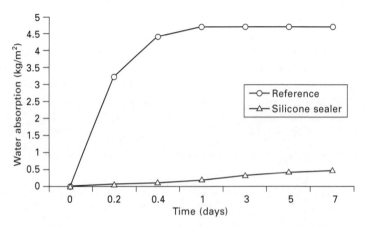

10.13 Water absorption of rammed earth impregnated with a silicone sealer.

10.14 Unaffected vapour transmission of a masonry substrate treated with a silicone sealer demonstrated by blowing air through the substrate.

of a masonry substrate treated with a silicone sealer by blowing air through the substrate. Figure 10.15 reveals the vapour permeability data of a rammed earth substrate treated with a silicone water-repellent sealer. The data indicate that the water vapour permeability of the rammed earth substrate treated with silicone sealer is almost the same as that of the untreated sample.

As a silicone water-repellent impregnant is a non-film forming material, it does not change the surface appearance of the rammed earth substrate.

We know that water penetration is one of the main causes of efflorescence and water-borne staining. Moisture is also a good medium for algae and other micro-organisms to grow on the surface. As discussed in the admixture section, it is understandable that efflorescence caused by water movement within the capillaries of masonry substrates will be significantly reduced if the substrate is treated with a silicone water-repellent sealer. Silicone water-repellent treatment keeps the surface dry and therefore protects the substrate against dirt pick-up, water-borne staining and the growth of micro-organisms.

The durability of silicone water-repellent sealers is excellent for rammed earth buildings. The polysiloxane hydrophobic molecular lining formed by the silicone sealer within the capillaries of the rammed earth substrate is not affected by common factors such as physical abrasion, UV radiation or harsh weathering. Crosslinking via strong chemical bonding between the substrate and polysiloxane within the capillaries makes the silicone sealer part of the substrate, which is hardly affected by normal climatic conditions. Although the surface water-repellent effect (surface beading) may be quickly reduced by UV or physical abrasion, the internal hydrophobic zone within the substrate surface is not affected. This internal hydrophobic zone provides a durable water-repellent function to the rammed earth substrate against water

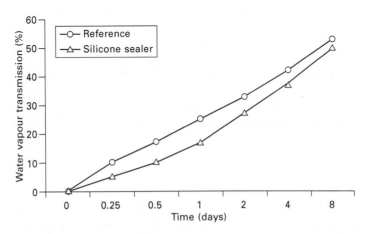

10.15 Water vapour permeability of rammed earth substrate.

penetration. The durability of the treatment with silicone sealer generally lasts over 10 years in practice.

10.3.2 Surface binders or dust suppressants

When sealing internal rammed earth walls, water resistance may not be important but dust binding or dust suppressing becomes the main aim. It is important that the original rammed earth finish should be maintained after the sealing. Water-based materials such as acrylics, latexes, PVA or other polymer emulsions are commonly used as materials for internal dust sealing. The sealer needs to be diluted to a low solid content in order to provide effective dust binding but impart minimal change in the surface appearance and the vapour permeability of the rammed earth building. For treating wet areas such as kitchens and bathrooms, water-resistant materials such as acrylic or latex emulsions should be used. PVA emulsion is not recommended for wet areas because PVA is a water-sensitive material.

The above-mentioned dust-sealing materials generally contain approximately 50% solids as a water-based emulsion. The material is required to be diluted with water at approximately 10–20% solids to form a working emulsion. The dilution rate may vary but should be controlled so that no film forming or a very light film is obtained over the treated surface with no significant change in the surface appearance. Spray application is preferred and two or more applications may be required depending on the permeability of the surface. It is important not to over-apply the sealer to avoid significant change to the surface appearance or the blocking of vapour permeability.

It is not recommended that dust sealing be applied to the exterior surface of rammed earth buildings. If sealing of exterior surfaces is undertaken then it requries a heavy application in order to provide sufficient dust binding against harsh weathering. However, the heavy application of dust sealer not only changes the surface appearance, but also blocks the vapour permeability of the rammed earth. This is because the sealing material such as acrylic or latex emulsion is a strong film-forming material that forms a continuous film over the substrate surface if applied heavily. Water can penetrate through the film via hairline cracks or small surface faults to wet the rammed earth behind the sealer film. Repeated water penetration will weaken the adhesion between the sealer film and the substrate, and water vapour behind the sealer film from the inside of the rammed earth wall will push the surface film off the substrate eventually causing delamination of the sealer. Figure 10.16 indicates the factors that affect the durability of the sealer.

If exterior dust sealing is required, a light dust-binding sealer containing silicone water-repellent may be used. In order to withstand harsh exterior weathering, a silicone water-repellent is highly recommended as an admixture to add into the dust sealer to make a combination of dust binding and water-

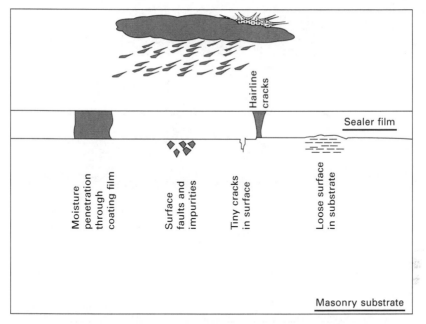

10.16 Factors that affect the durability of a dust sealer.

repellent sealer. Adding from 1–5% silicone water-repellent into an acrylic or a latex emulsion containg a resin solid of 5–10% is highly recommended for exterior dust-binding treatment. Silicone penetrates into the substrate forming a deep hydrophobic zone behind the surface film formed by the dust sealer. This deep hydrophobic zone formed by the silcone water-repellent covers faults, small cracks and loose surfaces in the substrate and acts as a second but much more effective water-repellent barrier. This barrier resists water penetration if the top sealer fails to stop water penetration due to hairline cracks and film faults. The silicone hydrophobic zone also provides a constant dry interface between the acrylic sealer and the substrate surface, resulting in a good adhesion between the dust sealer film and the substrate surface. The water vapour permeability of this combination sealer is not significantly affected due to the light dust sealing used. This negates the risk of film delamination under the harsh weathering conditions for earth buildings. The durability of treatment with this combination of dust binding and water-repellent sealer is significantly improved. Figure 10.17 indicates the advantages of a silicone admixture for an exterior dust sealer.

The data in Table 10.2 clearly indicate the advantage of a silicone water-repellent admixture added to an acrylic sealer for rammed earth substrates. The acrylic sealer contains 10% solid added with 5% silane/siloxane emulsion as water-repellent admixture. The water absorption of the treated rammed earth substrate is significantly improved compared to that of the sealer without

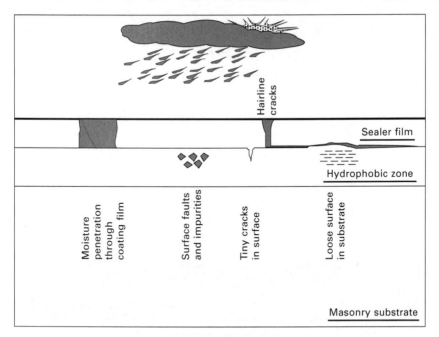

10.17 Advantages of silicone hydrophobic zone behind a dust sealer.

Table 10.2 Performance of an acrylic sealer containing silicone admixture

	Depth of penetration (mm)	Water absorption (% by weight in 24 hours)
Acrylic/silicone sealer	0.5	1.39
Acrylic sealer only	Nil	8.13

silicone water-repellent. The silicone water-repellent also achieves 0.5 mm penetration into the substrate, which is significant compared to that of pure acrylic emulsion sealer. Normal acrylic emulsion generally achieves almost no depth of penetration into rammed earth substrates.

10.3.3 Rendering and coating

Rammed earth wall surfaces are generally left untouched to give the wall the original colour and texture of natural earth. Regular horizontal lines from the wooden or steel formwork used in constructing the wall and subtler horizontal strata from the successive compacted layers of earth enhance the natural beauty of the rammed earth walls.

However, like any masonry wall, a rammed earth wall can be rendered if it is required. Renders such as traditional sand/cement render, polymer modified cement render and pure polymer (acrylic) render are all suitable for rammed earth walls. Render generally has a porous and permeable structure, which does not change the vapour permeability of rammed earth but provides no protection against water penetration under natural climatic conditions. Adding a water-repellent admixture into a render is a simple and cost-effective way to provide a significant protection against natural weathering for rammed earth buldings. Silicone water-repellent admixture not only increases reistance to water penetration, but also imparts resistance against water-based staining to the rendered surface. Figure 10.18 shows the water-repellent and anti-staining effect (against water-based ink) of an acrylic render containg a silicone water-repellent admixture. The render with silicone admixture shows a good water-repellent effect whilst the control shows complete absorption. After washing with water, the render containing silicone admixture shows no remaining ink mark whilst the control shows a permanent ink stain left on the surface.

The following is an example of a simple earth render mix that can be easily produced by rammed earth builders or earth building owners. The advantage of this render is in the use of local soil, which provides the same colour to the earth building. Further, this earth render system is simple and cost effective. This render has been practically used in Melbourne, Australia for the last 10 years and has proven to be a durable system for earth buildings in practice.

1   Dry soil mix (locally available)   10 parts
2   Polymeric emulsion   2 parts
3   Silicone water-repellent admixture   0.4 parts
4   Add water to form consistent slurry as final render mix.

The above render mix is made of locally available dry soil preferably with less than 10% clay content as the main render ingredient. Clean sand may be added into the render mix if the local soil contains a high clay content. A high clay content generally causes cracking after rendering. An acrylic, PVA, SBR latex or other polymeric emulsion is used as a binder which imparts a long-lasting binding and adhesion as well as flexibility to the earth render. Silicone water-repellent admixture imparts water repellency to the entire body of the earth render. The above render is mixed with water to a consistent slurry that can be applied to the rammed earth surface with a brush, roller or airless spray. To further improve the durability of the render, a silicone water-repellent sealer is applied as a top finish sealer to provide significant protection for both render and the rammed earth wall behind the render to form a complete protection system against weathering. Tests have shown that the above rendering system provided over 90% reduction in water absorption while still maintaining approximately over 80% vapour permeability for the earth substrate.

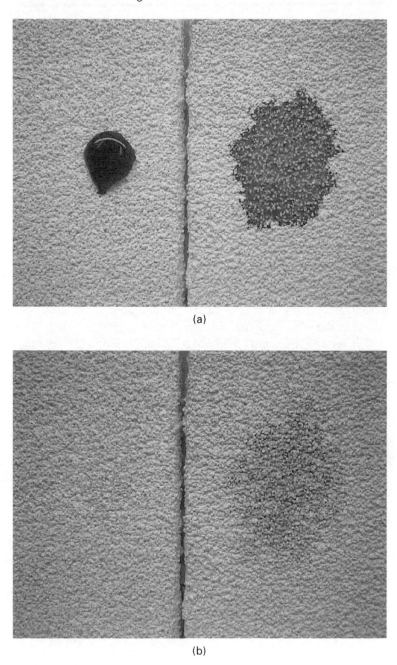

(a)

(b)

10.18 Water-repellent and anti-staining effect (against water-based ink) of an acrylic render containing a silicone water-repellent admixture. (a) Before washing, (b) after washing.

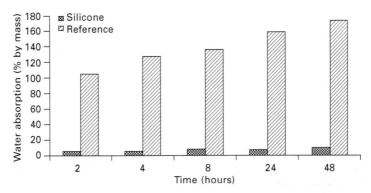

10.19 Water absorption of a highly filled acrylic coating with a silicone admixture.

If the earth wall requires surface coating with a paint, a vapour-permeable coating is highly recommended. Silicone masonry paint is useful for rammed earth buildings. Silicone paint contains a silicone resin as a binder, which not only provides binding for the paint body but also imparts significant water repellency to the paint film. The advantange of silicone masonry paint is that it has good vapour permeability due to its highly filled paint film. The silicone binder provides a good resistance to water penetration to provide significant protection against natural weathering. However, the high cost of silicone paint makes it less viable for rammed earth buildings.

Acrylic coatings with a high resin concentration or low pigment volume concentration (PVC) also have good water resistance and the coating is generally recommended for exterior application. However, such water-resistant acrylic coatings have been proven to exhibit poor vapour permeability due to their high resin concentration within the paint formula. The poor vapour permeability is not an ideal option for coating the exterior walls of rammed earth buildings. Highly filled acrylic coatings (e.g. with a high PVC) offer good vapour permeability but poor water resistance due to their porous and permeable film structure. By adding a small amount of silicone water-repellent admixture, highly filled acrylic coatings can achieve similar properties to those of silicone paints such as good water resistance and high vapour permeability. Figure 10.19 shows the water absorption of a highly filled acrylic coating with a 1% silicone water-repellent admixture. The water absorption results indicate that water absorption of the coating film was significantly reduced by adding the silicone admixture.

10.4   Future trends

Rammed earth buildings have been in existence for thousands of years. Recently the quality of rammed earth construction has been dramatically improved by the breakthrough of incorporating very small amounts of specific silicone water-repellent admixtures into the structure via nanotechnology to achieve vast improvements in durability. In this chapter, we have discussed this admixture and other integral functional admixtures as well as post treatment for rammed earth to impart some important properties to the rammed earth building.

The development in recent years of integral silicone water-repellent admixtures is considered to be a great invention for pressed concrete. Further, silicone admixtures dramatically improve the quality of rammed earth substrates. Other admixtures such as superplasticiser and set accelerators or retarders can also impart important functions to rammed earth, although these admixtures are not practically used for rammed earth. In the future, research should be undertaken into admixture materials that can further improve the quality of rammed earth buildings or that impart special functions to the rammed earth construction, or a combination of admixtures that can impart multiple functions to the rammed earth materials. For example, an admixture may be incorporated not only to provide a water-repellent effect, but also to provide oil-repellent properties to the rammed earth building. This would prevent water- and oil-based staining. Fluorinated chemicals such as fluorinated silane/siloxane or fluorinated acrylic polymers may be used for this purpose although these materials are currently too expensive as admixtures for rammed earth. Other admixtures may be incorporated to improve the thermal insulation properties of the rammed earth. For example, hollow glass beads made from recycled glasses recently developed for cementitious materials should also be useful for rammed earth to increase thermal insulation properies. These hollow glass beads decrease the density of the substrate without affecting the strength. They are also found to be effective in improving the workability of cementious materials. Anti-mould or anti-algae agents used in emulsion paints, for example, should also be effective for rammed earth buildings. The anti-mould function would be important for rammed earth buildings in areas where mould and algae growth is a problem. So there are enormous possibilities within existing or new additives to enhance the durability of rammed earth buildings.

Post treatments such as sealing, coating and rendering are equally important for improving quality or imparting more functions to rammed earth buldings. For example, fluorinated silicones or fluorinated acrylic polymers are already used as anti-staining sealers (oil repellent) for masonry substrates. These materials should also be effective for rammed earth buildings. The oil-repellent (stain resistant) sealer would be ideal for areas like kitchens, bathrooms or public areas where repeated cleaning is required. Repeated cleaning of a rammed earth surface is not practically recommended as the rammed earth surface is not strong enough for repeated harsh cleaning. Nano-acrylic emulsion, for example, is another new technology for paint primers to enhance penetration. This nano-acrylic emulsion would be an ideal material for sealing rammed earth for dust binding for either internal and external surfaces. Due to the small size of the particles, this nano-acrylic would have less impact on vapour permeability and the surface appearance of rammed earth than traditional acrylic or latex emulsions.

10.5   Sources of information

Unless otherwise stated, the information and data used in this chapter come mainly from internal test reports, technical data sheets and product brochures of Tech-Dry Building Protection Systems Pty Ltd. Some of the data have been previously published in professional magazines, conferences and seminars over the last 10 years. All the data and publications are available by contacting Dr Ren Kebao, Tech-Dry Building Protection Systems Pty Ltd at 177 Coventry Street, South Melbourne, Victoria 3205, Australia or visit the website: www.techdry.com.au.

10.6   References

1. Middleton, G. F., *Bulletin 5, Earth-Wall Construction*, CSIRO Division of Building, Construction and Engineering, 1987.
2. Russell, P., *Concrete Admixtures*, Viewpoint Publication, 1983, p. 14
3. Ohama, Y., Demura, K., and Wada, I., 'Inhibiting alkali-aggregate reaction with alkyl alkoxy silanes', The 9th International Conference on Alkali-Aggregate Reaction in Concrete 1992, pp. 750–757
4. Ren, K., 'Investigation of impregnants for low-cost buildings', PhD thesis, Victoria University of Technology, 1995
5. Silfwerbrand, J., *Water-repellent Treatment of Building Materials: Hydrophobe IV*, Proceedings of Fourth International Conference on Water Repellent Treatment of Building Materials, Stockholm, Sweden, Aedificatio Publishers, 2005.
6. Kaesler, K., 'Water-based siloxane emulsions as water repellents for masonry and concrete', Chapter 11, *Coatings for Masonry and Concrete*, conference papers, Brussels, 30 June–1 July, The Paint Research Association, Teddington, 2003
7. De Clercq, H. and Charola, A. E., *Water Repellent Treatment of Building Materials: Hydrophobe V*, Proceedings of Fifth International Conference on Water Repellent Treatment of Building Materials, Brussels, Belgium, Aedificatio Publishers, 2008

11
Weathering and durability of earthen material and structures

J-C. MOREL, University of Lyon, France, Q-B. BUI,
University of Savoie, France and E. HAMARD,
IFSTTAR, France

Abstract: Durability of earthen structures is mainly dependent on the action of water on the walls. In this chapter, the relevance of the current tests measuring the resistance of earthen material to water is analysed and some pertinent tests are suggested to corresponding domains of application. The second part of the chapter deals with the use of plaster to increase durability. The validation of the plaster is proposed thanks to two *in situ* tests: the shrinkage test and the shear test for surface coatings. The long-term performance of earthen material is then analysed through research that has been conducted on walls exposed to natural climatic conditions for several years.

Key words: durability, weathering, erosion, surface coatings, plaster shear test, in situ test, long-term performance.

11.1   Introduction

Durability of building materials can be defined as their resistance to functional deterioration over time. The durability of earthen structures is mainly related to the action of water on the walls. In this chapter, firstly, mechanisms of increasing water content in earthen walls are introduced. Next, we present current tests for assessing the durability of earthen materials in the laboratory consisting of the 'spray test', the 'drip test', the 'wire brush test', the 'saturated to dry strength ratio', the 'slake durability test', the 'rainfall test' and the 'stability in static water test'. The shrinkage test and the shear test for surface coatings are also presented. The results of research conducted on walls exposed to natural climatic conditions for several years are presented, which give an insight into the long-term performance of earthen material. Finally, we discuss the relevance of the current tests and suggest some pertinent tests for corresponding domains of application.

The durability of building materials can be defined as their resistance to functional deterioration over time (Heathcote, 2002). So, during an earthen wall life cycle, many factors can affect its sustainability:

• loss of mechanical strength due to a significant increase of moisture in

282

the wall (capillary rise, pipe damage, infiltration of water during the rains or floods)

- erosion on the surface of walls due to incident rainfall and rain splash at the foot walls
- damage affecting the wall material due to freeze–thaw owing to the presence of water in the wall
- damage to the wall caused by abrasion, insect and plants.

All these risk factors can be minimised with appropriate architectural design or technical solutions. For example, earth stabilisation (see Chapter 9) reduces abrasion, damage caused by insects or plants and erosion due to water. Coating (see Section 11.5) reduces the penetration of water and therefore the harmful effects of freezing.

Durability of earthen structures is mainly related to the action of water on the walls. This action is manifested primarily by increase of water content and erosion of earthen walls. The first problem is mainly due to capillary flow (from ground or surface). The second comes from incident rainfall. They are the two main subjects that will be discussed in this chapter. The points mentioned above will be detailed in the following sections.

11.2   Water content increase in earthen walls

11.2.1 The subsequent decrease in mechanical strength

Clay particles and siliceous particles are hydrophilic in nature (Morton and Buckman, 2008). This is the reason why, at microscopic scale, when earth material contains a large amount of water, mechanical bond strength between clay and grains (sand, gravel, etc.) decreases, leading to a reduction in the internal cohesion of the material. This translates to mechanical strength reduction of the material at macroscopic scale. Figure 11.1 shows an example of variation of compressive strength versus water content of unstabilised rammed earth specimens (9% clay content by dry weight).

For stabilised earth material, increase in water content also leads to a decrease in compressive strength. For cement-stabilised compressed earth blocks (CEBs) in studies by Walker (1995), Krishnaiah and Suryanarayana (2008) and Reddy and Kumar (2011), the ratio between average saturated and dry compressive strengths varied between 0.13 and 0.95, and was largely dependent on clay content (Fig. 11.2). The reason is that when clay plates are inserted between cement particles and aggregates, the bonds between these particles become hydrophilic. Therefore, cohesion decreases when water content increase.

When the ratio of cement to clay and total soil is sufficiently high (which is generally above 10% by weight), cement particles can cover all the clay particles and the granular particles, and the material becomes insensitive to water. These materials can be used even in the case of retaining walls.

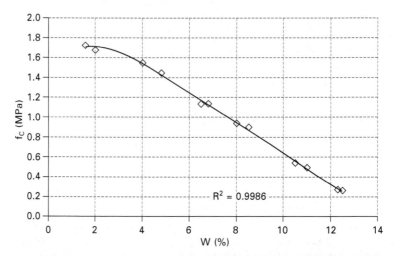

11.1 Decrease in compressive strength with an increase in the water content of unstabilised rammed earth samples, 9% clay content (ENTPE).

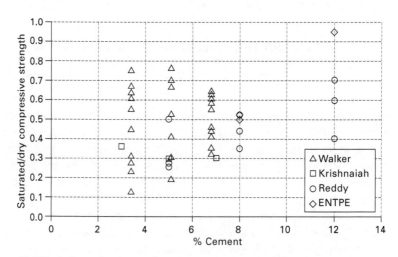

11.2 Variation of saturated to dry strength ratio following cement content.

11.2.2 Increase in freeze–thaw effects

On the one hand, freeze–thaw damage depends on the properties of the material itself, for example material manufactured with a sandy soil has a lower freeze–thaw risk compared to a clayey soil because larger micropores can more easily tolerate the expansion of water and it causes less damage (Minke, 2000). On the other hand, freeze–thaw damage also depends on the moisture content of the material during the freeze–thaw cycles: higher water content causes more harmful expansion in material micropores. The first element (material properties) depends on the choice of soil and construction techniques (rammed earth, adobe, CEB, etc.) which are presented in other chapters of this book. The second element (water content) is related to the ability of water to ingress into the material (by rain or capillarity from the soil).

11.2.3 Increase in shrinkage and efflorescence

An increase in water content tends to result in expansion of clay minerals. The amplitude of the swelling depends on the amount and the types of clays (montmorillonite, illite, smectite, kaolinite, etc.). After drying, clay shrinkage can create shrinkage cracks. These cracks will facilitate water access into the earth material during the next cycles of wetting and drying.

Moreover, soluble salts are often dissolved and redeposited on the wall surface where evaporation is occurring, and the resultant crystalline deposits are referred to as efflorescence (Hall and Djerbib, 2004, Grossein, 2009).

11.2.4 Causes of water content increase

Entry of water into earthen walls can be caused by a number of different mechanisms but is primarily due to wind-driven rainfall and absorption from the surrounding ground. Before water can penetrate a building enclosure, three conditions must exist simultaneously (Killip and Cheetham, 1984): there must be water on the wall, a route for it to travel on and a force to move it.

In earthen walls, the route for water migration is the open microporous network of the material. The force can be capillary suction, wind pressure or differential vapour pressure.

11.2.5 The mechanism of erosion

At the microscopic scale, the factor affecting erosion is the bond strength of particles, which is an intrinsic characteristic of the material. At macroscopic scale, the two predominant factors that determine the magnitude of erosion of the surface of earthen walls are the water content of the wall and the kinetic energy of incidental rain drops. Indeed, as mentioned above, with significant water content, internal cohesion of the material of the wall decreases. The material becomes more susceptible to erosion by rain. This is why, for the same amount of water, a strong but short rainfall is less erosive than prolonged rainfall (Heathcote, 1995). In the latter case, water has more time to penetrate the material, which reduces its resistance to erosion. This effect is more harmful than the increase in kinetic energy of incidental raindrops.

Kinetic energy of incidental raindrops depends firstly on the intensity of the rain – the stronger the force with which raindrops hit the wall, the more it is eroded – and secondly on the angle of the rainfall. The angle at which the raindrops beat the wall surface is determined by the speed of the wind. If there is no wind, raindrops will fall vertically and there will be no erosion. The kinetic energy is maximum for a 90° angle, but erosion will be favoured for a lower impact angle thanks to the digging effect (Fig. 11.3).

In the tropics, another important surface erosion factor of earthen walls is the water flow (Kerali and Thomas, 2004). Indeed, when water flow takes place on the wall surface, at grain scale, it creates a negative pressure, which increases the wall surface erosion, called cavitation erosion (ACI, 1998). This cavitation erosion is the reason why erosion is greater on a rough surface (e.g. adobes) than on a smooth one (e.g. rammed earth) (Heathcote, 1995).

11.3 Strategies to increase the durability of earth walls

Following what is mentioned in the previous section, to increase the durability of earth walls, they must be left at the 'dry' state and therefore it is necessary to prevent excess moisture sources in the walls.

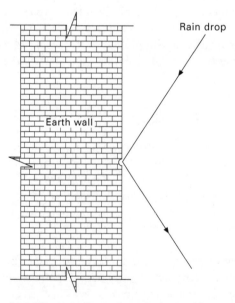

11.3 Digging effect of a raindrop with an angle smaller than 90°.

11.3.1 Cutting the capillary rise

A choice that is systematically applied in new earthen constructions is to add insulation layers to protect against capillary rise from the ground. A recommended option is to use a basement to firstly decrease capillary rise and secondly protect the wall against rain-splash. The basement can be made of concrete, stone masonry, bricks or stabilised earth materials.

A solution that is often used in France is 'canals' filled with gravel surrounding the exterior walls to facilitate the evaporation of water in the soil under the wall, which helps to reduce the capillary action to the wall (Fig. 11.4).

A damp-proofing barrier is often provided along the interface between the footing or plinth and the base of the earth wall. Heavy duty plastics-based damp-proof coursing materials (Walker *et al.*, 2005) or bitumen (Bui *et al.*, 2009a) are commonly used. However, in the case of rammed earth, the damp-proofing material should be capable of withstanding ramming without damage. To limit the possible risk of moisture build-up and damage at the interface between the rammed earth base and the permeable damp-proofing barrier, the barrier could be placed lower down the plinth beneath a course of bricks or similar.

11.3.2 Roof design

Wall erosion is often alleviated by good architectural design that 'cover' the walls as much as possible by roof overhang.

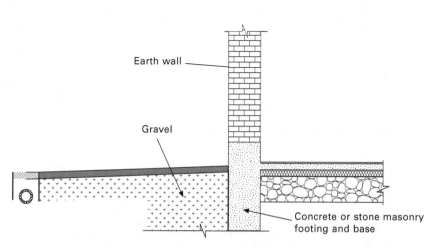

11.4 Example of a damp-proofing barrier.

11.3.3 Stabilisation or coating

There are usually some parts of a wall that are exposed to driving rain because the roof can protect only the upper parts of the wall (Bui *et al.*, 2009a). That is why, in the following sections, we shall present tests to assess the durability of earthen materials (stabilised or unstabilised) with various coatings.

11.4   Current tests for assessing the durability of earthen materials

11.4.1 Spray test

This test is also called the 'accelerated erosion test' (Fig. 11.5). It is often used to test stabilised CEBs, stabilised rammed earth and stabilised adobe.

In this test, the sample is sprayed over a period of 60 minutes or until the sample has completely eroded through. A jet of water projecting at 50 kPa from a standard nozzle is placed 470 mm from the sample. The exposed surface of the sample is a circle 150 mm in diameter. The maximum depth of erosion is measured after one hour of exposure and the sample is checked by eye to determine the extent of the penetration of moisture. Failure is judged to be when the maximum erosion thickness exceeds 60 mm or when moisture has penetrated to the back of the sample.

This test simulates two conditions of the erosion of earth walls due to rainfall: humidification (increasing the moisture content of the material, corresponding to a decrease in internal cohesion) and kinetic energy impinging on earth material (which will break the already weakened bonds of material particles). That is why this test is often used in practice (Walker *et al.*, 2005; Heathcote, 2002). However, these authors also noticed that the conditions of this test are more severe than actual climatic conditions observed onsite.

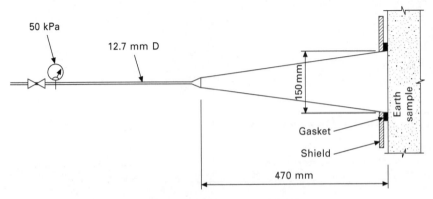

11.5 Spray test, following Walker *et al.* (2005).

It is valid only for stabilised earth materials. Therefore, optimisation of test conditions: water pressure, distance between the jet and the sample, the acceptable thickness of erosion will be required for each climate zone. In addition, the dispersion of results is often very important (Thomson *et al.*, 2008).

11.4.2 Drip test

This test was developed to provide a simple way for builders to determine soil suitability themselves. In this test, 100 mm of water is released via a wet cloth wick, which then falls 400 mm in height onto brick samples inclined at an angle of 27° from the horizontal (Fig. 11.6). This action is meant to simulate rain droplets. Frencham (1982) related the depth of pitting after the test to an *Erodability index* (Table 11.1).

With kinetic energy of droplets and increasing water content in the material during the drip test, this simple test can be acceptable in areas where annual precipitation is around 500 mm. Its applications to areas of higher rainfall are not yet confirmed.

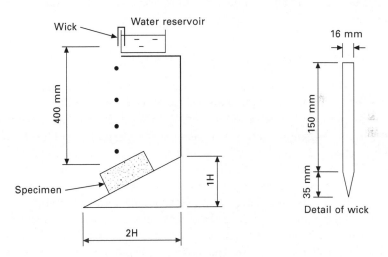

11.6 Drip test, following Heathcote (2002).

Table 11.1 Scale of assessment for 'drip test'

Erodability index (E_i)	Depth of pitting d (mm)	Note
1	0	Non-erosive
2	0 < d < 5	Slightly erosive
3	5 < d < 10	Erosive
4	d > 10	Very erosive

11.4.3 Wire brush test (ASTM D559, 1989)

In this test, earth is compacted in a 100-mm diameter mould to a depth of approximately 125 mm. After 7 days of storage, the samples are oven dried and weighed. They are then placed in water for 5 hours, oven dried and brushed with a firm wire brush (with a constant vertical pressure) to remove any materials loosened during the wetting–drying cycles. After 12 cycles of wetting and drying, they are oven dried and their final mass is recorded. The percentage loss in weight is calculated.

This test is similar to the 'abrasion test' which assesses the susceptibility of earth material to abrasion (Walker *et al.*, 2005). This test was originally developed to test the durability of soil–cement mixtures used in road construction. However, the test has similar conditions to heavy driving rain on the surface of the earth walls: first, there is a moisture content increase in the material and then a 'digging' energy is applied to create erosion on the material surface. This is why several authors have considered the test suitable to test the durability of cement stabilised CEBs (Heathcote, 1995). Fitzmaurice (1958) has proposed the limits of weight loss for CEBs in the case of permanent buildings in urban areas: 5% in regions with annual rainfall greater than 500 mm; 10% in regions with annual rainfall less than 500 mm.

11.4.4 Saturated to dry strength ratio

This test was developed to test stabilised CEBs. The specifications of this test are: a minimum compressive strength of 2 MPa in the 'dry' state and a minimum compressive strength of 1 MPa at the saturated state; plus a requirement that the ratio of saturated to 'dry' strength is not less than 0.5. It takes about 48 h for an earthen specimen soaked in water to be saturated.

This test is part of the family that uses the saturated compressive strength of the material to assess its durability, because if the material has sufficient strength in a saturated state, it will withstand the normal operating conditions of a building without a problem. However, several researchers have mentioned that this test is too severe and non-realistic compared to in situ conditions (Kerali and Thomas, 2004; Heathcote, 1995). According to the results of studies by Walker (1995), Krishnaiah and Suryanarayana (2008), Reddy and Kumar (2011) and our study at ENTPE on compressed earth blocks and stabilised rammed earth, many specimens that were stabilised at less than 4% by weight (by cement in these cases) could not meet the above criteria (Fig. 11.2).

Another criterion that should be used is that the compressive strength at saturated state must be greater than the maximum stress supported by the material.

11.4.5 Slake durability test

This test is usually used to test the durability of soft rocks (clayey rocks and mud-stones) (Fig. 11.7). Kerali and Thomas (2004) have proposed using this test to assess the durability of cement stabilised CEBs.

First, $30 \times 30 \times 30$-mm samples are cut from cement stabilised CEBs. For each test, four to five prismatic samples per drum are placed so that both specimen-to-specimen and specimen-to-mesh contact is occurring. Samples of oven dried prisms are initially weighed (w_i), and rotated at 20 turns per minute for 10 minutes in drums made of 2 mm steel mesh half immersed in tap water at 20°C.

A *slake durability index* is defined as the percentage of mass remaining after drum rotation:

$$Slake\ durability\ index\ (SDI) = w_f/w_i \qquad [11.1]$$

where w_f is the final dry weight.

Following the classification proposed by Franklin and Chandra (1972), samples with SDI above 50% are considered 'satisfactory'.

Following Kerali and Thomas (2004), the slake durability test gives results consistent with observations onsite. However, this test can only be applied to stabilised earth material. In addition, the preparation of $30 \times 30 \times 30$-mm samples by cutting CEBs will certainly change the initial block characteristics.

11.4.6 Rainfall test

This test was proposed by Ogunye and Boussabaine (2002b) to test the durability of stabilised CEBs. The schema of this test is illustrated in Fig. 11.8.

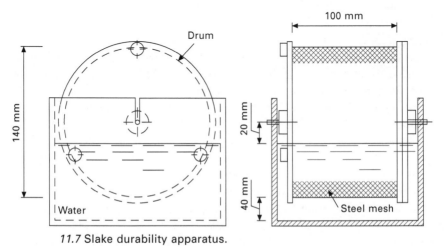

11.7 Slake durability apparatus.

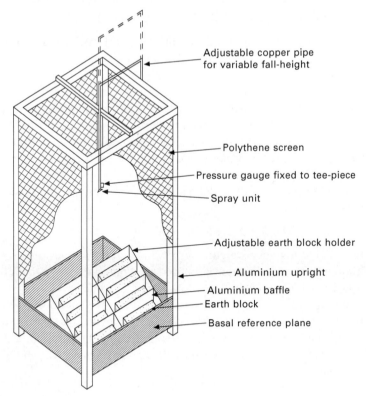

Adjustable copper pipe
for variable fall-height

Polythene screen

Pressure gauge fixed to tee-piece

Spray unit

Adjustable earth block holder

Aluminium upright

Aluminium baffle

Earth block

Basal reference plane

11.8 Schema of rainfall test, proposed by Ogunye and Boussabaine (2002b).

The distance between the nozzle and the basal reference plane is adjustable to permit variation in the fall-height of the drops thereby permitting variation of their energy at impact.

The (0.985 × 0.950 × 0.25 m) basal reference plane is fixed and inclined with a slope of 2.5° to permit free drainage to a 20-mm diameter copper outlet pipe. In this plane, there is an adjustable block holder (platform) on which samples are arranged for the exposure test. This (0.950 × 0.890 m) platform is pivoted and attached to a worm and screw mechanism that allows variable inclination of 15–45° to the spray water direction. This angle is based on the fact that rain always hits the building walls at an angle depending on the horizontal velocity component due to wind. The platform is partitioned into several parts by a 0.25-m height aluminium baffle to intercept soil splash and prevent possible interference with other samples. The samples are placed on a plate raised 3 cm above the level of the platform to avoid the possibility of water eroding the lower part of the sample.

In the study by Ogunye and Boussabaine (2002b), the platform was inclined 30° to the spray water direction. With the aim of simulating rain in Nigeria, the authors left a pressure of 0.5 kg/cm^2 and 2 m fall-height, corresponding to an intensity of 150 mm/h. The samples were exposed for 120 h, according to the average rainfall time in tropical countries. The weight of the lost material was determined after the test.

11.4.7 Stability in static water

Minke (2000) presents a test following the German standard (DIN 18952). A prismatic sample is immersed 5 cm deep into water and the time taken for the submerged part to disintegrate is measured. According to this standard, samples that disintegrated in less than 45 minutes are unsuitable for earth construction.

11.4.8 Freezing and thawing (ASTM D560)

The freezing and thawing test consists of placing a soil sample on an absorbent water-saturated material in a refrigerator at a temperature of –23°C for a period of 24 h and then removing it. The sample is then thawed in a moist environment at a temperature of 21°C for a period of 23 h and then brushed. The freezing–thawing cycles are repeated and then the sample is dried in an oven to obtain a constant weight. This test is considered too severe for earth material (Guettala *et al.*, 2006; Ogunye and Boussabaine, 2002a).

11.5   Surface coatings and finishes of earth structures

For modern stabilised walls, surface coatings are not always necessary, because the stabilised earth material can satisfy the conditions of durability without any coating and also the coating can detract from the appearance of the earth walls, which is aesthetically pleasing. However, in some cases, coating is still used such as the maintenance of existing walls or if occupants want to personalise their home.

The entry of moisture into an earthen wall is caused by rainfall, condensation, infiltration and adsorption from the surrounding ground and from general use of the building. Theoretically, a good coating would ensure full waterproofing and that external water cannot filter into the wall. However this solution is difficult to implement and requires very careful maintenance because any zone of weakness in the impervious layer can result in the concentration of moisture penetration in these regions, which lead to local areas of mechanical weakness in the wall (Bui and Morel , 2007; Hall and Djerbib 2006a).

For earthen walls, paint coatings often do not work well. A thin coating sticks well to earth but, over time, the paint coating develops cracks and peels off due to the differential thermal expansion of the earthen wall and the paint layer. Thus some portions of earthen wall are exposed in patches, facilitating water penetration into the wall. Due to the cracks in paint, the paint protection layer can be washed away by rain, which can lead to damage. Also, damage can occur due to accumulated water behind the paint layer. In a study by Bui and Morel (2007), the surface quality of the walls protected by paint was worse than the walls without any coating, which shows that paint protection is not appropriate for earthen media.

In general, for earthen walls, plaster is a better option than paint coating. With a permeable coating, water can penetrate through the coating but it stops after a few centimetres from the surface, because a saturated region is formed that provides a barrier to further water penetration (Hall and Djerbib, 2005a). This water is then evaporated in warmer or drier periods.

Plaster is composed of a granular skeleton, a binder and eventually of admixtures that improve its properties. The granular skeleton role is played by sands and silts, which are naturally present in earth, but generally in small amounts. The binder role can be played by clay or lime. Cement, is not recommended as a binder for plasters on earthen walls because a cement coating is usually impermeable to water vapour, which prevents the wall from 'breathing'. There are three main types of coatings for earthen walls: earthen plasters, lime plasters and gypsum plasters.

The role played by a coating is different depending on whether it is an interior or an exterior coating. An exterior coating has to protect the wall from weathering and impacts. An interior coating must contribute to the thermal, acoustic and aesthetic comfort of the room. Since exterior coatings are exposed to weathering, the exterior coating should be done using lime plasters. Interior coatings are not exposed to weathering, so can be done using lime, earth or gypsum plasters.

11.5.1 Implementation of coatings

Plasters are usually implemented in three layers: a scratch coat, a brown coat and a finish coat. The scratch coat is the bounding layer of the coating on the wall. Its binder/sand ratio is high and it must be wet enough to ensure a good migration of binder towards the wall. It strengthens the wall and the sand brings roughness to its surface. The brown coat corrects irregularities in the wall surface. The finish coat, thinner and less rich in binder than the brown coat, brings an aesthetic finish. The stabilisation of earth plasters can take place in all layers or only in the finish coat.

Unlike standardised building materials, earthen wall and plaster materials show considerable variability. Depending on the constructive mode and the origin of the materials of the wall, a plaster formulation may or may not be suitable. For each site, plaster formulation has to be adapted to the wall. Thus plaster formulation validation can only be done by on site tests in conditions as close to as possible to those of real work. The different earth/sand or lime/sand batching, possibly supplemented with admixtures, are tested for their shrinkage and shear behaviour on the wall to be plastered. A series of two tests is proposed to validate coating formulations on given earthen walls (Hamard *et al.*, 2013).

11.5.2 Shrinkage test

The shrinkage test of plasters on earthen walls consists of applying a 250 × 250-mm sample of brown coat of each formulation to be tested. After drying of the samples, when shrinkage is completed, it is possible to note the presence or absence of shrinkage cracks. Test results remains valid only if plaster is applied on a support in a state (brushing and humidification) similar to that of the test conditions.

Test schedule

- Scrape and wet the wall as for the plaster
- Implement the scratch coat as for the plaster
- Apply one sample (or more) of 250 × 250 mm for each formulation to test, using the number of layers and the thickness required for the final plastercoat,
  - Let the samples dry,
  - When samples are dry and shrinkage finished, examine plaster.

Validation of the formulations

To validate a formulation, the sample should not be cracked or bowed out. Crazing is acceptable. The validated formulations must undergo the shear test of earth plasters on earthen walls before being implemented.

11.5.3 Shear test

For earthen walls that consist of layers of earth (rammed earth, cob), elements of masonry (adobe, CEB) or of the filling of a timber-framed wall (wattle and daub), their heterogeneities must be taken into account in the testing. Therefore it is advisable to apply several samples testing different parts of the wall.

The purpose of the shear test for plasters on earthen walls is to ensure a sufficient bond between the plaster and the wall. The test consists of applying five samples of each plaster formulation to five different parts of the wall. Once dry, samples are submitted to a load of 2 kg. This ensures a safety coefficient greater than 10 since the thickest plasters are about 60 mm with a mass of 0.2 kg. This test must be carried out under conditions replicating those of real onsite implementation.

Preparation of samples

- Scrape and wet the wall as for the plaster
- Implement the scratch coat as for the plaster
- Apply, as for the plaster, five samples (or more) of 40 mm height, 50 mm width and 20 mm thickness. If the support is composed of different elements (mud blocks, adobe, layers of earth, etc.), at least two samples must be on an interface and at least two samples must be in the middle of an element, each sample testing a different element or a different interface,
- When samples are dry, it is possible to add a fine mortar layer to allow a good backing of the load device.

Test schedule

- First, place the load device on the upper part of the sample and be careful to minimise friction between the load device and the wall. Then, deposit the load of 2 kg on the load device and start the stopwatch (Fig. 11.9).
- If the sample does not fail after 30 s, we consider that it resists the load.

Validation of the formulations

- If the five samples resist the load, then the formulation is validated. If one or more of the five samples does not resist the load, then the formulation cannot be validated.

The shrinkage test eliminates formulations beyond a shrinkage threshold specific to the coating–wall combination, and the shear test eliminates formulations not offering sufficient resistance to traction and, therefore, sufficient shear strength, of coating–wall interface.

11.5.4 Wall preparation and weather conditions

Before coating an earthen wall it is advisable to let a full seasonal cycle pass, i.e. a year, to enable the wall to dry. For the same reason it is advisable to let a full seasonal cycle pass after the application of the brown coat to apply the finish coat.

Earthen walls must be properly prepared to receive a coating. They must

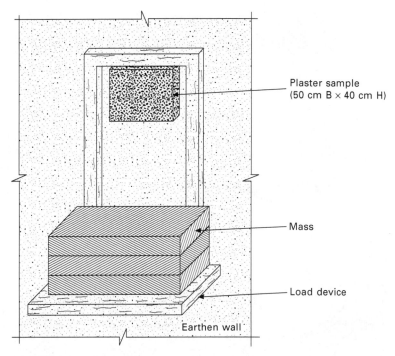

Plaster sample
(50 cm B × 40 cm H)

Mass

Load device

Earthen wall

11.9 Example of shear test.

be cleared of all non-consolidated elements to ensure a solid bond between the wall and the coating. They must be wetted to create a pool of water in the wall allowing a sufficient working time and a correct curing for the coating. Periods of heat and/or dry wind prevent proper wetting of the wall, they are therefore prohibited. During periods of freezing, the water released from the coating into the wall can cause damage. These periods are also not recommended.

11.6   Long-term performance testing of earth walls

11.6.1 Study by Guettala *et al.*

To assess the suitability of laboratory tests for the earth material durability, Guettala *et al.* (2006) have tested in two different ways on similar samples: laboratory tests and field tests. For the laboratory tests, *Saturated to dry strength test*, *Spray test*, *Capillary absorption test* and *Freezing–thawing test* were carried out on four CEB types with different compositions. For the field tests eight walls were built with the same CEBs as for the laboratory tests. These walls were exposed to real climatic conditions for 48 months (Fig. 11.10). Annual precipitation on the site was about 120 mm. Although

11.10 Walls exposed to real climatic conditions in the study by Guettala *et al.* (2006).

11.11 General view from the south of the in situ walls in study by Bui *et al.* (2009a).

quantitative measures on walls were still lacking, this study showed that all the laboratory tests mentioned above were too severe compared to what happens onsite for a dry climate.

11.6.2 Study by Bui *et al.*

Bui *et al.* (2009) presented a study on the durability of different types of stabilised and unstabilised rammed earth walls. These rammed earth walls were constructed and exposed to natural weathering for 20 years, in a wet continental climate (Fig. 11.11). The erosion of the rammed earth walls was measured using stereo-photogrammetry. The result showed that the mean erosion depth of the studied walls was about 2 mm (0.5% wall thickness) in the case of a rammed earth wall stabilised with 5% by dry weight of hydraulic lime, and about 6.4 mm (1.6% wall thickness) in the case of unstabilised rammed earth walls.

In general, the erosion of a rammed earth wall is not a linear function of time. Initially, the wall shows more erosion on the surface, and over time the erosion stabilises (Fig. 11.12 non-linear erosion). This non-linearity is due to the loss of compaction energy caused by friction on the formwork – the earth which is in contact with the formwork is less compacted and therefore more eroded. However, it is currently impossible to estimate exactly the lifetime of these walls using nonlinear functions. That is why two linear functions of time were used to assess the durability of unstabilised rammed earth (URE) and stabilised rammed earth (SRE) walls (Fig. 11.12). For URE walls, 63 years should be required before these walls are eroded to 5% thickness. For SRE walls, the time should be 204 years. The non-linear erosion curves in this figure are only examples of the possible non-linear behaviour of erosion that is often observed empirically in reality, but there are still no scientific data giving an exact function. However, because erosion with time is not linear, the lifetime of these walls may be much longer than 63 and 204 years, respectively, for unstabilised and stabilised rammed earth walls. This shows again that laboratory tests are generally too tough for a moderate climate.

11.7   Future trends and conclusions

For modern earthen constructions, the walls are often stabilised and are not coated, because the coating can detract from the appearance of the earth walls, which is aesthetically interesting. So, the stabilised earthen material itself must meet the durability requirements.

Spray tests are interesting because of their simplicity. However, this test simulates 'extreme' conditions and therefore gives harsh conclusions about

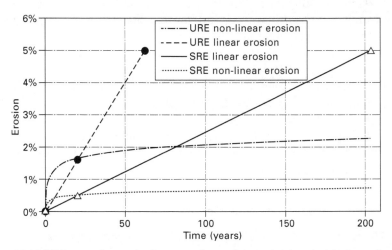

11.12 Difference between linear and non-linear erosions with time of unstabilised and stabilised rammed earths (Bui *et al.*, 2009).

the durability of earth material tested. Calibration of this test is required for each climatic zone.

The 'drip test' is suitable for areas where precipitation is moderate, while the 'wire brush test', the 'slake durability test' and the 'freeze–thaw test' are not particularly suitable to assess the durability of earth material in areas with moderate precipitation. They may be better suited to the 'extreme' climatic zones, for example areas there is the current risk of flooding.

The 'rainfall test' presents several advantages due to its complexity but this point is also a disadvantage for applying the test in practice. Also, the test needs to be calibrated for each climatic zone.

The 'saturated to dry strength test' is not a good indicator for unstabilised materials or materials stabilised at low dose (< 2% by weight). However it is simple and can be used for stabilised earth materials that suffer severe conditions such as high precipitation, flooding, retaining walls. For unstabilised materials or materials stabilised at low dose, the 'stability in static water test' can be applied.

In situ long-term testing gives valuable data because all factors depending on climate and time are combined. However, these tests are very expensive and very long. In addition, data are only valid for the climatic zone studied; it is difficult to 'extrapolate' for other climatic zones.

In the near future, pressure on sustainability of building will increase and, for example, embodied energy of construction will be taken in account rigorously in regulations as has happened recently with thermal building behaviour. In this context, there are three directions of development with regard to sustainability of earthen constructions.

Development of architectural design able to build with earth dug onsite without any admixtures (unstabilised), such as thick bearing walls or non-bearing walls. This would allow the use and reuse of earthen materials without causing pollution.

The second trend is the use of coatings to protect walls. This can provide acoustic and thermal qualities, and protection against moisture, while minimising the amount of high embodied energy material used. The use of coatings will be increased by establishing design guidance to implement mortars with local or able materials.

Finally, the use of low embodied energy stabilisers in walls or in coatings; for example, of vegetable origins (fibres or biopolymers), may be interesting because it may increase the durability of some earthen materials.

11.8   Acknowledgements

The authors wish to thank Ali Mesbah and Myriam Olivier from ENTPE, for all the research done since 1982 on engineering assessment of earthen architecture. The authors also wish to thank the professional, in earth construction at 'Réseau Ecobâtir', in particular Alain Marcom and Nicolas Meunier, for sharing their valuable know-how.

11.9   Sources of further information

Bui Q B, Morel J C (2009). 'Assessing the anisotropy of rammed earth', *Construction and Building Materials*, 23: 3005–3011.

Bui Q B, Hans S, Morel J C, Meunier N (2009b). 'Compression behaviour of nonindustrial materials in civil engineering by three scale experiments: the case of rammed earth', *Materials and Structures,* 42, 8: 1101–1116.

Hakimi A, Ouissi H, El Korbi M, Yamani N (1998). 'Un test d'humidification-séchage pour les blocs de terre comprimée et stabilisée au ciment', *Materials and Structures*, 31, 20–26.

Gelard D (2005) 'Identification et caractérisation de la cohésion interne du matériau terre dans ses conditions naturelles de conservation'. PhD thesis, Institut National Polytechnique de Grenoble.

Guelberth S R, Chiras D (2003) 'The Natural Plaster Book – Earth, Lime and Gypsum Plasters for Natural Homes', New Society Publishers

Hall M, Djerbib Y (2006b) 'Moisture ingress in rammed earth: Part 3 – sorptivity, surface receptiveness and surface inflow velocity', *Construction and Building Materials*, 20, 6: 384–395.

Jaquin PA, Augarde CE, Gallipoli D, Toll DG (2009) 'The strength of rammed earth materials', *Géotechnique*, 59, 5: 487–490.

Keefe L (2005) *Earth Building – Methods and Materials, Repair and Conservation*, Taylor and Francis, 197 pp.

Kenmogné E (1993) 'Dégradation des matériaux de construction: contribution à l'étude de la faisabilité des terres argileuses en construction', PhD thesis, Université Joseph Fourier.

McHenry PG (1984) *Adobe and Rammed Earth Buildings – Design and Construction*, The University of Arizona Press, 217pp.

Morel J C, Mesbah A, Oggero M, Walker P (2001) 'Building houses with local materials: means to drastically reduce the environmental impact of construction', *Building and Environment*, 36, 1119–1126.

Olivier, M (1994) 'Le matériau terre, compactage, comportement, application aux structures en blocs de terre', PhD thesis, INSA de Lyon.

Rubaub M, Chevalier B (1986) 'Etude de différents systèmes de protection superficielle du matériau 'terre' utilisé pour la construction – Expérimentation en vieillissement naturel sur murets au CSTB', for Centre Scientifique et Technique du Bâtiment.

Standards New Zealand (1998) *NZS 4298:1998, Materials and Workmanship for Earth Buildings*, Standards New Zealand, Wellington.

Taylor P, Luther MB (2004) 'Evaluating rammed earth walls: a case study', *Solar Energy*, 76, 79–84.

11.10   References

ACI 210R-93 (1998) *'Erosion of Concrete in Hydraulic Structures'*, 24pp.

ASTM D559 – 03 (1989) 'Standard test methods for wetting and drying compacted soil-cement mixtures',

ASTM D560 (2003) 'Standard test methods for freezing and thawing compacted soil-cement mixtures',

Bui QB, Morel JC (2007) 'Durability of rammed earth walls exposed for 20 years of natural weathering', International Symposium on Earthen Structures, Bangalore, August 2007 pp. 113–120.

Bui QB, Morel JC, Reddy BVV, Ghayad W (2009a) 'Durability of rammed earth walls exposed for 20 years to natural weathering', *Building and Environment*, 44, 912–919.

Fitzmaurice, R (1958) *Manual on Stabilised Soil Construction for Housing*. Technical Assistance Program, United Nations, New York.

Franklin JA Chandra R (1972) 'The slake durability test', *International Journal of Rock Mechanics and Mining Sciences*, 9, 325–341.

Frencham, GJ (1982) 'The performance of earth buildings', Deakin University, Geelong.

Grossein O (2009) 'Modélisation et simulation numérique des transferts couples d'eau, de chaleur et de solutes dans le patrimoine architectural en terre, en relation avec sa dégradation', PhD thesis, Université Joseph Fourrier, 218pp.

Guettala A, Abibsi A, Houari H (2006) 'Durability study of stabilized earth concrete under both laboratory and climatic conditions exposure', *Construction and Building Materials*, 20, 119–127.

Hall M, Djerbib Y (2004) 'Moisture ingress in rammed earth: Part 1 – the effect of soil particle-size distribution on the rate of capillary suction', *Construction and Building Materials* 18: 269–280.

Hall M, Djerbib Y (2006a) 'Moisture ingress in rammed earth: Part 2 – the effect of soil particle-size distribution on the absorption of static pressure-driven water', *Construction and Building Materials*, 20, 6: 374–383.

Hamard E, More JC, Salgado F, Marcom A, Meunier N (2013) 'A procedure to assess suitability of plaster to protect vernacular earthern architecture,' to be published in *Journal of Cultural Heritage*.

Heathcote KA (1995) 'Durability of earthwall buildings', *Construction and Building Materials*, 9, 3, 185–189.

Heathcote KA (2002) 'An investigation into the erodibility of earth wall units', Doctor of philosophy thesis, University of Sydney.

Kerali AG, Thomas TH (2004) 'Simple durability test for cement stabilized blocks', *Building Research and Information*, 32, 2, 140–145.

Killip IR, Cheetham DW (1984) 'The prevention of rain penetration through external walls and joints by means of pressure equalization', *Building and Environment*, 19, 81–91

Krishnaiah S, Suryanarayana RP (2008) 'Effect of clay on soil cement blocks', 12th IACMAG, Goa, India, pp. 4362–4368.

Minke G (2000) *Earth Construction Handbook – The Building Material Earth in Modern Architecture*, WITpress, 206pp.

Morton T, Buckman J (2008) 'Traditional cob wall: response to flooding', *Structural Survey*, 4, 302–321.

Ogunye FO, Boussabaine H (2002a) 'Diagnosis of assessment methods for weatherability of stabilised compressed soil blocks', *Construction and Building Materials*, 16, 163–172.

Ogunye FO, Boussabaine H (2002b) 'Development of a rainfall test rig as an aid in soil block weathering assessment', *Construction and Building Materials*, 16, 173–180.

Reddy BVV, Kumar PP (2011) 'Cement stabilized rammed earth – Part B: compressive strength and elastic properties', *Materials and Structures*, 44, 3, 695–707.

Thomson A, Pope D, Walker P (2008) 'Erosion characteristics of rammed earth', Kerpic08 – Learning from Earthen Architecture in Climate Change International Conference, Northern Cyrus.

Walker PJ (1995) 'Strength, durability and shrinkage characteristics of cement stabilised soil blocks', *Cement and Concrete Composites*, 17 301–310.

Walker P, Keable R, Martin J, Maniatidis V (2005) *Rammed Earth : Design and Construction Guidelines*, BRE Bookshop.

Part III

Earth building technologies and earth construction techniques

12

History of earth building techniques

P. JAQUIN, Integral Engineering, UK

Abstract: Building using subsoil is one of the oldest construction techniques, providing simple shelter using freely available material. Buildings made from soil are found in many parts of the world, in different forms, sometimes mixed with other traditional construction materials such as timber or stone, or with more modern inventions such as cement and steel. Moist soil is formed either as a monolithic wall that is then allowed to dry, or into independent units, such as bricks or blocks, which are allowed to dry before being placed as a wall. This chapter describes earth buildings in various parts of the world, and discusses how we are able to trace the development of earth building techniques through excavation and standing archaeology, current vernacular techniques and reviews of standing monumental architecture.

Key words: history, rammed earth, cob, adobe, international.

12.1   Introduction

Building using subsoil is one of the oldest construction techniques, providing simple shelter using freely available material. Buildings made from soil are found in many parts of the world, in different forms, sometimes mixed with other traditional construction materials such as timber or stone, or with more modern inventions such as cement and steel. Moist soil is formed either as a monolithic wall, which is then allowed to dry, or into independent units, such as bricks or blocks, which are allowed to dry before being placed as a wall. This chapter describes earth buildings in various parts of the world, and discusses how we are able to trace the development of earth building techniques through excavation and standing archaeology, current vernacular techniques and reviews of standing monumental architecture.

The poor resilience of earth compared to other building materials such as stone means that monumental archaeological sites may have been completely destroyed or remain undiscovered, and that it is difficult to infer the development of techniques. However, the ubiquity of earth as a vernacular construction material means that many ancient sites built using earth are being studied, and in time this will allow a much better understanding of the techniques used in antiquity.

It is probable that earth building techniques developed independently in different parts of the world, spreading with the movement of peoples. Early

307

man was constantly moving, following hunting and gathering patterns dictated by the surroundings. The earliest shelters utilised natural features such as caves, and the first earth buildings might have been extensions to natural features such as mounds of earth at cave entrances or pits dug into the ground. Agriculture first developed in fertile river valleys, and here the silt and clays provide excellent building materials for earth construction. The first technique to develop was likely to have been wattle and daub; construction of a façade or roof using timber or grasses, which is then covered in earth. A rammed earth type technique may later have developed, with earth placed against or between walls made from timber, and compacted into place, forming a thicker wall. The development of unit construction could have developed later, initially units were formed by hand and later more cuboid blocks made using formwork. When dry, these could be transported, allowing production of the materials to be separated from the location of the building, meaning for example that suitable earth could be taken from a river valley liable to flooding and used for construction of buildings at a higher level.

Agriculture and earth construction developed independently in the main cradles of civilisation. The development of agriculture beside major rivers led to people gathering together for the first time in towns. These fertile river valley civilisations provided the right types of soil for earth construction, and there is evidence for development of earth building independently in the valleys of the Tigris and Euphrates, Nile, Indus, Jordan, Murghab and Yellow Rivers. These cultures remained independent from each other, yet appear to have developed very similar earth building techniques. As civilisation and trade developed techniques were refined and updated.This is difficult to chart because earth building techniques can vary settlement to settlement and year to year. However, some patterns do emerge. It would appear that the transition from hand-moulded to cuboid bricks occurred in Mesopotamia around 5000 BC, and that rammed earth was not found in South America prior to its introduction by Europeans.

Earth is generally used in combination with other building materials when these are available, for instance wattle and daub houses combine earth construction with timber and are found in Japan and northern Europe. Where stone is readily available, earth is used as a plaster or a mortar, such as in Malton in the north of England, or turf used for roof or wall construction, such as in the Western Isles of Scotland. There are also instances of monumental architecture in one construction type, and vernacular construction in another, such as the stone cathedrals of England, contrasting with the surrounding wattle and daub vernacular houses.

In this chapter we chart the development of earth building, focusing on particular techniques, through specific examples from around the world. For brevity, many historic earthen architectural sites have been excluded.

12.2   Earth building techniques in Asia

12.2.1 China

Settled civilisation in China first developed around 2300 BC when nomadic peoples settled on the alluvial plans of the Yellow River, beginning the Lungshan civilisation. The soft soils here could be cut to form pit houses and heaped to form rammed earth type mound walls. This allowed the development of defensive settlements such as found at Lianyungang, Jiangsu, Taosi, Erlitou and Longwan. Evidence of formwork boards and ramming implements have been found at Pingliantai.

During the Warring States Period (475–221 BC) rammed earth was used for the construction of more elaborate rammed earth walls at larger settlements such as Langya, Anyang, Linzi and Xiadu. The Qin dynasty (221–206 BC) was the first to construct rammed earth defensive walls along their northern frontiers in western China. These walls were repaired and extended by the Han (206 BC–202 AD) and Jin (265 AD–420 AD) dynasties. The Tang dynasty (618–907) expanded Chinese borders and trade, but was harassed by tribes to the north, and as a result built fortified settlements in north western China along the eastern part of the Silk Route. These settlements, such as Jiahoe, Gaochang and Xi'an, are each encircled with large rammed earth walls, and the city of Kashgar in western China is built in adobe, while the fortress of Baishui, at the western end of the Great Wall is constructed wholly from rammed earth.

The Tang dynasty collapsed around 907 AD, which led to a period of major upheaval in China. The next major dynasty to produce major monumental earth architecture were the Ming (1368–1644) who pursued a policy of aggressive expansionism. The walls of the Ming capital Xi'an, originally rammed earth, were faced in stone, while along the Silk Road and to the northern borders the Great Wall was repaired and upgraded, and new forts at Jiayuguan and Hexibao were constructed in adobe.

In the Fuijan province of central China, the round houses of the Hakka people have recently been given World Heritage Site status. These large rammed earth buildings, called *Tulou* (literally earth structures) are defensive homes to many families and can be up to 60 m across and four storeys high. The oldest of these buildings was built in 1308 and their construction continued well into the 20th century.

Rammed earth and adobe are found on the Tibetan plateau and in parts of the Himalayas as both monumental and vernacular building techniques. At the west end of the Himalayas, rammed earth is found in the north Indian state of Ladakh, such as palaces at Shey and Leh and a fort at Basgo. In the Nepali kingdom of Mustang, much of the capital city of Lo Manthang is constructed from rammed earth, and a defensive wall, dating from 1380 surrounds the city. The country of Bhutan, at the east end of the Himalayas continues to promote traditional building materials with many homes and monumental architecture built in rammed earth.

12.2.2 Central Asia and the Indus valley

Settled civilisation developed in the Indus valley around 7000 BC. The small adobe settlement of Mehrgarh in modern Pakistan was a forerunner of the much larger Indus Valley civilisation, which developed around 3000 BC.

12.1 Rammed earth section of Kyichu Lhakhang Monastery, Bhutan.

12.2 Rammed earth fort at Basgo, India.

The civilisation spread along the Indus River, with two large settlements of Harrapa and Mohendjaro emerging around 2600 BC. Both settlements were laid out in a grid pattern, with adobe houses and individual streets.

Settled agriculture and the first buildings in Central Asia are related to the Bactria-Margiana archaeological complex, comprising around 300 discrete fortified adobe brick enclosures at sites such as Namazga-, Altyn- and Gonur-Depe, which have been dated to between 2200 and 1700 BC. Though the peoples of Central Asia were largely nomadic, in western Uzbekistan there are a number of forts or *Qalas* built in adobe and thought to date from around 300 BC. Few of these settlements remain to the present because of shifting trading patterns, passing armies and moving rivers. Settlements that did survive grew to become major trading centres such as Balkh and Merv.

Balkh (Bactria) in modern Afghanistan is called *Umm Al-Belaad* (Mother of Cities) because of its antiquity. Though the city dates from 2000 BC, the oldest standing structures are the rammed earth Takht-e Rustam and Top-Rustam attributed to the Buddhist or Zoroastrian religions. Balkh became a pre-eminent city in the region and a centre of trade and commerce. When Muslim traveller IbnHawqal visited the city around 950 AD he described it as 'built of clay with ramparts and six gates'.

The city of Merv in Turkmenistan is relatively unique amongst archaeological sites, with a number of different settlements being constructed adjacent to, rather than on top of each other, allowing archaeologists to uncover earlier structures without destruction of those built later. Almost all of the structures in Merv are built in earth, with the earliest settlements dated to around 2000 BC. The city was almost continually inhabited until its abandonment and destruction in 1787 AD and is now a major archaeological site.

Both Merv and Balkh lay on important trade routes that crossed Central Asia, and many other earthen settlements grew up on what has become known as the Silk Route. The adobe city of Panjakent in western Tajikistan is first mentioned around 500 BC, and was probably the highlight of the Silk Road before its decline in the 8th century. The site has been extensively excavated since the 1940s and is now a tourist attraction. Further east, the Uyghur empire capital and Silk Route city of Ordu-Baliqin modern Mongolia featured rammed earth defensive walls and buildings. The city was established in 745 AD, but abandoned in 840 AD. The armies of Alexander the Great around 330 BC, the Muslims around 720 AD and Genghis Khan in 1220 each ransacked or destroyed many earthen settlements, and as a result many of the sites in Central Asia are mere shadows of their former selves.

12.2.3 Middle East

The Euphrates and Tigris river valleys were home to nomadic civilisations that first developed settled agriculture and buildings around 9000 BC. These civilisations used hand-moulded oval bricks to form circular structures, found at the sites of Djade al-Mughara in Syria and Tappeh Ozbaki and Ganj Durrell in Iran. These oval bricks appear to have been used until around 6000 BC, at sites such as Jericho and Netiv Hagdud. From around 6000 BC onwards, square bricks are found at the Tell Hassuna site in Iraq, and at Jericho, the buildings change from being circular on plan to rectangular.

The settlement of Çatalhöyük developed independently on the alluvial plains of the Çarşambariver in central Turkey. This settlement is still being excavated but may have been the largest settlement in the world at the time, with 5000 inhabitants at its peak between 7300 and 6800 BC. The city was built from adobe with irregular plan buildings packed so tightly that access was via the roofs.

Settlements along the Euphrates and Tigris rivers grew in size and complexity, by 3500 BC the city of Uruk was the largest in the region, with rammed earth buildings and adobe temples. Closer to the Arabian Gulf, the Assyrian cities of Elba and Mari vied for influence, and excavations at these sites show each site had earthen city walls and adobe palaces. Technology developed such that when the Ziggurat of Ur was constructed around 2100 BC it was built with an adobe brick core and faced with fired

12.3 Kasbah in Asslim, Draa valley, Morocco.

12.4 El Badi Palace, Marakesh, Morocco.

brick set into bitumen. Such Ziggurats may have earlier been constructed in adobe and have not withstood the ravages of time, or remain unidentified to the present day.

Many settlements in this region feature a core of earthen buildings, which have been renewed and rebuilt over the centuries. In Iran the cities of Yzadand Isfahan contain a large number of historic adobe buildings. The city of Tousis is surrounded by rammed earth walls and the citadel of Bamdates from around the 7th century AD and was probably the largest adobe building in the world before its collapse in an earthquake in 2003. In Yemen, the city of Shibam is renowned for its particularly tall adobe buildings. From around 1700, the residents began to build up, and currently around 500 adobe 'skyscrapers' reach up to 30 m high. Close by is the town of Tarim, home to the Muhdhar Mosque. The adobe minaret of this mosque, completed in 1914, is probably the tallest earthen structure in the world at 53m.

12.3   Earth building techniques in Africa

Although Africa is known as the cradle of mankind, archaeology has not yet revealed a great history of earth building in Africa. The African mud hut, constructed from woven reeds or timber with an earth plaster may have remained unchanged for millennia. The earliest woven reed and branch earth-covered sites have been dated to 5000 BC at sites in the Nile Delta such as Mermid and Fayum. The Egyptian dynasties appear in the Nile valley around 2900 BC and the clay river silt mixed with desert sand and straw from cultivated grains allowed hand-shaped adobe brick manufacture. The large independent adobe structures at Shumet el-Zebib and Nekhen dating to 2750 BC show adobe was used as a monumental construction technique before the more well-known stone edifices were built.

Adobe continued to be used as a vernacular construction material in Egypt. The settlement of Deir el Medina (1550–1080 BC), which was home to the masons of the Valley of the Kings comprises square, single-room adobe houses laid out in a grid pattern. The city of Tel el-Amarna was a new capital city built by the Pharaoh Akhenaten around 1353 BC but abandoned soon afterwards. This city features single-storey rectangular adobe buildings with external stairs leading to a flat roof. Rameses II (1279–1213 BC) embarked on many building projects, and adobe bricks from his major construction projects were stamped with his seal. Egypt however remained isolated from the rest of Africa, and as a result had little influence on the building techniques found throughout the rest of the continent.

In North Africa, the Phoenician civilisation spread from the eastern Mediterranean founding settlements along the north coast. Their capital at Carthage (in modern Tunisia) was founded in 814 BC, and excavation reveals rammed earth walls used in homes there. The famous Carthaginian general Hannibal crossed into Europe in 218 BC, and the Roman author Pliny the Elder describes the rammed earth towers in Africa attributed to Hannibal. Around 700 AD Islam spread through North Africa, and the valleys of the Draa and Dades rivers in modern Morocco are filled with hundreds of rammed earth Kasbahs, such as Ait Ben Haddou and Tamnougalt, the earliest dated to around 1000 AD. The city walls of both Marrakesh and Fes are built in rammed earth, and it appears extensively in monumental Muslim architecture such as at the El Badi Palace in Marrakesh, built in 1578. Muslim rule in Egypt promoted the use of adobe brick, with the 10th century Fatmid tombs built in adobe.

Although complex societies have been present in West Africa since around 1500 BC, the first documented is the Ghanaian empire, ruling a large part of West Africa from around 830 AD. Although much of the monumental architecture is stone, it is assumed that current earth building practices found in Ghana, such as adobe and cob construction were used in vernacular architecture in antiquity. The demise of the Ghanaian empire around 1235 AD precipitated the development of the empire of Mali, with its famous earthen cities of Djenne and Timbuktu. The original great mosque of Djenne was probably first built around 1200 AD, but fell into disrepair before being reconstructed in 1907. The characteristic style is similar to other sites in West Africa such as the Sankore and Djinguereber Mosques at Timbuktu in Mali built around 1320, and the Grand Mosque of Agadez in Niger built around 1515. These buildings are unique, being decorated with bundles of palm stalk that project from the wall and serve as a scaffold for annual replastering of the buildings. Earth buildings are found as far east as

Cameroon, where the homes of the Musgum people are inverted catenary dome structures built in earth. Monumental earth buildings are not found in the forested and more humid regions of central and southern Africa, but earth construction in various forms continues to be used across Africa.

12.4   Earth building techniques in Europe

Earth building traditionally takes many forms in Europe, with adobe and rammed earth found in southern Europe, while in northern Europe, earth is used in conjunction with timber in wattle and daub and half-timbered techniques. The earliest use of adobe in Europe can be dated to around 5300 BC at the settlement of Sesklo in Greece with small homes built on stone foundations. The use of earth with timber in northern Europe means that many archaeological sites have decayed and only foundations remain, making assessment of the building materials difficult. Further east in Hattuša, central Turkey, remains of adobe buildings have been found dated to around 1600 BC.

Rammed earth may have been brought to Europe by the Phoenicians, who spread from the eastern Mediterranean, and founded settlements in Spain such as Morro de Mequitta. Rammed earth towers built by Hannibal are described by Roman historian Pliny, and architect Vitruvius describes rammed earth used in the French city of Marseilles and the Greece city of Athens being constructed entirely from adobe. Although much Greek and Roman monumental architecture was built in stone, it is likely that half-timbered and wattle and daub construction continued to be used in vernacular construction throughout Europe, and their use continued following the decline of these empires.

Islam came to southern Europe in 711 AD, bring with it building technologies from north Africa. Conflict at this time led to the construction of many rammed earth and adobe fortifications. Excavations of the fortifications of Calatayud and Plad'Almalain, Spain have been dated to 884 AD, and the Muslim defensive walls of historic cities of Cordoba, Seville and Granada are built in rammed earth. The World Heritage site of the Alhambra Palace in Granada was constructed from rammed earth around 1238. Though earth continued to be used as a building material, its use declined with the increasing penetration of fired brick from the 16th century onwards. In northern Europe, wattle and daub techniques developed as vernacular structures. Cob structures dated to around 1400 AD have been found in parts of the UK, and this building technique continued to be used as a vernacular technique until into the 19th century.

At the end of the 18th century, the political climate in Europe was turning towards freedom for the common man and revolution against the ruling classes. In this climate, rammed earth was 'rediscovered' and championed

12.5 The rammed earth Alhambra of Granada, Spain.

by Frenchman Francois Cointeraux. Cointeraux published a series of leaflets on rammed earth in Lyon in 1791, these were translated into English, French, German and Italian, allowing the technique to spread across Europe and to the United States.

Earth building again declined with the advent of the Industrial Revolution in Europe, meaning fired brick was more easily available, but was again rediscovered following the world wars. After the First World War, trials of rammed earth and chalk buildings were undertaken in the UK and, following the Second World War, rammed earth was used in East Germany leading the development of the first building standard for the material.

12.5   Earth building techniques in North America

Earth was used as a construction material by Native Americans in modern Mexico and the southern United States. The Aztec civilisation in Mexico constructed major monumental architecture in cut stone, but vernacular buildings are thought to have been adobe. The Hohokam culture of Arizona constructed adobe homes with slightly sunken floors cut into the alluvial soils, and remains of adobe Hohokam structures at the Casa Grande National Monument have been dated to around 750 AD. The Pueblo peoples of modern New Mexico built adobe structures, which were home to several families and were several storeys high, of which the most famous is the Taos Pueblo which is dated to around 1000 AD.

Europeans coming to North America continued to use adobe for the construction of many missions and frontier forts such as the adobe Tamacacori, Guevavi, and Calabazas Jesuit missions in Arizona built in 1691. In Albuquerque, the governor's house was built in adobe in 1706. As European settlement moved westward, the settlers required protection, and forts were established to protect settlers from Native American raids. Remains of the adobe Fort Union (1851) and Fort Selden (1865) are testament to the US army using the available materials to construct defences.

Many cities on the west coast, such as San Jose and Los Angeles may have originally been constructed in adobe, though continual expansion and rebuilding means that little remains of these original structures. A single adobe wall remains in Santa Clara University in San Jose, built in 1822 and part of the original lodges around which the university was founded. Casa de Estudillo in the Old Town of San Diego was built using adobe in 1829, and has recently been restored as a historic monument. Lured by gold mining, Chinese immigration to the west coast of the United States brought with it construction techniques such as rammed earth, which was previously unknown in the region. In Palo Alto, California, a business woman named Juana Briones built a rammed earth and cob house in 1845, and Chinese immigrants built a rammed earth herb shop in 1855 (the Chew Kee Store in Fiddletown, California). Later European immigrants are probably responsible for the around 150 rammed earth buildings clustered in the San Antonio Valley in Monterey County, which were built around 1896. German immigrants to the east coast of the United States brought the rammed earth technique from Europe. Hilltop House in Washington, DC was built in 1773 in rammed earth. Bushrod Washington (nephew of George Washington) built rammed earth lodges on his estate at Mount Vernon in Alexandria, Virginia in 1812. Future US president and architect Thomas Jefferson was aware of the technique, but it is unlikely that he personally constructed any buildings in rammed earth, although slave quarters at the Bremo plantation in Virginia, designed by Jefferson, were built by his friend General John Hartwell Cocke around 1819 in this material.

A rammed earth house was built in Trenton, New Jersey, by S.W. Johnson, drawing on the work of Francois Cointeraux in Europe. Johnson hoped to provide a model to newly arrived Europeans to settle farm land, and published a pamphlet in 1806 detailing rammed earth construction. This new construction technique was championed by John Stuart Skinner, editor of *The American Farmer* magazine who published many articles on rammed earth in the 19th century. Others began to experiment in rammed earth, and there are many articles in periodicals from the time referring to rammed earth. South Carolina academic William Anderson was a key proponent of rammed earth, and in 1850 built the Church of the Holy Cross near Stateburg, South Carolina. Keen to experiment with new building techniques, the new Marine hospital was built in New Orleans in 1867. This building was to be iron framed with rammed earth infill panels. Construction began, but the building was vastly over budget, never completed and eventually demolished. Use of rammed earth extended into Ontario, Canada, where St Thomas Church in Shanty Bay was built in 1838, and homes in Greensville in 1868.The expansion of the railroads at the end of the 19th century meant that it became much easier to transport heavy construction materials around the country and use of locally sourced building materials such as rammed earth and adobe declined.

Rammed earth saw another revival in the 1920s, following the interest generated in Europe by well-known English architect Clough Williams-Ellis. Karl Ellington published a book in 1924, and in 1926 an official from the Department of Agriculture published Bulletin No. 1500 detailing rammed earth construction. The depression and New Deal programme in the early 1930s saw several deliberately labour-intensive construction techniques tested, with Thomas Hibben building seven rammed earth houses at Gardendale, Alabama in 1935. This period also saw the first academic research, with Dr Ralph Patty and others publishing results of erosion testing at South Dakota Community College through the 1930s, leading to the publication of technical documentation for rammed earth construction.

Rammed earth advocate David Miller built his rammed earth home in Greely, Colorado around 1940, setting up the Rammed Earth Institute International, and inspiring a new generation of modern earth builders such as David Easton, Paul Graham McHenry and Bruce King.

12.6   Earth building techniques in South America

Archaeological evidence of earth building in South America is scant, with the richest area being the coastal regions of northern Peru. A recently discovered temple at the Ventarron site in northern Peru appears to be constructed from blocks cut directly from river sediment, and has been dated to around 2000 BC. The earliest recorded earth bricks relate to the Moche culture, which flourished in Northern Peru between 100 and 800 AD. The centre of this civilisation was the city of Cerro Blanco, with two pyramids dedicated to the sun and the moon. Huacadel Sol and de la Luna are adobe core pyramids around 50 m tall. Marks on the individual adobe bricks suggest that many different communities were involved in the construction of these structures. Contemporary to the Moche culture were the Lima culture (100–650 AD) of central coastal Peru. This culture also built adobe pyramids, such as the Huaca Pucllana and the Huaca Juliana, the latter being 25 m tall and formed using adobes laid vertically. In the south of Peru, the Nazca civilisation, most famous for the Nazca lines, built their capital at Cahuachi in adobe, which is still being excavated today. Although there is little archaeological evidence of vernacular architecture, it is likely that both monumental and vernacular constructions used adobe bricks. The collapse of the Moche culture around 750 AD led to the development of the Lambayeque culture who continued to build adobe pyramids at sites such as Batan Grande, Túcume and Apurlec. The largest civilisation to develop following the decline of the Moche was the Chimu, who emerged around 900 AD and built their capital of Chan Chan close to the modern city of Trujillo in northern Peru.

Chan Chan was probably the largest city on the continent at that time, home to up to 26 000 people and surrounded by adobe walls around 15 m high. Ten 'royal' enclosures are surrounded by 9 m tall adobe walls covered in relief patterns. The Chimu civilisation was conquered by the Incas whose monumental architecture utilised cut stone, although it is likely that the vernacular building continued in adobe.

The arrival of European settlers brought new building techniques from Europe, to develop missions and settlements. In 1549, a Jesuit missionary sent a request to Europe to send 'artisans able to handle loam, and carpenters, for the construction of a rammed earth wall' for the construction of the Colego da Campanhia in São Paulo. São Paulo became a focus of rammed earth building, with many monumental and vernacular buildings. The rammed earth Cathedral of Taubate was built in 1645, and the Church of Our Lady of the Rosary in 1720. Architectural styles followed those in southern Europe, and the rammed earth House of the Chamber was built in 1776 in a similar style to that found in Portugal around the same time. In 1850 major flooding in São Paulo made many buildings unsafe, precipitating a public campaign against earth buildings, leading to the demolition of much of the historic earthen architecture. Building with adobe continues to be popular in many Andean parts of South America.

12.7   Earth building techniques in Australasia

Earth building is not used by the nomadic aboriginal people native to Australia, but European settlers experimented with a wide range of building techniques from their home countries. An early reference to rammed earth in Tasmania is given in the Hobart town gazette of May 1823:

> Resolved that the mode of building in pise, or rammed earth, appearing to this Society to be both economical and expeditious, the Society earnestly recommend its adoption in Van Diemen's Land.

In 1839 the South Australian newspaper reported on 30 rammed earth houses being constructed, and rammed earth was often used as a quick construction technique in gold rush and frontier towns such as Penrith in New South Wales and Rushworth in Victoria. European settlers of New Zealand tried many forms of construction including rammed earth and adobe, but earthquakes in 1846 and 1855 meant that all forms of masonry fell out of favour. The best-known historic earth monument in New Zealand is Pompallier House in Russell, built in 1841.

Rammed earth building in Australia was rediscovered by an English trained architect, G.F. Middleton, who was employed by the Commonwealth Experimental Building Station. Middleton conducted a large number of tests, which were written into the famous Bulletin No. 5 in 1953, which until recently was the accepted standard reference in Australia and New Zealand.

12.8   Conclusions

This chapter has shown that earth building techniques have developed independently in different parts of the world, but there are some common traits that are found everywhere. Soil, and specifically the subsoils required for earth building are found in many parts of the world and, where they are close to the surface, it is usually because there is no timber or stone present. Were stone or timber present, they would be used as construction materials. Therefore, earth is generally used where timber or stone are not.

There are myriad different earth building techniques, and further research is required to allow accurate descriptions of the development and spread of these techniques.There are however two main categories of earth building, namely monolithic and unit construction. Unit construction requires soils that are particularly clay and silt rich, these are generally found in river valleys, and are generally combined with a binder material such as straw to produce small units. These units can be dried and carried short distances away from the production site. As a result, the earliest large settlements using earthen construction materials such as along the Tigris and Euphrates rivers, the settlement of Catalhöyük and those of Harrapa and Mohenjo-daro along the Indus river seem to have developed with settled agriculture, using the river valley soils combined with the cultivated crops now available.

The soft loess soils on the plains of northern China and in the southern parts of North America could easily be excavated and then piled. Thus, in China, the Lungshan culture, and the North American Hokokam people developed piled and earth sheltered techniques.

As civilisation developed, the building material became of secondary importance to the architecture, and thus we find angled adobe bricks, such as at Huaca Pucllana, Lima, or in the Draa valleys of Morocco or patterns cut into earth renders such as at Chan Chan in Peru. Rammed earth became decorated by the inclusion of decorative brick lines between each lift.

Earth has served as the construction material for many types of construction. Its primary use is usually vernacular construction, and the earliest settlements such as those on the Indus valley, and around the Tigris and Euphrates rivers were able to grow because of the ubiquity of the construction material. Earth was used for the construction of large religious and memorial monuments such as the Ziggurats in western Asia and the Pyramids of the Sun and Moon in modern Peru, and those in ancient Egypt before the stone construction. Earth is particularly notable for its use in defensive constructions, particularly city walls. In China these walls are perhaps the largest, with the city walls of Xi'an and Beijing being around 20 m thick at the base. In the Himalayas, the walls surrounding Lo Manthang in Mustang are built in rammed earth, and in north Africa and southern Europe, rammed earth was used for the city walls of many Islamic cites. The walls of Seville and Cordoba in Spain and Marrakesh in Morocco are all constructed in rammed earth.

Though earth continued to be used as both a vernacular and monumental construction material, its use has declined over time as other construction materials become available, both through improved production processes (first for timber and stone, and later for steel and concrete) and through improved transportation. These, coupled with the improved mechanical properties that other materials exhibit, mean that earth began falling out of use in some parts of the world.

By the 18th century, the Industrial Revolutions sweeping Europe and North America meant that other construction techniques could provide a viable alternative to earth building. Although earth building proved popular in the mid-west United States, the coming of the railways allowed more efficient construction techniques to develop. Likewise, the development of Portland cement in 1824 and the use of iron and steel in construction pushed earth building away from mainstream construction.

Earth building in the developed world has recently seen a resurgence as a sustainable construction material. The virtues that made it so viable to early builders, namely low transport distances, simple construction processes and easy availability, make earth the ultimate sustainable construction material.

12.9   Bibliography

Aaberg-Jørgensen, J. 2000. Clan homes in Fujian. *Arkitekten* 28: 2–9.

Alvarenga, M. 1993. *A architectura de terra no Ciclo do Ouro, em Minas Gerais, Brasil.* 7th International Conference of the Study and Conservation of Earthen Architecture, Silves, Portugal, Direcção Geral dos Edifícios e Monumentos Nacionais.

Arango Gonzalez, J. R. 1999. Uniaxial deformation-stress behaviour of the rammed earth of the Alcazaba Cadima. *Materials and Structures* 32: 70–74.

Azuar Ruiz, R. 1995. *Las técnicas constructivas en al-Andalus: el origen de la sillería y del hormigón de tapial.* V Semana de Estudios Medievales, Nájera, Logroño: Instituto de Estudios Riojanos.

Bazzana, A. 1993. *La Construction en terredans Al – Andalus: le Tabiya.* 7th International Conference of the Study and Conservation of Earthen Architecture, Silves, Portugal.

Bertagnin, M. 1993. *De Cointeraux a del Rosso: de la diffusion de la pensee technologique a la recherché des dernier stemoignages d'architecture en Pise deToscane.* 7th International Conference of the Study and Conservation of Earthen Architecture, Silves, Portugal, Direcção Geral dos Edifícios e Monumentos Nacionais.

Bowman, I. 2000. *Earth building in New Zealand, a little known heritage.* Terra 2000: 8th International Conference of the Study and Conservation of Earthen Architecture, Torquay, UK English Heritage, London.

Camarillo, A. M. 2005. *Juana Briones de Miranda House*. Stanford University, Palo Alto, California.

Chazelles, C. 1993. *Savoir-faire indigenes et influences coloniales dans l'architectures de terre antique de l'extreme-occident (Afrique du Nord, Espagne, France Meridionale).* 7th International Conference of the Study and Conservation of Earthen Architecture, Silves, Portugal, Direcção Geral dos Edifícios e Monumentos Nacionais.

Cody, J. W. 1990. *Earthen walls from France and England for North American farmers, 1806–1870.* 6th International Conference on the Conservation of Earthen Architecture, Las Cruces, New Mexico, Getty Conservation Institute, Los Angeles.

Crook, J. and Osmaston, H. 1994. *Himalayan Buddhist Villages*. University of Bristol, Bristol.

Easton, D. 2007. *The Rammed Earth House (revised edition)*. Chelsea Green Publishing Company White River Junction, Vermont.

Gallego Roca, F. J. and Valverde Espinosa, I. 1993. *The city walls of Granada (Spain), use, conservation and restoration.* 7th International Conference of the Study and Conservation of Earthen Architecture, Silves, Portugal.

Guntzel, J. G. 1990. *On the history of clay buildings in Germany*. 6th International Conference on the Conservation of Earthen Architecture, Las Cruces, New Mexico, Getty Conservation Institute, Los Angeles.

Houben, H. and Guillaud, H. 1994. *Earthen Architecture: a Comprehensive Guide*. Intermediate Technology Development Group, London.

Jaquin, P., Augarde, C. and Gerrard, C. 2008. A chronological description of the spatial development of rammed earth techniques. *International Journal of Architectural Heritage* 2: 377–400.

Jest, C., Chayet, A. and Sanday, J. 1990. *Earth used for building in the Himalayas, the Karakoram and Central Asia – recent research and future trends.* 6th International Conference on the Conservation of Earthen Architecture, Las Cruces, New Mexico, Getty Conservation Institute, Los Angeles.

Jiyao, H. and Weitung, J. 1990. *The protection and development of rammed earth and adobe architecture in China.* 6th International Conference on the Conservation of Earthen Architecture, Las Cruces, New Mexico, Getty Conservation Institute, Los Angeles.

Kleespies, T. 2000. *The history of rammed earth buildings in Switzerland*. Terra 2000: 8th International Conference of the Study and Conservation of Earthen Architecture, Torquay, UK, English Heritage, London.

Maxwell, G. 2000. *Lords of the Atlas. Morocco:The Rise and Fall of the House of Glaoua (text written 1966)*. Cassel & Co, London.

Michon, J. 1990. *Mud castles (Kasbahs) of south Morocco – will they survive?* 6th International Conference on the Conservation of Earthen Architecture, Las Cruces, New Mexico, Getty Conservation Institute, Los Angeles.

Oliver, P. 1997. *Encyclopaedia of World Architecture*. Cambridge University Press, Cambridge.

Palmgreen, L. A. 2005. *Rammed Earth in Sweden*. Rammed Earth Design and Construction Guidelines. Bath University, Bath.

Pecoraro, A. 1993.*The conservation of the church of Nossa Senhorado Rosario, Embu, Sao Paulo, Brazil*. 7th International Conference of the Study and Conservation of Earthen Architecture, Silves, Portugal, Direcção Geral dos Edifícios e Monumentos Nacionais.

Rizvi, J. 1996. *Ladakh. Crossroads of High Asia*. Oxford University Press, Oxford.

Schroeder, H., Schwarz, J., Chakimov, S. A. and Tulaganov, B. A. 2005.*Traditional and current earthen architecture in Uzbekistan*. Conference on earthen architecture in Iran and Central Asia: its conservation, management, and relevance to contemporary society, 12–13 November, UCL, London.

Stevens, A. and Talon-Noppe, C. 1983. Architecture de terre: monuments et sites de l'oasis de Turfan (Xinjiang) sur la route de la soie. *Momentum* 3(26): 46–69.

Swenarton, M. 2003. Rammed earth revival: technological innovation and government policy in Britain, 1905–1925. *Construction History Society Newsletter* 19: 107–126.

Walls, A. 2003.*The 300 year old history of an Arabian mud brick technology*. Terra 2003, Proceedings of the 9th International Conference on the Study and Conservation of Earth Architecture, Yazd, Iran, 29 November–2 December, Iranian Cultural Heritage Organisation, pp. 630–654.

Warren, J. 1993. *Earthen Architecture: The Conservation of Brick and Earth Structures*. ICOMOS, Paris.

13

Stabilised soil blocks for structural masonry in earth construction

B. V. VENKATARAMA REDDY,
Indian Institute of Science, Bangalore, India

Abstract: Stabilised soil blocks (SSBs) are energy efficient and low embodied carbon alternative materials for structural masonry. There is an upsurge of interest among building professionals in utilising low embodied carbon materials. This chapter deals with various aspects of SSBs applicable to structural masonry. A brief outline of soil classification and developments in SSB technology is provided. Different methods of soil stabilisation and production techniques for SSBs are discussed in detail. The influence of soil composition on SSB characteristics and optimum soil grading for the production of SSBs are discussed. The role of block density, moulding moisture content and stabilisers on various SSB characteristics including durability aspects is illustrated. The chapter ends with a discussion of the behaviour of SSB masonry, guidelines for SSB masonry design and some short case studies.

Key words: earth construction, stabilisation, stabilised soil block (SSB), compressed earth block, masonry, cement soil mortar.

13.1   Introduction

Soil has been the basic construction material ever since human civilisation learned to build structures for habitation. Earthen buildings can be seen across the world and earth construction techniques are still in vogue in many parts of the globe. Cob, adobe, rammed earth, and wattle and daub represent just some of the traditional soil-based construction techniques used for earthen buildings. The use of soil for wall construction has distinct advantages, such as the readily available local material, lower cost, and being recyclable and environment friendly. Also, earthen buildings provide better thermal comfort than buildings made from other conventional materials. Some of the major drawbacks of mud walls are the larger wall thickness, loss of strength due to saturation, issues relating to durability and erosion due to rain impact. These disadvantages can be minimised or eliminated by the use of stabilised soil products/techniques for the walls.

Burnt clay brick walls are highly durable and are most commonly employed for the construction of load-bearing buildings. Soils with a sufficient clay fraction are essential for the manufacture of burnt clay bricks. Processed

324

soil is shaped into a brick and then fired at 800–1000°C for manufacturing. The clay minerals in the fertile soil are permanently transformed into other types of mineral products during the firing process. Retrieving the original soil clay minerals from the burnt clay bricks is not possible. Burnt clay bricks are considered energy intensive and are not eco-friendly because of the destruction of fertile clay minerals. The embodied energy content of masonry using burnt clay bricks is about 2200 MJ/m^3, whereas for masonry using alternative materials such as cement-stabilised soil blocks (CSSBs), the embodied energy is in the range of 550–700 MJ/m^3 (Venkatarama Reddy, 2008b and 2009). There is clearly a need for an energy efficient, economical and eco-friendly technology to satisfy the ever-increasing demand for the construction of new buildings. SSBs represent one such alternative solution for the construction of structural walls in buildings. Manufacture of SSBs involves utilisation of the locally available soils in producing durable and eco-friendly materials for the construction of walls. The concept of SSBs for masonry was originated in the early 1940s with the introduction of hand-moulded SSBs. Later, machines came into the picture and the technology is now well matured and commercially exploited for the construction of load-bearing masonry structures.

13.1.1 Soil classification for stabilised soil blocks (SSBs)

Soil is the basic material for the manufacture of SSBs. It is therefore essential to understand the types of soils and methods of selecting the correct stabiliser additives. There are several soil classification systems based on properties such as grain size distribution, Atterberg's limits, etc. The basic purpose of such soil classifications is to facilitate an understanding of the properties and behaviour of a soil to be used for a specific engineering application (soil classification systems were dealt with in detail in an earlier chapter of this book). The type and percentage of clay mineral present in a soil dictates the selection of the stabilising additive needed for SSB production. Large varieties of clay minerals exist in nature. Kaolinite, illite and montmorillonite are the most commonly occurring clay minerals in soils. For SSB purposes, soils can be broadly classified into two categories: expansive and less expansive soils. Expansive soils are those with excessive swelling clay minerals such as montmorillonite. The presence of expansive clay minerals in soils can cause excessive swelling when the soil comes into contact with water and also shrinkage when it undergoes drying. Using such soils for SSB production demands the use of a sufficient quantity of lime (calcium hydroxide). Lime reacts with expansive clay minerals and forms cementitious hydrates of calcium–silicate, calcium–aluminates, etc., responsible for the development of strength in lime-stabilised SSBs. Less expansive soils, such as those containing minerals such as kaolinite and illite, do not swell and shrink as much as expansive soils. Cement stabilisation is best suited for less expansive soils and for the production of SSBs using such soils. Identification of a soil with respect to its swelling and shrinking characteristics is essential when selecting a suitable stabiliser additive for the production of SSBs.

13.1.2 Developments in stabilised soil block technology

Stabilised soils were successfully used for the construction of road and pavement sub-bases from 1920 to 1950 (Chaston, 1952; Lambe, 1962). Use of compacted stabilised soil (using cement/lime binders) for building construction can be seen after the 1940s. Initially, stabilised soils for building construction started with the concept of rammed earth walls. Later, stabilised soils were used to manufacture hand-compacted blocks, for use in the construction of load-bearing masonry. In the period from 1940 to 1950, and prior to the invention of machines for the purpose of compacted soil block production, hand-compacted SSBs were used for the construction of masonry walls. Cement-stabilised hand-compacted blocks were used for the construction of 260 houses in Bangalore during 1949 (Madhavan and Narasinga Rao, 1949), when wooden moulds were used to manufacture the blocks; load-bearing masonry walls (250 mm thick) were built with blocks containing 4–5% cement in mud mortar and the buildings had reinforced concrete flat slab roofs. Another Indian example from 1950–1951 was the construction of residential colonies in the northern Indian regions to house the workers employed in hydroelectric projects and dams during the period (Mitra, 1951).

The CINVA-RAM press, developed in 1952 by Paul Ramirez, was the first (manual) machine to produce compacted soil blocks for building construction. Its development led to the concept of machine-pressed SSBs. A number of groups in Europe, Australia and in other countries started working on SSB technology and began producing machines, and various buildings using SSB technology and employing machine-pressed SSBs were built in Columbia, Chile, Venezuela and Brazil during the early 1960s (UN, 1964). Social housing projects involving large numbers of houses of up to three-storey construction were built using cement SSBs and rammed earth in Isle D'Abeau, near Grenoble in France, during the early 1980s. Reports and other documents on stabilised earth construction by Middleton (1952), the Department of Housing and Urban Development (1955) and Fitzmaurice (1958) represent the beginning of compacted SSBs for building construction. Mukerji (1986) published a survey of soil block making machines developed in several countries across the world and a large number of machines (manual, semi- and fully automatic) are still commercially available in the market.

Investigations into SSB technology can be grouped into two major areas: (i) manufacturing and characterisation of SSB properties; and (ii) the behaviour of SSB masonry under compression, tension and shear. Cement and lime represent the most commonly used stabilisers for SSB production and hence discussions here have a bias towards cement and lime SSBs.

A good understanding of the various aspects of confined static compaction of soils is needed for the design and development of machines to manufacture compacted SSBs. The force/energy needed for the compaction of a soil block, the moisture, density and energy relationships and optimum moisture content are some of the issues investigated by Olivier (1994), Venkatarama Reddy (1983, 1991), and Venkatarama Reddy and Jagadish (1993). Production techniques, strength, dimensional stability, durability and stress–strain characteristics are some of the major parameters explored for cement-stabilised and lime SSBs. SSB technology has been addressed by Fitzmaurice (1958), Spence (1975), Lunt (1980), Middleton (1952), Olivier and Mesbah (1987), Heathcote (1991, 1995), Houben and Guillaud (2003), Venkatarama Reddy and Jagadish (1995), Walker and Stace (1997), Walker et al. (2000), Walker (2004), Venkatarama Reddy and Walker (2005), Venkatarama Reddy and Gupta (2005a), Venkatarama Reddy et al. (2007a), Venkatarama Reddy (2007), and in many other publications. Studies of masonry using SSBs have been undertaken by Venkatarama Reddy and Jagadish (1989), Shrinivasa Rao et al. (1995), Walker (1999), Walker (2004), Venkatarama Reddy and Gupta (2006, 2008), Venkatarama Reddy and Vyas (2008) and Venkatarama Reddy et al. (2009) amongst others.

13.2   Soil stabilisation techniques

Soil is used for the construction of roads, structural walls, masonry mortars, floors, roofs, etc. Natural soil has certain distinct characteristics or properties such as strength, plasticity, swelling, shrinkage, etc. These properties are controlled by its grain size distribution, and by the type and quantity of clay minerals present. Some of these properties may have to be altered in order to make the natural soil suitable for any specific construction. The alteration of existing soil properties to meet specific engineering requirements can be termed 'soil stabilisation'.

In SSBs, the bulk of material used is soil. Important requirements for an SSB masonry wall are adequate strength, resistance to moisture ingress, volume stability and durability, i.e. resistance to rain erosion and mechanical damage. Very rarely, the natural soils in their native state can impart the above requirements to the SSBs being produced. Normally, however, the natural soil has to be stabilised in order to make it suitable for the production of SSBs. Soil stabilisation techniques can be grouped into three broad categories: (i) mechanical stabilisation; (ii) stabilisation by compaction; and (iii) stabilisation by additives.

13.2.1 Mechanical stabilisation

Mechanical stabilisation is an inexpensive method of stabilising a soil, in which two or more soils with different characteristics are mixed, or sometimes inert materials such as sand and gravel are mixed in to get the desired grading for a soil. Generally, soils have either an excess clay fraction or are deficient in the sand fraction. In order to make them suitable for earth wall construction, it is essential to either dilute the soil by the addition of sand/gravel or mix it with another soil of different composition. Either way, the natural soil properties are altered to make them suitable for the production of SSBs.

13.2.2 Stabilisation by compaction

Compaction is the best-known soil stabilisation technique used to improve strength and reduce porosity. Here, the soil in its loose state is brought to a dense state by supplying compaction energy. There are two types of soil compaction techniques: static and dynamic compaction. Improvement in strength, reduction in porosity and other properties depend upon the soil composition, moulding water content and the compaction energy supplied.

13.2.3 Stabilisation by additives

In this type of stabilisation, both organic and inorganic additives in either solid or liquid form are added to alter the soil properties. Additives such as cement, lime, bitumen, polymers, certain salts, organic binders, organic and inorganic fibres are used. Selection of the type of additive depends upon the kind of alteration expected in the soil property and the ultimate product/application. For example, if control of shrinkage cracks on drying is required, the use of fibres is recommended; if there is a need to control the swelling of the soil, lime is added; if an improvement in the erosion characteristics of the earth blocks is required, bitumen can be added.

In the majority of soil stabilisation processes, more than one of the above-mentioned methods is adopted. In the case of SSBs, all three methods of stabilisation mentioned above are used. As mentioned previously, cement and lime are the most commonly used additives for the production of SSBs.

13.2.4 Cement stabilisation

Cement stabilisation is ideal for coarse-grained and gravely soils, and is used to improve strength (especially in saturated condition), resistance against rain erosion and mechanical damage of the SSBs. The principle involved in cement stabilisation is illustrated opposite.

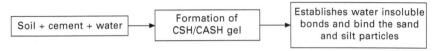

13.1 The principles involved in cement stabilisation.

The major factors affecting cement-stabilised soils include soil composition, percentage of cement, degree of compaction achieved for the product, and the curing period and curing temperature. The strength of cement SSBs increases with the increase in cement content, density, curing period and curing temperature.

13.2.5 Lime stabilisation

Lime used for construction can be classified as hydraulic lime and non-hydraulic (or fat) lime. Hydraulic lime can set and harden in the presence of water, whereas fat lime derives its strength from a reaction with a pozzolanic material or through carbonation. Hydrated lime is usually used for lime stabilisation.

Lime stabilisation techniques are generally adopted for clayey soils, in order to control swelling and shrinkage. When lime is mixed with the soil, the clay minerals present in the soil react with the lime. These lime–clay reactions lead to the formation of a water insoluble gel of silicate and silicate–aluminates, and (with time) this gel finally crystallises into hydrates of calcium silicate, calcium aluminates, etc. The cementing agents formed in lime–clay reactions are similar to those formed during Portland cement hydration. The cementitious gel formed coats the soil particles and establishes bonds. The pace of lime–clay reactions and formation of cementitious gel is slow. These reactions take place continuously, even in the presence of just a little moisture in the surrounding atmosphere.

With reference to stabilised soil based building products, it is difficult to exploit the lime stabilisation technique mainly because of the slow pace at which the lime–clay reactions take place and because of the insufficient quantity of binding material formed in such reactions to derive meaningful strength for the SSB to be used in structural applications. Lime stabilisation is a must while manufacturing SSBs using soils containing expansive clay minerals such as montmorillonite, in order to control the swelling and shrinkage problems associated with such soils. Lime–clay reactions can be accelerated by raising the curing temperature. For example, it is possible to achieve 3 months' ambient temperature curing results in less than 24 hours of curing at 80°C (Venkatarama Reddy and Jagadish, 1984; Venkatarama Reddy and Lokras, 1998; Venkatarama Reddy and Hubli, 2002). Steam curing methods can be adopted for such purposes but result in additional cost and energy expenditure.

13.3 Production of stabilised soil blocks (SSBs)

Stabilised mud blocks, or stabilised compressed earth blocks are other names for stabilised soil blocks (SSBs). Processed soil that is mixed with stabiliser and water compacted into a dense block (Fig. 13.2) can be termed an SSB. Compaction is carried out at optimum moisture content (OMC) (though not necessarily at Standard Proctor OMC) generally using a machine. The blocks produced in this way are cured and then used for masonry construction.

The characteristics of SSBs greatly depend on the clay fraction of the soil, block density and stabiliser content. Different types of SSB can be manufactured using different stabilisers and geometric configurations. Figure 13.3 illustrates the different types of SSB, whilst typical pictures of some of these SSBs are displayed in Fig. 13.4. Blocks of any desired shape and size can be manufactured using small attachments to the machine mould. SSB blocks are manufactured using machines employing a static compaction process. Block thickness in such a compaction process should be restricted

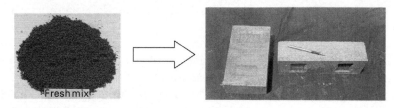

13.2 Processed soil converted to a dense block.

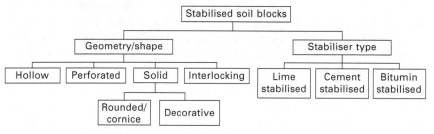

13.3 Classification of SSBs.

Cornice/corbelled/filler/rounded corner blocks

Interlocking blocks

13.4 Different types of SSBs.

to less than 100 mm in order to avoid stratification of block density across the thickness of the block.

SSBs are produced by a process, involving four critical steps: (i) soil selection; (ii) soil processing; (iii) pressing or compacting of the block; and (iv) curing.

13.3.1 Soil selection

Soils contain particles of different sizes such as gravel, sand, silt and clay minerals. Generally, soil is sieved through a 5-mm sieve, thus eliminating the gravel portion. Sand and silt particles are inert whereas clay is a mineral with a lot of affinity for water. Clay particles swell on absorption of water and shrink on drying; the amount of swelling and shrinkage is dependent upon the type and quantity of clay mineral present in the soil. Plasticity, shrinkage and strength, etc. of a soil are controlled by the quantity and type of clay minerals present.

Soils containing less expansive clay minerals (such as kaolinite) can be used for SSBs using Portland cement as a binder. Coarse-grained sandy soils containing predominantly non-expansive clay minerals are ideally suited for the production of cement-stabilised blocks. Soils containing an excessive silt fraction can lead to SSBs with very low green strength for handling during the block manufacturing process. In such situations it is advisable to add some coarse gravel or sand fraction to the soil. Acidic soils generally require the addition of hydrated lime to neutralise the acidity, as well as the use of Portland cement. Organic soils and soils containing sulfates are generally not suitable for stabilisation using cement/lime alone. Such soils require accelerating agents other than lime and cement. Expansive soils need lime to stabilise their swell–shrink properties and hence the use of lime is essential for producing SSBs from them. It is mandatory to test the cured blocks for strength and durability characteristics.

13.3.2 Soil processing

Soil processing involves different activities such as crushing/powdering of excavated soil, sieving, mixing with the sand and stabilisers in dry state, and then mixing with water. These processes bring the partially saturated mixture to a state ready for compaction in a machine. Generally, excavated soils contain lumps and the size of the lumps (in air dried state) mainly depends on the type and quantity of clay mineral present in the soil. Soil lumps have to be crushed and sieved (using a 4–5-mm sieve) in order to blend the natural soil with sand and stabilisers (such as cement or lime). Sieved soil is effective for arriving at a uniform mixture and better distribution of sand and stabilisers. Figure 13.5(a) shows a manually operated rotary sieve.

As we have already seen, natural soils in their original composition are rarely suitable for the production of SSBs. Hence, in the majority of cases the natural soils are blended with sand or equivalent inert materials such that the resulting mixture is suitable for SSB production. Basically, sand is used to modify the natural soil composition. For example a natural soil containing 30% clay can be reconstituted by mixing with sand in the ratio of 1:1 (soil to sand, by weight) such that the resulting mixture contains 15% clay, which may be the requirement for the production of SSBs with cement stabilisation. Soil and sand are generally mixed in a dry state. Dry mixing can be carried out in a rotating drum type of mixer (Fig. 13.5)(b). Mixing of soil and sand can also be done manually, where thin alternate layers of soil and sand are spread and mixed with a spade. Generally, mixing of powdery binders such as cement or lime with soil has to be carried out in a dry state. It can be performed either by using a mechanical mixer or manually. A mechanical mixer is more efficient in obtaining a uniform mixture. Adding the correct amount of moisture and mixing it uniformly with the soil–stabiliser mixture is a very important operation in the SSB production process. Here the soil is mixed with an optimum quantity of water (though not necessarily Standard Proctor OMC). This can also be carried out either manually or using a drum type mixer. In the manual process, the soil is spread into a thin layer (100–150 mm) and then water is sprinkled onto the layer. The wetted layer is mixed using a spade until a uniform mixture is ascertained. The mixing can also be carried out in a drum type or pan type mixer. The soil, sand, stabiliser and water mixture is brought to a semi-dry state. The partially saturated mixture (Fig. 13.5(c)) is then compacted into a high density block. The water content used in such mixes will be in the range of 10–14%.

13.3.3 Pressing or compacting the block

Processed soil in a partially saturated state is used for the manufacture of SSBs employing a static compaction process, which involves confined compaction of the wetted mixture in a mould using a piston, either from one side or from both sides, in order to densify the loose soil mix into a

13.5 (a) Manual rotary sieve; (b) rotating drum type mixer; (c) mix ready for compaction.

block. The compaction energy supplied is not a fixed value and depends upon the soil composition, moisture content and target density, and therefore the conventional standard Proctor density–moisture relationships have no relevance to the static compaction process employed in the production of SSBs using a machine. Various stages involved in the static compaction process employed in the production of SSBs are illustrated in Fig. 13.6.

The dry density of the compacted soil mass is an indication of the degree of compaction. Compaction curves for static compaction and dynamic compaction, such as the standard Proctor test, are shown in Fig. 13.7. In the Standard Proctor test, the peak density and corresponding moisture content are termed maximum dry density (MDD) and optimum moisture content (OMC), respectively. Moulding moisture content, energy supplied (or compaction effort) and soil composition all control the MDD and OMC values. Higher compaction effort leads to a higher value of MDD and lower OMC. In the Standard Proctor test, the energy supplied to the soil specimen is fixed.

In the static compaction process, the concepts of MDD and OMC do not arise, since the energy supplied is not constant here. While employing the static compaction process, it is preferable to achieve as much dry density as possible by using a suitable moisture content. Static compaction curves for constant energy input can be generated indirectly through energy–density and energy–moisture relationships (Venkatarama Reddy and Jagadish, 1993); such static compaction curves are shown in Fig. 13.7.

The dry density of a block has a controlling influence on the strength of the SSB and hence block weight has to be controlled to control dry density. The block compaction process involves: (i) weighing the wetted mix of soil; (ii) feeding it into the machine mould; (iii) closing the lid and compaction; (iv) ejecting the block; and (v) stacking. Figure 13.8 illustrates the various steps involved in block production employing a manual process (a time and motion study conducted by Venkatarama Reddy (1983) showed that the rate of block production using manual processes is not controlled by the weigh batching activity).

13.3.4 Curing

Cement SSBs need moisture for the development of hydration products responsible for establishing water insoluble bonds, and therefore they require curing. Such blocks are stacked one above the other and water is sprinkled three to four times daily for four weeks (Fig. 13.9). After completion of curing, the blocks are allowed to dry in the stack before being used for construction. In order to reduce the curing period and to achieve higher strengths, steam curing at low temperatures (80°C) is often undertaken, at atmospheric pressure for 10–12 hours. Lime–clay reactions are slow at ambient temperatures and hence need a longer curing period to achieve meaningful strengths; these

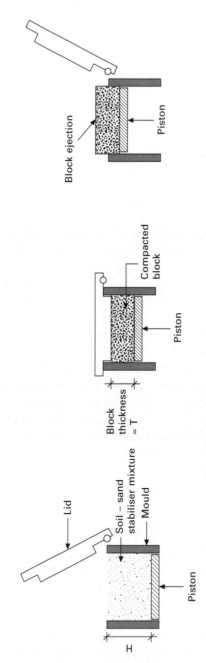

13.6 Various stages in the static compaction process (L–R): (a) mould filled with processed mix, (b) compaction through lid closure and piston movement, (c) ejecting a compacted block.

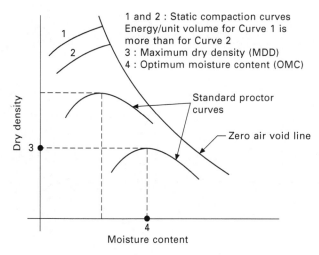

13.7 Compaction curves.

13.8 Various stages of compacted soil block production process (L–R: wetted mix, filling the mould, block compaction, block ejection).

blocks can also be cured using steam curing techniques (see Venkatarama Reddy and Hubli (2002) for more information on steam curing).

13.3.5 Machine requirements for block production

Manually operated, semi-automatic or fully automatic machines can be employed for the compaction and manufacture of SSBs. All these machines employ a static compaction process (the various steps of which are illustrated in Fig. 13.6) to achieve the requisite dry density for a block. The compaction ratio of the machines should not be less than 1.70. The compaction ratio = $(H)/(T)$, where H is the height of loosely filled mix in the machine mould (Fig. 13.6(a)) and T is the block thickness, which should be restricted to 100 mm or less. Block thickness in excess of 100 mm can lead to a variation in the density of the block across its thickness when the static compaction process is

13.9 Stacking and curing.

employed (Venkatarama Reddy, 1991), and thus production of thicker block results in poor compaction in the top portion and better compaction in the bottom portion (from where the maximum piston movement takes place).

13.4   Characteristics of stabilised soil blocks (SSBs)

Strength, durability, water absorption, development of bond with the mortar and stress–strain relationships are some of the important SSB characteristics needing to be considered when they are used for load bearing masonry. The major factors influencing theses characteristics are: (a) composition of the soil–sand mixture; (b) density of the block; and (c) the type and quantity of stabiliser.

13.4.1 Influence of soil grading and composition on the strength of stabilised soil blocks (SSBs)

The grain size distribution of the soil and the clay minerals present in the mix play a significant role in the strength of SSBs. The type and quantity of clay minerals in the mix can dictate and control the type and quantity of stabiliser required to achieve a particular strength of SSB. The compressive strength of an SSB is sensitive to the moisture content of the specimen at the time of the test: saturated or dry conditions are the two extremes under which the compressive strength of an SSB can be measured. Compressive strength measured under saturated conditions is termed wet compressive strength, whilst strength measured in totally dry condition is designated dry compressive strength. The ratio between wet and dry compressive strengths can range between 0.20 and 0.55 (Venkatarama Reddy and Gupta, 2005b; Venkatarama Reddy *et al.*, 2007a; Walker and Stace, 1997; Venkatarama Reddy and Walker, 2005) depending on factors such as block density, percentage of clay fraction in the mix, type of clay mineral present in the soil, stabiliser content, etc.

Soils with predominantly non-expansive clay minerals are ideal for cement stabilisation. The role of the clay fraction of the soil–sand mix on strength of CSSBs is illustrated in Fig. 13.10. The curves in Fig. 13.10 were generated for cement-stabilised compacted cylindrical specimens with constant dry density and similar moulding moisture content at each of the clay fractions in the mix. There is an optimum clay content yielding maximum strength for CSSBs (Venkatarama Reddy *et al.*, 2007a), optimum clay content being in the range of 10–15% (close to 15% for coarse-grained sandy soils and in the vicinity of 10% for fine-grained silty soils).

13.4.2 Density and strength relationships

The density of an SSB has significant influence on compressive strength. The compressive strength of a compacted cement-stabilised soil mix increases with the increase in density irrespective of cement content and moulding moisture content. Some of the typical strength–density relationships are shown in Fig. 13.11 for different cement contents. Compressive strength is very sensitive to variations in density, and marginal increase in density leads to a large increase in strength. Typical values of strength and density

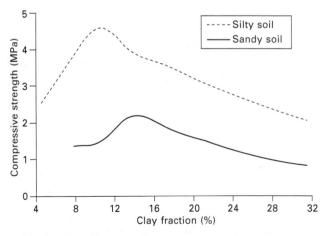

13.10 Strength versus clay content of the soil–sand mix.

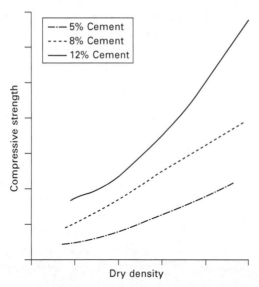

13.11 Typical density and strength relationships for CSSB.

Table 13.1 Strength and density for compacted stabilised soil specimens

Dry density (kg/m³)	Wet compressive strength (MPa)	
	5% cement	8% cement
1600	0.66	0.96
1800	1.37	3.00
2000	2.71	4.52

are given in Table 13.1 for compacted cement soil specimens with 5% and 8% cement (Prasanna Kumar, 2010), where it can be seen, for example, that a 10% increase in dry density from 1600 kg/m³ leads to a 200% increase in strength for both 5% and 8% cement. Venkatarama Reddy and Walker (2005) reported a sharp increase in strength as density increases. Increase in strength due to an increase in dry density can be attributed to a reduction in the porosity (hence a closer contact of the particles) of the compacted specimen, resulting in better bonding due to cement hydration products. Block density can be controlled by resorting to weight batching in the SSB production process.

13.4.3 Moulding moisture content and stabilised soil block (SSB) strength

Density has a controlling influence on SSB strength and the density achieved for an SSB depends on the energy supplied during compaction. Compaction characteristics of stabilised soil mixtures can be established by reference to the Standard Proctor test, under which the SSB density achieved could be either more or less than the Standard Proctor OMC, depending on the capacity of the machine used to supply energy for compaction. In such situations, the question of moulding water content to be used arises, in order to achieve best possible strength for the SSB. Prasanna Kumar (2010) examined strength–moulding water content relationships for compacted cement-stabilised specimens (see Table 13.2 for typical results); Standard Proctor OMC for the soil–sand mixture used in these studies was 11% and the corresponding Standard Proctor density was 2000 kg/m^3.

The results given in Table 13.2 show that the compressive strength of cement-stabilised compacted soil specimens increases with an increase in moulding water content. For 5% and 8% cement contents, the strength increase was about 30% when the moulding moisture content was increased from 8.5% to 14.5%, whilst strength increased by nearly 50% for the higher 12% cement content. A dry soil–cement mixture contains cement and clay particles, both with an affinity for water. When water is added to the dry soil–cement mixture, it is shared by both the cement and clay particles. When the samples are compacted using a small quantity of moulding water (say 8%), it is possible that there could be insufficient water for proper hydration of the cement to take place. As the moulding water content is increased (say to 14.5%), more water is available for cement hydration. Specimens compacted with a higher moulding water content can also suffer higher shrinkage, leading to a marginal increase in density, and this increased density can result in higher compressive strength (which might explain why there is an increased strength when a higher percentage of moulding water is used). These results clearly indicate the advantages of using a higher moulding water content during the manufacture of cement SSBs.

Table 13.2 Moulding water content-strength relationships for cement stabilised soil compacts (Specimen dry density = 1800 kg/m^3; Specimen size: 38 mm diameter and 76 mm height)

Cement content (%)	28-day wet compressive strength (MPa)		
	Moulding moisture content (by mass)		
	8.5%	12%	14.5%
5%	1.10	1.15	1.37
8%	2.31	2.38	3.00
12%	2.94	3.90	4.44

13.4.4 Type and quantity of stabiliser on the stabilised soil block strength

As already noted, a variety of stabiliser additives can be used for the production of SSBs. Lime and Portland cement are the most commonly used stabilisers, and the following discussion of stabiliser influence on strength is restricted to these two.

The selection of lime or cement for the production of SSBs depends on the clay minerals present in the soil used for such blocks. Ordinary Portland cement (OPC) is the most commonly employed stabiliser and is an ideal stabiliser for coarse-grained soils with less-expansive clay minerals. The strength of CSSBs increases with an increase in cement content. Generally, the cement content used for CSSBs is in the range of 5–10%. For a given block density, the percentage increase in strength as the cement content increases depends on the clay content of the mix used. Typical variations in strength as cement content is varied are illustrated in Fig. 13.12 for blocks using a sandy soil with 7% cement. As the cement content is increased, the percentage increase in strength (or the slope of the lines on the graph) can vary depending upon the clay fraction of the mix and also on the block density. Selection of the percentage of cement content for production of CSSBs is dependent on the compressive strength of the block desired.

Generally, lime in the form of calcium hydroxide is used for SSB production. Lime stabilisation becomes essential when handling soils containing expansive clay minerals, mainly to address the issue of high swelling and shrinkage characteristics, but it is also used for SSB production using soils with non-expansive clays. Figure 13.13 shows strength variations in lime-stabilised compacts when lime content is varied, and shows that there is an optimum lime content (OLC) yielding maximum strength, which varies from soil to soil. Table 13.3 summarises the OLC and strength for the four types of curve shown in Fig. 13.13. Type of clay mineral, percentage of clay, OLC

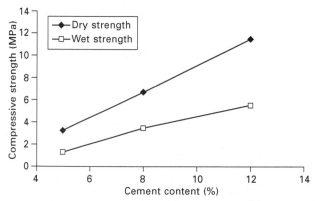

13.12 Strength versus cement content for CSSBs.

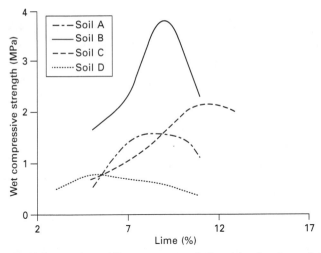

13.13 Strength and lime content relationships for lime-stabilised soil compacts.

Table 13.3 Details of OLC and strength of SSBs (Specimen size: 50 mm cube)

Soil designation	Type of clay	Clay content (%)	OLC (%)	28-day wet compressive strength at OLC (MPa)
Soil A	Montmorillonite	12.0	11.5	2.15
Soil B	Kaolinite	10.5	9.0	3.80
Soil C	Kaolinite and montmorillonite	6.0	8.0	1.60
Soil D	Kaolinite	5.0	5.0	0.80

and 28-day compressive strength values are also detailed in Table 13.3. It is interesting to note that there is a relationship between OLC and the clay type and its percentage in the mix, with OLC almost exactly corresponding to the clay content of the soil mix, and increasing with the increase in clay content of the soil.

The results of the parametric studies discussed above relate to the unconfined compressive strength of smaller size specimens. The wet compressive strength of CSSBs (size: 230 × 190 × 100 mm) with 7% cement (OPC) used for load bearing masonry is in the range of 5–7 MPa (Ullas and Venkatarama Reddy, 2007), whilst the wet compressive strength of lime-stabilised blocks (8–10% lime) is in the range of 1.5–2 MPa.

13.4.5 Absorption characteristics of stabilised soil blocks

SSB absorption characteristics include properties such as rate of water absorption, initial rate of absorption (IRA) and the saturated water content (called water absorption) of the blocks. The percentage of clay minerals present in the block, quantity of stabiliser, density, porosity and pore size distribution of an SSB control these properties. Investigations by Walker and Stace (1997), Venkatarama Reddy *et al.* (2007a), Venkatarama Reddy and Gupta (2005b), and Venkatarama Reddy and Vyas (2008) are just some of the studies that have been made on SSB absorption characteristics.

Figure 13.14 shows water absorption rates for CSSBs with various soaking durations in water. When dry blocks are soaked in water, the rate of water absorption is highest in the initial few minutes. CSSB blocks require 4–12 minutes to attain 75% saturation, and 15–200 minutes to attain complete saturation. The rate of water absorption is a function of density, soil grading and cement content. Soaking duration to achieve 75% and 100% saturation decreases with an increase in cement content and density of the blocks (Venkatarama Reddy and Gupta, 2005b).

The saturated water content (or water absorption) of CSSBs increases with an increase in the clay content of the blocks. Typical relationships between water absorption and block clay content (having a dry density of 1800 kg/m^3) are shown in Fig. 13.15 (Prasanna Kumar, 2010). Saturated water content values for CSSBs with a clay content of about 15% lie in a range of 10–16% for blocks with 7–10% cement content.

IRA for CSSBs decreases with an increase in the clay fraction of the block. The lower IRA values at higher clay content can be attributed to the low surface porosity of the blocks. As the clay content of the block increases, the percentage of fines in the block increases and can lead to lower surface porosity leading to lower absorption rates. Typical IRA values for blocks with 15% clay are about 2 kg/m^2/minute, as compared with 1–3 kg/m^2/minute for a variety of burnt clay bricks.

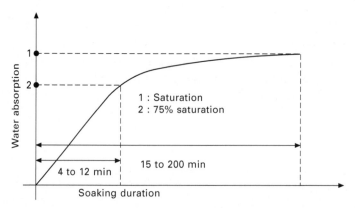

13.14 Variations in water absorption rate with soaking duration for CSSBs.

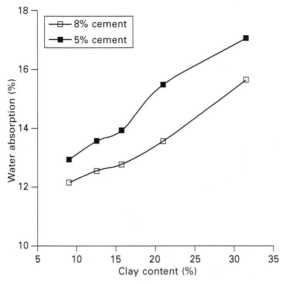

13.15 Variations in saturated water content with the clay content of CSSBs.

13.4.6 Stabilised soil block (SSB) durability

Durability of SSB buildings is a major concern expressed by the users of SSB technology. Several accelerated test methods have been devised to assess the durability of SSBs, including: (i) the spray erosion test; (ii) the drip test; (iii) the alternate wetting and drying test; and (iv) the linear expansion on saturation test. Each of these tests will now be considered in turn.

Spray erosion test

Several attempts have been made to standardise the spray erosion test (Middleton, 1952; Venkatarama Reddy and Jagadish, 1995 and 1987; Walker, 1998; Heathcote, 1995 and 2001, Walker, 2004). Walker (1998) highlights the different types of test procedures available to assess erosion resistance of soil blocks. A typical test procedure involves spraying a jet of water (at constant pressure) onto the surface of a stabilised soil block for a given time. After the test, the block surface is monitored for pitting, surface erosion (erosion depth) and weight of the material removed by the water jet. Depth of erosion and mass loss are quantified and correlated to the durability of stabilised soil blocks. Many investigators have attempted to make the spray erosion test represent more closely the damage caused by raindrops on a wall surface. Figure 13.16 shows a typical spray erosion test set-up. Accelerated spray erosion tests can at best give only some idea about the relative performance of an SSB. Many attempts have been made

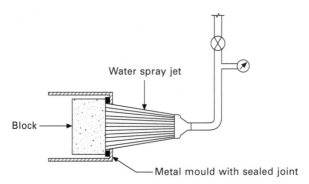

13.16 Spray erosion test set-up.

to correlate erosion rates with field erosion rates due to natural rainfall (Venkatarama Reddy and Jagadish, 1987); however, standardisation becomes a problem because the nature, intensity, amount and distribution of rainfall varies considerably from region to region.

The drip test

The drip test basically involves allowing water droplets or drips to impinge on an inclined block surface for a given time (Yttrup *et al.*, 1981). After the test, the block surface is monitored for pitting and depth of erosion. This type of test again gives a relative evaluation of the durability performance of different types of SSBs.

It has been observed that SSBs with adequate stabiliser content and manufactured using soils with optimum clay content perform exceedingly well when subjected to both spray erosion and drip tests. Evaluating the relative performance of SSBs with stabiliser content beyond a threshold limit becomes difficult because the surfaces of such blocks do not show any erosion, even after several hours of exposure to water spray (Venkatarama Reddy and Jagadish, 1987; Walker, 2004).

Alternate wetting and drying test

This cyclic wetting and drying test examines the durability of SSBs, with the test procedure and guidelines being supplied by the ASTM D559 code. The test involves monitoring weight loss after 12 cycles of cyclic wetting and drying and the brief test procedure is as follows:

1   Ascertain the initial dry block weight after drying it to constant weight in an oven
2   Soak the dried specimens in potable water at room temperature for a period of 5 hours and then dry them in an oven at 71°C for 42 hours.

Apply two firm strokes on surfaces of the partially dried specimens with a standard wire brush. Eighteen to 20 strokes are applied to cover all the sides of the specimens (see ASTM D559 code). This constitutes one cycle of wetting and drying

3   Repeat the same procedure and complete 12 cycles

4   At the end of 12 cycles, oven dry specimens again at 60°C until they attain constant weight and record the final dry weight

5   Estimate the weight loss to ascertain mass loss after 12 cycles of the wetting and drying test.

This test, involving scratching the specimen with a metal wire brush, is considered too severe a test to evaluate the durability of SSBs (Lunt, 1980).

The Portland Cement Association (1956) specified acceptable limits for weight loss ranging from 7 to 14% for various types of soils when used with cement for road construction. These limits have been modified and adopted by several authorities over the years, according to the nature and type of construction and climatic conditions. For example Fitzmaurice (1958) recommended limits for maximum weight loss as 5% for permanent buildings and 10% for rural buildings in any type of climate; Walker and Stace (1997) attempted to correlate mass loss with the clay content of the cement–soil mix used for block production.

Tests performed on a variety of SSBs and also on blocks collected from different buildings with age distributions ranging between 5 and 14 years show useful results to specify a limit for weight loss. After investigations by Venkatarama Reddy et al. (2003), Venkatarama Reddy (2007), Venkatarama Reddy and Jagadish (1995), and observations of the performance of real time SSB buildings, a limit of 3% weight loss has been specified by Venkatarama Reddy (2007) and is being adopted by the revised Indian code of stabilised soil blocks.

Linear expansion on saturation

An accurate length comparator is an essential tool for performing any test for linear expansion on saturation. A typical experimental set-up is shown in Fig. 13.17. The test involves measuring the initial length of an oven dried (at 50°C) block, soaking the oven dried block in water for 24 hours, and then measuring the length of the saturated block. The difference in the lengths of the dry and saturated block gives the linear expansion of the block on saturation.

Venkatarama Reddy (2002), Ullas and Venkatarama Reddy (2007) and Venkatarama Reddy et al. (2003) examined the performance of existing CSSB buildings with exposed block-work walls (without protection), with age

13.17 Experimental set-up for linear expansion on saturation.

distributions ranging between 5 and 14 years. They monitored linear expansion on saturation for the blocks collected from several of these constructions, and established a correlation between the performances of these buildings (the extent of damage) with linear expansion on saturation of the CSSBs collected. These results indicate that the use of SSBs with < 0.10% linear expansion on saturation generally leads to satisfactory performance of SSB buildings when exposed to alternate wetting and drying caused by rain and sunlight over a period of several years.

Figures 13.18 and 13.19 show the relationships between mass loss and strength, and mass loss and linear expansion as reported by Venkatarama Reddy *et al.* (2003). Figures 13.20 and 13.21 show CSSB wall surfaces after exposure for 14 years (with linear expansion on saturation of 2.32%) and 11 years (with linear expansion on saturation of 0.03%), respectively. A large value for linear expansion on saturation indicates severe damage to the block and wall surface after repeated wetting and drying caused by the natural weathering process. If an SSB has a smaller value for linear expansion on saturation, the wall surface is likely to be intact even after several years of weathering. Hence, keeping linear expansion less than 0.10% and mass loss after the wetting and drying test below 3% could assure stable long-term performance for the SSBs. The test for linear expansion on saturation can also be conducted in the field to assess block quality, and blocks having

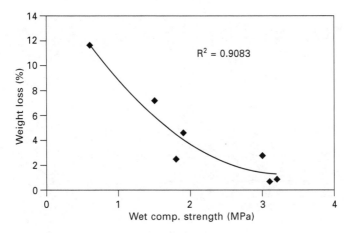

13.18 Mass loss versus strength.

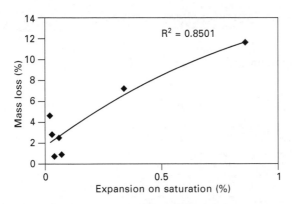

13.19 Mass loss versus linear expansion.

< 0.10% linear expansion have performed satisfactorily in such field tests. The linear expansion on saturation value can be bench marked to indicate the dimensional stability of an SSB under natural weathering conditions.

13.4.7 Stress–strain characteristics

Stress–strain relationships and stress–strain characteristics such as modulus, Poisson ratio, strain at peak stress and ultimate failure strains for SSBs depend on the following parameters:

1 composition of the soil–sand mixture, especially the percentage of clay mineral
2 type and quantity of the stabiliser
3 block density
4 moisture content of the block.

13.20 Damage after 14 years of exposure to natural weathering (linear expansion on saturation = 2.32%).

13.21 Wall surface after 11 years of exposure to natural weathering (linear expansion on saturation = 0.03%).

Comprehensive studies on the stress–strain relationships and elastic properties of SSBs are limited. Investigations by Venkatarama Reddy and Jagadish (1989), Venkatarama Reddy and Gupta (2005b), Venkatarama Reddy *et al.* (2007b), Venkatarama Reddy and Vyas (2008) and Walker (2004) throw some light on the stress-strain relationships for CSSBs. Investigations by Kouakou and Morel (2009), and Maniatidis and Walker (2008) provide information on stress–strain relationships for unstabilised compacted soil blocks. The initial tangent modulus (in saturated state) for CSSBs is in the range of 3000–4500 MPa for 7–8% cement blocks with a density of 1800 kg/m^3. Typical stress–strain relationships for CSSBs (with an optimum clay content of about 15%) in dry and saturated conditions are illustrated in Fig. 13.22. The stress–strain behaviour for CSSBs shows a large softening portion beyond peak stress. The ultimate strain at failure is in excess of 1.5% for CSSBs in dry condition, whereas in a saturated state the blocks have an ultimate failure strain of about 0.5–0.6%. Strain at peak stress is in the range of 0.002–0.004. The Poisson ratio increases with the increase in clay content of the block, and can be as high as 0.24 for low cement content and high clay fraction. Generally, for blocks with 6–8% cement (and dry density > 1800 kg/m^3) with an optimum clay content of 15% and using coarse-grained soils, the Poisson ratio is in the range of 0.10–0.18 in the saturated state (Venkatarama Reddy *et al.*, 2007b; Gupta, 2004; Vyas, 2007).

13.5   Cement–soil mortars for stabilised soil block masonry

Clearly, an important role for any mortar in masonry is to develop a good bond with the masonry unit and prevent moisture ingress through the mortar joint. Cement mortar and cement–lime mortar are the commonly used conventional mortars for masonry. Pure cement mortars, especially

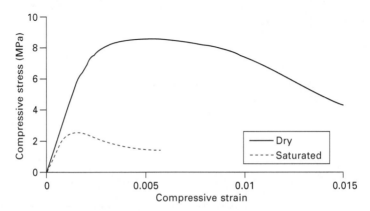

13.22 Typical stress strain curves for cement SSB.

the leaner mortars, have poor workability characteristics coupled with low water retentivity and develop poor bond strength. Cement–lime mortar possesses good workability, water retentivity and better bond strength when compared to pure cement mortars. Replacing lime with soil in cement–lime mortar results in cement–soil mortar. Cement–soil mortars are economical and possess better properties when compared to conventional mortars. Such mortars are commonly used for SSB masonry in many countries.

Cement–soil mortars are composed of OPC, soil and sand with the requisite quantity of water. Such mortars can be prepared on-site by carefully mixing the raw materials to a desired consistency. Soils containing less expansive clay minerals such as kaolinite can be used. Both natural river sand and manufactured sand (conforming to the general requirements mentioned in standard codes of practice for mortars) can be used. Advantages of cement–soil mortars are: (i) superior workability and better water retentivity, and no segregation even at very high water-cement ratio; (ii) higher masonry bond strength; (iii) lower cost; and (iv) they possess natural soil texture. A detailed discussion of cement–soil mortars can be found in Walker and Stace (1997) and Venkatarama Reddy and Gupta (2005a).

Table 13.4 gives details of some typical characteristics of commonly used cement–soil and cement-lime mortars (Gupta, 2004; Venkatarama Reddy and Gupta, 2005a). Workability of mortars can be quantified by measuring the flow value. Gupta (2004) measured the flow values of fresh mortars used by masons for the construction of SSB masonry and found that a mortar with a flow of about 100% was preferred. To achieve a flow of 100%, the water–cement ratio differs amongst mortars with different compositions. Water retentivity of cement–soil mortars is more than that of conventional cement–sand mortar. Whilst the compressive strengths of cement–lime mortar (1:1:6) and cement–soil mortar (1:2:5) are comparable, the masonry bond strength of cement–soil mortar is superior to that of cement–lime mortar. Figure 13.23 shows stress–strain relationships for conventional cement–lime mortar and cement–soil mortar.

Table 13.4 Characteristics of cement–soil mortars

Mortar proportion (by volume) *C : L : So : Sa	Flow (%)	W/C ratio	Water retentivity (%)	Cube strength (MPa)	Initial tangent modulus (MPa)	Bond strength (MPa)
1 : 1 : 0 : 6	100	1.79	82	5.94	4800	0.18
1 : 0 : 2 : 5	100	1.68	87	5.40	5400	0.20
1 : 0 : 2 : 8	100	1.96	84	3.42	4000	0.14

*C: cement, L: Lime, So: Soil, Sa: Sand.

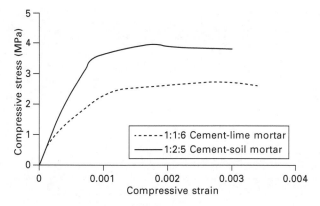

13.23 Stress–strain relationships for cement–lime mortar and cement–soil mortar.

13.6 Stabilised soil block masonry

Masonry is a layered composite structure consisting of masonry units and mortar. Structural behaviour of SSB masonry is influenced by the strength and deformation characteristics of the blocks and the mortar. Generally, the two materials have different strengths and elastic properties. The strength of SSB masonry is influenced by a large number of factors, such as block strength, block height, mortar strength, mortar joint thickness, modulus and Poisson ratio of the mortar and direction of loading.

13.6.1 Stabilised soil block masonry strength and design guidance

The compressive strength of masonry is less than that of the compressive strength of the individual brick or block. This is attributed to the development of splitting tensile cracks developed in a masonry unit due to the differing deformation characteristics of the masonry unit and the mortar. Typical failure patterns of a brick or block and masonry prism are shown in Fig. 13.24. These failure patterns clearly show that the brick or block fails in shear, as a result of platen friction, whereas the masonry prism and wall fail due to the development of vertical splitting cracks. Vertical splitting cracks indicate that the bricks or blocks in the masonry wall are subjected to tensile stresses even though they are loaded in compression. This is due to the differing modulus values for the bricks and the mortar. In addition to the magnitude of compressive loads, slenderness ratio and load eccentricity also influence the compressive strength of masonry walls. For a given mortar–block combination, the compressive strength of a wall is much less than that of a block.

(a)

(b)

13.24 Failure patterns in (a) an SSB and (b) masonry wallette under compression.

Investigations by Venkatarama Reddy and Jagadish (1989), Shrinivasa Rao *et al.* (1995), Walker (2004), Venkatarama Reddy *et al.* (2007b), Venkatarama Reddy and Vyas (2008) and Venkatarama Reddy and Gupta (2008) throw some light on the strength of cement SSB masonry. Table 13.5 gives a compilation of the compressive strength (wet) of cement SSB masonry prisms extracted from the above mentioned investigations. A range of block strengths (2.5–11.5 MPa) and mortar strengths (0.75–9.40 MPa) have been considered to arrive at the masonry prism strengths. The prism strength to block strength ratio is in the range of 0.43–0.52, except in a few cases where it is more than 0.60. Masonry prism strength is not very sensitive to mortar strength. There is a large scope for understanding the block aspect ratio (height to width) on masonry compressive strength. Given the limited data on masonry strength, the relationship between masonry prism strength and block strength can be used to assess the compressive strength of masonry walls. Assuming a conservative value of prism to block strength ratio of 0.40, the masonry prism strength can be assumed to be 0.40 times the block strength. This value of prism strength can lead to an assessment of either characteristic compressive strength or basic compressive stress of SSB masonry. If working stress design principles are used, the basic compressive stress of SSB masonry can be obtained, being the prism strength divided by a factor of 4 (as per the IS 1905 code). A brief summary of masonry design procedures using working stress design principles is given below.

Procedure for the design of a masonry wall under vertical gravity loads

1   First, assume the masonry wall thickness
2   Estimate the total load realised on the wall and compressive stress developed at the base of the wall
3   Determine the slenderness ratio and eccentricity (using masonry code guidelines) and obtain the stress reduction factors from the code
4   Allowable compressive stress = (basic compressive stress) × (stress reduction factors)
5   Equate allowable compressive stress to the compressive stress developed at the base of the wall and obtain the basic compressive stress value
6   Masonry design codes give relationships (either in tabular form or graphical form) between the masonry unit strength, mortar type and basic compressive stress. For any computed value of basic compressive stress, select brick or block strength and mortar type from such relationships, under concentric and eccentric loads.

Similar design procedures can be evolved using limit state design principles provided a procedure is established to compute the characteristic compressive strength of SSB masonry using masonry prism strength values. There is still scope for further comprehensive investigations to be undertaken into the behaviour of full-scale SSB masonry walls.

13.7   Long-term performance, repair and retrofitting of stabilised soil block buildings

Satisfactory performance of an SSB masonry building is the building owner's most important concern. Deterioration/weathering of SSB walls can be

Table 13.5 Strength of stabilised soil blocks and masonry

Block characteristics		Mortar characteristics		Masonry prism characteristics		(B)/(A)
Size (mm)	Strength (MPa) (A)	Proportion	Strength (MPa)	Size (mm)	Strength (MPa) (B)	
305 × 146 × 82	2.51	1:6 cement mortar	3.38	305 × 146 × 368	1.52	0.61[@]
305 × 143 × 100	8.34	1:2:5 Cement–soil mortar	3.45	305 × 143 × 550	3.66	0.44[*]
305 × 143 × 100	8.34	1:1:6 Cement–lime mortar	2.93	305 × 143 × 550	3.64	0.44[*]
305 × 143 × 100	7.19	1:6 cement mortar	5.40	305 × 143 × 436	4.55	0.63[**]
305 × 143 × 100	7.19	1:1:6 cement–lime mortar	5.94	305 × 143 × 436	5.27	0.73[**]
305 × 143 × 100	4.94	1:6 cement mortar	6.07	305 × 143 × 345	2.14	0.43[+]
305 × 143 × 100	4.94	1:2:8 cement–soil mortar	1.51	305 × 143 × 345	2.38	0.48[+]
255 × 122 × 80	5.09	1:1:6 cement–lime mortar	3.42	255 × 122 × 440	2.30	0.45[++]
255 × 122 × 80	5.09	1:0.5:4 cement–lime mortar	9.40	255 × 122 × 440	2.50	0.49[++]
255 × 122 × 80	11.46	1:1:6 cement–lime mortar	3.42	255 × 122 × 440	5.75	0.50[++]
295 × 140 × 120	3.3	Cement–soil mortar	0.75	295 × 140 × 640	1.70	0.52[#]

[@] Venkatarama Reddy and Jagadish (1989); [*] Venkatarama Reddy et al. 2007b; [**] Venkatarama Reddy and Gupta (2008); [+] Rao et al. (1995); [++] Venkatarama Reddy and Uday Vyas (2008); [#] Walker (2004) attributed to various factors relating to material composition of the block, and to disintegration caused by mechanical means. Clay minerals present in the blocks (and with a lot of affinity for water) are a major contributing factor for deterioration of SSBs. Unstabilised clay pockets present in the blocks can absorb water and undergo swelling, and then shrink on drying; this process can manifest as surface cracks, and further cycles of wetting and drying can disrupt the established cementitious bonds between various particles causing deterioration of the SSBs with age. Frost action can also damage and cause deterioration. Hence, the quantity of clay present in the SSBs, the percentage of stabiliser used, and the extent of exposure to cyclic wetting and drying can all control the long-term performance of SSBs. Figure 13.25 shows cracking and deterioration of SSB walls. Blocks present at the corners and edges of walls, window sills and jambs are more prone to such types of failures, which can damage walls considerably and lead to collapse of a structure. All such types of damage can be attributed to inadequate stabilisation of clay minerals in the soil used for the SSBs.

Damaged and deteriorated SSB walls can be repaired and retrofitted either by patch plastering or by progressively replacing the damaged parts. Patch plastering repairs the cracks, pitting, spalling and damaged surface of SSB walls whilst using a compatible mortar. Here, the compatibility is in terms of shrinkage characteristics of the mortar and the SSB, mainly to avoid differential shrinkage between the block and the patched up mortar on the damaged surface. Cement–soil mortars with a clay fraction of about 10% and cement content of 10–12% are compatible, and hence such mortars can be used for the patch plastering of damaged SSB block surfaces. Figure 13.26 shows a patch plastered surface performing satisfactorily.

Severe damage to walls can jeopardise the safety of a building and could lead to collapse. Progressively replacing a badly damaged or partly collapsed

13.25 Different types of SSB wall damage and failure (L–R: Splitting cracks; erosion and pitting; spalling).

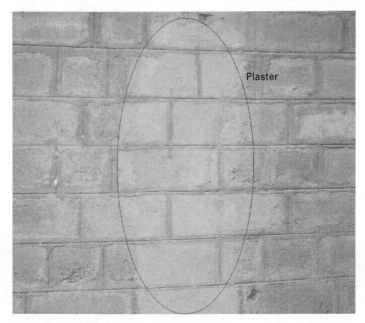

13.26 SSB wall with patch plaster.

SSB wall can repair it without disturbing the neighbouring portions. Figure 13.27 shows an SSB wall that has been retrofitted without disturbing the first floor wall (Venkatarama Reddy, 2008a).

13.8   Case studies of cement-stabilised soil block (CSSB) buildings

A variety of load-bearing masonry CSSB buildings have been built across India and a number of case studies are highlighted here.

13.8.1 Residential building, Bangalore, India (Fig. 13.28)

This two-storey building (with a provision for one more floor) has load-bearing SSB masonry walls (230 mm thickness) in cement–soil mortar and was built in 2000. The blocks of size 230 × 190 × 100 mm were manufactured with 9% Portland cement on-site, using a portable manual machine. Soil processing and block manufacturing was a completely manual operation. The wet compressive strength of the block was 6.5 MPa (courtesy: Design and construction by Dr M. R. Yogananada, Bangalore, India).

13.27 Retrofitted SSB wall.

13.28 Case study – Residential building, Bangalore, India.

13.8.2 Seminar hall complex, Indian Institute of Science, Bangalore, India (Fig. 13.29)

This is a two-storey load-bearing CSSB masonry wall building (with provision for one more floor) and was built in 2004. The ground floor consists of a seminar hall (clear span = 6.25 m) attached to a staircase room. The walls are 230 mm thick built using $230 \times 190 \times 100$ mm blocks and a cement–soil mortar. Exposed block-work outside and the interior wall faces are plastered with cement–soil mortar. The wet compressive strength of the CSSBs with 8% cement was 7.0 MPa. The floors and roof for the main hall are made up of composite masonry jack-arch using CSSB blocks. The staircase portion has a CSSB filler slab floor/roof (courtesy: Design and construction by Prof. B. V. Venkatarama Reddy, Indian Institute of Science, Bangalore, India).

13.8.3 Cement-stabilised soil block (CSSB) masonry vault and CSSB decorative features (Fig. 13.30)

This CSSB masonry vault was built using cement–soil mortar in 1995. The vault span is 3.5 m, with a thickness of 143 mm. CSSB corbelled blocks and

(a)   (b)

13.29 (a) and (b) Case study – Seminar hall complex, Indian Institute of Science, Bangalore, India.

13.30 Case study – CSSB masonry vault and CSSB decorative features.

decorative jali block masonry was used for the balcony. This is a three-storey load-bearing CSSB masonry building with 190-mm thick walls (courtesy: Design and construction by Dr M. R. Yogananada, Bangalore, India).

13.8.4 Vikas Project, Auroville, India (13.31)

This four-storey load-bearing masonry residential building complex was built using CSSB blocks in cement–soil mortar during 1991–1998. The complex has a cellar portion with cement-stabilised rammed earth foundation. The soil excavated in the cellar was utilised for the production of CSSB blocks. The masonry walls are 240 mm thick. The CSSBs and rammed earth were stabilised with 5% Portland cement. The floor and roofing system consists of unreinforced masonry vaults. This project was a Finalist for the *2000 World Habitat Award, UK* (courtesy: design and construction by Mr Satprem, Auroville, Pondicherry, India).

13.8.5 Multi-storey residential building, Bangalore, India (Fig. 13.32)

This multi-storey (basement plus three floors) load-bearing masonry residential building using CSSB masonry in cement–soil mortar was built during 2006.

13.31 Case study – Vikas Project, Auroville, Pondicherry, India.

13.32 Case study – Multi-storey residential building, Bangalore, India.

The walls are 230-mm thick. CSSB compressive strength (wet) was 9 MPa for the blocks used in the basement floor and 7 MPa for blocks in subsequent floors. The block strength was conveniently adjusted by using 12% Portland cement for the blocks used in the basement floor and 9% cement for the blocks in the other floors. The CSSB blocks were manufactured on-site using a portable manually operated machine (courtesy: design and construction by Dr M. R. Yogananada, Bangalore, India).

13.9   References

ASTM D 559 – 03, (2003), 'Standard test methods for wetting and drying compacted soil-cement mixtures', American Society for Testing and Materials, Pennsylvania

Chaston, F. N., (1952), 'Soil-cement progress in Australia', *The Indian Concrete Journal*, 26(12), 354–356

Department of Housing and Urban Development, (1955), 'Earth for homes', Division of International Affairs, Washington DC, 20140

Fitzmaurice, R. F., (1958), 'Manual on stabilised soil construction for housing', UN Technical Assistance Program, New York

Gupta, A., (2004), 'Studies on characteristics of cement–soil mortars and soil-cement block masonry', MSc (Engg.) thesis, Dept. of Civil Engineering, Indian Institute of Science, Bangalore, India

Heathcote, K., (1991), 'Compressive strength of cement stabilised pressed earth blocks', *Building Research and Information*, 19(2), 101–105

Heathcote, K., (1995), 'Durability of earth wall buildings', *Construction and Building Materials*, 9(3), 185–189

Heathcote, K., (2001), 'An investigation into the erosion of earth walls', PhD thesis, University of Technology, Sydney

Houben, H. and Guillaud, H., (2003), Earth Construction – A Comprehensive Guide, CRATerre-EAG, Intermediate Technology Publications, Belgium

I S 1905, (1987), 'Code of practice for structural use of unreinforced masonry', Bureau of Indian Standards, New Delhi, India

Kouakou, C. H. and Morel J. C., (2009), 'Strength and elasto-plastic properties of non-industrial building materials manufactured with clay as a natural binder', *Applied Clay Science* 44, 27–34

Lambe, T. W., (1962), 'Soil stabilisation', *Foundation Engineering*, edited by G.A. Leonards, McGraw Hill Book Co. Inc., New York, 351–437

Lunt, M. G., (1980), 'Stabilised soil blocks for buildings', *Overseas Building Notes* No. 184

Madhavan, R. and Narasinga Rao, C. N., (1949), 'Report on Labour Housing Scheme', The City Improvement Trust Board, Bangalore, India

Maniatidis V. and Walker, P., (2008), 'Structural capacity of rammed earth in compression', *J of Mat. in Civil Engineering* (ASCE), 20(3), 230–238

Middleton, G. F., (1952), (revised by Schneider, L. M., 1987), 'Earth-wall construction', *Bulletin 5*, 4th ed., Commonwealth Scientific and Industrial Research Organization, Australia

Mitra, J. N., (1951), 'Suitability of soil for stabilised soil houses for Rangawan dam colony', *Indian Concrete Journal* 15, 234–238

Mukerji, K., (1986), 'Soil block presses', report on Global survey, GATE, Dag-Hammerxjold-weg, 1, 6236, Eschborn, Germany

Olivier, M., (1994), 'Le matériau terre-compactage, comportement, application aux structures en blocs de terre', PhD thesis, Institut National des Sciences Appliquées, Lyon, France

Olivier, M. and Mesbah, A., (1987), 'Influence of different parameters on the resistance of earth, used as a building material', Proceedings of the International Conference on Mud Architecture, Trivandrum, India

Portland Cement Association, (1956), 'Soil-cement laboratory handbook', Portland Cement Association, Skokie, Ill

Prasanna Kumar, P., (2010), 'Stabilised rammed earth for walls: materials, compressive strength and elastic properties', PhD thesis, Dept. of Civil Engineering, Indian Institute of Science, Bangalore, India

Shrinivasa Rao, S., Venkatarama Reddy, B. V. and Jagadish, K. S., (1995), 'Strength characteristics of soil-cement block masonry', *The Indian Concrete Journal*, 69(2), 127–131

Spence, R. J. S., (1975), 'Predicting the performance of soil-cement as a building material in tropical countries', *Building Science*, 10, 155–159

Ullas, S. N. and Venkatarama Reddy, B. V., (2007), 'Characteristics of soil-cement blocks from different construction sites', *Proceedings of the International Symposium on Earthen Structures*, Interline Publishers, Bangalore, India, 141–146

UN Report, (1964), 'Soil-cement – its use in building', Dept. of Economics and Social Affairs, United Nations, New York

Venkatarama Reddy, B. V., (1983), 'On the technology of pressed soil blocks for wall construction', MSc (Engg) thesis, Dept. of Civil Engineering, Indian Institute of Science, Bangalore, India

Venkatarama Reddy, B. V., (1991), 'Studies on static soil compaction and compacted soil-cement blocks for walls', PhD thesis, Dept. of Civil Engineering, Indian Institute of Science, Bangalore, India

Venkatarama Reddy B. V., (2002), 'Long-term strength and durability of stabilised mud blocks', *Proceedings of the 3rd International Conference on Non-conventional Materials and Technologies*, Construction Publishing House, Hanoi, Vietnam, 422–431

Venkatarama Reddy, B. V., (2007), 'Indian standard code of practice for manufacture and use of stabilised mud blocks for masonry', *Proceedings of the International Symposium on Earthen Structures*, Interline Publishers, Bangalore, India, 194–202

Venkatarama Reddy, B. V., (2008a), 'Retrofitting of damaged stabilised earth block buildings', *Proceedings of the 5th International Conference on Building with Earth* (LEHM 2008), 10–12 October, Koblanz Germany, Dachverband Lehm e.V, Germany, 4–111

Venkatarama Reddy, B. V., (2008b), 'The stabilised mud block: a low carbon emission alternative for masonry buildings', *Proceedings of the First International Conference on Building Energy and Environment* (COBEE-2008), Dalian, China, 13–16 July, 1397–1404

Venkatarama Reddy, B. V., (2009), 'Sustainable materials for low carbon buildings, *International Journal of Low Carbon Technologies*, 4(3), 175–181

Venkatarama Reddy, B. V. and Gupta, A., (2005a), 'Characteristics of cement–soil mortars', *Materials and Structures (RILEM)*, 38(280), 639–650

Venkatarama Reddy, B. V. and Gupta, A., (2005b), 'Characteristics of soil-cement blocks using highly sandy soils', *Materials and Structures (RILEM)*, 38(280), 651–658

Venkatarama Reddy, B. V. and Gupta, A., (2006), 'Strength and elastic properties of stabilised mud block masonry using cement soil mortars', *J. of Mat. in Civil Eng.*, 18(3), 472–476

Venkatarama Reddy, B. V. and Gupta, A., (2008), 'Influence of sand grading on the characteristics of mortars and soil-cement block masonry', *Construction and Building Materials*, 22(8), 614–1623

Venkatarama Reddy, B. V. and Hubli, S. R., (2002), 'Properties of lime stabilised steam-cured blocks for masonry', *Materials and Structures (RILEM)*, 35, 293–300

Venkatarama Reddy, B. V. and Jagadish, K. S., (1984), 'Pressed soil-lime blocks for building construction', *Masonry International*, 3, 80–84

Venkatarama Reddy, B. V. and Jagadish, K. S., (1987), 'Spray erosion studies on pressed soil blocks, *Building and Environment*, 22(2), 135–140

Venkatarama Reddy, B. V. and Jagadish, K. S., (1989), 'Properties of soil-cement block masonry', *Masonry International*, 3(2), 80–84

Venkatarama Reddy, B. V. and Jagadish, K. S., (1993), 'The static compaction of soils', *Geotechnique* 43(2), 337–341

Venkatarama Reddy, B. V. and Jagadish, K. S., (1995), 'Influence of soil composition on the strength and durability of soil-cement blocks', *The Indian Concrete Journal*, 69(9), 517–524

Venkatarama Reddy, B. V. and Lokras, S. S., (1998), 'Steam-cured stabilised soil blocks for masonry construction', *Energy and Buildings*, 29, 29–33

Venkatarama Reddy, B. V. and Vyas, U. C. V., (2008), 'Influence of shear bond strength on compressive strength and stress strain characteristics of masonry', *Materials and Structures (RILEM)*, 41(10), 1697–1712

Venkatarama Reddy B. V. and Walker, P., (2005), 'Stabilised mud blocks: problems, prospects', Proceedings of the International Earth Building Conference, Sydney, Australia, 63–75.

Venkatarama Reddy, B. V., Richardson, L. and Nanjunda Rao, K. S., (2007a), 'Optimum soil grading for the soil-cement blocks', *J. of Mat. in Civil Eng.*, 19(2), 139–148

Venkatarama Reddy, B. V., Richardson L. and Nanjunda Rao, K. S., (2007b), 'Enhancing bond strength and characteristics of soil-cement block masonry', *J. of Mat. in Civil Eng. (ASCE)*, 19(2), 164–172

Venkatarama Reddy, B. V., Richardson, L. and Nanjunda Rao, K. S., (2009), 'Influence of joint thickness and mortar-block elastic properties on the strength and stresses developed in soil-cement block masonry', *J of Mat. in Civil Eng.*, 21(10), 535–542

Venkatarama Reddy B. V., Williams, T. and Walker, P., (2003), 'Durability of stabilised mud block buildings in Southern India', *Proceedings of the 9th International Conference on the Study and Conservation of Earthen Architecture (Terra–2003)*, Yazd, Iran, 492–503

Vyas, U. C. V., (2007), 'Studies on shear bond strength – masonry compressive strength relationships and finite element model for prediction of masonry compressive strength', MSc (Engg) thesis, Dept. of Civil Engineering, Indian Institute of Science, Bangalore, India

Walker, P., (1998), 'Erosion testing of compressed earth blocks', *Proceedings of the 5th International Masonry Conference*, London, UK, 264–268

Walker, P., (1999), 'Bond characteristics of earth block masonry', *J. Mater. Civ. Eng.*, 11(3), 249–256

Walker, P., (2004), 'Strength and erosion characteristics of earth blocks, and earth block masonry', *J. Mater. Civ. Eng.*, 16(5), 497–506

Walker, P. and Stace, T., (1997), 'Properties of some cement stabilised compressed earth blocks and mortars', *Mater. Struct.*, 30, 545–551.

Walker, P., Venkatarama Reddy, B. V., Mesbah, A. and Morel, J.–C., (2000), 'The case for compressed earth block construction', *Proceedings of the 6th International Seminar on Structural Masonry for Developing Countries*, Bangalore, India, 27–35

Yttrup, P. J., Diviny, K. and Sottile, F., (1981), 'Development of drip test for the erodibility of mud bricks', Deakin University, Geelong, Australia

14
Modern rammed earth construction techniques

D. EASTON and T. EASTON, Rammed Earth Works, USA

Abstract: This chapter provides an overview of modern rammed earth construction techniques, with a particular emphasis on current practices in the United States. It describes all of the key steps in constructing rammed earth wall systems, recognizing the myriad alternative approaches that exist for every part of the process. The chapter ends with a reflection on likely future trends and a discussion of factors that could contribute to growth in the field.

Key words: rammed earth, amending mix designs, machine mixing, manufactured form systems, conveyor delivery.

14.1 Introduction

The earliest examples of rammed earth were made by compacting one damp soil layer at a time into temporary wooden formwork. Soil could be dug by hand from the site and, when the moisture was suitable, installed directly into the formwork. Time, simple formwork and a compacting tool were essentially the only resources required for constructing these modest but long-lasting structures.

Over the course of centuries, advancements in the field have been significant. The aesthetic associated with modern rammed earth is hardly comparable to its humble ancestry (see Fig. 14.1); engineering specifications may require that rammed earth be stabilized with cement, reinforced with steel and designed to reach compressive strengths in excess of 3000 pounds per square inch (psi) (20.68 MPa). In its most complex applications, rammed earth is an engineered, high-strength, environmentally responsible building material installed with sophisticated machinery. The low-tech, labor-intensive version of rammed earth construction still has great value, however. In developing nations, labor is inexpensive and conventional building resources scarce. There, the less sophisticated method of building rammed earth can prove appropriate and effective.

Rammed earth construction has a wide range of approaches. In one scenario, an installation can involve a single form made of reuseable, hand-sawn lumber, shovels, buckets and heavy wooden rammers. Materials for

364

14.1 Rammed earth with modern aesthetic, Las Vegas, NV.

construction can be harvested locally, forming lumber can be reused for roofing or other construction applications, and unskilled volunteer labor can be utilized. The modern method involves laminate plywood, aluminum form supports, conveyor belts, air compressors, backfill tampers, tractors, mixing machines and laboratory testing.

In California, engineering requirements demand that rammed earth be stabilized with cement to improve compressive strength and other durability characteristics, reinforced with steel to protect wall systems from failure in the event of seismic activity and amended with imported aggregate materials to optimize quality. High labor rates require that many workers with non-mechanical tools be replaced with capital-intensive, labor-lean systems. Rammed earth's acceptance into the United States building industry, especially seismically active California, has naturally been biased towards reinforced and stabilized methods.

The resulting range of approaches to building with rammed earth can be highly variable in (i) cost, (ii) labor requirements, (iii) production times and (iv) associated ecological impacts.

Although techniques for the application of rammed earth vary considerably from region to region, the following chapter has been organized and written to provide a broad working knowledge of the various aspects of the process, from selection of raw materials to treatment of finished walls. It is heavily weighted towards the authors' direct experience – which has been primarily gained in California.

14.2   Material sourcing

The strength and look of a finished rammed earth wall is directly governed by the raw materials in the mix design, and material sourcing and processing have significant financial and environmental impacts on a project. The final selection of bulk materials will involve making trade-offs between cost of materials and wall quality. Geotechnical evaluation in conjunction with trial mix designs and pre-construction testing is highly recommended.

14.2.1 Site-sourced materials

In the ideal situation, suitable soil materials required for construction can be found on the building site, and required quantities can be matched to excavations associated with site work. On small projects, site sourcing will be limited to the footprint of the structure. On larger parcels it is often possible to encounter different soil profiles, one or more of which may have the required characteristics.

In many cases, before a site-sourced material can be used in a mix formulation, screening or other forms of processing may be required. For smaller projects, a screening operation may be as simple as shoveling raw material through a static screen. Larger projects requiring greater quantities may warrant front-loaders and vibratory screening plants. Our experience has indicated that the expenses associated with on-site processing can in some cases equal the price per yard of an imported quarry product.

14.2.2 Amending and blending

In most cases, a suitable mix design can be developed using some percentage of site-excavated soil in combination with one or more imported amendments. Optimum proportioning between site soil and amendment is determined through materials testing. Cost and intended wall quality will govern the decision. Imported materials can be expensive but their inclusion can enhance wall characteristics and reduce stabilization requirements. Common field and laboratory testing for rammed earth material inputs can be found in Chapter 6 of this text, as well as in Section 14.7 of this chapter.

Importable amendments typically come from quarries. Occasionally excess soil material from some close or adjacent construction project such as a pipeline, road cut, basement or pool excavation might be obtained for the cost of transport. Be aware that free imports may require on-site screening to reduce gravel size and to break-up clods to improve the materials' suitability.

The second source for amendment is virgin aggregate from a rock, sand or gravel quarry. The advantages to importing from a quarry are that the selected material is usually uniform, consistent and requiring no on-site processing. Additionally, materials can be delivered in manageable, incremental quantities, which can be of benefit on small sites.

14.2.3 Quarry waste and recycled materials

Quarries may prove an even greater resource if they have what can be classified as post-industrial aggregates that are suitable for use as soil amendments. Most quarry operations are geared to the manufacture of clean building products with characteristics that adhere to industry standards. In many cases, the production of these materials results in the creation of lesser value by-products. These by-products may be 'overburden', 'crusher fines' or 'tailings' – essentially waste products generated during primary processing. These types of products are in many cases suitable for use in rammed earth mix designs, with the nature of their creation allowing them to be classified as post-industrial recycled. This can work to double advantage for the environmentally conscious rammed earth builder – they are sold at reduced price and can reduce the ecological impacts associated with construction. Furthermore, the LEED rating system awards points for the use of post-industrial recycled materials.

Recycled brick, tile and concrete can also act as amendments for a balanced rammed earth mix formulation. Gradation varies depending on the input and the processing. Advantages include (i) color enhancement: (brick and tile can introduce strong colors into the mix) and; (ii) increased strength (recycled concrete in some cases retains free cement, which can allow for a reduction in stabilization requirements).

14.2.4 Transportation expenses

Heavy trucks and diesel fuel account for a high percentage of the cost of soil amendment. It follows that the further the amendment source is from the project, the more significant are the environmental (and monetary) costs associated with the importation of that amendment. The choice of a final mix design, therefore, will involve decision-making in three key areas: (i) wall quality and performance; (ii) raw material costs and (iii) ecological impacts of construction. An argument might be made that the more durable the wall (i.e. improved wall quality) the more years of service it can provide, thus offsetting potentially higher initial construction costs (both monetary and ecological). Conversely, a rammed earth wall built entirely of site-sourced material with low to no stabilization will consume fewer non-renewable resources but may not have the same durability. Life-cycle costing takes into account both the energy inputs and the life expectancy of the finished product.

14.2.5 Stockpiling and moisture control

Site logistics are an important factor in staging any rammed earth construction project. Stockpiling raw materials, proportioning and blending, delivering the prepared mix to the formwork – all require substantial room on the site and around the building perimeter. When a portion of site-soil is to be used in the formulation, this will have been excavated in previous stages of construction and should be stored near the work area but out of the paths of other trades. Imported material must be off-loaded and stored so that it is both accessible and convenient. Work areas for carpenters to construct formwork must be maintained. Careful pre-construction planning during the early stages of construction can reduce wasted man-hours and equipment costs.

Once the selected soil and aggregate materials have been stockpiled on site, it is important to ensure that the stockpiles are protected from rainy and windy weather. If such conditions are anticipated, covering the piles with tarps or plastic is critical. All materials should be relatively dry at the time of proportioning and blending. This will help to achieve correct material ratios, simplify mixing, reduce clods and clay balls, and achieve homogeneity in the final mix.

14.2.6 Summary

The creation of a high-quality mix design usually requires a minimum of two material inputs. In situations where suitable soil materials are found on-site, their composition will commonly include higher amounts of silt and/or clay-sized particles than is optimum. Additionally, they may require screening or other on-site processing. Sand and gravel are normally readily available from nearby quarries or landscape supply yards and can be used to improve a high-fines site-soil. Waste and recycled, both post-consumer and post-industrial, amendments should be preferable to virgin aggregates because of the associated environmental benefits, though performance testing should be the qualifying factor. Because of the high costs associated with transporting bulk soil and aggregate materials, informed decision-making throughout the process of selecting raw materials is critical. After those stockpiled materials are assembled they should be protected from the elements.

14.3   Proportioning and mixing

The finished quality of a rammed earth wall is governed by the characteristics and proportions of its material inputs. On one hand, a completely serviceable wall may be constructed of native soil excavated from the site, placed directly into the formwork, and compacted to density. However, the future success of the rammed earth industry demands that procedures for the precise control over quality and consistency of ingredients continue to be established and improved. To date, this control has largely existed in the form of engineering performance specifications defining acceptable ranges for: (i) input material gradations and (ii) the finished product's ultimate compressive strength. Where gradation requirements remain relatively consistent within the industry, compressive strength requirements vary by as much as a factor of ten. Upper limit compressive strengths (e.g. > 2000 psi (13.79 MPa)) are achievable only with careful formulation, high stabilization ratios and precise quality control. History and current practice in some parts of the world leave no room for debate as to the potential for success and long term serviceability of unstabilized rammed earth. With proper soil selection and compaction, such rammed earth walls can meet all realistic structural expectations for low-rise construction. However in New Zealand, Australia and North America, where much of the recent revival of rammed earth has taken place, building codes have specified compressive strengths that are unattainable without cement stabilization. For this reason, the formulations and processes to be discussed in the following sections will be related with the assumption that cement is included in the mix design.

14.3.1 Proportioning

In basic formulations, where only one raw aggregate source and one stabilizer are involved, proportioning is a simple volumetric ratio between soil and cement. A 7% cement stabilization ratio is roughly 13 parts soil to 1 part cement. Measured with shovels, it is 1 shovel of cement to 13 shovels of soil.

In larger batches, proportioning is accomplished with the buckets on front-loading tractors or skid-steer tractors. Cement is measured in sacks. One sack equals one cubic foot. Tractor buckets are typically 1/4 yard, 1/3 yard or 1/2 yard (roughly 7 cubic feet, 9 cubic feet or 13 cubic feet). The challenge when measuring with tractor buckets is to develop the ability to maintain a consistent fill line. A heaping bucket may contain as much as 50% more material (by volume) than a level bucket. When two or more equipment operators are sharing proportioning responsibilities, it's essential that they agree on and maintain consistent fill lines. Take for example, one operator using a 1/3 yard tractor bucket deposits 6 level buckets onto the mix pad, the result is roughly 2 cubic yards. For a cement ratio of 7% he would then empty 4 sacks of cement onto the pile. A second operator deposits 6 heaping buckets onto the mix pad, resulting in roughly 3 cubic yards. Four sacks of cement onto this pile would yield a stabilization ratio of 5%, a reduction in total cement content of nearly 30%.

Increased control over proportioning can be achieved using any of several volumetric mixing machines manufactured for the concrete industry. These machines are designed to give precise control over the ratio between cement, aggregate, water and admixtures. In addition to the major equipment manufacturers such as Cementech, Elkin, Zimmerman, Airplaco and Strong, there are a few smaller companies marketing soil-blending machinery.

In formulations where two or more raw material sources are being blended, pre-proportioning materials prior to mixing is advantageous. Pre-blending the full quantity of soil and aggregate calculated for the entire wall system will yield more consistent results.

14.3.2 Mixing

As with proportioning, there are a variety of approaches to mixing materials in preparation for installation, with varying levels of precision and capital investment. Determination of the most suitable method to employ should involve deliberation on the following aspects of the project: (i) scope, (ii) budget, (iii) available time, (iv) labor skill-set and (v) desired finished wall quality. The approaches can generally be divided into two groups: ground mixing and machine mixing.

Ground mixing

This is by far the most common mixing technique (historically and in current use), and requires the least capital investment. After proportioning the soil, aggregate and cement materials on a hard, clean mixing area, the ingredients are incorporated with one another in their dry state. Once thoroughly combined, the materials are continually mixed as water is added to attain the proper moisture content. The optimum moisture content for a particular mix design is based on maximum dry density and maximum achievable compressive strength, and should be determined prior to construction through laboratory testing. Heavily affected by stabilization ratios, typically it will be in the range of 8–12% of the mix by weight.

The process of blending dry materials on a mixing pad can be accomplished in one of four ways (presented here in order of increasing capital investment): (i) by hand with shovels, (ii) with a garden type rototiller in combination with shovels, (iii) with a tractor bucket or (iv) with a rototiller mounted on a tractor used in combination with the tractor bucket. The use of a rototiller in mixing operations can yield significant improvements in mix quality, especially when dealing with soils with clods. The desired pace of the installation will influence which strategy is optimal.

With small unit set-ups, a single two- to four-yard batch can be mixed at one time using any of the above techniques, with proper moisture being maintained throughout installation either with tarps or misting. The prepared mix should be placed into the forms and compacted as quickly as is feasible, with not more than two hours elapsing between initial hydration of the cement and installation. Special care should be taken and increased haste used in especially hot or windy conditions.

For larger form set-ups, where material requirements exceed four yards, smaller batches are preferable, with each batch mixed, moistened and delivered to the formwork as it is needed.

Machine mixing

As rammed earth projects have increased in size and complexity, structural engineering and project oversight have also increased. This is especially true in California where earthquake safety and litigation aversion are paramount. Building officials and special inspectors question the quality control when proportioning with buckets, mixing on the ground and moistening with a garden hose.

In 1990, the authors began the conversion from ground mixing to machine mixing in order to gain better control over mix consistency and directly respond to the quality control concerns of design professionals.

In machine mixing, the dry soil and aggregate are loaded into one or more stationery bins, with either a conveyor belt or a dry material auger pulling material out of the bin at a set rate and into a mixing auger. At the same time, cement is metered into the dry material. Once soil and cement are combined, the mixing auger blends the material and moves it towards the output end of the machine. Part way along the auger, water is added to the dry mix. Volumetric mixers can be adjusted to precisely control all input proportions, and to deliver the mixed material at the rate at which it is being installed in the formwork. They can be stopped and started within seconds. Small mixers are capable of producing four to five cubic yards per hour. Large mixers can produce ten or more cubic yards per hour.

The capital expense of a mixing machine will have a significant impact on operating costs for a rammed earth builder. Costs of volumetric mixers range from $50,000 – 100,000.

There are several challenges to competitively pricing rammed earth using expensive equipment. It is important that the improvements in mixing and delivery result in better wall quality and higher strengths. Second, the forming system and crew capabilities must be such that they utilize the mixing machine's high rate of production. Third, it is advantageous if construction opportunities follow one after another to allow the rammed earth contractor to maximize equipment use thereby recouping investment costs.

14.4   Formwork

If material sourcing represents the single greatest impact on cost and environmental benefits associated with a rammed earth project, then the choice of forming system represents the second greatest. As with cast-in-place concrete, there are numerous strategies for designing and erecting suitable formwork for rammed earth. Which of the many systems to use on any given project will be influenced by several factors: cost, structural design, reusability, desired finish and available manpower.

Historically, formwork for rammed earth comprised a cumbersome composite of planks and ties, assembled in place and used to define small wall units. The advantages were low cost and reusability. Disadvantages included difficult assembly, low rates of production and a lack of precision. Ragged edges and rough joint lines are inherent in this method. Relatives of this forming system are still in use today in many developing countries, where neither speed of installation nor wall surface quality are of governing importance.

In contrast, on projects where wall quality is a top priority, where panel seams and cold joints must be as inconspicuous as possible, the selection of forming systems is of significant importance. This is especially true because the rammed earth aesthetic has evolved over recent decades from a low-cost utilitarian wall system to a highly prized architectural finished wall.

14.4.1 Small panel formwork

In 1976, David Easton developed a forming system using the traditional rammed earth form as a starting point. This lightweight forming system, quick to assemble and disassemble, could be relied on to produce walls that were true and plumb, having straight tight seams. Called the California form, the system was primarily developed in response to engineering uncertainties about the earthquake safety of rammed earth construction. The early applications of the California form were to build free-standing panels of rammed earth around the building perimeter and to later cast reinforced concrete posts to lock adjacent panels together, all capped with a reinforced concrete bond beam (see Figs 14.2 and 14.3). Since that time, as test data have continued to document the strength of stabilized rammed earth, design engineers have come to rely less on the concrete frame.

The components of the California form system are: (i) plywood form faces backed by (ii) 2″ × 12″ walers spaced at close intervals, with (iii) full-height 'end-boards' (shutters), all held together with (iv) pipe clamps as form ties. The use of wide walers to resist pillowing of the plywood form faces horizontally allows for clamps to be spaced up to nine feet apart. The walers also provide both a ladder and a comfortable working platform at the top of the wall. The wide spacing of the clamps creates an open box form, which allows for uninterrupted and thorough compaction. The finished wall panel is uniform, dense and free of tie holes. Adjacent panels lock into place in a tongue and groove configuration after finished rammed earth panels have

14.2 Panel to panel rammed earth construction using the California form. Finished rammed earth wall with concrete posts and bond beam.

14.3 Continuous form.

been allowed to cure for at least 24 hours. Chamfer strips can be used at form panel junctures if desired.

The California form is well adapted to owner-builders and to applications in developing countries. It is inexpensive, reusable, takes up little room in storage and, being small, establishes realistic daily production milestones. Once a form is fully filled and the top layer of rammed earth is compacted, the form can be immediately disassembled and re-set. Indeed, one of the great advantages to rammed earth construction in general is that a wall or wall section does not rely on its formwork for support during curing (though performance characteristics will improve if protection from the elements is provided at this stage).

14.4.2 Continuous form systems

In contrast to the small form, panel-to-panel approach, continuous forming involves forming long sections of wall at a time (see Fig. 14.4). Broadly, there are two approaches to continuous form systems: site-built and manufactured. Advantages to continuous form systems are (i) finished walls are joint-free for longer expanses than with a panel system, (ii) daily production quotas increase and (iii) enhanced performance characteristics result from extended curing times within the formwork. Disadvantages are: (i) initial cost, (ii) potential for lumber waste and (iii) storage requirements in between uses. For single use form systems, the framing lumber and plywood can be recycled into other elements of the structure.

14.4 Installation using conveyor delivery.

Site-built forms

Site-built forms typically comprise sheets of forming plywood, wooden framing members and extractable form ties. One approach is to build 4′ × 8′ panels of plywood and 2″ × 4″s. The panels are set adjacent to one another along the foundation and 2″ × 12″ walers are used horizontally on opposing sides of the wall, spaced at roughly two-foot intervals. Form ties are located at roughly eight-foot intervals to pull against the walers and support the form system against expansion. If desired, additional support members, called strongbacks, can be placed vertically (perpendicular to the walers) at roughly eight-foot intervals. Strongbacks will provide added insurance that the wall will stay aligned and will allow for increased spacing of the form ties in the vertical direction. Aluminum strongbacks can be rented from several concrete supply houses or they can be fabricated of timber or double 2′ × 12′.

A second approach to site-built formwork is to construct a frame wall along the foundation and attach the plywood sheets to the framework once it is erected, plumbed and braced. The support system of walers, ties and strongbacks is the same as described above. With a simple frame wall, plywood sheets are nailed from the face side, with nails small enough that framing lumber can be pulled away from the sheets once the wall is compacted.

For an easier disassembly, the frame wall can be built with 2″ × 4″ Ts. This way the plywood sheets can be attached to the frame from the backside, through the flange of the T, which allows for a cleaner disassembly. The advantages to this system over the panel system are (i) the layout of plywood seams can be adjusted to the architectural plan, (ii) no screw heads or nail heads showing on the finished wall and (iii) safe disassembly. The disadvantage to the T system is simply that it uses more lumber.

The site-built methods described above are by no means the only options, simply two with which we have had success. Carpenters and concrete form-setters will commonly feel more comfortable erecting formwork of their own design, although it should be emphasized that the forces exerted on the forms from compaction are greater than those from poured concrete.

Note that building forms for curved walls will in nearly all cases require the expertise of a knowledgeable carpenter. Depending on the radius and the symmetry of the curve, constructing, aligning, restraining and bracing a curved wall can consume many more man-hours than forming straight walls. Architects and clients should be made aware of the challenges involved and time that must be committed to building curving rammed earth walls.

Manufactured form systems

Form systems designed and manufactured for the concrete industry can be readily adapted to rammed earth construction. They are typically sold or rented as complete packages, with supports, ties and bracing. In some cases, plywood is attached to panels, in other cases, it is to be installed on site. For large-scale projects, the supplier of rented formwork will provide take-offs and installation diagrams.

Two popular manufactured forming systems are Symons and Atlas. Symons forms are plywood panels wrapped with a steel frame. Panels are typically four feet high, held together with clips and flat ties spaced roughly two feet on center. Walls built using the Symons system are typically set and compacted to four feet in height, after which point a second level of panels is set on top of the first to allow for construction of the rest of the wall. The main advantages of the Symons system are the ease and speed of panel installation and disassembly. The disadvantages are the proximity of form ties and the creation of holes in the finished wall that will need to be patched after tie removal.

The Atlas system comprises aluminum waler beams, aluminum strongbacks and removable form ties – either taper ties or she-bolts. Forming plywood is attached to the aluminum beams on site, either once the framework is erected or on a working deck from which it can be set in place with a grade-all or crane. The advantages to the Atlas system are that wall forms are set to full wall height for one-storey structures, form ties are widely spaced to provide room for thorough and timely compaction, there are a limited number of tie holes to patch and large areas of formwork can be set and re-set quickly. Disadvantages include capital costs and the fact that equipment is required to move the forms, further adding to overall wall cost.

Both Symons and Atlas, as well as all the other manufactured systems available for rent or sale, provide special components designed to create and brace inside and outside corners.

When it comes to forming plywood: the denser the veneers, the stiffer the plywood and the straighter the wall. Quality is proportional to cost. Plywood comes in several face designations: CDX, AC, BBX, MDO and HDO and barrier film. CDX, AC and BBX will display the grain of the wood veneer. They will hold up for fewer uses and will get dinged during ramming. MDO (medium-density overlay), HDO (high-density overlay) and barrier film (high-density with a superior overlay) have the outside veneer coated with a thin film to obscure the grain. They will provide dozens of satisfactory uses. Barrier film results in the tightest finished surface.

Other systems

At least two rammed earth builders have developed unique, patented forming systems specific to their installation methods. These are: the SIREWALL form from Terra Firma Builders in British Columbia, and the Stabilform from the Associated Earthbuilding Group in Western Australia.

Permission to use the form requires that an agreement be in place with the patent holder.

14.4.3 Summary

A range of forming options are available to the rammed earth builder, with decisions regarding method of forming (continuous vs. panel-to-panel), materials selection (plywood type, manufactured vs. site-built forms), level of precision, etc. all affecting the cost of construction and the aesthetic of the final product. Furthermore, decisions made at this stage will affect production rates and other stages of wall construction, including method of material delivery and installation.

Setting and disassembling formwork takes a great deal of time (especially a continuous form that incorporates corners, door and window block-outs, and mechanical services). Depending on the forming system selected and time expended in filling the forms, the formwork phase of a project can take up to twice as many man-hours as the installation and ramming.

Two recently completed projects by the REW team confirm this calculation. Using the Atlas system to construct three $20' \times 24' \times 10'$ high modules where soil material was mixed volumetrically and delivered to the formwork by conveyor, a total of 750 man-hours was invested in form setting and 500 man-hours on compaction. (2640 total square feet of wall for a combined total of 1250 hours or roughly two square feet per man-hour labor investment.)

14.5   Installation

The process of installing the rammed earth mix into the formwork to create a finished wall can be separated into two steps: delivery and compaction. Each of these can be approached in a variety of ways. The decisions regarding which methods to use are influenced by several factors: type of forming system, site access, manpower resources, labor rates, budget, schedule and finish quality expectations.

14.5.1 Delivery

Delivery of rammed earth to the forms is a process that has seen much experimentation and innovation throughout the evolution of the industry. The most economical and historically widely used method is the bucket or basket, carried up a ladder or passed hand to hand. Innovative builders have developed a range of low-cost improvements to the hand-carried bucket, such as bottom-dumping backpacks, wheelbarrows and ramps, ropes and pulleys. These are all slow and labor-intensive methods requiring lower capital expenses. They are most appropriate to owner-builders and in developing countries.

The majority of the rammed earth industry today uses the front-loading tractor to carry material from the mixing area to the forms and a worker with a shovel standing either on top of the forms or in the tractor bucket to distribute it. A variation of the tractor bucket delivery is a hopper on a forklift. The tractor loads material on to the hopper, the forklift drives to the form and tilts the mast forward, and material is 'raked' off of the hopper and into the forms. Having a forklift in the operation frees up the tractor for additional mixing time, provides a superior working platform and facilitates delivery to taller formworks.

The REW team utilizes tracking conveyor belts to move prepared material from the mixing machine directly into the formwork in regulated course depths. Conveyors can be linked to extend filling ranges (see Figs 14.5 and 14.6).

The rate at which material can be moved from the mix area to the formwork is usually directly proportional to the capital investment in delivery equipment. Cranes and conveyors are extremely fast, but they can be expensive.

14.5.2 Compaction tools and techniques

There are two ways to compact soil: manually and mechanically. Manual compaction (by hand) is slow and laborious, although it can be every bit as effective as mechanical compaction.

14.5 Installation using conveyor delivery.

14.6 Concrete posts and bond beam yet to be poured.

Manual compaction

Historically, all rammed earth was compacted by hand, using heavy wooden rammers. Hand compaction is still the most viable method in any economy where labor is inexpensive and widely available. Steel rammers made of 1″ diameter pipe and 4″ square heads are an improvement over heavy wooden rammers. In all but the smallest form there will be room for extra people. Compacting in a group can remove much of the tedium from the process.

Mechanical compaction

Nearly all rammed earth builders in industrialized countries use pneumatically powered backfill tampers (see Fig. 14.7). Some experiments have been undertaken to assess the effectiveness of gas-powered whackers and electric-powered vibratory rollers, but to date the backfill tamper has proved the most versatile and effective.

Backfill tampers possess a number of attributes, the most important being: (i) weight, (ii) foot (or butt) size, (iii) strokes (blows) per minute and (iv) air pressure requirements. Manufacturers include Jet, Ingersoll-Rand and Chicago Pneumatic. Backfill tampers typically cost between $800 and $1200.

While large, heavy backfill tampers will yield fast and effective results, there are frequently conditions (especially where there are a large number of imbeds or prolific amounts of reinforcing steel) where less cumbersome and more maneuverable tools are preferable. In situations where even the

14.7 Pneumatic compaction at the top of a rammed earth wall.

smallest backfill tampers cannot access every area within the formwork, steel or wooden hand tampers will prove useful. Having a variety of tools available for use throughout the entire installation is ideal.

When walls are 18″ or wider, it is often possible for installers to work within forms set to full wall height. In instances where wall thickness or an abundance of imbedded elements prohibit installers from getting inside the forms, installers will have to work from the top of the formwork. Small backfill tampers custom fit with long handles can be used to install walls of heights up to 12 feet, though quality control at the bottom of the wall becomes more difficult as the height of the formwork increases.

Air compressors

Pneumatic backfill tampers are powered by compressed air. The weight and total number of tampers employed on the project will dictate the required output of the air compressor, which is measured in cubic feet per minute (cfm). Where one small Jet 2T rammer will require roughly 20 cfm, a big Ingersoll Rand 441 requires nearly 40 cfm. A large nail gun compressor can marginally power a single Jet 2T. Running six Jets or two 441s will tax a small towable 100 cfm compressor. Rental yards commonly stock 185 cfm compressors, which are adequate for most installations.

The rate of soil delivery will dictate how many rammers to put into use. Delivery from tractor buckets may require one or two rammers; forklift hoppers two to four; conveyor delivery six to ten rammers.

The production quotas on the projects referenced above were 40–50 cubic yards of placed material per shift with six people on Jet 2T backfill tampers.

Strata and compaction techniques

The signature of architectural rammed earth is the horizontal lines (reminiscent of geologic strata lines) that reflect the courses (or *lifts*) of compaction. Control over the depth, density and uniformity of these lines reflects the skill of the builder. Some architects and clients prefer widely diverse line and depth, infused with different soil colors. Others specify subtle, consistent, horizontal lines. It is essential that all parties agree on the wall characteristics well before construction begins.

Soil characteristics, lift depth and lift homogeneity are the principle factors that affect the finished wall appearance. Clay soils demand more effort during compaction than sandy soils or installation using smaller lifts – depth should not exceed 7 inches to prevent the risk of under-compaction. Walls made with heavy clayey soils will have improved durability characteristics if course depths are kept down to 5 inches or less. On the other hand, sandy soils will in some cases reach satisfactory compaction at much larger lift depths (12 inches or greater). Varying lift depths in test wall panels is the best way to understand a specific material's tolerances and establish guidelines for the project.

Sharp, distinct interfaces are obtained by fully compacting each lift before adding more soil. Blended interfaces are obtained by slightly under-compacting courses and allowing the course above to integrate with the top surface of the course below.

Each course up the wall, with the exception of the top course, experiences two types of compaction: the first is the direct force of compaction from the impact of the rammer; the second is a sort of surcharge resulting from the compaction of subsequent courses on top of it. The final course at the top of the wall never benefits from the surcharge effect and, while compacting it more thoroughly is helpful, it will typically be less dense and may weather differentially. Bond beams and wall caps can alleviate this problem.

The incorporation of different colors into the rammed earth wall is a matter of both taste and technique (see Figs 14.8 and 14.9). Colors can be dramatically different from one another and used in multiple bands, or changes can be subtle and used sparingly. Coloring can be achieved naturally, with clay pigments, or with chemical dyes. Perfecting the art of the stratified earth wall is a challenge to be taken seriously and with patience.

14.8 Rammed earth wall with color changes.

14.9 Rammed earth wall with color changes.

For designs where changes in color between wall courses are intended, ground mixing necessitates the creation of separate batches on the mix pad. Distinct color strata are achieved in the wall by careful separation between batches and complete compaction of courses within the formwork. For subtle color changes, batches are allowed to mingle and shorter wall courses may be left slightly under-compacted so that there will be some blending with subsequent courses. With careful planning and pre-construction testing, a number of approaches can be used to successfully produce a variety of aesthetics.

14.5.3 Stripping and curing

With small panel construction, forms are disassembled soon after completion (in some cases within minutes). Freshly built rammed earth walls are not yet strong. Corners can easily be broken and faces marred. Rapid drying from wind or sun will have adverse effects. Walls should be kept moist and covered with curing blankets or plastic if the full value of the stabilization and construction effort is to be realized. With stabilized rammed earth, it can be assumed that significant performance increases will continue to be realized if curing conditions are maintained for up to 28 days. Like concrete, stabilized rammed earth will be serviceable without undertaking any special effort to prevent moisture loss and exposure to the elements, but enhancement of performance characteristics is possible with the application of good curing techniques. With continuous form systems, formwork is normally left in place for several days at least, while reinforcing steel and bond beams are completed. This time can contribute significantly to the ultimate strength and durability of the finished wall.

14.5.4 Summary

Installation of a rammed earth wall consists of two steps: delivery and compaction. As in other parts of the rammed earth process, delivery can be achieved in a variety of ways, depending on the capital investment. Traditionally, delivery was by bucket or basket. Modern delivery is by tractor bucket, forklifts or conveyors – equipment to expedite installation, reduce labor requirements and, in some cases, enhance wall quality. Compaction in the forms can be achieved manually or mechanically, with mechanical compaction requiring more capital and less labor. Techniques for creating rammed earth with color striations are varied, though achieving a tasteful aesthetic takes experience and patience. Care should be taken to not damage walls during form disassembly, and they should be kept moist for a minimum of seven days.

14.6   Future trends and conclusions

How well the construction industry responds to an expanding use of rammed earth is in many regards a factor of local building practice. In regions of the world where wood frame construction is common, solid walls – whether of stone, brick, block, concrete or rammed earth – will remain at a competitive disadvantage. The construction industries in these regions have evolved to support frame walls. Tradesmen and suppliers invest their time and resources developing skills and products to support the majority demand. On the other hand, in regions of the world where construction favoring solid wall systems predominates, rammed earth can more quickly gain recognition.

Within the US, the use of rammed earth has achieved a very small market penetration, appealing principally to clients and architects with an appreciation for the unique and compelling qualities of raw earth and of natural building systems.

What changes could help rammed earth gain traction in the construction industry? (i) expand the pool of trained installers, (ii) reduce the cost of construction, (iii) compile test data to define formulations and to document structural performance and (iv) develop a program of subsidies or incentives supporting low-carbon thermal storage wall systems.

14.7   Sources of further information

Easton, David (2007), *The Rammed Earth House – Revised Edition*, VT, USA, Chelsea Green Publishing

Houben, Hugo and Guillaud, Hubert (1994), *Earth Construction – A Comprehensive Guide*, London, Intermediate Technologies Publications

Minke, Gernot (2006), *Building with Earth – Design and Technology of a Sustainable Architecture*, Basel, Switzerland, Birkhauser

Walker, Peter, Keable, R., Marton, J. and Maniatidis, V. (2005), *Rammed Earth – Design and Construction Guidelines*, Bracknell, BRE Bookshop

15

Pneumatically impacted stabilized earth (PISE) construction techniques

D. EASTON, Rammed Earth Works, USA

Abstract: This chapter describes an industrialized method for constructing monolithic reinforced cement-stabilized earth walls. The process utilizes high-pressure air to impact a carefully selected blend of mineral aggregates against an open form. Pneumatically impacted stabilized earth (PISE) differs from rammed earth in that only one face of the formwork is required and the material is placed at higher water content. The method utilizes equipment designed and manufactured for the shotcrete industry.

Key words: PISE, alternative earth construction, shot earth, gunite, single-sided form.

15.1   Introduction

PISE, an acronym for pneumatically impacted stabilized earth, is a process for constructing stabilized earth walls using high-pressure air for delivery and placement (see Fig. 15.1); procedures and equipment are similar to dry-mix shotcrete (or gunite). A carefully selected, tested and proportioned mixture of soil and cement is blended and then conveyed via high-pressure air to the point of placement. Water for hydration is added to the dry materials at the nozzle. The force of impact at the point of placement in conjunction with stabilization contributes to ultimate strength. Like traditional rammed earth, the end product is a monolithic structural wall (see Figs 15.2, 15.3 and 15.4). PISE differs from rammed earth in that only one side of the formwork is required and the finished surfaces have a different texture.

This chapter will cover several aspects of PISE installations for new construction including materials, equipment, crew qualifications and organization, preliminary preparation, proportioning, placement and quality control. The techniques and procedures described herein are based on 20 years of technological development and construction implementation by the author in conjunction with experienced gunite installers. It should be remembered however that as the technology gains in popularity and migrates to different geographical regions, procedures may vary from one region to another, and adjustments may be required to meet the needs of the particular project. Most importantly, mix designs will significantly differ from one geographical region to another, depending on available mineral soils. Due to

385

15.1 PISE being installed.

15.2 PISE structure.

15.3 PISE structure.

15.4 PISE structure.

this range of suitable soil types, the reader must be warned that substantial pre-construction testing will be necessary in all cases before undertaking a project of any significant size.

15.2   Materials used for pneumatically impacted stabilized earth (PISE) construction

PISE mix designs are composed of mineral soil, supplemental aggregate, cementitious binders, water and, in some cases, one or more admixtures such as fly ash, fibers or other components.

15.2.1 Mineral soils

All soil materials selected for PISE construction must be free from organic matter and contain a relatively uniform distribution of coarse and fine aggregates, ranging from 1/2″ gravel to fine silt and clay.

A wide variation in particle size improves the strength, durability, and resistance to shrinking of the in-place PISE. The maximum size aggregate should be 1/2″ and, for best results, no more than 15% should pass through a 200-mesh sieve. A grain size distribution table of a soil that has demonstrated good results for PISE work is shown in Table 15.1.

Particular attention must be paid to the expansive characteristics of the fine particle constituent in order to reduce shrinkage cracking. In general, soils with a plasticity index of less than 10 will achieve better in-place results. The plasticity index of the aggregate blend can be reduced through the addition of sand or small aggregate.

Laboratory testing of the proposed mineral soils will provide information on the gradation and plasticity index. Based on these results, the type and percentage of aggregate admixtures can be determined. Clean coarse sand is a commonly available and suitable admixture, providing both strength and crack control.

It should be mentioned that the mix design requirements for PISE are much more restrictive than for traditional rammed earth. Maximum aggregate size is restricted to 5/8″ and clay content to 15%. Water content is significantly higher, resulting in greater likelihood of shrinkage cracking. Pre-construction

Table 15.1 Sieve analysis

Sieve size	% Passing
1/2″	100
3/8″	97
#4	77
#8	54
#16	38
#30	29
#50	22
#100	18
#200	15

testing and careful mix design formulation are of paramount importance in order to achieve optimum results.

15.2.2 Supplemental aggregates

In situations where either the gradation or the plasticity index of the selected mineral soil does not fall within the range of acceptability, supplemental aggregates will be required. These aggregate amendments are most often coarse sand or small gravel. They are selected based on size, shape, color, cost and proximity to the construction site. In some cases a combination of sand and gravel from separate sources will provide the best results. Predicting results based on laboratory formulations is possible although not nearly as reliable as shooting pre-construction test panels.

15.2.3 Cement (binder)

Cements used for stabilizing mineral soils are the same as those used in concrete mix designs: Portland cement types I, II and III; white cement and block cements. In PISE construction, the selection of which cement to use is normally based on the desired color of the finished wall, rather than the structural qualities, as ultimate strength requirements are generally low compared to concrete. Fly ash may be combined with cement in some mix designs to reduce the dependence on virgin cement and to improve the workability and resistance to cracking of the finished PISE. At the time of writing, only limited experimentation had been undertaken with cement/fly ash mix designs.

15.2.4 Water

Mixing water should be clean and free from substances which may be injurious to concrete or steel. It is recommended that potable water be used. Curing water should be free from substances that may be injurious to concrete. Water for curing of architectural PISE should be free from elements that will cause staining.

15.3   The forming system

Formwork for PISE differs significantly from that for traditional rammed earth in that the form is open on one face (see Fig. 15.5). The principle component of the PISE form is the backer. This establishes the dimensions, plane and elevation of the wall. The backer may be of any rigid material, such as wood, steel, rigid insulation board, expanded metal lath, straw bales or an excavated cut bank.

15.5 PISE formwork.

In conjunction with the backer (or panel), shut-offs establish the end of a wall and block-outs create openings for doors and windows. The top of the form is left open, with a thin wooden ledger (or screed) attached as a guide to define the finished wall height. Shut-offs and block-outs are typically cut to the width of the finished wall. In all cases, the forms must be adequately braced and secured to prevent excessive vibration or deflection during placement of the PISE. All formwork must be designed to provide for the escape of compressed air and rebound during installation. For column construction, two sides can be formed or the four corners can be formed using light narrow wood lath. Similarly, in beam construction, the soffit and one side may be formed leaving the other sides open. Beams should be braced or shored so that no deflection will occur due to the weight of fresh PISE.

15.3.1 Simple forming

There are many different methods for constructing a plywood form. The most common are wood framing, pre-fabricated form panels, manufactured aluminum concrete forming systems, sheet metal studs or custom framing components. The method shown in the accompanying photographs illustrates conventional wood framing, in which a 2 × 4 frame wall is built on the slab, fixed into place along the inside wall line, then sheeted with the backing material (in this case 3/4″ HDO plywood). Forms for the entire perimeter of the roughly 30 × 50 foot room were set to full wall height, eliminating the need for shut-offs. Block-outs for doors and windows were built and braced with 2 × 4s and 2 × 6s. The project illustrated took a crew of three men ten days to erect. One advantage to setting forms using a frame wall approach is that plywood sheets can be cut and fit into a seam pattern that can correlate to interior room details or to the placement of windows or other significant architectural elements. Another advantage is that the lumber used to build the frame can be reused elsewhere in the house. A disadvantage to setting plywood against a frame is that disassembly is more time-consuming. The 2 × 4s are levered away from the plywood, protruding nails nipped off and the sheets taken down carefully one at a time.

15.3.2 Alternative forming systems

Alternative forming systems such as pre-constructed plywood panels, manufactured concrete forming panels, or custom built forms have their own set of advantages and disadvantages. These relate to cost, speed of assembly and disassembly, and likelihood of reuse. In a large project where the PISE walls will be shot in several sections, a reusable form will normally prove more economical.

In order to make an informed decision regarding which form system to use, four factors should be considered: experience and skills of the crew, cost of a rented system versus purchase price of new materials, opportunity to reuse framing lumber and plywood in the project and finally opportunity to reuse form panels in a future project.

15.4   Reinforcement of pneumatically impacted stabilized earth (PISE) walls

Reinforcing bars or welded wire mesh are generally required in all PISE structures subject to lateral loading, although in some instances, low width to height ratios (less than 1:4.5) may justify a mass wall (unreinforced) design. Where the risk of seismic activity is low, reinforcing can be significantly reduced. Generally reinforcing can be lighter in a PISE wall than in a concrete wall because the high mass of PISE (as in all earthwall structures) provides an inherent resistance to lateral movement. PISE walls have the ability to absorb energy without breaking.

For thin (3″ or less) applications such as veneers over concrete or straw bale, reinforcement in the form of welded wire fabric or stucco wire is recommended. Wire mesh limits the development and depth of cracking resulting from shrinkage and temperature stresses.

To produce uniform walls, reinforcing should be designed and positioned to cause the least interference with the placement of the PISE. Normally no bars larger than 5/8″ in diameter (number 5 rebar) should be used. If larger sizes are required by the design, exceptional care must be taken to properly encase them with PISE. In all cases, reinforcing should be sized, spaced and arranged to facilitate the placement of PISE and to minimize the development of voids or shadowing. Lapped bars should be spaced apart at least three times the bar diameter. For most PISE applications, one layer of reinforcing is usually sufficient. Where additional layers are specified, spacing and positioning of the bars must allow for penetration of the PISE to the deepest recesses. The minimum cover over reinforcing should comply with the job specification, but should never be less than 2″.

Reinforcing should be rigidly secured in place to prevent its movement or deflection. Vibrations in the reinforcing steel can cause sagging and dropouts, reducing the in-place strength and adversely affecting the cosmetic quality of the PISE. Reinforcing should be free of oil, rust, dirt or any substance that could impair good bonding.

15.4.1 Embedded ledgers, hardware and conduits

Wooden or steel ledgers, hardware such as J-bolts, beam brackets, purlin anchors and straps or conduits, chases, electrical boxes and plumbing mounts may be installed directly in PISE walls. In every case the embedded devices must be securely fastened in place prior to installation so as to prevent movement or misalignment due to the force of impact. Electrical boxes and plumbing ends should be taped shut or otherwise protected against filling with overspray. Exposed threads on bolts should be taped or sleeved. Conduits and plumbing lines should be installed to provide for a minimum of 3″ of cover.

15.5   Equipment for proportioning, mixing and placement

The successful installation of PISE requires properly operated and maintained equipment. The contractor should choose the equipment for a project only after careful evaluation of the specifications, size and character of the work, job site conditions, the availability and quality of local materials, labor, and time available. A basic complement of equipment for PISE usually consists of, but is not limited to, the mixing equipment, air compressor, delivery equipment (gun) and the required hoses and nozzles.

15.5.1 Mixing equipment

Most PISE operations utilize portable mixing equipment set up on the job. This can be either batch (or drum) mixers or continuous auger mixers. Auger mixers are more capable of providing a continuous and consistent supply of material. Small continuous mixers are generally comprised of a single 1–3 cubic yard soil hopper, fed by tractor bucket, and a smaller hopper for cement, supplied one sack at a time. Larger volumetric mixers normally have two hoppers, each 3–5 cubic yards, and a cement hopper with a 3000-pound (30-sack) capacity. In both types, the soil and cement are metered into the mixing trough where the auger combines the dry ingredients as it moves material to the outfall. Water required for pre-dampening is added in the mixing trough. The equipment should have adequate controls and gauges to allow the operator to adjust mix ratios as required to assure compliance with the specified mix design. Some volumetric mixing machines are equipped with liquid admixture systems.

15.5.2 Air requirements

A properly operating air compressor of ample capacity is essential to a satisfactory PISE operation. The compressor should maintain a supply of clean, dry, oil-free air adequate for maintaining required nozzle velocities while simultaneously operating all air-driven equipment, and a blowpipe for clearing away rebound. Operation of compressors at higher elevations requires increased volumes of air. Compressed air requirements vary depending on the type of equipment, its condition and mode of operation. Check the gun manufacturer's recommendations for required compressor capacity. The compressor capacities shown in Table 15.2 are a general guide for PISE applications using air-motor driven continuous feed guns. These air capacities must be adjusted for compressor age, altitude, hose and gun leaks, and other factors that reduce the rated capacity of the air compressor. In addition, hose length, unit weight of material, bends and kinks in the hose, height of nozzle above the gun and other air demands, will all affect the air requirements of a particular equipment layout.

The operating air pressure is the pressure driving the material from the gun into the hose and is measured at the material outlet or air inlet on the gun. The operating pressure varies directly with the hose length, the specific weight of the materials mix, the height of the nozzle above the gun, the number of hose bends and other factors. A rule of thumb is that operating pressures should not be less than 40 psi (275 kPa) when 100 ft (30 m) or less of material hose is used, and the pressure should be increased 5 psi (35

Table 15.2 Compressor capacities

Hose ID (″)	Hose ID (mm)	CFM @ 100 psi	M3/min @ 700 kPa
1–1/2	38	600	17.0
2	51	750	21.0
2–1/2	64	1000	28.0

kPa) for each additional 50 ft (15 m) of hose and 5 psi (35 kPa) for each additional 25 ft (8 m) the nozzle is above the gun.

15.5.3 Delivery equipment

Although there are several types of delivery equipment used for dry-mix shotcrete, the rotary gun is the most common for PISE construction. It is designed to provide a continuous supply of dry mixed material through the delivery hose to the nozzle. Basically, a feed bowl comprising a number of U-shaped pockets rotates beneath a material hopper. As each pocket in turn passes below the air supply, material is forced out of the pocket, through the outlet neck and into the hose. Continuous rotation of the bowl and continuous discharge of material under pressure provides the force required to move material through the delivery hose to the nozzle. The internal configuration of the gun may vary somewhat between manufacturers, but all types rely on the same principle and utilize a machined steel wear plate and replaceable rubber wear pads. Proper functioning of the gun depends on proper service and adjustment of the plate and pads.

15.5.4 Hoses

All of the hoses used to convey air, water and dry material should be of the highest quality and inspected regularly for worn spots. Material hoses are particularly susceptible to wear and should be replaced well in advance of failure. Water hoses and small air hoses for auxiliary uses should be commercial grade. Hoses should be properly sized to minimize friction loss in the line. Material delivery hoses and the air hose connecting the compressor to the gun typically have internal diameters of 2".

15.5.5 Nozzle

The nozzle attaches to the end of the delivery hose and is composed of the body, water ring, water valve and nozzle liner. The most efficient nozzle type for PISE construction is the standard gunite nozzle (distributed by Ridley as the 700-902). The water ring should be checked periodically to ensure uniform flow, and the nozzle liner should be replaced as needed.

15.5.6 Auxiliary equipment

Other components in a complete PISE equipment package include an in-line water booster pump for situations where water pressure is inadequate; a blow pipe for removing rebound and overspray; an air-driven impact chisel for cleaning the bowl and gooseneck; a small sledge hammer for freeing clogged lines; and a full complement of hose and valve repair parts.

15.6   The pneumatically impacted stabilized earth (PISE) method

The quality of PISE installation depends on the successful integration of several factors and individuals: the gun operator, nozzleman, control of mixing water, nozzle velocity and nozzle technique. In each case, the expertise and experience of the responsible crew member determines the quality of operation.

15.6.1 Control of mix water

In placing PISE just enough mix water is added at the nozzle so that the surface of the in-place material has a slight gloss. The nozzleman can change the water content instantaneously by as little or as much as needed. Depending on the position of the work, too much water can cause the PISE to sag, slough, puddle or drop out. Dropouts can occur in overhead work where too much material is gunned or 'hung' in one location at one time. Too little water leaves a dry, dark, coarse surface with no gloss. This condition increases rebound, creates dry pockets, makes finishing more difficult and can produce weak and laminated PISE. For effective water control, the water pressure at the nozzle should be 15–30 psi (103–207 kPa) over the air pressure.

15.6.2 Nozzle velocity

The velocity of the material at impact is an important factor in determining the ultimate properties of PISE. For most installations where standard nozzle distances of 2 to 6 ft (0.6 to 1.8 m) are used, material velocity at the nozzle and impact velocity of the material particles are almost identical. At longer nozzle distances they may differ and it may be necessary to increase the nozzle velocity so that the impact velocity will suit the requirements of the installation. Consideration must also be given to the fact that increasing velocity means increasing rebound.

The factors that determine material velocity at the nozzle are volume and pressure of available air, hose diameter and length, size of nozzle tip, type of material and the rate it is being gunned. These factors allow for great flexibility and versatility in that large, intermediate or small volumes of material can be gunned at low, medium and high velocities according to the immediate needs of the application. Small or large variations in flow, water content and velocity can be made by the gunman on instruction from the nozzleman.

15.6.3 Gun operation

Proper gun operation is critical to ensure a smooth, steady flow of material through the hose and nozzle. If a suitable balance of air and material flow is not maintained, slugging, plug-ups or excessive rebound may occur. Pulsating and intermittent flow causes under- or overwetting of the mix and requires the nozzleman to quickly adjust the water, manipulate the nozzle, direct it away from the work or stop.

Whenever possible, sections should be gunned to their full design thickness in one layer to reduce the possibility of internal cold joints and laminations. The exception to shooting full thickness would only be in cases where a thin (less than 1″) flash coat is required to achieve cosmetic uniformity. The distance of the nozzle from the work, usually between 2 and 6 ft (0.6–1.8 m) should be such that it gives best results for work requirements. As a general rule, the nozzle should be held downward toward the work at an angle between 30 and 45°. Steeper angles result in excessive rebound. When gunning tops of walls, steeper angles are acceptable to reduce overspray, and when gunning flash coats and other veneer type work, a nozzle angle perpendicular to the wall will be most effective.

To uniformly place the PISE and achieve good compaction, the nozzle is directed slightly downward and rotated steadily in a series of small oval or circular patterns. Waving the nozzle quickly back and forth changes the angle of impact, wastes material, increases overspray and results in a rougher surface that creates more work for the finisher. An exception to small circular patterns would be when gunning to encase heavy rebar configurations, in which case the nozzle must be moved from side to side to direct material behind the bars. Also, the mix should be wetter than normal to aid the PISE in flowing to the backside and to prevent build-up on the front face of the bar.

When shooting PISE for thick structural walls, application should begin at the bottom against the form face. The first layer should fully encase the reinforcement and care should be taken to prevent overspray building up higher on the form face than the defined work area. An effective technique for shooting thick walls is 'shelf' or 'bench' gunning. Instead of gunning the full thickness of the wall with a horizontal top, a thick layer of material is built up, the top surface of which is maintained at approximately a 45° slope. With this technique, rebound is less likely to be trapped in the wall as the loose material is free to fall downslope, and compaction is improved as the nozzle angle will be perpendicular to the surface receiving PISE.

When inside corners, door or window block-outs, or other projections are part of the area to be shot, they should be gunned first and continuously built up as the layers become higher. This will prevent rebound and overspray from filling the corners and being covered up. Slight overwetting of the initial layer helps bond and reduces rebound. On top of door or window headers it is particularly important that rebound and other loose material be cleaned off before shooting. Prior to gunning, use a jet of air from the nozzle or a blowpipe to clean horizontal surfaces.

15.6.4 Rebound and overspray

Two of the unwanted by-products of PISE construction are rebound and overspray. These can be minimized with proper mix formulation and nozzle expertise. Overspray is the material disbursed away from the receiving surface. It adheres to guide wires, forms, reinforcing steel and other projections, leaving an unconsolidated thickness of low-quality PISE. It should be removed, preferably before it hardens, especially in areas to be covered with fresh PISE. Shooting fresh PISE on top of overspray left on rebar or anchor bolts will decrease the bond strength. Overspray left on formwork can cause delamination or other unwanted cosmetic defects when forms are removed.

Rebound is the term used both for the portion of mix that ricochets off the surface during shooting and the excess material that is shaved off the wall during screeding (trimming). The amount of rebound that results depends on several factors: percentage of fines in the mix design, percentage of cement stabilization, moisture content, thickness of the work, air pressure, nozzle angle and nozzle technique. Nozzle technique is especially important to minimize the overshot material that must be cut back by the finisher.

Rebound should not be reused in the structural wall, but can be salvaged and used for other applications on the job site, such as rammed earth garden or patio walls, heavy blocks for low retaining walls, paths, paving or other applications. Because the amount of rebound generated can be as much as 25% of the total volume of material placed, it is financially and environmentally imperative that some plan for utilization of rebound be in place prior to beginning wall construction. With a rebound utilization program, valuable by-products can result as opposed to the generation of a waste material needed to be off-hauled and disposed of.

15.6.5 Finishing

The most common technique for finishing a PISE wall is to shoot a fraction of an inch beyond the guide wires and then shave the excess material off to a straight and uniform finish (see Fig. 15.6). Care must be taken by the finisher to maintain a consistency to his rodding technique. Although there are other finishing techniques, the rough rod finish has the advantages of hiding hairline cracking and other surface inconsistencies.

Other finishes for PISE walls include the gun or natural finish and the

15.6 PISE being rodded (finished).

smooth or steel trowel finish. The natural finish is that left by the nozzle after the wall has been brought to approximate finish grade but not rodded. A gun finish is textured and uneven. In some cases small craters created by the aggregate burying itself in the wet mix can leave the wall with an unattractive 'moon surface' finish. The expertise of the nozzleman is essential in order to achieve an acceptable natural finish.

A smooth finish is obtained by first shooting a slightly wetter 'flash' coat, approximately 1/4″ thick, over the wall surface after it has been rodded flat. The fresh surface should first be finished with a wood float to achieve a uniform granular finish, then finished again with steel plastering or pool trowels. Because of the small aggregate in the PISE mix, a high-quality smooth finish is difficult to achieve and requires considerably more effort than the rodded wall. It is also subject to cracking and checking. It is recommended that small samples of smooth wall be created for approval prior to undertaking a large wall area.

15.6.6 Curing

Wall forms should be left in place for a minimum of 12 hours prior to stripping. Door and window jambs should remain in place for a minimum of 36 hours. Headers and beam forms should be left in placed and braced for a minimum of seven days.

All PISE should be properly cured so that its potential strength and durability are fully developed. This is particularly true for thin sections and veneers. For thick walls, curing will naturally take place more slowly as the moisture within the mass of the wall will contribute to the curing process. When conditions are cool or damp, the mass of the wall may be sufficient to retard moisture loss from the wall with no additional curing measures. In hot, dry or windy conditions, the walls should be sprayed with water several times per day or wrapped with plastic or other tarps to trap moisture. Curing conditions should be maintained for a minimum of seven days.

15.7   Conclusion

The nature of PISE construction affords a dynamic new approach to traditional earth building. Walls are faster to construct than with other methods, and the open formwork allows easy access for installation of mechanical services and steel reinforcing. In general, it conforms more closely with the work flow on a modern construction site than does rammed earth. The hurdles in the path of wider acceptance of the process are the limitations on acceptable soil gradation and the high capital and fuel consumption of the mixing and shooting equipment.

15.8   Appendix

Estimating the costs of PISE construction:

Simple rectangle $24' \times 40' \times 8'$ high and 18" thick

> Forming components
> $80 - 2 \times 4 \times 16'$ wall frame
> $32 - 4 \times 8 \times 3/4"$ form sheeting
> $32 - 2 \times 4 \times 12'$ wall bracing
> stakes, 1 5/8" and 2 1/2" drywall screws

Aggregate materials (assume 2 parts site soil 1 part imported sand)

> 60 yards site soil
> 30 yards imported coarse sand

Cement @ 10% stabilization rate
> 150 sacks Portland cement

Labor

> 120 man-hours form setting

160 man-hours shooting
48 man-hours stripping and clean-up

Equipment

3 days skid-steer for mixing and installation
2 days gunite machine
2 days air compressor

16

Conservation of historic earth buildings

G. CALABRESE, Architect, Australia

Abstract: The first part of the chapter outlines the initial analysis and study of the earth-walled city of Diriyah in the Atturaif region of Saudi Arabia. It discusses what lessons can be learnt from this example, including the different climatic issues and the importance of site maintenance. The second part of the chapter discusses the restoration of 'Casa Patacca' in the Abruzzo region of Italy.

Key words: earth buildings, Atturaif, earth conservation, adobe, massone, earthen plaster, earthen techniques, traditional earth architecture, sustainable architecture.

16.1 Introduction

Historic cities need to be able to adapt to modern living if they are to continue to flourish and grow. If they do not adapt the cities can start to stagnate, becoming run down, eventually leading to disuse. Therefore the restoration of historic buildings is often necessary to keep them in a usable condition, even though this can sometimes conflict with the main aim of heritage preservation.

The subject of authenticity is often brought up when discussing conservation of heritage. This is because authenticity is considered to be the way in which historic buildings and artefacts can be 'valued'. In the Atturaif region, the authenticity of the buildings is expressed in their architectural form and function, the materials adopted and the techniques used to build them, as well as through the site where the buildings are placed.

As an example of how contentious authenticity can be, consider the problem of sacrificial layers. These are solid layers of soil that are applied to the base of the walls of earth buildings to protect them from being affected by moisture uptake. These layers require continuous maintenance. The question that is often raised is will the building still be authentic after the new sacrificial layers are added to protect the original layers below? And furthermore, will the building still be authentic after it is restored and replastered? Earth buildings require periodic maintenance, as well as new renders and sacrificial layers. The World Heritage Committee has recognised this. One opinion is that the authenticity of a historical building lies in the specific techniques used to create and maintain it, and not exclusively in the materials themselves.

401

Parts of the city of Diriyah are to be restored and adapted in order to be used to house museum structures. An example of how authenticity can be preserved is through the use of new modern reversible buildings on pad footings, which are designed to be inserted inside the restored ruins. Thus the historical buildings can be both preserved and utilised.

16.2   Common causes of deterioration on historic earth buildings

Often, the greatest threat to an earth building is moisture and poor site drainage conditions. Moisture that accumulates at the base of walls migrates through the walls through capillary action, causing the earth to lose cohesion, subsequently weakening the wall. Continued exposure to moisture can degrade the internal structure of the earth wall by allowing the clays to expand. Moisture that is channelled into the base of the earth walls through poor site conditions can begin to undercut the wall through mechanical erosion, causing the structure to collapse in the long term.

Another threat is rainwater, which is often directed onto the wall where there is poor roof protection and drainage. Snow is also a great threat, as it accumulates at the base of the buildings, keeping that section of the wall permanently wet, which can, over time, cause the loss of the entire wall.

Other significant threats are caused by human and animal activities. Human activities can cause differential decay of the walls. This is often caused by poor building practices and systems that are incompatible with the existing systems. Vibrations from motor vehicles and excavations can also cause numerous problems if there is not appropriate site management in place. Animal activities, such as the nesting of insects and other animals, microgrowths, rodents boring galleries and livestock brushing the base of the walls, can all cause direct and indirect damage the earth structures.

16.2.1 Erosion mechanisms

Water is the greatest enemy of an earth building. With persistent dampness the walls move and pass from a solid state to an elastic one. Because the water is prevented from moving and evaporating from the wall, it causes condensation on the cold surface. This can in turn cause serious deterioration that in most cases is not directly visible from the outside. It is only noticeable once the majority of the wall has severely deteriorated. Owners of earth structures often cover the base of walls with cement render in order to make the wall more resistant to water penetration; however, this can cause much more severe erosion over time.

Standing water can be caused by placing objects against walls, plants that need to be watered, and also by earth falling towards the wall base. This water is then subsequently absorbed and transmitted into the wall through capillary action. The base course and the top of the wall require continuous maintenance in order to fight erosion.

16.2.2 Structural defects

Structural defects can be caused by a number of things. Initially, defects may appear as a minor crack in the wall, but in the long term defects can lead to the collapse of the entire wall. The major causes are unsuitable soils used as construction materials, poor construction techniques and unsuitable damp protection. The use of soil as a building material requires good practice, respect of building codes and systems to be in place from the very beginning, to the completion, of the structure. Structural defects can lead to structural cracking, by which point the earth wall is no longer resistant to stress. Structural defects can also be triggered by vertical shrinkage cracks caused by carelessness in quality control of the earth used for construction. The sourcing of suitable materials is of equal important for repair and reconstruction of heritage earth walls, where an optimum balance of sand and clay is necessary. This can be attained through analysis of the material in a laboratory before construction begins.

Another cause of structural defects is poor site choice, for example, areas where water pools or where the building may be exposed to strong winds or natural disasters.

16.2.3 Water study and protection principles

The cycle of wetting and drying with the movement of salts in the building material can cause severe erosion at the base of the wall when the cohesion of the earth particles is compromised. Sealing the perimeter of the building with a cement footpath, with the intention of protecting the wall base, will only increase the causes of erosion. It is important for water to be able to evaporate from the walls, as it will move through the foundations in the wall and through the base coure where it will evaporate, causing exfoliation of the layers on the wall surface. The base of the wall is highly vulnerable to water damage and for that reason it is an area that requires constant scheduled maintenance and attention.

A rise in ground level can also cause several problems. Activities such as excavation for pipe work, road maintenance and waterproof footpaths all diminish the height of the base course of the wall, therefore adding to the exposure of the wall base to moisture. The basic technique of adding a wearing layer over the base course is dependent on the continuous repristination of the layer itself over time. The idea being that this layer of soil will be affected by moisture erosion before the base course.

The height of the base course, commonly made of brick, stone or concrete, is also of primary importance in protecting against erosion. It needs to be adequate to handle the wet season, as well as water carried by strong winds. Of fundamental importance is the provision of a waterproof membrane, which is laid over the base course of the earth wall to stop water rising through capillary action throughout the wall surface. Moisture barriers offer great protection against water in both stabilised and unstabilised walls. They can be laid vertically to both the exterior and interior surfaces of the wall or together horizontally above the base course, therefore stopping the capillary action of water movement, although they must be uninterrupted for this to work effectively. Moisture barriers can be made out of bituminous products or water-repellent cements. Chemical binders are also commonly used amongst earth builders. The danger in using these products is that, although they have been tested, they have only recently been introduced to the market, therefore their performance over the coming years is not guaranteed. New water-repellent additives can waterproof walls throughout, thus making earth walls more suitable for very exposed conditions, including retaining walls. However these additives may inhibit the breathability of the material. All water, no matter how small the quantity, needs to be able to evaporate from the walls and these additives, if not used in moderation, might cause more harm than good. Care should be taken with regards to choice of waterproofing in order to avoid adding toxicity to the surfaces of the walls, and consequently movement of toxic particles throughout the building.

It is important that all protection principles are used together with the correct site maintenance programme. The effectiveness of a site maintenance programme is often dependent upon the scheduling of the maintenance activities themselves. Activities such as cleaning gutters and downspouts and removing accumulated debris and objects at the base of the walls are all part of a good preservation maintenance programme.

16.3   Conservation of earth architecture

Earthen sites make up 10% of the World Heritage List of UNESCO. Many of these sites are currently under threat. Modern earth construction has seen a consistent increase in the industrialisation and standardisation of the industry, whereas the conservation of earth architecture has unfortunately undergone much slower progress. However, over the past 40 years, the field of conservation has received some much-needed interest, with a sequence of international conferences aimed exclusively at earth conservation that were started in Iran in 1972. Another wave of interest was caused by a series of educational activities known as the PAT courses or Pan-American Courses on the 'Conservation and Management of Earthen Architectural and Archaeological Heritage' offered from 1989 to 1999. A series of workshops, seminars and courses throughout the world have also advanced the field of study and the conservation of earthen architecture. The reality is that earthen architecture is finally being recognised as part of our heritage. The conferences made one thing clear: it is impossible to reduce conservation to a 'treatment' that is simply aimed at stabilising or consolidating a surface. Conservation is also about education, research, public awareness and professional practice. In 1983 the international symposium held in Peru on earthen architecture reiterated the need for education; however, it was not until 1987 that the International Centre for Earth Construction – School of Architecture of Grenoble (CRATerre-EAG) assumed responsibility for such programmes. In 1989 the International Centre for the Study of the Preservation and Restoration of Cultural Property (ICCROM) began to share the responsibility for educational programmes. The Gaia Project came from the 1989 agreement between CRATerre-EAG and ICCROM on educational programmes, which stressed the importance of on-site education. Later came Project Terra, another agreement of collaboration, between the Getty Conservation Institute GCI, CRATerre-EAG and ICCROM. In May 2004, the GCI participated in 'Unfired Clay Construction in Italy: Toward a National Building Standard', an event organised by Citta' dell Terra Cruda with support from UNESCO (Italian Commission) and other Italian institutions working towards development of a national standard for earthen architecture.

Earth architecture is the oldest and most widespread architectural expression of our monumental heritage, for this reason its conservation is a priority. There is a growing interest in preserving our cultural heritage, hence the plethora of international courses and workshops on the subject. As architectural acculturation becomes more acute, innovative ways of conserving earthen architecture will need to be found.

16.4   Case study of the UNESCO heritage site of Diriyah in the Atturaif region of Saudi Arabia

This section discusses the evaluation of the archaeological site of Diriyah, its historical evolution, the physical fabrication of the buildings and the social environment in which the site developed.

16.4.1 Analysis

The cultural significance of the site was assessed in terms of aesthetic, social and scientific significance. Whilst examining the collapsed walls of the city of Diriyah, it was found that the collapse was not directly due to the failure of the foundations, as the city was built on a rocky basement of limestone which came to the surface in some points. The collapse was caused by erosion; the product of salts crystallising at the wall base above the limestone stem wall and the separation between material layers, causing further severe erosion. The introduction of service infrastructure along the curved roads, coupled with poor site management, meant that the excavated earth and walls collapsed on one side, directing water to pool in several other areas and channelling it through the standing walls towards detached areas. This caused severe erosion to numerous buildings that subsequently required propping. This demonstrates that even local site work that can appear insignificant can be disastrous when coupled with a large amount of rainfall and bad site management.

16.4.2 The earth city of Riyadh and the first Saudi state

During the second half of the 18th century, through to the first quarter of the 19th century, the Atturaif region became the centre of the first Saudi state. The development of this area expanded across the seasonal river Wadi Hannifah, from the existing Addiriyah quarters, with the construction of an administrative centre, a treasury and the palaces of the Saudi Princes. Saud Bin Muhammad took over as chief of Addiriyah in 1720 and became the founder of the House of Saud. His son was responsible for the establishment of the centre of the first Saudi state in Atturaif. In 1818, and again in 1824, the Ottoman Empire destroyed the site. Afterwards, the people who had resettled in Atturaif and reconstructed some buildings resettled in Riyadh. Following this event, the House of Saud was relocated to the city of Riyadh and the earth city was left to decay, neglected for over 125 years.

In the mid-20th century some 200 families resettled in the east part of the site, building new houses on the debris and ruins of the first Saudi state. The destruction of the site and years of total neglect took a heavy toll on the site. The reoccupation of the site itself meant the removal of several structures and detrimentally affected the integrity of others (Crosby, 2007). Today, this historical area is considered to be an important part of the metropolitan area of Arriyadh. People from all over the country can come to visit the home of their forebears and it is planned to become a major tourist attraction once it has been stabilised and restored.

16.4.3 Climatic conditions in the Najd region

The Najd region in Saudi Arabia has a long, hot and almost dry summer with large diurnal temperature differences. Winter is almost non-existent, and it rains for a total of one month per year. The reason for this is that the air masses are exhausted before they reach the internal areas. The high mountains also create an almost impenetrable wall that the rain clouds cannot easily pass over. During the day there is intensive solar radiation and at the night a rapid cooling occurs under the cloudless sky. The maximum air temperatures range between 30 and 46°C in the months of June, July and August in the city of Riyadh. The highest temperature during the day is about two hours after midday. Rainfall follows irregular patterns and is very minimal, the yearly average being around 100 mm. Although total rainfall is low, when it rains, it comes in very large quantities all at once. This kind of intense rain is an obvious enemy of earth buildings, as it causes severe erosion and runoff. The rainfall is so intense that even the modern buildings now built in the city centre cannot withstand the rain and occasionally leak. This is also due to the enormous amount of sand and dust that continuously accumulates in gutters and downpipes – if these building elements are not part of a continuous maintenance programme they will be obstructed when the rainfall comes and will not work efficiently.

Due to these climatic conditions, with the sky virtually cloudless, the solar radiation reaches the ground very quickly resulting in very high temperatures. The winds are charged with dust as a consequence of the temperature contrasts between day and night. The winds from the north-east carry dust and debris due to the absence of vegetation and sand instability. This also is not good news for earth buildings as sand storms accelerate the erosion of the buildings, hitting the delicate earth plasters at enormous speeds.

16.1 Aerial view of the historic Diriyah site, showing the inserted buildings organised around a central courtyard and the only vegetation growing around the wadi (seasonal river). The prevailing winds blow from the wadi (top left of image) and this the major erosion explains and missing roofs.

16.4.4 Architectural and planning features

Diriyah is located approximately 20 m above Wadi Hanifah, the seasonal river; the site was strategically chosen and achieves what we would call today a sustainable climatic environment. The wadi is the corridor of the seasonal river. It is dry most of the time, with an occasional water stream that becomes a fast-flowing river once the rainy season brings a substantial volume of water. With its fertile soil the wadi was a perfect location to build a settlement.

The site choice and the climate are strongly linked. The walled city of Diriyah is located on the top of a hill; an area not suitable for farming. It is surrounded by palm trees and athel trees, an evergreen tree that can grow up to 18 m tall, commonly found along watercourses in arid areas. The prevailing wind direction is north-north-west. The site has an intimate relationship with the surrounding vegetation and the climatic conditions discussed previously. One interesting aspect of this is that when the hot, dry air starts blowing through the vegetation of the palm trees and the athel trees, the humidity content of the prevailing wind increases, and this in turn leads to a decrease in temperature. By the time the wind enters the buildings through the windows and ventilation holes, it has lost most of the dust and has also cooled by a few degrees. Because the site is around 600 m above sea level, it is not affected by large dust particles as it would be if located lower, near to the seasonal river. If the wind passed over a nearby river or lake before reaching the buildings, the reduction in temperature would be further amplified.

Vegetation is of high importance, not only in the natural air conditioning process, but also due to its ability to reduce the dust and lower the wind speed, which are both great forces which can cause erosion of earth buildings.

16.4.5 Settlement patterns

The compact layout of the site, in a fortified cluster of settlements, works strongly with the single units when analysed with regard to heat gain, street orientation, grid pattern and architectural features. In terms of layout, most of the dwellings have up to three walls in common, this strategy allows for minimal heat gain, as well as loss, through the external walls. The orientation of the streets is another important factor, as it minimises the amount of radiation from the sun that the buildings receive during the day. There are lots of continuously shaded areas all throughout the day, as the main pattern of the site is in the form of a grid, diagonal from east to west. The air in these shaded areas is a few degrees cooler than the surrounding environment, and it cools the surrounding dwellings when it penetrates through the triangular ventilation holes as well as through the windows. Because the streets are so winding and narrow, the wind velocity drops a little, thus reducing the severity of the erosion caused by the wind itself.

A characteristic of the western section of the site is differential weathering. It is easily noticeable how numerous walls have extensive weathering on their east face, where over half of the entire wall has been eroded away

(a)

(b)

16.2 (a), (b) and (c) Detail of ventilation holes that formulate decorative patterns, mainly triangular in shape. These motifs were organised in different patterns based on the position of the family in the society, but had the key role of initiating a breeze throughout the building.

(c)

16.2 Continued

in contrast to the west face. Stone masonry was often used for this more aggressive orientation, where stones in herringbone patterns can still be seen as the sacrificial layers have been eroded away from them.

16.4.6 Dwellings and climatic factors

The Islamic style house is introverted, with most rooms lacking a specific function so that their use can be changed according to family components and necessity. Intrusion of the outside world seems to have been discouraged in this settlement, with women only observing public life from behind the screens or through very small windows of up to 600 × 600 mm. The percentage of openings on the façade ranges up to a maximum of 10% of the total façade area. An important element of these buildings is the high location of the windows for privacy from the street. This is also to allow warm air to escape and reduce dust infiltration. White lime render is used around the openings of the windows to reflect as much heat as possible before it enters the building. Green planted courtyards, possibly used in conjunction with a pool of water, can be invaluable for cooling. The courtyard was seen as the heart of these traditional buildings, moderating the climatic extremes. They are normally square or rectangular in shape, a traditional response to the harsh environment, supplying light to the verted rooms surrounding it. The air circulates through a continuous breeze via a natural convection current created by the courtyard. At night, the courtyard behaves like a sink of fresh air, attracting the cool air down from the roofs (sleeping areas), whilst during daytime, as the sun heats up the courtyard, the air temperature rises, acting as a thermal chimney setting up a breeze from outdoors into the courtyard. Because the courtyard is located at the heart of the building, not facing the prevailing wind, it works in creating a low pressure so that when the wind passes, it initiates a continuous breeze throughout the day. The bigger the ventilation holes on the building opposite to the direction of the prevailing winds, the stronger the breeze that is created.

The location of the courtyard and its proportionate size with regards to the size of the dwelling appears to have come from centuries of experimentation. The courtyard easily modifies the microclimate of the dwelling, reducing temperatures and increasing comfort for its inhabitants, insulating them from dust and noise. Shading elements were also used in the courtyards, when there were few or no trees present, the colonnades would shade the walls themselves. Remnants of cooking tools have also been found in the courtyards, from which we can infer that cooking was carried out outdoors, at least in the hot season. Here, the urban unit can be described as a cluster, with each cluster composed of a series of dwellings. There are random divisions between each cluster some of which are wider semi-private streets.

16.4.7 Materials, construction methods and building components

The athel tree was often used as a building material on the site of Dariyah. Waterspouts were made from athel trees, and they were also used as floor beams covered with palm branches, palm leaves and finally compressed with soil to create a watertight layer. The end of the beams in each room would sit on a stone layer embedded in the wall, as this would facilitate the weight distribution to the adobe brick wall.

The beams were often chosen to have a bend, when the beam was laid horizontally forming a floor joist. The bend would face upwards in contrast to the weight. Such beams would then be intensively decorated with brightly coloured and simple geometrical motifs, as well as the doors the windows and the window shutters, by the women of the family. Each decoration would identify a particular family; no two were ever the same.

Adobe bricks made of earth were the main building material used in Diriyah, along with earth mortar and earth plaster. These adobe bricks were made by pre-mixing earth, water and straw, then compacting these elements

(a)

(b)

16.3 (a) and (b) Examples of how the buildings share common walls between them, with a courtyard (often shaded) used for lighting the rooms, keeping animals and plants, for cooking and as a circulation space.

in a mould. Straw was generally added to improve the tensile strength, acting as a three-dimensional rebar in the single adobe brick and reinforcing the entire structure when the building was finished. Clay binds all the elements of the brick together and was easily available from the wadi river. Bricks measured on site ranged from 16–25 cm with a height between 9 and 12 cm. External walls ranged between 40 and 50 cm while the internal walls ranged from 30–40 cm. The outer walls in most of the walls still standing have been tapered towards the interior to improve stability. It is curious that of the buildings still standing today, the toilet structures called long drop privies still survive while the rest of the building is in ruin. These wet areas of the buildings stand like skyscrapers and tapering of the wall is clearly visible from a distance as these structures reach over 15 m in height.

The adobe bricks were not directly laid on the ground but were separated by a strip stone foundation. No other foundations were present, as the site sits over pure limestone. Limestone drums, joined with earth mortar, were used to create columns. The drums had a height and diameter of 25 cm with a coarse external finish and render applied. The capital was created using two larger stones, over which a wood beam would then be placed to distribute the weight over each column.

16.4.8 Site erosion

The most visible erosion to this site has been caused by basal erosion. Every rainfall is severe, and it causes water to be rapidly channelled at the base of the earth wall, especially just above the limestone base stem wall. This initiates a movement of salt crystals throughout the wall, which causes erosion and undercuts just above the stone base. Many of the tall walls are still standing on site. They are reasonably stable in the lower sections but not so much in the higher sections.

16.4.9 Future outlook

Unfortunately, most of the population of the modern city of Diriyah do not have a good knowledge of building with earth. Some of the buildings in the city centre have shapes and forms of the old earth buildings, but these are generally made of concrete panels, resulting in a false architecture with no relationship to the surrounding environment or the people. Many of the shops in the city centre have an interior paint that is made of mud and straw over the fired brick or concrete walls, rather than ordinary painted interiors. This shows they do still have an association with the past. Modern architecture, just like the architecture of the past, must respond to the climate and site, utilising science and technology in order to find a solution to the major issues of our environment, and of global warming. Modern architecture should not only concentrate on façades, but should make the entire building work with its surroundings. By understanding the environment and by fully comprehending the traditional design criteria and methods, new solutions can be found.

(a)

(b)

16.4 (a) and (b) Examples of columns that were left standing at the Diriyah site.

Innovative solutions today will be regarded in years to come as 'traditional'. Some of the elements from the historic architecture that are still in use today are tree planting, which can help to maximise shade and microclimatic effects with high and low-pressure zones, as well as reduce the intensity of sunlight whilst reducing the heat reflection of uncovered ground; narrow and winding streets; no asphalting or dark coloured surfaces to minimise heat radiation; grid pattern in relation to sun orientation; minimum presence of external openings; green and water elements working in combination; buildings in clusters with compact planning, the use of courtyards with a ratio of 1:1 up to 1:3 compared with the main building creating external spaces used as circulation areas; and finally, use of pergolas, light roofs and tent structures to create shaded courtyard areas adequate for recreation.

16.5   Case study of earth buildings in Italy: Loreto Aprutino in the Abruzzo region

The houses analysed in this case study are in the Abruzzo region of central Italy, in the town of Loreto Aprutino, in the province of Pescara. Buildings have been studied, analysed and catalogued. One building of major significance, with a kitchen and stalls at ground floor for the animals, and bedrooms above grouped around a central fireplace and staircase, was chosen as the main subject of restoration and study for a degree thesis in architecture. The technique commonly used in the area is that of the massone, an irregular

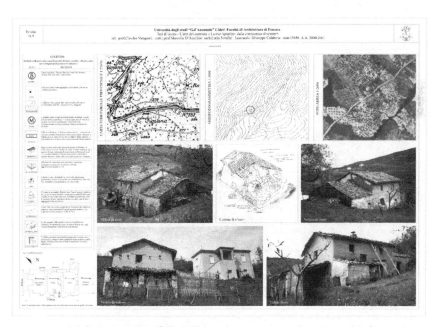

16.5 An example of the tables that were used to record information for each earth building, including a detailed list of elements that characterised it, aerial photographs of the location on maps of different scales and an axonometric sketch analysis.

block made of earth and straw in the shape of a bread loaf, which is stacked in a herringbone pattern in this locality.

16.5.1 The Patacca house

Once the home of a peasant, the Patacca house now corresponds to modern living standards. Some parts of the restoration project of this building were started during the thesis stage, with the rest being completed later by the owners. In the 1970s, as a new oversized concrete building had been built just a few metres from it, the Patacca house began to be used for storage. Once routine maintenance of the house was halted, several issues arose. In the 1990s the deep excavation of a road bank metres from the main façade created a discontinuity in the equilibrium of the underground water table, which took its toll on the stability of the structure itself.

An earth building is constantly linked with its surroundings. If something changes in the surrounding area, this can create varying levels of readaptation

16.6 This is an example of an earth building that has been consolidated using concrete. The base of the building has rejected the concrete, and it in turn has rotated away from the earth wall. This is caused by severe erosion due to trapped moisture below the concrete itself.

of the structure, more often for the worse. Over the years, incorrect maintenance of the Patacca house had seen placement of cement render over the stone base up to the level of the first block courses, creating a waterproof seal that resulted in more detrimental issues.

Cement renders and earth blocks have different thermal expansion coefficients, which can sometimes result in cracks forming along the façade, often leading to accelerated erosion. By cladding the building in concrete with a view to making it more resistant to water, rather than taking the approach of keeping the water away from the building via good site maintenance, more rather than less erosion was caused.

16.5.2 Field recording

The methodology of this recording was to conduct an initial survey of several undocumented country sites on the outskirts of the town of Loreto Aprutino, Pescara, Italy. Samples of earth from walls, together with plasters, were collected, analysed and displayed.

The first approach to analyse the buildings was through photography. All photos were taken with a scale of 30 cm, clearly identifiable at any distance. The scale was of fundamental importance as it allowed further analysis and comparisons at later stage. A dismountable framing timber square representing 1 m^2 was hung on every side of the buildings, and photographs were taken perpendicular to the surface in order to be able to compare the fabrics. Photographs were especially taken of areas where there was a change in material, where the plaster was eroded or where the size of the bricks changed from the average size of the rest of the façade. The timber square also had intervals of 10 cm clearly indicated on each side. This method allowed comparison of construction methods in the subject building, as well as between distant buildings and ones where fired bricks were used.

During the field recording sessions, several people working in the fields where the buildings were located were interviewed. The buildings were for the most part built after the Second World War and were now seen by their children and themselves as a symbol of poverty associated with very difficult times; certainly not an image of progress. Many buildings still standing today are being gradually replaced in their exact location by a modern concrete building. This is because this is classed as an alteration or an addition to the existing building, rather than a new construction, which would mean some taxes payable to the Council can be saved. This strategy is becoming more and more common in Italy, as people are trying to save money.

After the Second World War, earthen building techniques were seen as a symbol of poverty. Following this period, there was a loss of interest in the know-how of earth building and therefore techniques were not passed onto

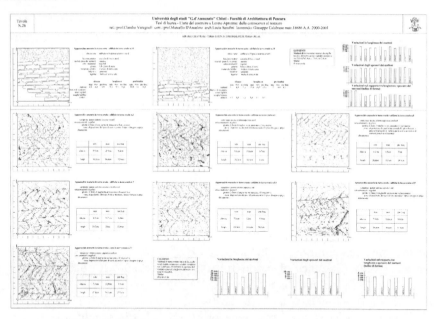

16.7 This table shows how m² from every wall orientation of the buildings was recorded. Both the brick and mortar dimensions were recorded. This analysis showed which areas of the building were most eroded, it was then used by the builders during construction to protect the surfaces that were known to be more exposed.

the following generation. Common labourers need to be reacquainted with the construction methods but do not want to invest energy in a process they see as outdated and with almost no interest from the consumer. Today, thanks to a renewed interest during the 1980s, authorities are starting to recognise the importance of this heritage. Buildings are starting to be scheduled and recorded, including their construction methods and parts studied.

16.5.3 Condition documentation

All conservation action must begin through scientific studies and a thorough understanding of the problems involved. Graphic symbols and pre-established descriptive terminology have been employed to aid recording methods, leading to a better understanding of each site. Prior to field recording, aerial photographs were collected by the state library, as well as maps of the territory at various scales. This allowed each building to be located in its context. Unfortunately historical photographs were never available for any of the buildings.

Once on site a 'cartone di rilievo' was produced and dated. This is a sketch of the building, in plan and elevations, recording as much information as possible during the first visit to site. As well as beam orientation, north point orientation, room space, surface finish, prevailing wind directions, repairs, overlays and any major erosion was recorded.

A range of graphic symbols was utilised to describe the building conditions as illustrated in Fig. 16.9. Each graphic symbol corresponds to a breakdown of the conditions and variables which allows correlation amongst those of the various buildings. Some variables recorded were the proximity to a road, number of buildings, presence of a fireplace, state of preservation, construction technique, levels, type of roofing and presence of foundations. These records allow us to postulate correlation amongst the buildings and further understand the cause and effect relationships through this initial reading of the physical evidence.

In order to define classes of condition and layers of materials applied to the surfaces of the buildings, graphic recording systems were developed in both linear and overall pattern symbols. No colour was used for ease of reproduction.

16.5.4 Digital documentation

All façades were photographed in sections and information was put together and scanned.

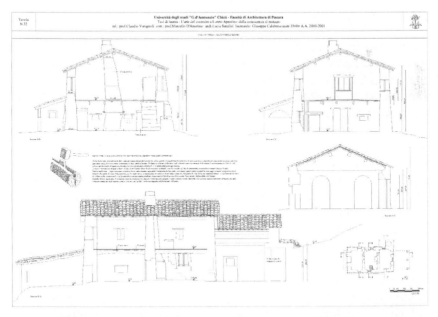

16.8 This figure shows the way in which the buildings were recorded graphically, using symbols to differentiate different aspects of the buildings.

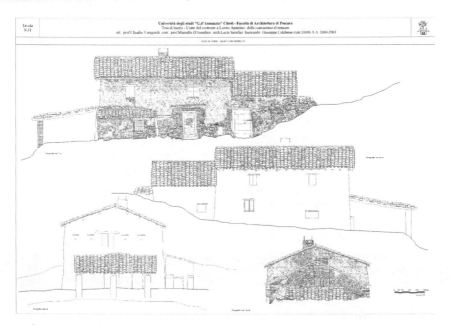

16.9 An example of how CAD was used to create accurate representations of the buildings and the earth around them.

After thorough measurements of the façades, the field recorded information was imported into both photographic and drawing computer applications to transfer the field information into practical drawings. Several trips back to site were necessary as this type of building is never rectilinear, and one cannot assume a straight line between two points as all elements have moved and settled over time; the building itself as well as the ground. Following each rain season minor discrepancies could be detected in the measurements due to the movement of the various parts, and so it was decided to measure each building every few days and to transfer all the information to the computer immediately. The end product is a series of plans, sections and elevations with layers of information.

16.5.5 Types of building

In his investigations, Baldacci divides the earth buildings in Italy into three functional types: the farmer's, the labourer's and the craftsman's house. The type of building has evolved as a consequence of its function. The functional demand in an area has therefore initiated a certain typology (Baldacci, 1958).

16.5.6 Structural instability

Structural instability is certainly the most serious of all the conditions affecting a historical building, as it can result in a sudden loss of all the fabric. Of the various buildings studied, structural instability was a major concern in all structures. Often the structural condition is the result of a number of variables; previous vandalism, environmental exposure, wall material, construction methodologies or subsequent stabilisation and incorrect maintenance interventions over time. It is often a combination of these factors that creates a range of conditions that we refer to as structural instability. Earthquakes and water table subsidence appear to have been the major causes of structural instability in the buildings analysed here. It has been found that seismic activity remains a major threat to loss of the building fabric.

16.5.7 Orientation and exposure

In the buildings analysed, it was impossible to establish to what degree the deterioration was caused by orientation, exposure or incorrect maintenance over time. The walls that appeared to be particularly vulnerable were those that were exposed to the prevailing winds. In several cases these had been covered with plastic sheeting by the owners, which can create a cycle of condensation and deterioration of the building behind the plastic film.

Where the wind has a higher speed it is inevitable there will be a higher level of deterioration. This was recorded along the eaves of the buildings, where the plaster had been totally eroded away over the years. Vegetation had also contributed to the decay of the buildings, creating stresses and differences in temperature along the walls through overshadowing and prevention of proper evaporation of the water trapped around the building after heavy rainfall. Also, drifting snow was seen to be collecting and melting at the wall base, which can cause further erosion. It was decided to remove the trees and their roots, which had entered the buildings, foundations contributing further to the erosion and loss of fabric.

16.5.8 Appropriate repair methodologies: plaster stabilisation

In the case of the earth buildings in Italy, a mechanical cleaning and removal of debris was necessary using small compressed air pumps and picks. Some areas were stabilised by spraying Japanese tissue facing paper, which was found to be very efficient. This system of rehydration was found to be the best system as it did not introduce any foreign materials to the walls, relying on the clays to re-establish lost cohesion. The areas that were removed and dampened were then reset into their initial location using the Japanese tissue

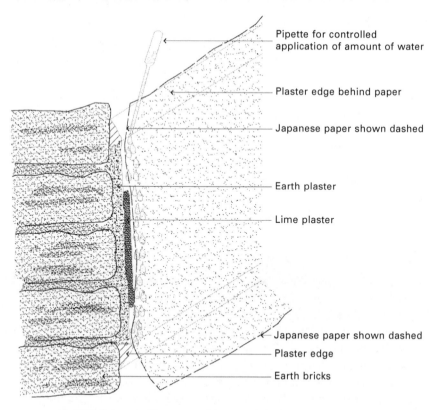

Pipette for controlled application of amount of water

Plaster edge behind paper

Japanese paper shown dashed

Earth plaster

Lime plaster

Japanese paper shown dashed

Plaster edge

Earth bricks

16.10 This image illustrates the use of Japanese paper and the close relationship with the plaster and mortar edge.

paper and applying some light pressure via cosmetic sponges. Some areas were treated by injection grouting, as the detachment of the plaster from the substrate was major. These areas were initially flushed with a 3:1 water and alcohol solution. Soil and water were mixed prior to use. The mix was determined for each application on the site and approved only after test panels were employed. The area of plaster to be stabilised was sprayed using a gentle mist. The Japanese paper was gently positioned over the section of plaster. A pipette was used so that only a small amount of water would reach the paper and not the wall directly. In this way the water that soaked through the Japanese paper pulled the paper towards the surface of the plaster creating a bandage effect. The Japanese paper with the section of plaster was then lightly pressed against the rest of the wall. Once the area had cured and the paper totally dried, the Japanese paper was removed. All cracks were then sprayed damp and mortar compatible with the existing plaster was inserted via a syringe.

16.5.9 Restoration

The first intervention was to remove the concrete sealing of the ground around most of the building, which was trapping water below ground and moving it from the foundations to the stem walls. It was also decided to remove all trees around the building, to be replaced later, after a careful solar study and analysis of the prevailing winds to find out what effect these would have on the vegetation. Inside, at ground floor, the terracotta tiles had been laid directly on the bare earth. It was decided to remove the tiles and introduce insulating materials and vents, so that any trapped humidity could escape. Along the eroded surfaces of the wall, small grids made of olive wood were hooked to the wall in order to make it more compact, facilitating a connection between the existing and the new wall layer. Several tests were carried out in order to study the effects of the various plasters and mixtures before final application to the surface of the wall.

16.6   Conclusions

One of the major goals of the projects described here was to understand the earth buildings as much as possible before any intervention. The variables of each earth building are greatly dependent on location, which means there isn't a common solution that can be applied worldwide, in every climate and for all wall compositions, orientations and exposures. The subsequent stabilisation and maintenance interventions are all different.

It is hoped that the research discussed in this paper will be of some value and assistance for the restoration of earth buildings worldwide, and that this understanding will allow for more effective preservation strategies in the future. There is a need for more funding to maintain buildings and ruins alike, and in order to allow a cost-effective use of those funds, prioritising work is imperative. Historic sites and cities need be adaptable in order to survive; to conserve them is to continue using them, otherwise they will be condemned to be seen only as an exhibit from another era.

The key point is to learn from climatic regionalism; in this case the use of wind towers, ventilation holes, chimneys and courtyards to work with the local environment. Indigenous or traditional building styles often use climate responsive designs and are appropriate in relation to other environmental factors, such as weather and natural hazards. Regrettably, architecture in our age has become increasingly mechanised, and earth is very rarely used as a resource for sustainable housing projects.

On the fringes of environmental issues, a range of disciplines, in research and practice, associated with earth building have re-emerged in the modern era. Earth is a 100% eco-friendly building material, which doesn't require manufacturing or transporting. Earth is non-toxic, non-polluting and it 'breathes'. Clays within the earth soils are hydroscopic, releasing and absorbing moisture in response to changing local atmospheric conditions, improving air quality, removing asthma triggers and reducing other respiratory diseases – this alone is an important factor to be considered when promoting and re-launching earth structures as a viable option for today's living.

16.7   Sources of further information

Websites

Eartharchitecture.org
Casediterra.it
Getty.edu

Book and reports

Facey, W., *Back to Earth, Adobe Building in Saudi Arabia*. Al-turaif, Riyadh, Saudi Arabia, 1997

Houben, H. and H. Guillaud *Earth Construction, A Comprehensive Guide*. Originally published by Editions Parentheses as *Traite de Construction en Terre de CRATerre*. Intermediate Technology Publications, London, 1994

'Summary report, Project Terra research meeting', Report of meeting held in Torquay, England, 14 May, 2000

'Terra Consortium: Guidelines for institutional collaboration', guidelines for program of Project Terra and the UNESCO Chair of Earthen Architecture, Constructive Cultures, and Sustainable Development, 2000

Associations

Cedterra, Earth Building Association of Casalincontrada in Italy

16.8   References

Baldacci O. 'L'ambiente geografico della casa di terra cruda in Italia', *Rivista Geografica Italiana*, LXV, 14–43

Crosby, A. 'Atturaif Living Museum, conservation manual', unpublished report

Part IV
Modern earth structural engineering

17

Earth masonry structures: arches,
vaults and domes

J. F. D. DAHMEN, University of British Columbia, Canada
and J. A. OCHSENDORF, MIT, USA

Abstract: This chapter discusses design and construction of vaulted roof
systems using earthen masonry. Arches, vaults and domes can be used to
create unique architectural spaces whose aesthetic appeal is matched by
their environmental performance. The structural behavior of vaulted roofing
systems is explained by way of contemporary and historic case studies. The
chapter includes an overview of design criteria and other considerations
when designing vaulted roofing systems with earthen materials, and
concludes with future trends for the design of structural forms in earthen
construction.

Key words: masonry, arches, vaults, domes, graphical analysis.

17.1 Introduction

Builders have constructed arches, vaults and domes using earth masonry for
centuries. When designed and detailed properly, these ancient forms create
striking architectural spaces that can match the longevity of any contemporary
building material. Existing earthen masonry structures can be found that are
over 1000 years old, offering ample evidence that earthen construction is
capable of considerable durability when properly designed.

The chapter begins with contemporary motivations for designing vaulted
roof systems with earth masonry, which include environmental as well as
aesthetic, economic and structural considerations. A general discussion of
the theory explaining the structural behavior of arches, vaults and domes
follows, which provides background for historic case studies of existing
arches, vaults and domes. More recent examples will show that these forms
are capable of creating compelling contemporary architectural spaces. The
chapter includes an overview of design criteria and other considerations
when building with earthen materials, and concludes with future trends for
the design of structural form in earthen construction.

427

17.1.2 Motivation for design of earth masonry vaults and domes

There are a host of reasons for the renewed interest of architects and designers in the ancient methods of building with earthen masonry, but the primary advantage of earth masonry over other materials is its low environmental impact. In contrast to most other masonry materials, earth masonry is minimally processed and low in embodied energy. Often the building material can be sourced at or nearby the building site, lowering the energy profile even further. While earth masonry does not have the compressive strength of concrete or fired masonry, with proper design it can be used to create safe, efficient and enduring structural systems that require only a small fraction of its available material strength. The thermal mass of earthen building materials can also be incorporated into design to reduce the energy required to heat and cool buildings. The non-toxicity of natural earthen building materials, combined with the ease with which they can revert to their preconstruction state, address issues of disposal and waste. Finally, and no less important, using earth masonry offers the architect an opportunity to express the natural attributes of a site in a unique way. Building ambitious structures with materials sourced on-site offers architects greater design possibilities than many contemporary processed materials that are often high in embodied energy.

The environmental advantages of earthen masonry come into sharpest focus when it is compared to the material it most often replaces: concrete. The production of cement and concrete accounts for 6–7% of CO_2 emissions worldwide and each ton of cement manufactured releases a roughly equivalent amount of CO_2 into the atmosphere (Chaturvedi and Ochsendorf, 2004). In addition to greenhouse gas emissions, concrete products contribute roughly 50% to the 140 million metric tons of construction waste stream generated annually in the United States. Fired clay products are also responsible for high carbon emissions. Replacing concrete and fired clay masonry materials with minimally processed earthen masonry offers significant environmental advantages.

Despite these advantages, earthen arches, vaults and domes present unique challenges to the designer and builder alike. Designers must accept the limitations inherent to the material, including relatively low strengths and susceptibility to weathering. In addition, the methods used to design structurally efficient geometries for arches, domes and vaults are largely unknown in the contemporary structural engineering community, and most building codes do not include earthen vaults as a viable structural system. Proper detailing is absolutely essential to the longevity of earthen structures, as small incursions of water can have disastrous effects on earthen masonry. Finally, working with earthen masonry can be a challenging experience for builders who are accustomed to highly predictable standardized materials.

Despite these limitations, earthen masonry arches, domes and vaults can be used to produce sustainable and dramatic structure and spaces, as shown in Fig. 17.1.

17.2   Structural theory for arches, vaults and domes

The structural design of earthen masonry arches, vaults and domes requires a different approach than is used for conventional framed structures. Structural engineering practice has generally focused on three main structural criteria: strength, stiffness and stability (Heyman, 1995). A building must support the applied loads without exceeding the strength of the material, it must not deflect more than established acceptable limits and the structure as a whole must be stable under all expected loading conditions. Although these criteria also apply to masonry structures, only the third criterion, stability, plays a central role in the design of earth masonry structures.

Due to the high self-weight and low compressive stresses, unreinforced masonry vaults are typically a problem of stability rather than strength. The compressive strength of unreinforced masonry is generally far stronger than the loads placed on it, and the tensile strength can conservatively be considered as zero. Unbaked bricks routinely achieve 5 MPa in compressive strength, which would allow a hypothetical tower to be constructed roughly 200 m-high in earth before the bricks at the bottom begin to crush. Applying a safety factor of 10 against crushing would allow for 20 m-high earthen walls. Thus the compressive strength of earth is likely to be at least an order of magnitude more than is required by most structures. Conversely, the tensile strength can be assumed to be zero because earthen bricks and

17.1 Vaults at vaulted interior at Mapungubwe Interpretation Center, Limpopo, South Africa, Peter Rich Architects 2009 (photo: Peter Rich).

mortar have virtually no resistance to tensile forces. Stiffness is rarely an issue in masonry structures, because the low stresses account for very small displacements due to elastic deformation of the material. While strength and stiffness are of negligible importance in the design of masonry structures, the third criteria, stability, is of paramount importance. In particular, foundation movements or seismic activity can lead to instability of unreinforced masonry. In general, it is not advisable to build earthen vaults in seismic regions.

Instability can best be understood in geometrical terms: if the proportions of an arch, vault or dome are correct, the internal compressive forces will be maintained within the masonry, with relatively low compressive stresses (stress is defined as force per unit area). The task of the designer of an unreinforced masonry structure, whether of stone, fired brick or earth, is to find a shape for it such that the structure will remain in compression under all foreseeable loading conditions.

Take, for example, the simple two block structure shown in Fig. 17.2. The load P necessary for collapse is clearly a question of stability, not stress. Of course stress plays a role when the entire load is focused at point A at collapse (Fig. 17.2b), but in most cases of historical masonry construction, the working stress is two orders of magnitude below the compressive strength. The potential cause of collapse is therefore far more likely to be the instability of the structure rather than insufficient compressive strength. This simple example illustrates the value and necessity of stability analysis for unreinforced masonry structures.

17.2.1 Elastic versus plastic design understanding

Conventional methods of structural analysis, known as linear elastic analysis, are not capable of predicting the load, P, which will cause the overturning of a masonry block. Therefore, identifying a proper shape for masonry presents a problem for conventional elastic design techniques. These techniques are effective for determining local stresses in building elements such as steel

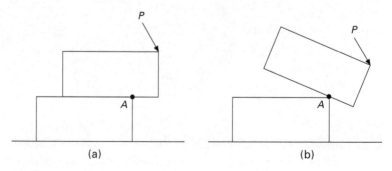

17.2 Example of stability analysis for simple two-block structure.

frames, but do not offer a reliable picture of the overall flow of forces in a structure across a range of support conditions. By idealizing the material as linear elastic, any elastic calculation will incorrectly predict tensile stresses that the material is not capable of supporting. According to an elastic analysis, typically performed using the finite element method (FEM), a masonry arch could not support its self-weight without incurring tensile stresses that would crack the arch. Thus, according to elastic analysis, a Roman stone arch bridge can be declared unsafe due to insufficient tensile strength, despite the fact that it has been standing for 2000 years. Furthermore, an elastic analysis will predict high tensile stress concentrations in some locations, such as the corners of a window in a brick wall. Conventional linear elastic methods are incapable of predicting the actual behavior or the failure state of a masonry vault and are not useful for determining a good form for the structure.

In contrast to elastic design techniques, the plastic design approach, typically called limit analysis or the 'equilibrium approach', allows for the existence of many possible load paths within a structure. Rather than attempting to calculate the stresses in a structure in an overly idealized perfect elastic continuum, the safe theorem of plastic design dictates that if one equilibrium state can be found for a structure given a set of loads, the structure can be demonstrated to be safe. There are many different methods of carrying out an equilibrium analysis. Of these, graphic statics is the fastest and most intuitive method for designing masonry vaulting. The graphical method relies on thrust lines to represent the paths of resultants of compressive force flowing through the structure (Ochsendorf and Block, 2009).

17.2.2 Robert Hooke's 'hanging chain'

English scientist Robert Hooke (1635–1703) identified 'the true Mathematical and Mechanical form of all manner of Arches for Building,' which he summarized with the phrase: 'As hangs the flexible line, so but inverted will stand the rigid arch'. A hanging chain, which forms a catenary in tension under its own weight, can be inverted to find the ideal form of an arch, which stands in compression (Fig. 17.3). Both the hanging chain and the arch must be in equilibrium, though the chain can only work in tension and the arch can only work in compression. More generally, the shape that a string or chain takes under a set of loads, if inverted, is an ideal shape for an arched structure to support the same set of loads. This simple idea, sometimes called a funicular form, produces a line of thrust that can be used to design and analyze vaulted structures in earthen masonry.

To apply the idea of a hanging chain to an actual masonry structure, consider the semi-circular stone arch shown in Fig. 17.4a. Each *voussoir*, or stone, in the arch has a known mass and centroid. Because masonry can resist only compressive forces and is very weak in tension, the arch can

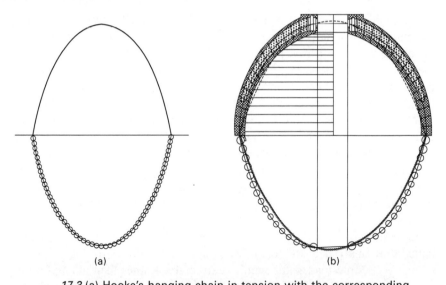

17.3 (a) Hooke's hanging chain in tension with the corresponding inverted arch in compression; (b) the safety of the cracked dome of St Peter's in Rome demonstrated by Hooke's principle.

stand only if a path of compressive forces is shown to lie entirely within the masonry. A model of the arch could be constructed with a piece of string supporting masses that correspond to the weight of each block, and the resulting shape of the string would be the path of the tensile forces necessary to support these weights. By pulling on the ends of the string with different values of horizontal force, an infinite number of equilibrium solutions can be found for the same loading condition. By inverting this string, it can be used to calculate the line of compressive forces lying within the masonry. Though builders in the past have sometimes used physical models made of weights on hanging strings to derive the form of masonry arches and vaults, the same solution can be found using graphic statics as shown by the force polygon (Fig. 17.4b). More detail on graphic statics and how it can be used to calculate the form and forces of masonry structures can be found in the book by Allen and Zalewski (2009).

Equilibrium in a masonry arch is visualized by means of a *line of thrust*, a theoretical line that represents the path of the compressive forces through the structure. If the thrust line strays outside the section, compressive forces must travel through the air, which is only possible if the material can resist tension. The stone arch of Fig. 17.4a has an infinite number of possible thrust lines that lie within the masonry, though only the minimum and maximum values of horizontal thrust are illustrated. The masonry arch is statically indeterminate, with an infinite number of solutions between the minimum and maximum values. In short, multiple equilibrium solutions are possible,

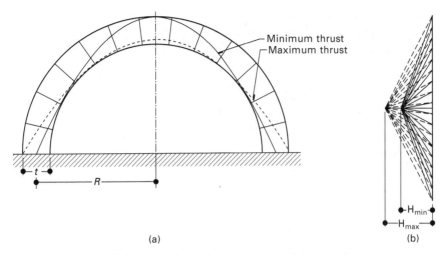

17.4 (a) Minimum and maximum horizontal thrust values in a semicircular masonry arch; (b) the force polygons corresponding to the lines of minimum and maximum thrust for this arch according to graphic statics (image: Katherine Hriczo).

with each solution corresponding to a different compressive line of thrust inside the masonry.

The geometry of the arch determines the minimum and maximum horizontal force that is possible within the arch. To assess the safety of a historical arch, a range of possible thrust values can be determined, though the exact internal value of force depends on the exact support conditions, which are known only to the arch. If support movements occur, then the arch will adjust to these movements by forming hinges. Such hinges make the arch statically determinate because a single thrust line can be found that passes through the hinges. A small outward movement of the supports will lead to the state of *minimum thrust* and a small inward movement of the arch supports will lead to the state of *maximum thrust*. The minimum (or passive) thrust state represents the least amount the arch can push horizontally on its neighboring elements, as a function of its self-weight and shape. The maximum (or active) state of thrust represents the largest possible horizontal force that this arch can provide. Vaulted masonry structures must bring their applied loads safely to earth within the fabric of the masonry. A more general example of a random stone arch and one possible compressive line of thrust as calculated by graphic statics is demonstrated in Fig. 17.5.

Each individual portion of the arch is held in place by the compressive forces applied to it from adjacent portions. In this manner, the weight of each segment of the arch is carried down to the supports in compression. The equilibrium of a single block in the arch is represented by the bold triangle shown in the funicular polygon of Fig. 17.5c. Because the load

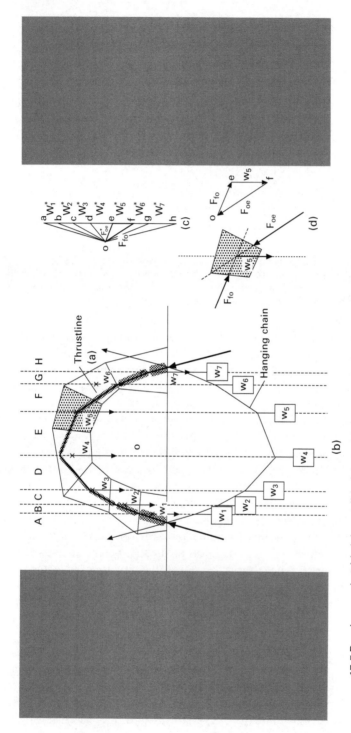

17.5 Random arch with (a) possible internal line of thrust; (b) its inverse, a hanging string with weights proportional to the weights of the voussoirs; (c) the corresponding funicular polygon; (d) one of the arch voussoirs with the resultant forces on it and a closed triangle visualizing the equilibrium of forces in it (image: Philippe Block).

accumulates from the center of an arch down, the keystone at the center of the arch supports lower forces than the blocks nearer the base.

17.2.3 Funicular versus geometrical design in masonry

Masonry structures may follow traditional geometry or more structural geometry. *Classical* geometry is composed of vertical columns and walls, which support arches, vaults and domes that are circular arcs of constant radius. *Funicular* geometry is defined by the path of forces through structures. Classical geometries are easier to lay out and build, but they use more material because the forms are determined by geometry rather than lines of thrust. In contrast, funicular geometries place material nearest the dominant thrust line, which is the path of forces taken due to self-weight only for most masonry structures. Funicular geometries use less material, but are more difficult to define and construct than classical arches.

17.2.4 Hanging models

Many designers have used hanging models to derive the form of funicular structures, by inverting the geometry found from using weights on strings, or sections of bare chain (Fig. 17.6). Uniform loads can approximate the form of an ideal funicular structure due to dead load, and additional weights can be added to account for other load cases, or to define a larger envelope for the masonry structure. Because masonry cannot resist tension, the designer must create a form that can contain the complete envelope of compressive thrust lines for all expected loadings, including live loads caused by wind,

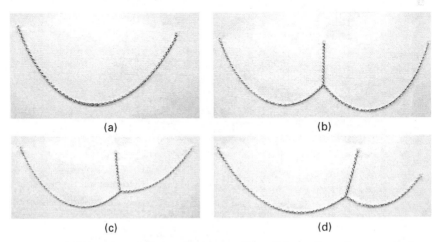

(a)   (b)

(c)   (d)

17.6 Hanging chain explorations of possible sections for a vaulted masonry building (photos: John Ochsendorf).

snow and other sources. A chain model allows for rapid exploration of the infinite possibilities of structural form and gives insight into the importance of the support conditions. Such physical models are particularly powerful during initial design explorations. More precise graphical methods can be used to refine the design during later stages.

17.3   Earth masonry arches

The arch is the fundamental structural form for spanning in unreinforced masonry. Even the simple two-dimensional arch has many possible variations, and these can be combined to create countless other structural shapes. The structural action of complex vaults and domes can be understood by analyzing them as a series of two-dimensional arches contained within these shapes. When analyzed this way, the additional structural integrity resulting from the three-dimensional aspect of vaults and domes provides a further margin of safety.

17.3.1 Historic case study: Taq-I Kisra arch

The arch of Ctesiphon is the surviving portion of a barrel vault that originally enclosed the Iwan-i-Khosrau, the throne room commissioned by Khosrau I, ruler of the Sassanid Empire, around 540 AD located in the ancient city of Ctesiphon in modern-day Iraq. (Fig. 17.7). During construction, the horizontal

17.7 Taq-I Kisra arch from 540 AD, as photographed in 1932, Ctesiphon, Iraq (photo: Photo Department of American Colony, Jerusalem).

courses of bricks above the imposts were progressively corbelled inwards so that the two walls approached each other at slight curves, which reduced the large area that had to be vaulted (Scarre, 1999). The remaining portion of the vault was then constructed against the rear wall using a method of pitched brick vaulting to avoid centering, which was especially advantageous in Mesopotamia where wood was scarce (Heywood, 1919). The vault was constructed with fired clay bricks adhered to one another with gypsum mortar. Gypsum mortar was ideal because it was quick setting and gypsum deposits were locally available (Pope, 1977). In the 1500 years that following its completion, part of the vault collapsed, leaving behind a monumental ruin. The arch of Taq-I Kisra is one of the most impressive masonry buildings in the world today.

The surviving arch measures approximately 35 m high at its crown, 25 m wide from the spring points and 50 m long. The arch is approximately 7 m thick at its base, diminishing in thickness to approximately 1 m at the crown of vault. The dimensions of the burnt bricks are approximately $30 \times 30 \times 7.5$ cm thick. The Ctesiphon vault was constructed as an extruded series of arches, and it can therefore be considered to act structurally by dividing it into two-dimensional arches.

17.3.2 Structural analysis of Taq-I Kisra arch

A simple thrust line analysis of the arch of Taq-I Kisra can be used to estimate the magnitude of the internal compressive forces and to demonstrate its safety (Fig. 17.8). The forces are considered for a 1-m-wide strip of the arch, which is divided into segments whose weight and centroid must be found by multiplying the volume of masonry by the density of the material (typically around 2000 kg/m^3). A possible equilibrium solution for the arch is determined from trial and error using graphic statics, and the resulting thrust line is transferred to the ground entirely within the masonry. Of particular interest for estimating the stability of the remaining arch is the location at which the thrust line falls within the base of the supporting wall. This analysis demonstrates that the arch is well-equilibrated, with the resultant force of roughly 145 metric tons falling within the middle-third of the wall at the base.

17.4   Earth masonry vaults

Though the arch had been used in Egypt much earlier, master builders of the Roman Empire expanded the use of the arch in architecture to create a wide range of barrel vaults and domes. Drawing on the Roman tradition, builders in medieval Europe created simple vaulted churches by using cylindrical barrel vaults, eventually employing pointed vaults to increase interior heights.

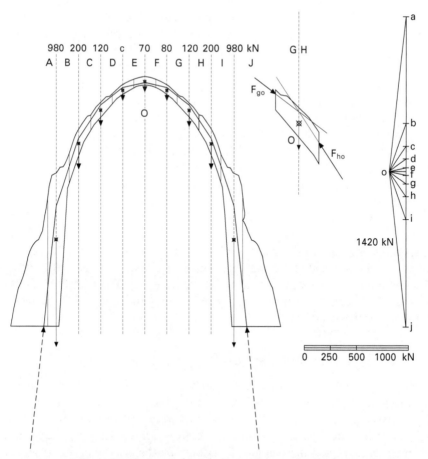

17.8 Thrust lines for Taq-I Kisra arch (image: Joseph Dahmen).

Gothic builders created increasingly complex arrangements of arches and vaults, using less material than Romanesque barrel vaults. In the Islamic world, master masons used remarkable combinations of arches vaults and domes to invent new spatial possibilities. These historical precedents can serve as inspiration for future designs.

17.4.1 Equilibrium analysis of vaults using thrust lines

To determine possible equilibrium states for a masonry vault, we may divide the vault into a series of intersecting arches (Fig. 17.9). Most vaults can be divided into an infinite arrangement of different arches, though it makes most sense to use symmetry in dividing up a vault. A *quadripartite vault*, defined by masonry ribs across its diagonals, may be divided into parallel arches that span between the ribs. The ribs then collect the forces to the support at

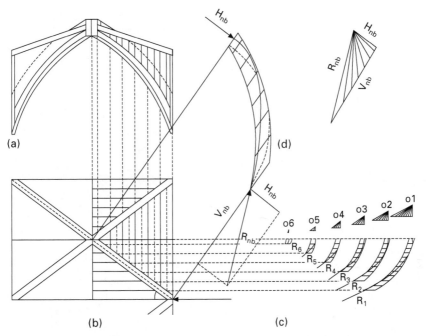

17.9 Graphical analysis of the forces in a ribbed quadripartite vault
(a) cross section; (b) plan view; (c) arch sections span between
the main ribs; and (d) the rib profile with a safe thrust line and its
corresponding force polygon. The loads on the main rib come from
the resultants of the series of web arches (image: Philippe Block).

each corner, where the thrust of the vault must be resisted by an adjacent
vault or by a buttress. In this way it is possible to estimate the forces in a
range of vault geometries, with or without ribs.

17.4.2 Historic case study: Guastavino vaulting, ca. 1900

Rafael Guastavino Jr (1872–1950) used graphic statics to design and analyze
numerous thin shell masonry domes, bringing a historical Mediterranean
construction method into 20th-century American buildings (Ochsendorf,
2010). Such equilibrium calculations can help to define the form of the dome,
or even suggest locations for flying buttresses, as proposed for the church of
St. Francis de Sales in Philadelphia (Fig. 17.10). Their load-bearing domes
spanned long distances with minimal material, by making good use of efficient
structural forms. In addition, decorative tile finishes and openings for natural
light, as in the Elephant House of the Bronx Zoo, add to the visual impact
of Guastavino vaults (Fig. 17.11).

Guastavino vaults are strong in compression and weak in tension, and
therefore behave similarly to other masonry vaults. By using doubly curving

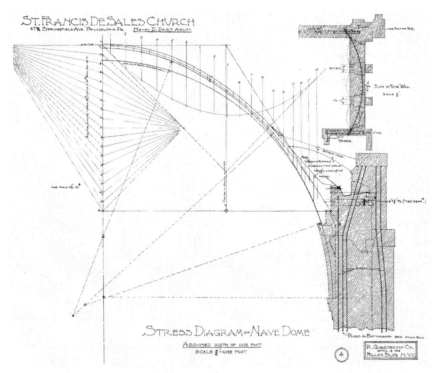

17.10 Graphic analysis for St. Francis de Sales Church Dome by the Guastavino Company, Philadelphia, 1908 (image: Guastavino/ Collins Collection).

17.11 Inner dome of the Elephant House in the Bronx Zoo, 1909, with numerous openings for natural light (Avery Library; photo by Michael Freeman).

shells in masonry, Guastavino vaults have numerous load paths, giving rise to their legendary strength. Of the more than 10,000 vaults constructed in approximately 1000 buildings throughout North America between 1881 and 1962, no Guastavino vault has ever failed in service.

17.4.3 Contemporary design case study: Mapungubwe Interpretation Center

The Mapungubwe Interpretation Center (Fig. 17.12) was commissioned by South African National Parks to provide an interpretive center devoted to the Mapungubwe culture in Southern Africa. The unique design was a collaboration between Peter Rich Architects and vault designers John Ochsendorf and Michael Ramage. James Bellamy led the construction team and trained local workers to build vaults. The structurally efficient earthen vaults were designed using equilibrium methods to act in compression, with no internal steel for reinforcing. The thin bricks were made from cement-stabilized soil, and the vaults were constructed by training minimally skilled workers to build in the 700-year-old method of Mediterranean tile vaulting. Training workers in the method addressed budget, time and materials constraints while providing jobs for unemployed workers in the region. By using fast-setting mortar to build the first layer of masonry, the vaults were constructed with minimal formwork. The result is an elegant building that is economically, socially and environmentally sustainable.

The construction process consisted of making tiles, constructing work guides and laying tile. Stabilized earthen tiles measuring $140 \times 290 \times 20$ mm

17.12 Vaulted interior at Mapungubwe Interpretation Center, Limpopo, South Africa, Peter Rich Architects 2009 (photo: Robert Rich).

were fabricated from local soil by modifying a standard Hydraform block press to produce thin tiles instead of blocks. Locally trained masons constructed the vaults following the traditional layered tile technique, laying tiles with their thin edges joined by fast-setting gypsum mortar (Fig. 17.13). Two edges of the tile are in contact with mortar, giving enough purchase to support the tile in a matter of seconds as the gypsum plaster sets. The technique allows masons to construct the vaults with minimal formwork, which reduces the amount of material and increases the speed of construction. After the first layer of tiles is installed, masons increase the thickness of the vault by adding successive layers of tiles interspersed with Portland cement mortar until the entire construction is three to four tiles thick. To be stable during construction, it is essential for each section of the vault to contain compressive arches within the thickness. The completed vaults are capable of impressive spans, with a large vault spanning 21 m, and other vaults spanning 14.5 m. The vaults are designed to ensure that all forces in the structure act only in compression, minimizing the need for steel reinforcement, which is expensive and energy intensive. The Mapungubwe

17.13 Mapungubwe Interpretation Center under construction (photo: James Bellamy).

Interpretation Center revives a centuries-old technique in tile vaulting with local materials and labor, with dramatic reductions in embodied energy due to the use of minimally processed soil bricks. Among other international design awards, this building was named the 'World Building of the Year' at the 2009 World Architecture Festival in Barcelona.

17.4.4 Contemporary design case study: Start Festival Vault

The Start Festival Earth Pavilion was a temporary earthen structure built in the gardens of Clarence House, the London home of HRH Prince Charles (Fig. 17.14). The pavilion was constructed for *Start*, a national initiative created by HRH Prince Charles' Charities Foundation to promote and celebrate sustainable living. Architects Peter Rich and Tim Hall worked with vault designer Michael Ramage to design the structure, which was sponsored by The Earth Awards to demonstrate sustainable building techniques. Mason Sarah Pennal led the construction team.

Like the Mapungubwe Visitor Center, the Start Festival Earth Pavilion was constructed of earthen tiles created from local materials. The builder used a modified Hydraform manual press to create the tiles from a mix design consisting of 90% east London soil combined with 10% cement and a small amount of sand. The tiles, which were approximately 20 mm thick,

17.14 Start Festival Earth Pavilion (photo: Michael Ramage).

were assembled over a lightweight plywood formwork after curing for a minimum of four days. The total assembly was 60 mm thick, which included two layers of tile and a mortar layer between them in which a nylon triaxial geotextile was embedded to help resist any tensile forces. The temporary foundation for the project was constructed of plywood, and in keeping with the theme of sustainability, the floor consisted of salvaged plywood planks, which were previously used for pressing the tiles. The screen between the vaults was chestnut, harvested a day before it was installed.

The area covered by the vault is approximately 75 m^2, which results in a usable floor area of approximately 40 m^2. The structure was built in ten days using fast-setting gypsum mortar and earthen tiles. Because of the short duration of the festival, no waterproofing was used on the structure, which instead relied on effective detailing to shed water. The vaults of the Start Festival Earth Pavilion were designed to have low stresses throughout the structure. Vertical forces were distributed through large areas of masonry while the horizontal thrust was minimized by making the vault nearly vertical at the foundations. The resulting design develops a maximum stress of only 0.1 MPa, ensuring that the compressive strength of the bricks, which was greater than 5 MPa, was 50 times greater than the highest compressive stress in the structure.

The unique form of the vault presented challenges in design and construction. The parabolic shape of the latitudinal axis was designed according to the flow of forces of the vault, but the longitudinal curve was chosen largely for its architectural appeal. Although the compound curvature of the vault develops considerable stiffness, the saddle shape resulted in tensile forces that required the addition of a geotextile between tile layers to ensure that safe envelope of thrust lines was sufficient. The geotextile also provided tensile strength to withstand asymmetric wind loads, effectively allowing the thrust lines to pass outside of the masonry if necessary.

17.4.5 Contemporary design case study: Freeform Thin-tile Vault

This freeform unreinforced masonry vaulting prototype was designed and built by the BLOCK Research Group at the Swiss Federal Institute of Technology (ETH) in Zurich, Switzerland (Fig. 17.15). Although the experimental vault was constructed with commercially available fired masonry tiles, the method of design used to find the compressive form and the manner of construction pertain to earthen vaults. The undulating form was designed using Thrust Network Analysis (TNA) – a novel computational form finding approach for exploring compressive form in three dimensions. The resulting freeform shells presented new challenges in tiling patterns, sequence of building and especially the guidework. A complex vaulted structure is capable of

17.15 Masonry vault prototype, BLOCK Research Group, ETH Zurich (photo: Lara Davis).

generating local forces that are significantly higher than those occurring in the final structure. Temporary means of support must be found to manage these forces to ensure that the partially completed structure does not collapse during construction. The BLOCK Research Group developed a continuous expandable cardboard guidework system using CAD-CAM (computer-aided design and computer-aided manufacturing) cutting and gluing processes to ensure structural stability of the vault during construction (Fig. 17.16).

The thin shell prototype was designed by Matthias Rippmann, Lara Davis and Philippe Block and constructed over a period of six weeks by Lara Davis. The custom CAD-CAM guidework was fabricated by Tom Pawlofsky, and the project was sponsored by ZZ Wancor AG and Rigips.

17.4.6 Pitched brick vaults

In contrast to vaults employing double curvature, the pitched-brick vault, sometimes called the 'Nubian vault', is an extruded arch whose simplicity of construction offers great appeal. While a typical barrel vault requires formwork, or centering, during construction, the pitched brick vault requires no support from below. When constructing such a vault, masons build one permanent end gable wall with brick to brace the subsequent construction. The bricks are laid in a roughly parabolic shape at an angle of approximately 20° to perpendicular along the longitudinal axis of the vault. This angle allows prior courses of masonry to support successive courses such that no centering is required (Fig. 17.17). Once the vault is complete, buttresses are sometimes formed by raising the side walls by 8 to 10 courses of bricks

17.16 Continuous expandable cardboard guidework system during construction (photo: Lara Davis).

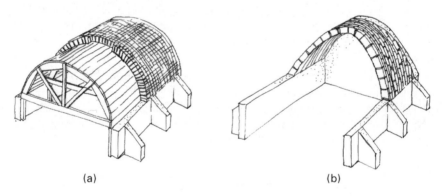

(a)   (b)

17.17 Vaults under construction (a) Typical barrel vault with centering; and (b) pitched brick vault (drawing: Joseph Dahmen after Edward Allen).

and filling in the volume between the buttress and vault with plain earth to provide resistance against the horizontal forces, though this is only effective if the supporting walls are buttressed as well. The first examples using this efficient construction technique date back at least 3500 years.

Several organizations have been established to train masons in pitched brick vaulting techniques to address low income housing. The Association La Voûte Nubienne (AVN) in France applies the traditional Nubian vault technique to produce contemporary earth buildings throughout Africa. AVN trains local masons to build the vaults with sun-dried adobe bricks created on-site or nearby, creating dwellings whose thermal mass is well-adapted to the harsh climate. The Nubian technique is especially well-suited for this application: construction is straightforward and does not require wood centering, which is a precious resource. The finished structures do not require metal roofing, which can be expensive in developing countries.

The Adobe Alliance in Presidio, Texas, along the border between the United States and Mexico, was established by Simone Swan to preserve and continue the rich tradition of building with sun-baked adobe bricks. Swan founded the organization after apprenticing with her mentor Hassan Fathy (1900–1989), a noted Egyptian architect who worked to re-establish the use of earthen vaults in Egypt during the 1940s (Steele, 1997). With the help of local masons, Swan creates earthen homes on both sides of the border (Fig. 17.18). The Adobe Alliance teaches the design and construction of hand-made adobe vaults and domes built without centering. The vaults and domes are created with hand-molded earthen adobe blocks without the addition of cement-stabilizing materials (Fig. 17.19). Completed vaults are covered in an earthen plaster composed of clay, sand, water, nopal juice, horse manure, lime and finely chopped wheat straw. The natural materials breathe, regulating interior humidity while moderating the harsh temperatures characteristic of the desert climate by absorbing heat in the daytime and radiating it at night.

17.18 Nubian vault under construction in Texas by masons trained by the Adobe Alliance (photo: Adobe Alliance).

17.19 Adobe vaulted interior in far west Texas by Adobe Alliance (photo: Simone Swann).

Outdoor courtyards create pleasant micro-climates that provide additional living space. The Adobe Alliance has trained many masons in pitched brick vaulting methods and is building several adobe houses each year.

17.5   Earth masonry domes

Masonry domes have created some of the most grandiose spaces ever conceived. Masonry domes can be designed and analyzed by dividing the dome into a series of wedge-shaped arches that lean against each other along radial lines (Fig. 17.20a). This is a safe but overly conservative approach, because it ignores the circumferential *hoop* forces which can develop in a dome (Figure 17.20b). For a hemispherical dome, the hoop forces act in compression in the upper portion, and then change to tension below an angle of approximately 52° from the vertical, which may sometimes require the addition of tension reinforcement near the base of a dome. Wolfe (1921) analyzed masonry domes graphically for a range of hypotheses (Fig. 17.21).

A two-dimensional arch will collapse if any voussoir is removed, since there would be no other path in compression within the structure. Because of their double curvature, domes allow for many additional load paths in three dimensions (Lau, 2006). Small rounded openings can be introduced in almost any location, provided that compressive forces can flow around the opening.

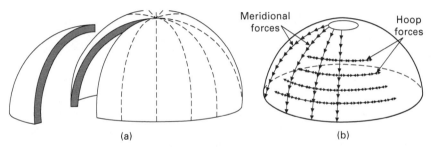

17.20 Idealization of masonry domes (a) considered as series of independent wedges that lean on each; and (b) Combination of meridional (arching) and hoop forces (image: Wanda Lau).

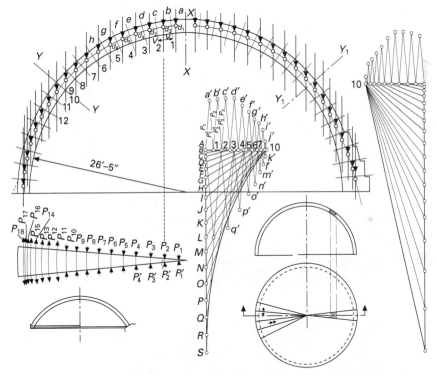

17.21 Graphical analysis of a masonry dome (source: Wolfe, 1921).

A horizontal circular hole at the top of a dome, creates an opening known as an *oculus*, which can allow for natural light and natural ventilation. Forces can flow around the oculus in a compression ring, and multiple openings can be made in a dome, as in the case of the Bronx Zoo building by the Guastavino Company (Fig. 17.11). The Pantheon in Rome, whose dome was cast in unreinforced concrete and thus acts as an unreinforced masonry dome, contains an 8.3 m oculus at the center of its 43-m span.

17.5.1 Historic case study: Syrian 'beehive' houses, ca. 1700

The semi-nomadic people of northern Syria have constructed corbelled dome dwellings of unbaked earth bricks since the 1700s (Fig. 17.22) (Mecca and Dipasquale, 2009).

After outlining the square perimeter of the house on the ground, the mason excavates the earth until a solid layer of soil is encountered on which to construct a foundation wall of local stone 60–85 cm thick. The mason then constructs a stone wall stabilized by small stones used as wedges. The dome is created by laying bricks in a continuous helicoidal spiral (built without centering) and is blocked at the top (Fig. 17.23). The construction method results in the trademark 'beehive' shape, which appears more like a cone than a traditional dome. Because the structure shifts from a square stone foundation to a circular dome, it can be challenging to bring the dome forces down to the supports. The fragile nature of the unbaked earthen bricks makes the beehive dwellings susceptible to damage due to weathering, and they are often plastered with mud–straw mortar to provide some measure of protection from the elements. The beehive houses require frequent maintenance due to erosion.

The square base of these dwellings varies from 3–3.5 m with heights generally between 4 and 6 m. Larger structures known as 'Sultan Domes'

17.22 A village of Syrian beehive dwellings (photo: Saverio Mecca).

17.23 Syrian beehive dwelling ceiling detail (photo: Saverio Mecca).

have bases that measure as large as 4.5 m with heights similar to the smaller structures. The dimensions of the unbaked bricks formed from local soils vary from village to village.

17.5.2 Structural analysis of typical Syrian 'beehive' house

The beehive dome can be approximated initially as a series of arches leaning against each other (Fig. 17.24), though for some geometries, this simplification will not allow for the thrust line to be contained within the dome (Fig. 17.25). Domes gain greater stability from the three-dimensional flow of forces and, in particular, from compressive hoop forces, which prevent the dome from falling inward under dead load. The effect of hoop forces is most pronounced in the upper region of a dome, where circular compression rings maintain equilibrium. This gives a resultant reaction on the meridional, or arching forces, which can deviate the thrust line to keep it within the section of the dome. Many different geometries are possible for domes, provided that compressive hoop forces are capable of developing to assist the arching action.

17.5.3 Contemporary design case study: dome of the Dhyanalinga Meditation Shrine, India

The Temple of the Dhyanalinga, located in the foothills of the Velliangiri Mountains of Southern India, was commissioned to provide a space for

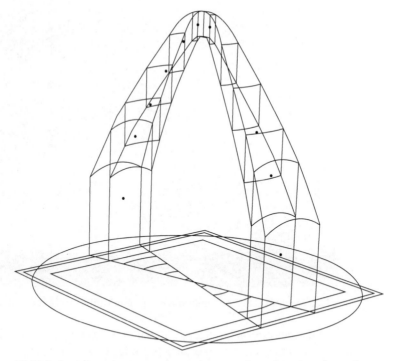

17.24 A dome can be approximated as a series of wedge-shaped arches leaning toward the center of the dome (image: Joseph Dahmen).

meditation (Fig. 17.26). The temple dome was constructed in eight weeks by 300 local unskilled laborers and volunteers. The dome is constructed of fired clay bricks that vary considerably in size because they were supplied by approximately 20 different kilns due to the large numbers required. Because no reinforcement was used anywhere in the dome construction, the engineers who designed the structure had to ensure that only compressive forces acted on it. The completed dome which is elliptical in cross-section, measures 10.1 m high and 23.2 m in diameter. The thickness of the dome is 50 cm at the base, which tapers to 20 cm at the crown, roughly the thickness of a single brick. The dome exterior is parged with cement mortar to provide a weather-resistant finish.

The dome rests on a foundation of granite rubble masonry set in lime mortar that measures roughly 3 × 3 m. The first course of the segmental dome is laid at an angle of 13° to the horizontal, increasing to 82° to the horizontal at the last course. As the dome was rising, the specification of mud mortar stabilized with lime and sand was adapted by adding more soil to achieve the ideal adhesion based on the angle of the brick layers. To give the mortar its strength and fast-setting properties, 13% cement and 19%

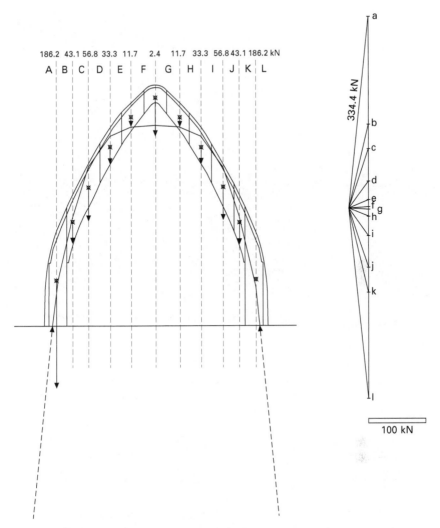

186.2 43.1 56.8 33.3 11.7   2.4   11.7 33.3 56.8 43.1 186.2 kN

A B C D E   F   G   H   I   J K L

334.4 kN

100 kN

17.25 Simplified arch analysis through section of Syrian beehive dwelling demonstrates that hoop forces are required for the dome to stand (image: Joseph Dahmen).

lime were added (Auroville Earth Institute, n.d.). The dome shelters a silent atmosphere conducive to spiritual growth and meditation, and is one of the longest unreinforced masonry spans built in recent years.

17.6   Material properties of earth masonry structure

The case studies above provide a general overview of a range of vaulted masonry designs. A series of historical and contemporary examples, together

17.26 Dhyanalinga Meditation Shrine, India, 1999 (photo: Auroville Earth Institute).

with structural analysis and construction details, allow the designer of an earthen vault to understand the broad considerations for design. However, it is worthwhile to go into more detail for various other aspects of earthen vault construction.

17.6.1 Bricks

It is not possible to speak of one particular size or shape of earthen brick suitable for earthen masonry arches, vaults and domes because of the different demands of each project and location. Bricks used for vaulting may vary widely, with specific brick geometry dictated by performance demands placed on the individual units by the design of the structure and the characteristics of local soils. Adobe blocks such as those used for pitched brick vaults by the Adobe Alliance typically measure 45 × 60 cm (18 × 24 in), whereas tiles for Guastavino-style vaults often measure a mere 25 mm (1 in) thick, with compressive forces acting along the thin axis of the tile. It is necessary that the geometry of bricks be suited to the construction method, and that design strengths be capable of meeting loading with appropriate safety factors, though axial compressive stresses are generally quite low. For environmental and economic reasons, cement content should be kept to the minimum necessary to meet performance requirements, and local soils should be used wherever possible.

17.6.2 Mix designs

Mix designs governing the composition of soil bricks depend on the specific performance characteristics required by a design. Where clay binders are used, only non-expansive clays should be sourced to avoid problems with expansion and contraction. Soils should be well graded with even distribution of aggregate sizes and an equal proportion of fines to produce tightly packed tiles. Soil for brick making should be screened to ensure that individual particles are limited to 5 mm maximum, though this can be increased to as much as 15 mm in the case of larger bricks. Sample bricks should be load-tested to failure to ensure that they are capable of meeting design loads with an acceptable factor of safety. For environmental reasons the minimum amount of cement should be used that is capable of providing bricks of sufficient strength. Moreover, bricks should be capable of withstanding moisture likely to be encountered during construction and the service life of the building.

17.6.3 Bond patterns

Bond patterns vary according to specific construction techniques and desired appearance, but in general, bond patterns should be staggered to avoid weak lines along mortar joints. Stack bonds should be avoided for the same reason, as they can result in weakness along mortar joints. Vaults are sometimes constructed with tiles laid in herringbone pattern, or in multiple layers, which allows masons to improve the cohesion between bricks. Geotextiles, or other reinforcement, can be embedded between masonry layers to address tensile forces due to asymmetrical loading.

17.6.4 Mortar types and properties

The composition of mortar depends on the strengths required by specific designs for arches, vaults and domes. Furthermore, mortar composition will depend on the mix design used for bricks as the two must interface effectively for durable construction overall. Extreme differences between durability of mortar and brick can accelerate weathering, as will occur in an adobe wall with a Portland cement mortar between bricks. For tile vaults, gypsum plaster offers fast setting for the initial layer of tiles but it must be protected from any precipitation during and after construction. Until the initial layer of tiles is parged with cement mortar, construction should be protected from moisture with appropriate measures. Subsequent layers of tile are laid on the initial layer using cement mortar. A final parging coat of rich cement mortar is applied, preferably covered by a layer of impermeable material to prevent water infiltration.

17.7   Design and construction criteria for earth masonry structures

Arches, vaults and domes are capable of creating spaces that are unique in contemporary architecture. These techniques offer efficient use of resources, requiring few or no additional structural elements. Moreover, the earth masonry can be the final finish, reflecting elements of process and the site in a way that few other materials and techniques can match. Design of earth masonry arches, vaults and domes should abide by the design criteria outlined below to ensure longevity, and safety.

17.7.1 Design criteria

Earthen arches, vaults, and domes should be designed such that tensile forces in the masonry do not occur. Compressive force resultants should remain in the middle third of the masonry at the base of the structure. Masonry joints should be approximately perpendicular to the calculated compressive forces to minimize the risk of sliding wherever possible. Point loads on shells should be avoided to prevent local failures. Safety factors must account for variable soil composition, which can cause considerable variation in individual brick strength. Steel reinforcing should be minimized or eliminated entirely with earth masonry as it is difficult or impossible to attain an effective bond between earth and steel and it can cause significant problems in earthen structures due to corrosion. Generally, metallic reinforcement, such as steel reinforcing bars, is prone to corrosion and will shorten the life of a masonry structure if not protected from oxidation. Moreover, this expensive and energy-intensive material is unnecessary when proper design ensures that forces act only in compression.

Structural elements should be designed to resist loads during the construction process as well as in the finished structure. Because arches, vaults and domes derive their strength from their geometry, individual elements often experience greater loads during construction, when geometry is incomplete, than in the finished structure. Foundations should be adequately sized to resist the often considerable horizontal thrust of earth masonry arches and vaults, in addition to vertical loads. Finally, the finished project must be protected from precipitation and other forms of moisture, which can severely compromise the strength of earthen masonry. Durable waterproof finishes should be specified and proper detailing provided to ensure precipitation is channeled away from structure and provision for site drainage provided. Design should provide isolation from foundations to address rising damp in climates where that is an issue.

Undertaking the design and construction of an earthen vault is a serious matter, and is not an amateur activity. Even for experienced vault builders, it is often a good idea to build a smaller-scale version first, to ensure that the construction principles are sound. Not only should the structure be safe under all expected loading conditions, but it must also be safe during each phase of construction. Building an earthen vault or dome can be dangerous, as a collapsing vault can easily kill a person. Finally, unreinforced masonry vaults are not suitable for seismic regions.

17.7.2 Construction criteria

At the outset of any earthen masonry project, a testing program should be established for materials and assemblies prior to the start of construction. Such a program is best established during the design stage to give a sense of the performance characteristics of bricks made of local soils. Such a program should include testing structural components to failure in the event that new designs or unfamiliar soil types are utilized. In addition, random samples of bricks should be tested to failure throughout the project to ensure variability is within acceptable limits of required design strengths. Where final construction will be exposed to natural elements, weathering blocks should be constructed to gauge the effect of precipitation on actual materials. Geometry of arches, vaults and domes should be checked frequently during construction to ensure that they conform to design specifications as even small variations in geometry can cause drastic variation in the structural behavior. This is especially important for unreinforced masonry, which is incapable of withstanding even modest tensile forces. Arches, vaults and domes should be braced as necessary during construction to account for local loading that can be considerably greater under construction than in the finished structure. Finally, projects should be carefully protected from precipitation and other forms of moisture during construction, unless design calls for the project to be exposed to the weather and material selection has been made accordingly.

17.8   Future trends

Earth masonry has been in use constantly for nearly 10,000 years and is one of the oldest building materials on the planet. Arches, vaults and domes have been employed to enclose space for several millennia. Current interest in these traditional structural shapes and materials is not merely historical, however. They offer the contemporary designer rich possibilities for achieving radical new designs. Future improvements to design tools coupled with an increase in environmentally sustainable building techniques will contribute to the continued relevance, innovation and growth of earthen masonry arches and vaults. Sophisticated software will render masonry design in accordance with flow of forces more intuitive. Computational design tools will extend equilibrium methods to three dimensions, increasing our understanding of forces and allowing for complex forms in the future. As the sophistication of three-dimensional modeling software increases, architects are challenged to find materials capable of realizing non-standard shapes. Double curvature that can be difficult or impossible with many planar materials is possible with traditional earth masonry. Finally, as environmental concerns grow more pressing, using locally sourced materials low in embodied energy to build structurally efficient forms will only increase in popularity. Thousands of years after these materials and methods first appeared, earthen vaults still have untapped potential to create exceptional buildings.

17.9   Acknowledgments

The authors would like to thank Michelle Morales, who assisted with the background research for this chapter as part of the Undergraduate Research Opportunities Program (UROP) at MIT in 2010. Michael Ramage kindly offered details and images of the Start Pavilion. Philippe Block, Lara Davis and Matthias Rippmann provided information and photographs of the vaulted structure by the BLOCK Research Group at the ETH. Simone Swann provided the images for the Adobe Alliance.

17.10   Sources of further information

Books

Heyman, J., *The Stone Skeleton: Structural Engineering of Masonry Architecture*, Cambridge: Cambridge University Press, 1995.

Heyman J., *The Masonry Arch*, Chichester: Ellis Horwood, 1982.

Huerta, S., *Arcos, bóvedas y cúpulas: geometría y equilibrio en el cálculo tradicional de estructuras de fabrica*, Madrid: Instituto Juan de Herrera, 2004.

Lancaster, L., *Concrete Vaulted Construction in Imperial Rome*, Cambridge: Cambridge University Press, 2005.

Mecca, S. and Dipasquale, L., Eds. *Villages of Northern Syria. An Architectural Tradition Shared by East and West*, Pisa: Edizioni ETS, 2009.

Ochsendorf, J., *Guastavino Vaulting: The Art of Structural Tile*, New York: Princeton Architectural Press, 2010.

Sondicker, J. *Graphic Statics*, Charleston, SC: Bibliolife, 2009.

Steele, J., *Architecture for People: the Complete Works of Hassan Fathy*, New York: Whitney Library of Design, 1997.

Internet resources

http://web.mit.edu/masonry
Research group in masonry structures at the Massachusetts Institute of
  Technology.
www.guastavino.net
Resources on Guastavino tile vault construction.
www.block.arch.ethz.ch/tools
Thrust Network Analysis developed by the BLOCK Research Group at the
  Swiss Federal Institute of Technology in Zurich (ETH Zurich).
www.block.arch.ethz.ch/equilibrium/
Web hosted program developed at the BLOCK Research Group explains the
  behavior of structures and allows users to making their own drawings for
  their structural analyses and design explorations.
http://acg.media.mit.edu/people/simong/statics/data/
Interactive demonstrations of graphic statics that involve the user in
  experimentation with the relationship between structural form and
  forces,
www.designexplorer.net/newscreens/cadenarytool/applet/index.html
3D Hanging chain modeler developed by Axel Kilian.
www.adobealliance.org/
Simone Swan's Adobe training workshop in Presidio, Texas.
www.lavoutenubienne.org/
French/African NGO devoted to training masons to build Nubian vaults.

17.11   References

Allen, E. and Zalewski, W., *Form and Forces: Designing Efficient Expressive Structures*,
  Hoboken, NJ: Wiley, 2009.
Auroville Earth Institute, 'Dome of the Dhyanalinga Meditation Shrine', last accessed
  2010, www.earth-auroville.com/maintenance/uploaded_pics/12-dhyanalinga-dome-
  en.pdf, n.d.
Chaturvedi, S., and Ochsendorf, J., 'Global environmental impacts due to cement and
  steel', *Structural Engineering International*, 14 (3), pp. 198–200, 2004.
Heywood, E., 'Ctesiphon and the Palace of Khosroes', *The Geographical Journal*, 53
  (2), pp. 105–108, 1919.
Heyman, J., *The Stone Skeleton: Structural Engineering of Masonry Architecture*,
  Cambridge, Cambridge University Press, 1995.
Lau, W., 'Equilibrium analysis of masonry domes', [Master's Thesis] Cambridge, MA:
  Massachusetts Institute of Technology, 2006.
Mecca, S. and Dipasquale, L., Eds. *Villages of Northern Syria. An Architectural Tradition
  Shared by East and West*, European Institute of Cultural Routes, 2009.
Ochsendorf, J., *Guastavino Vaulting: The Art of Structural Tile*, New York: Princeton
  Architectural Press, 2010.
Ochsendorf, J. and Block, P., 'Designing unreinforced masonry', in Allen, E. and

Zalewski, W., *Form and Forces: Designing Efficient Expressive Structures*, Hoboken, NJ: Wiley, 2009.

Pope, A. U., *A Survey of Persian Art Vol. II*, Tehran: Soroush, 1977.

Scarre, C., *The Seventy Wonders of the Ancient World: the Great Monuments and How They Were Built*, London, Thames & Hudson, 185–186, 1999.

Steele, J., *Architecture for People: the Complete Works of Hassan Fathy*, New York: Whitney Library of Design, 1997.

Wolfe, W., *Graphical Analysis: a Textbook on Graphic States*, New York: McGraw-Hill, 1921.

18

Structural steel elements within stabilised rammed earth walling

R. LINDSAY, Earth Structures Group, Australia

Abstract: This chapter deals with how structural steel elements can enable stabilised rammed earth (SRE) to be used to greater architectural and structural effect. It includes descriptions of lintel systems, portal frame cladding systems, attaching of steel members to walls and the embedment of structural steel posts within the wall panels. It also describes potential problems of using structural steel within SRE walling.

Key words: structural SRE, lintels, portal frames, elevated SRE, stability, SRE design, SRE openings.

18.1   Introduction

Modern stabilised rammed earth (SRE) walling may require the inclusion of steel members to resist specific point loads, horizontal span loads and, in some cases, general wind loads. This chapter will provide an overview of how steel is being used to extend the architectural usefulness of modern SRE. The chapter will also provide designers with some measure of warnings about how SRE walls can react to the inclusion of structural steel, and hopefully highlight the best options for various applications. The chapter deals with structural applications that have been used successfully on over 3000 SRE projects predominantly in Australia, South Korea, Thailand, the UK and, most recently, the USA.

We strongly suggest structural designers from countries with specific seismic codes use ideas within this chapter with the understanding that they are considered for seismic applications. See Section 18.7 of this chapter.

18.2   Structural steel for stabilised rammed earth (SRE) walling

In the design of a typical single-storey SRE house the only structural steel required is for the lintels over door and window openings. Vertical steel posts are only needed in the event of a massive point load-bearing on a wall face or at the very end of a wall where there is little mass to resist a given weight. As a rule 300- or 400-mm-thick SRE or 400-mm-thick insulated

461

stabilised earth (ISE) built with a compressive resistance of 5 pascals (MPa) can withstand common roof loads assuming these loads are spread across wall plates fastened down onto the tops of the walls.

The three variables commonly used by engineers to assess the capability of SRE to withstand loads are:

1   wall thickness – 300 or 400 mm
2   wall density – 2080 kg/m^3
3   unconfined compressive strength – design to 5 MPa. Increased MPa can be achieved with careful soil blending and increased stabiliser. Site testing of batch samples can further assist an engineer to use SRE to greater structural effect.

The tensile resistance of SRE is normally calculated as 10% of the compressive strength.

18.2.1 Window and door openings

Window and door openings are constructed in situ between two opposing wall panels, or 'block-outs' within the confines of larger panels. These 'over-fill' sections require lintels, which in most cases can be bolted into the panels on either side or cast over the top of 'block-out' openings. For openings 1500 mm or greater, steel tee lintels are used. For smaller openings, layers of horizontal steel bars are used to create solid masonry lintels. For a detailed description of these lintel systems see Section 18.4 of this chapter.

18.2.2 Point loads on wall faces

Where a beam needs to be attached to a wall face there are two means of providing an effective contact point:

1   For beam loads requiring moderate resistance, the end of the beam can be attached using a plate with fastening holes drilled through it. The plate can be anchored to the wall face using (preferably) chemical anchor bolts embedded two-thirds of the thickness of the wall
2   For beam loads with a serious point load, a steel column can be embedded in the centre of the wall with a cleat attached to fasten the end of the beam to at a later time. For further details on the use of columns, see Section 18.5 of this chapter.

18.2.3 High wall stability

The architectural parameters for SRE are forever pushed to greater heights (literally). The use of embedded steel sections can provide lateral support

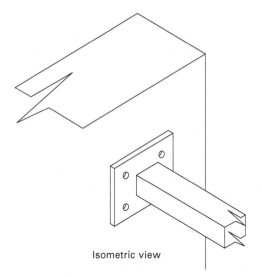
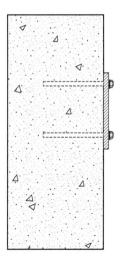

18.1 End-plate bolted to SRE wall.

for walls beyond 6-m heights. The requirement for embedded steel columns or portal frames has been especially prolific where there are few intersecting floor or roof frames to act as bracing, or few intersecting walls to act as buttressing. For more details on embedded steel columns, see Sections 18.5 and 18.5.5 of this chapter.

18.2.4 Cantilevered wall stability

Many modern buildings use SRE as a 'spine wall'. This wall will dissect the building, often as a corridor or 'gallery' with the living areas to the side, which captures winter sun, and the sleeping and utility areas to the other side.

These long walls can provide an excellent structural brace through the guts of the building. In many cases the walls extend beyond the constraints of the building and out into the garden as striking cantilevered elements.

These cantilevered walls may require steel columns within them (see Section 18.5 of this chapter). The engineer will determine wind loads and what resistance is required. As an added stiffener Earth Structures have used inverted 100-mm steel angles with the vertical leg cut into the centre of the wall top and the horizontal face bolted at 600-mm centres onto the wall top. These steel elements can be capped over using folded steel or cast masonry render capping systems.

Another precaution for preventing rocking action between SRE cantilevered panels is to cast in horizontal dowels at 600-mm increments between each panel.

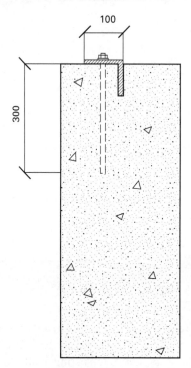

18.2 Inverted angle for stiffening cantilevered walls.

18.2.5 Precast walls

The use of precast SRE is becoming increasingly used for two particular applications:

1 to construct sign or entry walls for remote locations
2 to clad SRE buildings where the usual in situ construction methods are not viable.

Precast SRE walls are lifted from a steel tee 'lintel' base and can be either transported to site on flat-bed trucks or elevated from a casting site adjacent to the building. For details about pre-cast steel elements see Section 18.6 of this chapter.

18.2.6 Retaining walls

Structural engineers need to be aware that SRE has the capacity to retain ground loads to a certain extent. Successful applications of SRE retaining walls have employed structural systems described in Section 18.6.1 of this chapter.

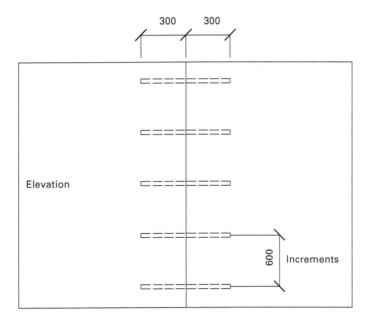

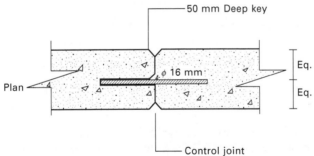

18.3 Horizontal dowels for cantilevered walls.

18.3 Design parameters for using structural steel within stabilised rammed earth (SRE) walling

The purpose of this section is for the author to impart to the reader some of the hard-earned lessons of using steel elements within SRE walling. In many ways it is the most important section of this small chapter.

18.3.1 Shrinkage cracking

The most important factor to consider when designing SRE with embedded steel sections is shrinkage. SRE walls shrink on both the horizontal and vertical axis. By contrast, steel does not shrink, therefore it can be used to restrain shrinking walls. However, these embedded steel sections have the capacity to cause shrinkage cracks.

How to avoid shrinkage cracking:

- Use a mix design that eliminates wall shrinkage. Initial soil testing will determine what the linear shrinkage rate is of the selected material (see Chapter 22, Section 22.4.3, 'linear shrinkage testing'). If the linear shrinkage is above 2% then add sharp, washed sand to the matrix until a linear shrinkage rate of 1% or less is achieved. For construction using SRE to clad portal frames the linear shrinkage rate should be *nil*. This is known as a non-plastic material
- Ensure all steel sections are placed with a minimum of 20% of the wall length between the embedded steel section and the ends of the wall panels. Thus within a 2500-mm-long panel, ensure the steel post is a minimum of 500 mm from the ends of the wall
- Ensure embedded steel sections are wrapped using closed cell foam to provide some space for the walls to shrink before they meet the resistance of the steel
- Ensure all horizontal gussets, cleats or stiffeners that are welded as attachments to the vertical steel section are covered using low-grade polystyrene with a minimum 25 mm thickness
- Ensure any anchor-bolts at the base of steel columns are wrapped using polystyrene or 'filler-foam'
- Ensure that the very top of embedded steel posts are clad with a minimum of 25 mm polystyrene. The tops of steel posts are often overlooked as shrinkage-crack points with devastating consequences.

It is important to have adequate cover over embedded steel within SRE walls. Earth Structures Group policy ensures a minimum cover of 100 mm SRE between the steel and the external face of the wall. This applies to the side surface of the wall only. Needless to say the ends of the walls require a minimum 20% of the wall length as cover.

Engineers will determine whether embedded steel will require galvanising. Rusting steel sections within the walls will in time 'explode' wall surfaces as the corrosion expands the profile of the steel. The increased embodied energy created by galvanising is normally offset by the extended lifespan of the wall.

18.3.2 Working within the parameters of the SRE formwork

The construction sequence of SRE needs to be taken into account so cast-in steel elements don't hinder the formwork erection. Much of the skill of good structural design lies in working within the confines of the SRE formwork.

Unlike concrete formwork, SRE forms cannot be cut and discarded. The forms are designed to last indefinitely which contributes to the remarkably low embodied energy of the product. An example: a steel post needs to be cast into a wall that will have an intersecting horizontal beam attached. The attached horizontal beam will be an obstruction to the SRE formwork erected on either side of the wall. The engineer allows for a cleat to be welded to the side of the post that protrudes only as far as the inside face of the SRE formwork. All sides of the cleat are clad using 100 mm of polystyrene to enable the beam to be bolted to the post later. The use of 'extender cleats'

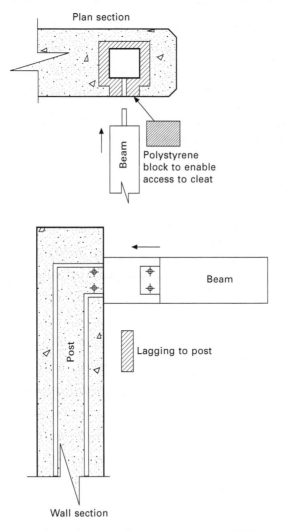

18.4 Sympathetic design to avoid cutting SRE formwork.

is a common way to overcome the installation of beams and trusses at a later time.

Modern SRE design and construction allows for some degree of movement to occur between each panel. This is called fully articulated design, and it ensures that slight sagging in the concrete footings or raft slab will not be transferred into the fabric of each panel, but rather will be taken up as movement in the control joints between each panel.

Thus, when designing lintel systems it is important not to 'embed' lintels from one panel to the next, but rather to allow for adequate shear load resistance that will reduce the likelihood of cracking.

18.4   The use of steel lintels for stabilised rammed earth (SRE) applications

There is no end to how engineers might design steel lintels for SRE applications. Here are the systems we have used for efficient and structurally sound outcomes over the past 30 years.

We avoid the use of timber lintels as even the most seasoned timber will eventually twist with dire consequences.

18.4.1 Steel tee lintels – for openings 1500 mm and greater

It is important that the attachment point of the lintels to each opposing wall allows for articulation between each panel (see Section 18.3.2 of this chapter). Thus do not 'embed' the lintels in the opposing walls. Rather, use an end plate welded to each end of the tee section with sufficient holes to allow for adequate shear resistance of the bolts embedded in each opposing wall.

18.4.2 Y bar reinforced masonry lintels – for openings 1500 mm or less

The first row of Y bars need to be one course of SRE above the horizontal soffit. The following courses are normally laid at every second course thereafter. Each course of SRE is normally 150 mm deep. To save time drilling the holes for the Y bars in the opposing wall ends, sections of plastic electrical conduits can be placed and rammed into the wall ends.

18.4.3 Massive steel lintels – for openings 5000 mm or greater

In the event of the architect requiring a very wide opening, the steel lintels can become massive, and generally require the lintel ends to be mounted

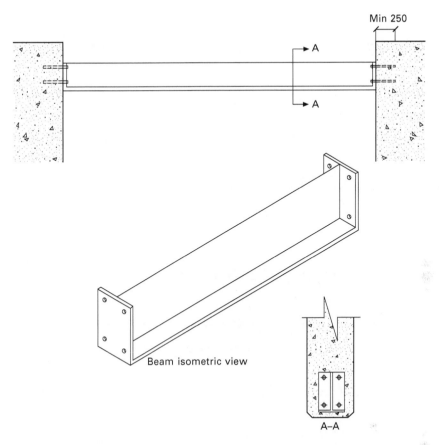

18.5 Tee lintel design.

at each end on steel posts embedded within the walls at each end. This is problematic as the builder will need to construct customised end shutters to fit around the lintels.

18.4.4 Reinforced masonry lintels for 'block-out' openings

At times a smaller opening is required within the envelope of a larger SRE panel. These openings are called 'block-outs' as they comprise a shape 'blocked out' of the wall using a collapsible timber 'box'. Reinforcing Y bars are laid in sequence within the SRE above the opening, usually with a 300 mm bearing on either side of the opening.

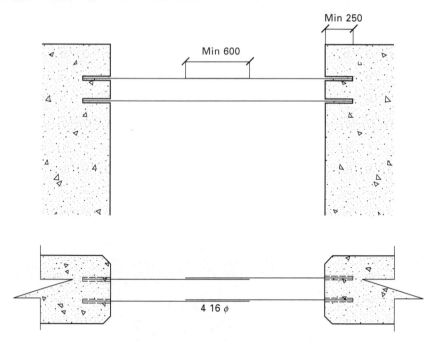

18.6 Reinforced masonry lintel design.

18.4.5 Header-beam style reinforced masonry lintels – for over-fills 1500 mm or greater

In the case of an architect not wanting to see a steel lintel expressed at the window head, a header-beam system of masonry lintel can be used. This system incorporates vertical ligatures, which are slow to construct and require the use of small-head pneumatic rammers. They should be avoided as they are a cumbersome and expensive option.

18.5   Steel columns embedded within stabilised rammed earth (SRE) walls

When a point load is too great for a bolted end-plate fixture, an engineer may need to pitch the load from a column embedded within the SRE wall. There are three conditions that will ensure the column will not cause problems with the wall.

18.5.1 Column placement

Ensure the column is placed either at the very end of the wall (so one face of the column is flush with the end of the wall) or at least 20% of the total

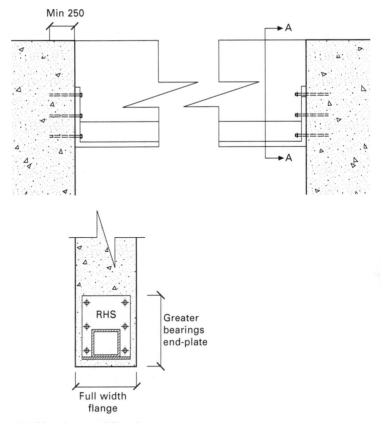

Min 250

A

RHS

Greater
bearings
end-plate

Full width
flange

18.7 Massive steel lintels.

wall length from the end of the wall. Both these precautions will minimise the risk of the wall cracking as a result of the column acting as a resistance to shrinkage.

18.5.2 Cleats and intersecting beams or trusses

Ensure any cleats for beam or truss attachments welded to the column do not extend past the face of the SRE formwork. The cleat can be clad using polystyrene, which can later be stripped out allowing access for attachment of the beam or truss. If the cleat protrudes past the face of the formwork, the SRE contractor has to cut expensive formwork or custom-make forms to fit around the cleat.

18.5.3 Lagging steel columns

Ensure steel columns are lagged using closed cell foam with a minimum thickness of 15 mm. this will allow the steel to expand or contract and will

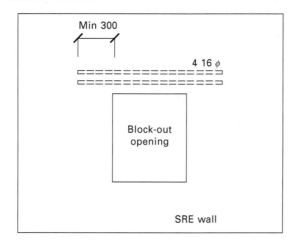

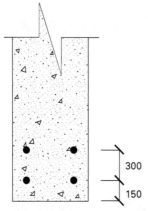

18.8 Reinforced masonry lintels above 'block-out' openings.

also further minimise the risk of wall shrinkage cracking along the face of the column.

18.5.4 Steel reinforcing bar

As a rule SRE contractors condemn the use of vertical steel reinforcing within walls. SRE shrinks on the vertical plane, while the embedded reinforcing steel acts as a hanger. The result is minor but unsightly cracking along the horizontal form lines.

Vertical steel reinforcing bar also impedes efficient production as it becomes an obvious obstacle for the rammers and for the safe feeding of the walls. Engineers need to make every effort to minimise or exclude the use of vertical reinforcing bar in SRE walls.

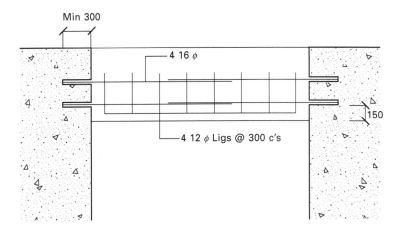

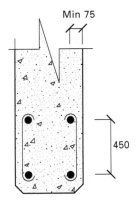

18.9 Header-beam style reinforced masonry lintels.

18.5.5 Steel portal frames

Portal frames can provide a last resort solution for stabilising SRE structures, especially for multi-storey projects. Portal frames create a highly stable structure around which SRE walls can be used as a cladding material. They are an expensive option both economically and in terms of embodied energy. One challenge with designing a portal frame-SRE system is ensuring minimal obstacles for the SRE cladding to be built up to and around the intersecting trusses and beams.

Building SRE up to the base of intersecting steel beams is quick and relatively simple. The vertical posts of the frame can act as props for the walls, which can be a huge advantage. If the beams can then be placed at heights that 'work' with SRE formwork increments, the challenge of building around them is significantly reduced. Intersecting beams and trusses need to be placed at large enough spaces to allow elevating machinery such as

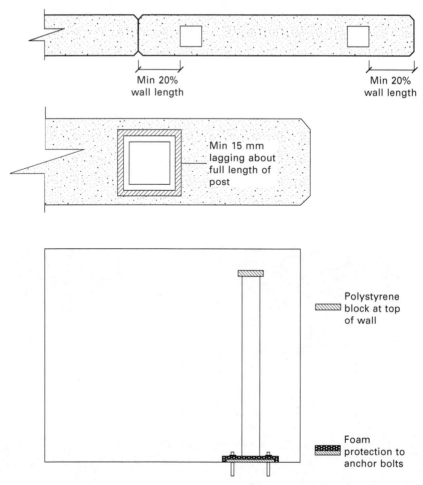

Min 20%
wall length

Min 20%
wall length

Min 15 mm
lagging about
full length of
post

Polystyrene
block at top
of wall

Foam
protection to
anchor bolts

18.10 Treatment of columns within SRE walls.

scissor-lifts to pass between them. The message here is for the structural engineer, architect and SRE contractor to sit down very early in the design process to sort out the beam heights and to imagine the process of efficient SRE production from the word go.

18.6   Structural systems for elevated or 'precast' stabilised rammed earth (SRE) panels

Elevated SRE panels, or 'precast' SRE is an increasingly popular method to clad buildings that have difficult or no access for in situ construction. SRE precast requires a different approach to structural design from precast concrete. Precast SRE requires a lifting restraint from the very base of the

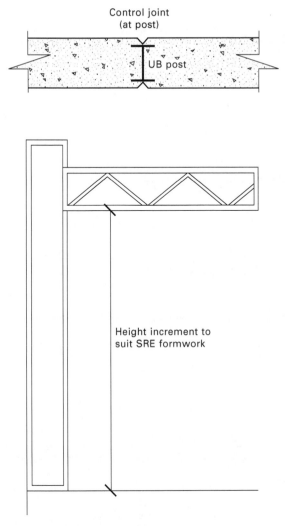

Control joint
(at post)

UB post

Height increment to
suit SRE formwork

18.11 Steel portal frame ideas.

panel. A simple solution is to build the panel on a steel tee 'lintel' with a full-panel-width flange at the base. Two lifting bars are welded to the web of the lintel set one-sixth of the panel length distance in from the end of the panel. The lifting bars need to be sleeved using a plastic conduit to minimise the risk of the SRE shrinking along the length of the steel. A spreader bar is attached to the lifting rods and the panel can be hoisted in place by crane or transported to site on a flat-bed truck prior to placement. Steel ferrules for locating the panel back to a portal frame can be threaded down the lifting rods to allow for greater resistance, thus using the actual lifting rods a load restraint.

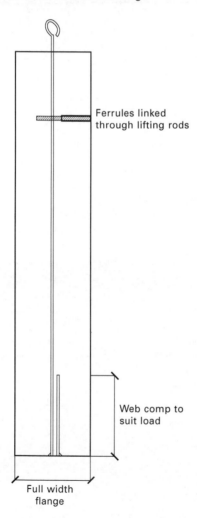

18.12 Lifting systems for precast SRE panels.

18.6.1 Retaining walls

SRE retaining walls require some measure of additional structural resistance for backloads of heavy, wet ground loads.

Wall thickness

SRE retaining walls can be designed using thicker sections to withstand loads. In some cases wall thickness can be staggered to create a tapered structure with the greatest base loads bearing on the thickest sections of the wall.

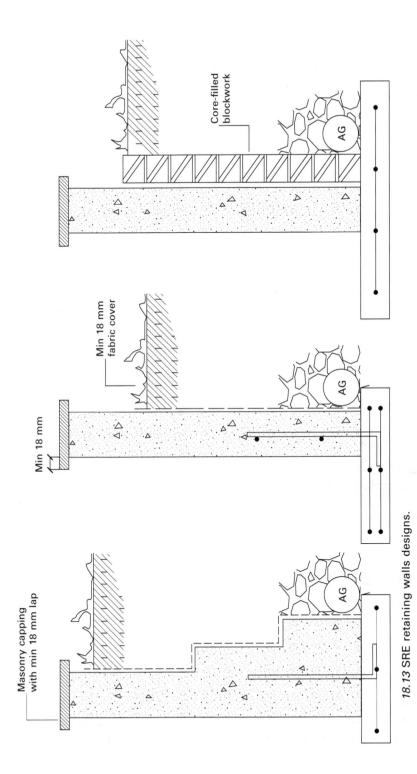

Masonry capping
with min 18 mm lap

Min 18 mm

Min 18 mm
fabric cover

Core-filled
blockwork

AG

AG

AG

18.13 SRE retaining walls designs.

Footing design

The concrete footings beneath the walls can have extended toe bearing to resist tipping action. These extended footings need to be used in conjunction with steel reinforcement within the SRE wall.

Vertical steel reinforcement

Vertical steel reinforcement bars can be used to restrain loads on SRE retaining walls. Use caution to prevent horizontal cracking of the SRE (see Section 18.5.4 of this chapter). Ensure the steel bars do not exceed one-third of the height of the wall. Also ensure the bars are lagged at the very top of the bar using a polystyrene 'cap' so the wall can shrink into the cap rather than creating a resistance point.

Waterproofing and drainage

The most important thing is to drain the ground water away from the retained area to minimise the load on the wall by using slotted drainage pipes and coarse, loose drainage aggregate.

All SRE walls are prone to rising damp. Walls must be thoroughly protected from ground water to a level 75 mm above the surrounding ground level using both painted membranes and plastic geo-fabric membranes.

SRE in conjunction with other masonry support

At times a structural or civil engineer will require further support to restrain large retaining loads against an SRE wall. Reinforced core-filled block-work or reinforced 'shot-crete' can be used to mitigate the retaining loads against SRE walls. Ensure SRE walls are waterproofed regardless if other masonry elements are employed.

18.7   North American structural steel

North American rammed earth has developed in a very different way from the Australian methodology, which enjoys an entirely non-seismic environment and uses different soil blends. North American rammed earth compressive strength requirements are typically a minimum of 6 MPa compared with the Australian 3.5–5 MPa minimum compressive strength.

Linear shrinkage in the vast majority of North American soil blends is so low as to not be a consideration. The delightful simplicity of rammed earth walls without steel in them at all is something North American earth building contractors dream of. The standard reinforcing in North American rammed earth is one or two mats of 12 mm or 16 mm rebar in a 600 mm grid. Structural steel and rebar mats seldom work well together, either in terms of buildability or with regards to thermal bridging.

Designers of North American rammed earth buildings should use structural steel very cautiously as the combination with rebar mats can create designs that are expensive and difficult to build.

A couple of final notes:

- cover over rebar is typically 75 mm
- given the typical thickness of North American footings (250–300 mm) under rammed earth walls, there is no danger of 'sagging'.

18.8   Conclusion

The use of structural steel, in particular portal frames, has enabled modern SRE to achieve some remarkable architectural outcomes. However, as a contractor it is always preferable to build walls with minimal steel sections – purely from an efficiency point of view, but also knowing the structural capacities of the SRE alone are being used by the architect and engineer to the best of their ability.

Steel contains considerable embodied energy. Given our current environmental challenges, an inevitable shift towards simpler design will require less structural steel within solid masonry walling, including SRE. The challenge for ideal architecture is to minimise the embodied energy created within the building while maximising the structural and life cycle efficiencies of materials such as earth walling.

18.9   Acknowledgements

The author would like to thank the following for their imaginative and courageous help over the years designing workable structural systems for the SRE industry in Australia: Giles Hohnen, Phil Taylor, John Bahoric, Simon Swaney, Brad Overson, Alan Brooks, Bill Smalley, Peter Chancellor, Dale Simpson, John Brock and the many other structural engineers who have had a crack at solving seemingly unsolvable challenges.

18.10   Sources of further information

Little has been written about the structural capacities of contemporary SRE. The following are useful guides to be used in conjunction with the suggestions made within this chapter.

Australian Standards

AS 4100   Steel Structures Code

AS 3700   Masonry Code
AS 3600   Concrete Code

CSIRO Bulletin No 5, edition 4 'Earth wall construction'. (Note this is a relatively outdated document. The design recommendations are based on an assumption of a maximum bearing resistance for SRE of only 2.5 MPa. Authorised site MPA testing can elevate the engineer from designing walls within this low recommended MPA parameter.)

19

Natural disasters and earth buildings: resistant design and construction

H. W. MORRIS, University of Auckland, New Zealand

Abstract: Earth buildings are particularly vulnerable to earthquakes, tropical storms, major wind storms and floods; the risks, causal mechanisms and failure modes for these natural hazards are described. Thousands die each year due to unreinforced earth building collapses in earthquakes; designers must assess the hazard, ground conditions and seismic lateral loads and use suitable structural forms and reinforcement. Unstabilised earth materials are highly susceptible to flood inundation requiring site consideration; in severe winds the low tensile strength requires special attention especially to the roof tie-down anchorage. The risks from other natural hazards are overviewed and the design principles are discussed.

Key words: natural hazards, earthquake origins, seismic resistance of earth walls, wind resistant design.

19.1   Introduction

19.1.1 Chapter overview

Natural hazards to which earth buildings are particularly vulnerable are earthquakes, tropical storms, major wind storms and floods. The risk profile of each of these natural hazards varies around the world.

Conventional earth construction has very heavy mass so buildings with earth walls attract large loads during seismic events. The low tensile strength and brittleness of earthen materials makes them susceptible to damage and life-threatening collapse. In low-seismic-risk areas it is essential that buildings are well tied together and great care is taken with connection details. In moderate and higher risk areas the earthquake hazard must be quantified, seismic lateral loads determined, and the design must include suitable reinforcement to prevent sudden collapse and take best advantage of the earth material strength. A regular structural form should be selected to respond well dynamically.

Severe winds and torrential rain are generated by tropical storms that originate over warm seas. They intensify as they move away from the equator then reduce in temperate latitudes. They attenuate as they pass over land so the winds and storm surges are at their most devastating in coastal areas and on islands. The most frequent wind damage to buildings is the peeling

481

away of roof claddings or removal of whole roofs. Special attention must be paid to the tie-down anchorage into low-tensile-strength earth materials. Tall buildings, long or high unsupported lengths of wall, and lightweight earth construction are more vulnerable to wind loads and require specific design detail. Flooding can be from rainfall, meltwater or sea level rise. Most earth materials are highly susceptible to inundation so the most significant mitigation is to identify the possible flood risks and to ensure that earth walls are well above possible rising flood waters. When there are slopes nearby, it is necessary to identify if there is any catchment or drainage channel that could flow towards the building. This may require a diversion or that the building is moved to a better position.

Preparation for earth construction requires that designers identify the natural hazards for the site, establish the level of risk and building vulnerability from each hazard and select the appropriate design response to achieve a safe earth building.

19.1.2 Types of hazards and their distribution

Natural hazards such as typhoons, tornados, storm surges, blizzards, earthquakes, tsunamis, floods, landslides, land subsidence, droughts, wildfires and volcanoes pose dangers to most types of buildings. This section briefly overviews the overall context of natural hazards with a focus on hazards to which earth buildings are most vulnerable. As illustrated in Fig. 19.1,

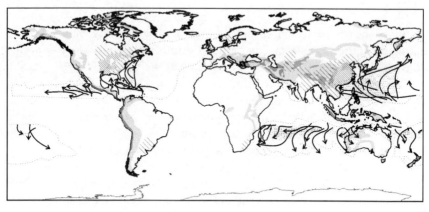

Tropical storms	Very high wind speeds	High speeds	Risk zone
	Typical storm tracks		
Tornados	Very high wind speeds	High speeds	
Earthquake	High ground accelerations	Moderate ground accelerations	

19.1 Distribution of natural hazards of significance to earth buildings (Munich Reinsurance Company, 2011; UNEP UNSDR, 2011).

zones of high hazard cover large parts of the world. Tropical storms move from the outer tropics away from the equator into vulnerable coastal areas particularly in south east Asia, the Caribbean and southern USA, and Oceania. Extra-tropical gales bring major winds across northern Europe and tornadoes occur where warm and cold air masses meet and are most severe in areas with open plains. Floods are most frequent in low-lying areas or mountainous areas, whereas earthquakes and volcanoes are most prevalent on tectonic plate boundaries.

Figure 19.2 graphically illustrates the relative numbers of significant disasters, which includes biological, extreme temperature and drought events over a 25-year period. Table 19.1 provides disaster data that are relevant to structures listed by decade from last century. Earthquakes, tsunami and volcanoes – events with geological origins – make up approximately 14% of physical events from 1950 to 1999 when compared with hydrological windstorms and floods. The table also shows that the numbers of disasters are statistically very variable. Increasing numbers of events can additionally be attributed to recording becoming more comprehensive with time.

19.1.3 Hazard, vulnerability and risk

The expected damage to life and property, or risk, can be determined from a combination of hazard, vulnerability and value. Hazard can be defined as the potential threat from any particular phenomenon that will impact on human activity. Vulnerability is the level of exposure and susceptibility as a proportion of the value of an item. This can be reduced by community preparedness and response during and after an event. Risk is the quantifiable probability that social or economic consequences will exceed a specified

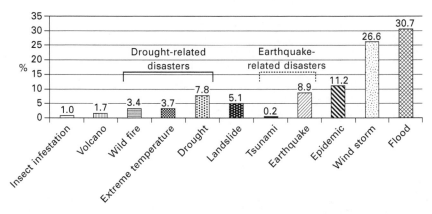

19.2 Proportions of types of natural disasters 1970–2005 (redrawn from UNSDR, 2006).

Table 19.1 Distribution of hydrological and geological disasters worldwide

Decade	1900–1909	1910–1919	1920–1929	1930–1939	1940–1949	1950–1959	1960–1969	1970–1979	1980–1989	1990–1999	2000–2005
Hydro-meteorological	28	72	56	72	120	232	463	776	1498	2034	2135
Geological	40	28	33	37	52	60	88	124	232	325	233

value at a particular site. In terms of property, equation 19.1 evaluates risk for a given hazard.

$$\text{Risk} = \text{Hazard} \times \text{Vulnerability} \times \text{Value} \qquad [19.1]$$

Engineers, architects, designers and builders minimise vulnerability by designing and constructing structures prior to an event to resist that hazard, and incorporating design features to mitigate the damage and social impact during and after the event.

19.2   Earthquakes and earth buildings

19.2.1 Vulnerability of earth buildings to earthquakes

Low-quality traditional earth buildings are very vulnerable to earthquakes and have been responsible for thousands of fatalities per year. This is due to a combination of ignorance of good structural practice, as these houses are frequently built in low socio-economic societies where people lack education and resources; and the inherent properties of earth construction. Conventional earth building materials are heavy, brittle and have low tensile strength while the best materials for good seismic performance are lightweight and ductile (not brittle). The sobering fatality statistics in Table 19.2 are dominated by poor countries and highlight the need for good engineering and detailing.

In February 2010 a great earthquake of moment magnitude (M_w) 8.8 offshore from Maule in Chile combined with a tsunami for which there was minimal warning and resulted in 577 deaths. There were a large number of older adobe and unreinforced masonry structures in the Talca area with perimeter walls as the primary load resisting system that were damaged (EERI, 2010a). The absence of reinforcement or bond beams, and weak connections between adjoining walls led to the collapse of walls and roofs in many buildings, and resulted in some fatalities. Many unreinforced buildings collapsed and an out-of-plane failure example is shown in Fig. 19.3 about 80 km from the epicentre. Conventional reinforced masonry suffered less, with damage reported as predominantly diagonal cracks and cracks around openings (EERI, 2010a).

In September 2010, a M_w 7.1 earthquake occurred near Darfield, 40 km from Christchurch, New Zealand (population 362,000). A combination of fortunate timing at 4:38AM, and resilient house construction meant there were no casualties. A replica historic building of unreinforced earth in the area of maximum earthquake shaking suffered typical wall damage (Fig. 19.4). The end wall shows a diagonal in-plane tension crack, sliding near the base, and out-of-plane flexure cracks near the edges. A longer panel in the same wall line in the adjacent room collapsed during the aftershocks. A

Table 19.2 Twentieth-century earthquake fatalities

Decade	1900–1909	1910–1919	1920–1929	1930–1939	1940–1949	1950–1959	1960–1969	1970–1979	1980–1989	1990–1999	2000–2009
Fatalities (thousands)	150	65+	570+	175+	145+	15	50+	440+	65+	100	465+*

1900–1999 Dowrick (Dowrick, 2009) *2000–2009 from USGS (USGS, 2010a) includes 228,000 from 2009 tsunami

19.3 Out-of-plane wall failure of a house in Parral, Chile with no ceiling diaphragm, poor connection to the cross wall and lack of connection at the near corner. (photo: Exequiei Araos)

19.4 Damage in 2010 to an unreinforced replica historic building constructed in 1977 of unstabilised rammed earth near to Darfield, New Zealand.

M_w 6.3 aftershock in the middle of the day in February 2011 that had much higher ground accelerations in the populated areas resulted in some building collapses and 185 fatalities. Damage surveys are discussed in Section 19.4.5 including earth buildings that performed poorly and those that performed well. Observation of earthquake damage is very useful in identifying failure modes, and demonstrates the importance of understanding earthquakes.

19.2.2 The origin and location of seismic events

Plate tectonics

Errors are often made in interpreting earthquake damage in relation to earthquake magnitude. An introduction to the fundamentals of earthquakes assists in understanding factors that influence a local site response.

Seismic activity occurs because the thin crust of the earth is made up of a number of rigid tectonic plates that move relative to each other. Infrequent and difficult-to-predict 'intraplate earthquakes' occur within the tectonic plate away from the boundaries, they account for about 10% of the total seismic energy. The vast majority of earthquakes are generated at the boundaries of tectonic plates when the build-up of strain is suddenly released. Other geological features of volcanism and mountain building occur on plate boundaries, which is well demonstrated in the Pacific 'ring of fire' shown in Fig. 19.5.

The plate movement is due to the convection-type behaviour of the semi-molten interior of the earth and the sinking of the subducting crust. The main forms of plate movement and boundary interaction are shown in Fig. 19.6. The lithosphere moves differentially on the less viscous asthenosphere, which is in the upper mantle from a depth of about 50 km.

The major types of plate boundary interactions are:

- Divergent zones, such as the mid Atlantic ocean ridge, where new plate material is added from the interior of the earth
- Subduction zones are where plates converge and the dense underthrusting plate pushes beneath the continental plate and is reabsorbed as it sinks into the mantle. The vast majority and the strongest of all earthquakes occur in regions where tectonic plates meet and occur after a period of strain build-up. Subduction causes the largest earthquakes, often near populated coasts such as on the Pacific rim (Fig. 19.5). This was the mechanism of the earthquakes, M_w 9.2, in Alaska 1964, and the great earthquake (M_w 9.1) that caused tsunamis in Indonesia in December 2004 (off the coast of Sumatra), M_w 9.0 of the earthquake in Tohoku, Japan in March 2011, the largest earthquake ever recorded in Chile in 1960 (M_w 9.5) and the M_w 8.8 earthquake in February 2010 near Maule, Chile mentioned earlier

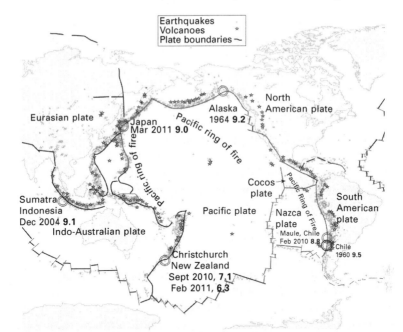

19.5 Tectonic plate boundaries showing major seismic events and volcanoes with the Pacific ring of fire clearly evident (based on USGS, 2008a). Some major or relevant earthquakes are noted with their magnitude.

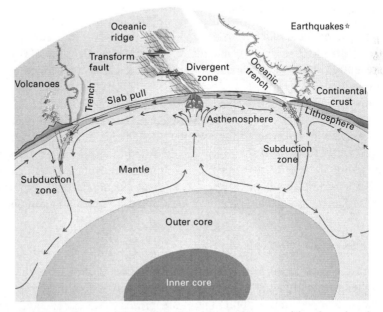

19.6 Tectonic movement and plate interactions resulting in volcanic and seismic effects.

- Transform plate boundaries are where plates move past each other horizontally; for example, the 1906 San Francisco earthquake on the San Andreas fault in California. Transform boundaries have the next most severe earthquakes when compared to subduction boundaries.

Earthquake faults

The focus of an earthquake is deep beneath the earth where the rocks under stress rupture at a specific point and the rupture rapidly propagates over a whole surface. This rupture surface forms a fault that is at a particular orientation to the surface of the earth. The types of faults are classified as strike slip faults, normal faults and reverse faults as shown in Fig. 19.7.

Depending on the depth of the rupture, its size and the overlying soils, faults may reach the surface and be expressed as a surface trace as illustrated in Figs 19.8 and 19.9 in Sept 2010 in Darfield, New Zealand.

Known active faultlines are localised but are seldom a concern for houses. For large structures, design consideration is needed to avoid or span these significant movement zones.

19.2.3 Energy release, wave propagation and measurement

Deep and shallow earthquakes

Earthquakes occur when rock ruptures at the 'focus' or 'hypocentre', usually at a depth of 5 to 50 km depth but sometimes as deep as 600 km. The 'epicentre' is the surface location above the focus or hypocentre as shown in Fig. 19.10.

In Fig. 19.6 earthquakes are shown occurring near the interface of the subducting plate at varying depths. Most damaging earthquakes occur at depths of less than 70 km where the travel path distance is moderate. Intermediate depth earthquakes at 70–300 km are less common but can be moderately damaging and are felt over a wide area. One such event was the earthquake in Bucharest in March 1997 (M_w 7.2) with a focal depth of 90 km (Dowrick, 2009). Deep earthquakes greater than 300 km are more uncommon.

Seismic waves

When the rock ruptures at depth the displacements propagate through the mantle and rock in all directions as a series of waves, similar to ripples from a stone dropped into a pond. These are primary waves (P-waves), which propagate the fastest in the same direction as the compression and dilation as

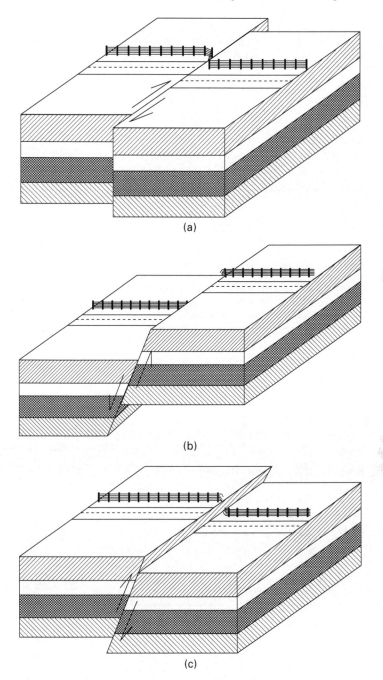

19.7 (a) Strike slip fault – horizontal movement along the fault direction; (b) normal fault – due to extension; (c) reverse fault, or thrust fault due to compression.

19.8 Earthquake fault trace displacements of about 1.5 m vertically and 4 m horizontally. Road previously straight and level, New Zealand 2010 (photo: M. Pender).

19.9 Horizontal strike slip trace with 4 m offset evident in fenceline in New Zealand 2010 (photo: M. Pender).

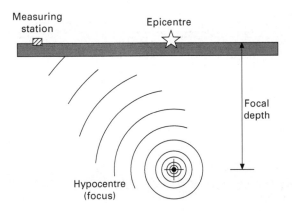

19.10 The epicentre is a point on the ground surface directly above the hypocentre. Seismic waves radiate in all directions, at a measuring station the arrival times for different seismic waves are used to establish the epicentre and focal depth.

shown in Fig. 19.11a, and secondary (shear) waves (S-waves), which move particles perpendicular to the direction of propagation (Fig. 19.11b).

S-waves are shear waves that cannot propagate through fluids and travel at a slower speed (around 60% of P-waves). It is the difference in arrival time of these two types of body waves that is used to determine the distance from a measuring station to the epicentre. S-waves are usually the most damaging to structures.

Love waves, Fig. 19.11c, are surface waves that are similar in surface motion to shear waves but reduce in amplitude with depth. Surface waves follow the surface of the earth and therefore travel a larger distance and usually travel more slowly than shear waves and arrive after S-waves. Rayleigh waves as illustrated in Fig. 19.11d are also surface waves that are slightly slower than Love waves with a smaller amplitude motion that reduces with depth. Like ocean water waves, a particle below the surface will move in an elliptical motion, these waves can sometimes be seen on flat open ground or felt as a rolling motion. Surface waves attenuate more rapidly than body waves and are more significant close to the epicentre.

Earthquake magnitude

Magnitude is best known from the historic Richter Scale which gives local magnitude, M or M_L, and is widely reported after earthquakes. It correlates with the amount of energy released at the hypocentre for medium-sized local events. (It is important to note that this scientific measure of magnitude may not relate to damage or felt intensity due to distance, depth and other

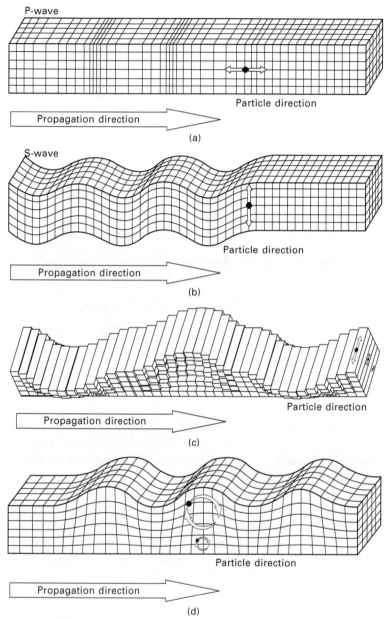

19.11 (a) P-waves – primary body waves that compress and dilate in the direction they propagate. A soil particle moves back and forth parallel to the direction of propagation. (b) S-waves – secondary shear body waves that move perpendicular to the direction of propagation. (c) Love waves – surface waves that have a similar motion to shear waves but travel along the surface and reduce with depth. (d) Rayleigh waves – surface waves that have a rolling motion that reduces with depth. A soil particle has an elliptical motion.

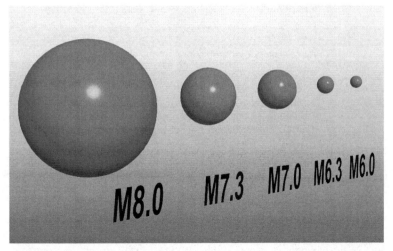

19.12 The relative size of earthquakes at the focus, a magnitude 7 event releases 32 times the energy of a magnitude 6 on the Richter scale or the moment magnitude scale.

factors.) The moment magnitude scale (M_w) is more recent than the Richter Scale and is directly related to the size of the rupture surface and length, and is a much more accurate measure of energy release for events greater than M_w 6.5. These scales of magnitude are logarithmic; the energy content increases by a factor of 32 with each complete integer on the scale as shown in Fig. 19.12.

One magnitude 8 earthquake releases around one million times the energy of a frequently felt magnitude 4 earthquake. In the last 20 years there has been an annual average worldwide of about 140 earthquakes of magnitude 6–6.9, 14 of magnitude 7–7.9 and one of magnitude 8 or greater.

Macroseismic intensity

Intensity is an expression of the extent and distribution of felt shaking intensity and damage reported as caused by an earthquake. It is related to ground velocities and accelerations and is of significant social and economic interest. The surface effects of an earthquake vary depending on the type of rupture, the distance to the epicentre, the focal depth and the geology. The scale most commonly used worldwide is the Modified Mercalli Intensity Scale (MM), which has a comprehensive list of damage states but is given in a commonly abbreviated form in Table 19.3. In Europe the very similar European Macroseismic Scale (EMS-98) is used.

Both the Modified Mercalli and European Macroseismic Scales use building damage as a measure of intensity. The EMS 98 Damage States are well

Table 19.3 The Modified Mercalli scale – short version (USGS, 2009)

Intensity	Description
I	Not felt except by a very few under especially favourable conditions.
II	Felt only by a few persons at rest, especially on upper floors of buildings.
III	Felt quite noticeably by persons indoors, especially on upper floors of buildings. Many people do not recognise it as an earthquake. Standing motor cars may rock slightly. Vibrations similar to the passing of a truck. Duration estimated.
IV	Felt indoors by many, outdoors by few during the day. At night, some awakened. Dishes, windows, doors disturbed; walls make cracking sound. Sensation like heavy truck striking building. Standing motor cars rocked noticeably.
V	Felt by nearly everyone; many awakened. Some dishes, windows broken. Unstable objects overturned. Pendulum clocks may stop.
VI	Felt by all, many frightened. Some heavy furniture moved; a few instances of fallen plaster. Damage slight.
VII	Damage negligible in buildings of good design and construction; slight to moderate in well-built ordinary structures; considerable damage in poorly built or badly designed structures; some chimneys broken.
VIII	Damage slight in specially designed structures; considerable damage in ordinary substantial buildings with partial collapse. Damage great in poorly built structures. Fall of chimneys, factory stacks, columns, monuments, walls. Heavy furniture overturned.
IX	Damage considerable in specially designed structures; well-designed frame structures thrown out of plumb. Damage great in substantial buildings, with partial collapse. Buildings shifted off foundations.
X	Some well-built wooden structures destroyed; most masonry and frame structures destroyed with foundations. Rails bent.
XI	Few, if any (masonry) structures remain standing. Bridges destroyed. Rails bent greatly.
XII	Damage total. Lines of sight and level are distorted. Objects thrown into the air.

illustrated for masonry (Fig. 19.13) (European Seismological Commission, 1998).

19.2.4 Ground accelerations, velocities and displacements

Ground accelerations measure earth motions at a specific site and are fundamental in determining design load levels. Any point of the ground surface will follow a complex three-dimensional path with rapidly changing accelerations and displacements. In some locations the direction of the

| Grade 1 | Grade 2 | Grade 3 | Grade 4 | Grade 5 |

Grade 1 – negligible to slight damage, 2 – moderate damage, 3 – substantial to heavy damage, 4 – very heavy damage, 5 – destruction

19.13 Damage states illustrated for two-storey masonry used in EMS-98 (European Seismological Commission, 1998).

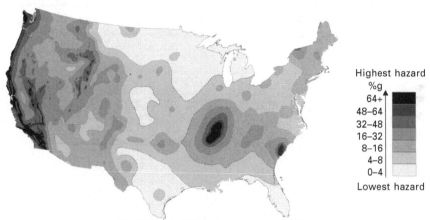

Ground accelerations with 2% probability of exceedance in 50 years

19.14 Hazard map showing probabilities of lateral ground accelerations for mainland USA (extracted from USGS, 2008a).

most severe shaking will be strongly in one direction, but no two events are the same and another earthquake is likely to be strongest in another direction. Strong motion seismographs record full acceleration records on accelerograms. From these the velocity and displacement are determined and response spectra are derived.

Seismic wave attenuation characteristics of the intervening mantle and rock are developed with attenuation equations that relate to distance from the source. In seismic regions these equations are used for a range of probable seismic events at locations along known active faults to predict maximum ground accelerations.

The overall statistical data of past events can be compared with the expected likelihood of possible events in order to generate hazard maps similar to that shown for mainland USA in Fig. 19.14. These maps are used as the basis for seismic zones in loading standards.

Local site effects

The way the rupture propagates within the fracture itself and the form of the fault will influence the earthquake characteristics including the seismic waves and their directional components. Exaggerated near-fault directional effects can occur at specific locations in relation to a large earthquake propagating for some distance along a fault. When a fault rupture progresses towards the site of interest there can be a cumulative effect as the seismic wavefronts superimpose or arrive in a compressed timeframe. Specific geological features may focus or disperse earthquake energy, and for very large or high-risk structures a site-specific hazard analysis should be undertaken by seismologists. A specific response prediction would take account of the range of possible earthquake sources and include a detailed site soil profile.

Near fault effects

The way the rupture propagates can result in significantly increased accelerations and deformations due to earthquake 'fling', and standards will have additional factors for sites very close to known faults.

The influence of soils

Ground accelerations are strongly modified by the soil conditions. Rock sites will have high frequency shaking, while on soft soil sites high frequencies (short period) will be reduced or filtered out, but low frequencies will be amplified. In the Mexico City earthquake of 1985, the deep soft soils of an old lake bed meant that tall buildings with long period motions were more severely damaged than many of the less engineered low-rise masonry structures.

Figure 19.15 shows the different site responses depending on soil type and depth.

The New Zealand standards classify soils into:

(A) strong rock
(B) rock
(C) shallow soil (e.g. < 20 m of 12.5–25 MPa or < 40 m of stiff cohesive soil or 40 m of loose dry sand)
(D) deep or soft soils
(E) very soft soil.

19.3   Earthquake engineering

'Earthquakes don't kill people – buildings do' – this decades-old adage is important to remember. Some people are killed by rockfalls and earthquake-

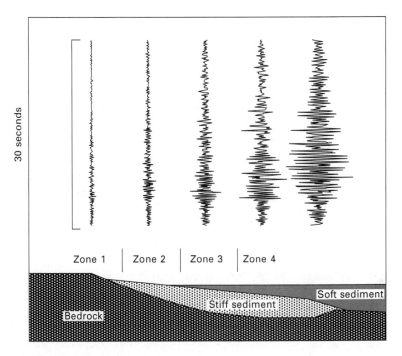

30 seconds

Zone 1 | Zone 2 | Zone 3 | Zone 4

Soft sediment

Stiff sediment

Bedrock

19.15 Earthquake response on sites with different soil depths (John Taber, Ansell and Taber, 1996).

initiated landslides, many are killed by tsunamis, but the bulk of casualties are due to failures of poorly designed and/or poorly constructed buildings. Over the last 50 years an average of over 15,000 people per year have died in earthquakes and the majority are in unreinforced stone, rubble, burnt brick or earth masonry dwellings.

Engineers, designers and builders of modern earth buildings in seismic zones need to design and construct high-quality structures that are well-engineered and set the benchmark for what needs to achieved worldwide.

19.3.1 Seismic response of buildings

Engineers are interested in the dynamic response of a structure to the strong ground motions at a particular site. This is usually dominated by lateral motion but vertical accelerations can occur and may significantly exceed 1 g. Vertical accelerations should be considered for structures that are vulnerable to increased vertical load and for unreinforced masonry that relies on gravity load to maintain lateral shear resistance. As a minimum a lateral strength reduction of 30% should be incorporated to account for the vertical accelerations in masonry walls that rely on gravity for shear and do not have full height vertical reinforcement.

Building response and resonant frequency

Structures have the largest dynamic response at their natural resonant frequency, or fundamental period. Like a pendulum gently moved backwards and forwards, buildings will respond with varying amplitude to changing input frequencies. The maximum displacement will be at its first mode fundamental period. For very tall and flexible buildings the fundamental period may be several seconds, for short stiff earth masonry buildings the period is likely to be 0.05 to 0.2 s.

Design earthquakes and design loading

Buildings standards use hazard maps to determine a zone factor (Z or similar) based on probabilistic hazard as illustrated earlier in Fig. 19.14. Earthquake characteristics are compiled into simplified design spectra that define equivalent accelerations related to building period as shown in Fig. 19.16. Soil characteristics are grouped into categories and define the amplification factors for rock through to deep soft soils and, in combination with the design spectra, determine a C factor. Acceptable levels of community

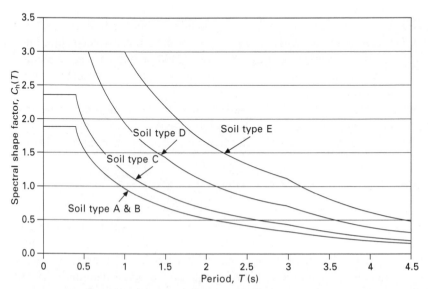

19.16 Typical design spectrum based on fundamental period for the building. (Standards New Zealand, 2004).

risk for different types of building use are given a factor, R for example, based on the return period. The acceptable risk is much higher for a single family dwelling than for a hospital, which must be fully functional after a severe earthquake. The equivalent static design approach for low-rise and regular single-storey building uses simple lateral load coefficients, which will represent the inertial earthquake loads as a proportion of the equivalent gravity force acting sideways. This can be represented by a simple equation often of the form:

$$\text{This equivalent elastic horizontal load is} = C\,Z\,R \qquad [19.2]$$

The lateral load that a building is designed to resist is strongly modified by the ductility of the structure. Buildings that can reach their maximum strength and then deform well beyond that level of deformation possess ductility, and that reduces the lateral demand. If a structure is able to move four times the equivalent yield deformation without losing strength, it has ductility of four. Earth buildings that are unreinforced, or have nominal reinforcement, will lose strength after first cracking, so should be designed for elastic response, a ductility of one. Walls reinforced within the fabric of the wall should maintain strength for modest deformations and have a limited ductility of around 1.25.

19.3.2 Load resisting systems for earth buildings

Inertial loading

Earthquakes rapidly move the structure by moving the ground under the foundations in all directions. These accelerations produce inertial loads within the structure that are directly related to the mass of each component of the structure.

Assuming structures are developed from a plan with straight walls, the loads may be conceptualised based on their orientation to the walls in two principal ways: 'in-plane' and 'out-of-plane' as shown in Fig. 19.17. If the accelerations are in-plane (parallel to the wall) then the structure tends to be stiff and diagonal tension cracks occur in wall panels, which then lose stiffness and strength. If the accelerations are out-of-plane (perpendicular to the wall) then the wall is more flexible and prone to collapse but is supported at the edges. The wall is supported at the sides (by the stiffer in-plane walls), at the base (by the foundations) and at the top (by a stiff bond-beam or heavy top plate with a ceiling or roof diaphragm).

After one direction of load has been evaluated the load direction perpendicular to that needs to be considered. The walls that were initially analysed as loaded out-of-plane are then loaded in-plane and vice versa.

19.17 Earthquake loading nominally along one axis of a building showing in-plane loads and out-of-plane wall orientations.

Out-of-plane wall actions and an analytical approach

Adobe and other discrete block masonry walls are prone to fail when seismic loads act out-of-plane (perpendicular to the wall). This is the more flexible direction as illustrated in Fig. 19.17. Adobe bricks or blocks of bricks can be shaken out of the wall, and due to their size and weight can be life threatening. For these walls to perform well it is very important that single-storey walls are:

- restrained at the top by a diaphragm or bond beam
- well connected to the in-plane walls
- well tied together as a wall matrix using some form of reinforcement.

Adobe walls and cob walls benefit from being very thick and from the inherent dynamic stability of a good height to width ratio.

Out-of-plane performance of unreinforced earth and conventional masonry under seismic loads can be calculated based on empirical approaches and simplified models, although research is still ongoing into understanding and analytically modelling the failure mechanisms. This type of wall behaviour at high seismic loads is sensitive to the dynamic response of the system. Given the limits of analytical understanding, laboratory shake table tests are very useful in verifying conceptual understanding and observing failure modes.

A simple approach is to assume that the top support is inadequate and the face-loaded walls act as an elastic bending cantilever from the foundation as shown in Fig. 19.18a; or as a vertical span from foundation to a fully effective bond beam (or top plate restrained by a diaphragm with minor flexibility) as shown in Fig. 19.18b. This behaviour assumes only a vertical span, which applies to long walls and is a reasonable approximation for typical building configurations with large widths relative to height. The stress must also be in the elastic range with bond between layers fully effective. The elastic analysis is therefore only applicable at moderate lateral loads and the flexural tensile strength would indicate the initial crack strength and likely crack location.

Once cracks extend into the wall matrix the wall behaves as discrete elements and the next simple check is to assume cracks are full depth and perform a simple force stability check for the load at which these elements collapse as shown in Fig. 19.18c. If the force was constant such as for strong wind then the wall collapse load can be determined.

Seismic loading is dynamic with many load reversals so the load will not continue in the same direction as equivalent, static loading would suggest. Elastic analysis is useful to avoid cracking of the wall at serviceability levels of earthquake, but for collapse of the wall panel elastic stress analysis no longer applies.

The collapse of a wall, considering only the vertical span direction, can be considered only to occur when the cracked sections of a wall move far enough that they become unstable. This mechanism was proposed for conventional unreinforced masonry by Priestley (1985), using an energy

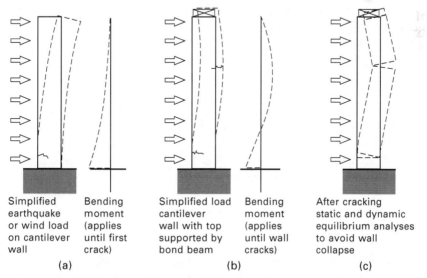

Simplified earthquake or wind load on cantilever wall	Bending moment (applies until first crack)	Simplified load cantilever wall with top supported by bond beam	Bending moment (applies until wall cracks)	After cracking static and dynamic equilibrium analyses to avoid wall collapse
(a)		(b)		(c)

19.18 Simple out-of-plane performance for walls of single-storey construction.

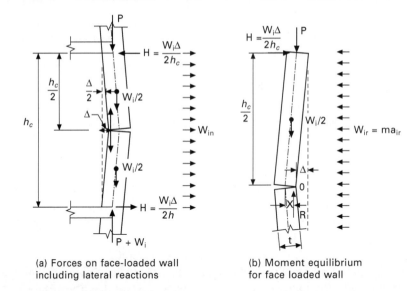

(a) Forces on face-loaded wall including lateral reactions

(b) Moment equilibrium for face loaded wall

19.19 Out-of-plane analysis lateral reactions and moment equilibrium diagrams for unreinforced walls as outlined in NZS 4297 (Standards New Zealand, 1998a). P = load per unit length, W = self weight of wall section, *h* = height between restraints, R is the vertical reaction at the crack.

method of analysis, and was included in a draft document by the New Zealand Society for Earthquake Engineering (NZSEE, 1996) before being implemented in the New Zealand Earth Building Standards, as shown in Fig. 19.19 (Standards New Zealand, 1998a).

The vertical span assumption is conservative because the actual span is in two directions, as is evident from the out-of-plane crack patterns of failed walls. This assumption is even more conservative for short walls unless the supporting return wall connections have low stiffness or have failed

In Australia the static calculation methods incorporate both horizontal and vertical span considerations in the masonry standards (Standards Australia, 2001). This is nearer to the reality of how walls perform, but would need development to account for dynamic effects.

Out-of-plane failure modes

Some typical out-of-plane failure modes for unreinforced adobe walls are shown in Fig. 19.20. In these cases cracks initiated in the horizontal span direction lead to major wall collapse, which highlights the need for top support along the wall and reinforcement at return wall connections.

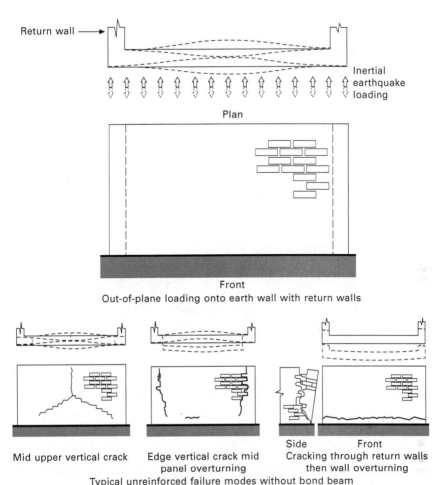

Plan

Out-of-plane loading onto earth wall with return walls

Mid upper vertical crack   Edge vertical crack mid   Cracking through return walls
                    panel overturning   then wall overturning
Typical unreinforced failure modes without bond beam

19.20 Typical failure modes for unreinforced out-of-plane walls.

In-plane loaded walls and load transfer

In-plane walls in unreinforced systems are stiff relative to out-of-plane walls so they carry the inertial load of their own mass plus the inertial load of any directly connected out-of-plane walls. It is strongly recommended that all buildings have bond beams or diaphragms to transfer roof and out-of-plane wall loads to the in-plane walls. Cantilever walls have higher vulnerability and require a low height-to-width (h/w) ratio and special reinforcement and foundation consideration.

Top horizontal support can be provided to out-of-plane walls by a floor or ceiling diaphragm (see Fig. 19.21). A diaphragm is usually made from plywood or engineered timber sheeting in a single horizontal plane, or in a

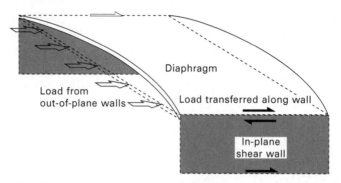

19.21 Structural diaphragm transferring load to in-plane walls.

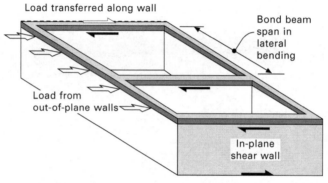

19.22 Bond beam supporting out-of-plane walls and transferring load to in-plane walls.

plane with a very low pitch slope. It must be well fastened to timber members that are substantially anchored into the fabric of each wall. The diaphragm needs to be designed with adequate strength to fully support the out-of-plane walls and with sufficient stiffness to avoid major mid-span deflections. Sheet joints in ply diaphragms need to be nailed at very close centres around the edge of each sheet (less than 150 mm) with specific design to transfer adequate load from sheet to sheet, and provide adequate stiffness.

Bond beams may be reinforced concrete or timber, and again need to be well anchored into the earth walls with plugs or dowels at close centres. Bond beams need to have adequate reinforcement and stiffness to restrain the out-of-plane walls as shown in Figure 19.22.

Where bond beams or diaphragms are present they transfer the load to the in-plane walls. As shown in Fig. 19.23, in-plane walls need to resist (i) the inertia force of the in-plane wall itself, (ii) the loads from any low strength or in-plane panels with openings, (iii) around 50% of the out-of-plane wall loads and (iv) the roof or upper floor inertia forces.

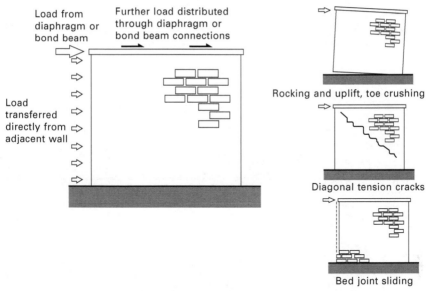

19.23 Typical in-plane failure modes.

Three common failure modes for in-plane walls are: firstly overall wall panel rocking, which causes uplift at the heel and crushing at the toe if the compressive load exceeds the wall material compressive strength; secondly, the most frequently observed seismic damage of diagonal tension cracks (where compression along the diagonal induces perpendicular tensile stress); thirdly, for adobe or pressed earth brick type masonry, slipping can occur along the bed joint or near the bottom course.

Wall configuration for lateral load resistance

Regular form and well distributed walls are important for seismic performance of earth buildings (see Section 19.3.3). In the New Zealand Earth Building Standards (Standards New Zealand, 1998c) the approach for buildings not specifically designed by an engineer is to divide the building into nominal bracing lines in the two orthogonal directions. In Fig. 19.24 the effective mass and lateral load in each direction is determined and then the available resistance of the shear walls is totalled. There are limits to the spacing between bracing lines, and the offset of walls cannot exceed more than 1 m from the nominal bracing line.

Reinforcement and detailing

In-plane walls need to be well anchored to transfer loads into the foundation and to prevent uplift. For adobe and block type construction vertical load due

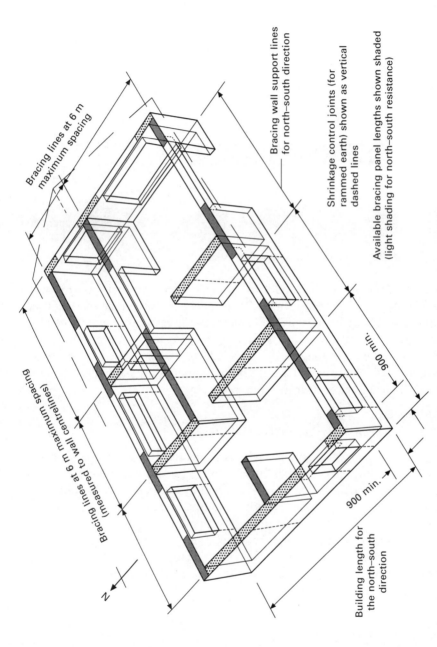

Bracing lines at 6 m maximum spacing

Bracing wall support lines for north–south direction

Shrinkage control joints (for rammed earth) shown as vertical dashed lines

Available bracing panel lengths shown shaded (light shading for north–south resistance)

Bracing lines at 6 m maximum spacing (measured to wall centrelines)

900 min.

900 min.

Building length for the north–south direction

N

19.24 Bracing lines where both directions of loading are considered. Resisting walls are well distributed (Standards New Zealand, 1998c).

to self weight will keep the lower part of the wall compressed to maintain shear performance, particularly at the mortar joint interfaces. Full height vertical reinforcement will also maintain this axial load at elastic load levels and for modest further deformations. Horizontal reinforcing is needed in discrete masonry construction to prevent diagonal tension cracks at low loads, to provide stiffness and wall continuity once cracking starts, and to provide ductility, which will prevent sudden failure (see Fig. 19.25).

In the colder areas of North America, a proprietary cantilever wall system is in use. This system has two skins of rammed earth with the major benefit of an intermediate layer of insulation. This system increases the effective h/w ratio of the wall by providing full physical shear transfer across the insulation, and the top is well integrated with a beam and cross ties. Cantilever walls require large foundations to provide lateral stability and stiffness with reinforcement to provide lateral moment rotation strength.

Because adobe has a low bond strength to steel it has been New Zealand practice to use clipped steel reinforcing mesh for horizontal reinforcing. The steel longitudinals are used with the cross mesh at 150 mm centres cut midway to the adjacent longitudinal rod, the clipped cross pieces provide direct physical embedment in addition to bond. Polysynthetic geogrid horizontal reinforcement is now regularly used in the mortar bed as an alternative to steel. It has high tensile strength and is easily cut to width and has flexibility that allows it to mould well into the mortar.

19.3.3 Essentials for seismic survival

If earth buildings are constructed in earthquake regions, even if the risk is low to moderate, then seismic design principles must be applied with special attention to structural detailing.

Summary guidelines, explained in more detail below, are:

- Build single storey only or, if two storeys high, use a lightweight timber upper level
- Use a bond beam that is well reinforced to resist moment at intersections and has 135° hooks in shear ties, or use a roof or upper floor diaphragm with adequate strength and stiffness
- Use regular forms in plan that have a good aspect ratio
- Use walls with a low h/w (height to thickness) ratio
- Provide adequate wall lengths in each direction to provide adequate in-plane strength
- Limit the lengths of window and door openings
- Provide high-quality materials and maintain high-quality work-manship
- Use reinforcing to tie all the components of the building together

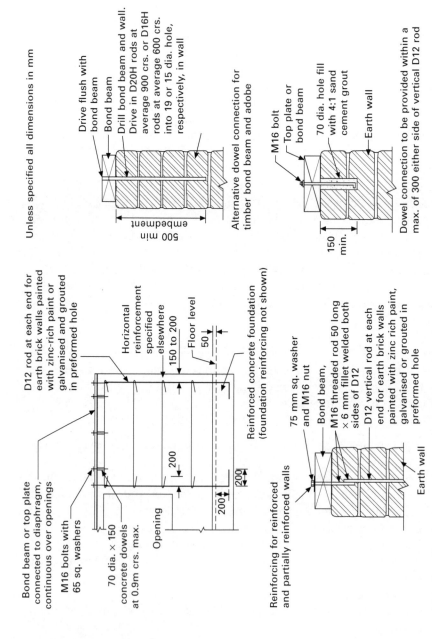

Unless specified all dimensions in mm

Drive flush with bond beam

Bond beam

Drill bond beam and wall. Drive in D20H rods at average 900 crs. or D16H rods at average 600 crs. into 19 or 15 dia. hole, respectively, in wall

500 min embedment

Alternative dowel connection for timber bond beam and adobe

M16 bolt

Top plate or bond beam

70 dia. hole fill with 4:1 sand cement grout

Earth wall

150 min.

Dowel connection to be provided within a max. of 300 either side of vertical D12 rod

D12 rod at each end for earth brick walls painted with zinc-rich paint or galvanised and grouted in preformed hole

Horizontal reinforcement specified elsewhere

150 to 200

Floor level

50

Reinforced concrete foundation (foundation reinforcing not shown)

Bond beam or top plate connected to diaphragm, continuous over openings

M16 bolts with 65 sq. washers

70 dia. × 150 concrete dowels at 0.9m crs. max.

Opening

200

200

200

75 mm sq. washer and M16 nut

Bond beam, M16 threaded rod 50 long × 6 mm fillet welded both sides of D12

D12 vertical rod at each end for earth brick walls painted with zinc rich paint, galvanised or grouted in preformed hole

Earth wall

Reinforcing for reinforced and partially reinforced walls

19.25 Details of wall reinforcing for seismic resistance from New Zealand standard NZS 4299 (Standards New Zealand, 1998c).

- Do not use thin single skin infill panels or thin cavity wall construction
- Do not use heavy masonry in the upper part of gable ends.

Earth houses in seismic regions experience high inertial loading, because the walls are very heavy they pose a major risk at an upper level. If a second storey is needed, it should be constructed of lightweight materials that impose only a minor increase in capacity on the lower level.

Bond beams and diaphragms, described earlier, are essential in tying the walls together. The connections must be adequate to transfer the loads. The bond beam in Fig. 19.26 only had a friction connection to the walls. Dowel connections along its length would have made a far better connection and prevented major damage to this building.

Buildings that are regular in shape perform most reliably during an earthquake. The simplest starting point for good seismic performance is a regular plan form without re-entrant corners (Dowrick, 2009). If the plan is eccentric, with the centre of mass offset from the centre of stiffness, then torsional vibrations will occur as shown in Fig. 19.27.

Buildings can be regular in plan but have asymmetric wall layout. This creates high demand on the more flexible walls (see Fig. 19.28). A series of good and poor plan forms are given in Fig. 19.29. Very long buildings can experience differential movement along their length and also lack cross walls. It is recommended to have seismic separation gaps that have specific allowance for differential movement between two buildings. In very low

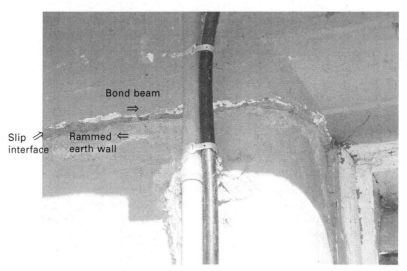

19.26 A reinforced concrete bond beam that has been poured onto the rough surface of a rammed earth wall but without positive connection. The earthquake has displaced the bond beam relative to the wall, near Darfield, New Zealand, 2010.

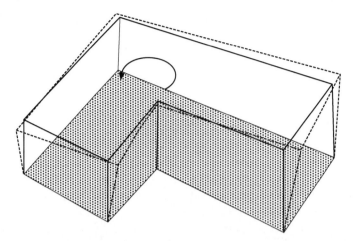

19.27 A building with an asymmetric plan, the centre of stiffness is not aligned with the centre of mass causing a torsional response. The re-entrant corner has to resist loads due to different performance in the two directions.

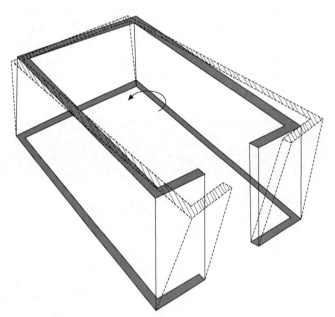

19.28 A building with a symmetric plan form, but very different wall stiffnesses at opposite ends create eccentricity and place high demand on the near-end wall panels.

seismic risk areas if two-storey earth wall construction (not recommended) is used then structure regularity in elevation must also be used. A regular profile would have walls and openings in the lower storey predominantly

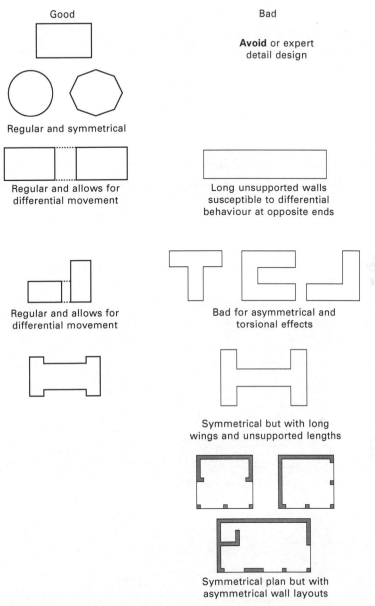

19.29 Plan forms of structures illustrating long unrestrained walls, building eccentricity and re-entrant wall effects. For low-rise buildings the eccentricity and re-entrant effects are best avoided but can be mitigated by high-quality reinforcement detailing.

aligned with openings in the upper storey. A lower storey with significantly larger and additional openings reduces the stiffness but must be avoided, even though it is likely to be architecturally desired.

Wall stability in low strength materials is significantly improved with wall thickness. Recommended maximum h/w ratios are 6 for cantilever walls and 10 for walls with top restraint. Where material tests have demonstrated a reliable tensile strength, this ratio could be reduced.

In elevation, regularity and good spacing between openings is another simple rule of thumb.

Materials and workmanship are described in other chapters. Seismic loading provides the greatest test for low strength masonry and will find the weakest link in a structural system.

Reinforcing needs to tie all the components together so the structure works as an integral whole. In New Zealand it is common to use vertical reinforcing within the wall that is continuous from the foundation to the top plate. This is very effective in giving integrity to wall panels. It prevents uplift and maintains the shear capacity of the wall. Horizontal reinforcing prevents the opening of diagonal tension cracks and provides corner continuity. Figure 19.30 illustrates vertical reinforcement that is full height in an adobe wall;

19.30 Horizontal geogrid reinforcement between courses in adobe. For effectiveness the ends of the reinforcement needs to be well anchored (photo: Richard Walker).

horizontal geogrid reinforcement is laid into the mortar and anchored to avoid slip and to supplement reinforcing bond to prevent diagonal tension cracks. It provides a high-quality corner connection at every third course.

Figure 19.31 illustrates the behaviour of a poorly constructed unreinforced house. Figure 19.32 illustrates a number of the key features in a reinforced and well-constructed seismic resistant earth building in a moderate seismic area.

19.3.4 Secondary earthquake effects

Earthquakes induce other secondary effects and one is ground deformation such as lateral spreading and liquefaction. Liquefaction occurs in course silts or sandy soils when the water table is high and the soils are not compact. As the soil is vibrated the particles move to a more compact arrangement and release the void water that builds pore pressure severely reducing bearing capacity and, as shown in Fig. 19.33, sometimes ejecting sand at the surface. Liquefaction can occur in lenses and caused major ground disruption when it occurred under inhabited slopes in Alaska in 1964 (Bolt, 2004; Booth and Key, 2006). On near-level sites, differential movement and lateral spreading can occur under a building (see Fig. 19.34).

Earthquakes can also cause landslides and rockfall as covered in Section 19.8.2, and can also cause tsunamis. The great M_w 9.3 earthquake of 2004

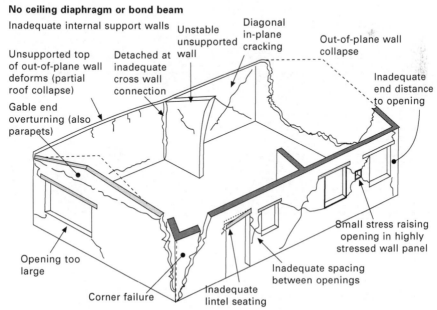

No ceiling diaphragm or bond beam

Inadequate internal support walls

Unstable unsupported wall

Diagonal in-plane cracking

Out-of-plane wall collapse

Unsupported top of out-of-plane wall deforms (partial roof collapse)

Detached at inadequate cross wall connection

Gable end overturning (also parapets)

Inadequate end distance to opening

Opening too large

Small stress raising opening in highly stressed wall panel

Inadequate spacing between openings

Corner failure

Inadequate lintel seating

19.31 Typical earthquake damage in an unreinforced or poorly detailed building.

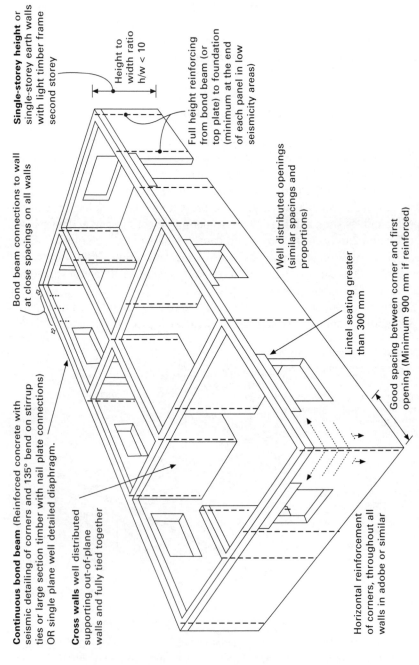

Single-storey height or single-storey earth walls with light timber frame second storey

Height to width ratio h/w < 10

Full height reinforcing from bond beam (or top plate) to foundation (minimum at the end of each panel in low seismicity areas)

Bond beam connections to wall at close spacings on all walls

Well distributed openings (similar spacings and proportions)

Lintel seating greater than 300 mm

Good spacing between corner and first opening (Minimum 900 mm if reinforced)

Continuous bond beam (Reinforced concrete with seismic detailing of corners and 135° bend on stirrup ties or large section timber with nail plate connections) OR single plane well detailed diaphragm.

Cross walls well distributed supporting out-of-plane walls and fully tied together

Horizontal reinforcement of corners, throughout all walls in adobe or similar

19.32 Key features that characterise a good reinforced building in a low to moderate seismic area.

19.33 Sandboils due to ground liquefaction in New Zealand (photo: M. Pender).

19.34 Major ground disruption around a house due to lateral spreading initiated by liquefaction. Not an earth house – timber frame with fired brick masonry veneer (photo: M. Pender).

caused a devastating tsunami in Sumatra and around the Indian ocean, and was followed by the deadly and economically devastating M_w 9.0 earthquake in Japan in 2011. Tsunami damage is caused by several mechanisms: firstly the surge impact, which could be three to six times the simple hydrostatic pressure for the depth of water. After the impact there is flow around the structure causing hydrodynamic drag and scour of the foundations, and there is the potential for major debris impact loads. Earth buildings, like most others, are unlikely to survive a direct wave impact so should be sited away from potential inundation and impact.

19.3.5   Earthquake reconnaissance surveys of building damage

Over many decades earthquake engineers have visited sites around the world immediately following major earthquakes. These studies make a major contribution to understanding of structural performance, and many are well documented (EERI, 2010b). Reconnaissance reports usually have excellent technical reviews of commercial structures, but seldom have detailed analysis of earth houses. The Earthquake Engineering Research Institute (EERI) methodology was modified by Tolles et al., when they surveyed historic adobe buildings damaged by the 1994 Northridge earthquake in California (Tolles et al., 1996), and was used for a survey in New Zealand (Morris et al., 2010) with the key categories listed in Table 19.4. When future earthquakes occur, modern earth buildings should be evaluated using a similar methodology.

An example of a modern adobe house that performed well in the M_w 7.1 2010 Darfield earthquake in New Zealand is shown in Figure 19.35a, reinforced with continuous vertical reinforcing at each end of every wall panel and with a structural ceiling diaphragm. It was built in 1997 and was within 800 m of the fault trace shown in Figs 19.8 and 19.9. The house had a 0.5-mm crack through the reinforced concrete foundation and a 3-mm crack in the concrete floor due to differential ground movement. Damage in the area was consistent with Modified Mercalli intensity (MM) VIII, and an accelerometer 1 km away registered a vertical peak ground acceleration of 1.26 g and horizontal peak ground acceleration of 0.82 g. The damage to the adobe was minimal with several cracks and opened joints similar to that shown in Fig. 19.35b, and illustrates that the reinforcing and integrity of the structure was effective.

Another modern adobe house that performed well was at a site that experienced lower accelerations but was on liquefied soils (Figs 19.36a and 19.36b). Further earth buildings are discussed in the December 2010 NZSEE Bulletin, issue 34 for the Darfield earthquake (Morris et al., 2010). Preliminary assessment of the more severe February earthquake is outlined

Table 19.4 Standardised damage states used for the earth building reconnaissance surveys (modified from Tolles *et al.*, 1996)

Damage state	EERI Description	Commentary on damage to historic and earth buildings
A None	No damage, but contents could be shifted. Only incidental hazard.	No damage or evidence of new cracking.
B Slight	Minor damage to non-structural elements. Building may be temporarily closed but could probably be reopened after minor clean up in less than 1 week. Only incidental hazard.	Pre-existing cracks have opened slightly. New hairline cracks may have begun to develop at the corners of doors and windows or at the intersection of perpendicular walls.
C Moderate	Primarily non-structural damage; there also could be minor but non-threatening structural damage. Building probably closed 2 to 12 weeks.	Cracking damage throughout the building. Cracks at the expected locations, and slippage between framing and walls. Offsets at cracks are small. None of the wall sections are unstable.
D Extensive	Extensive structural and non-structural damage. Long-term closure could be expected due either to amount of repair work or uncertainty on feasibility of repair. Localised, life threatening situations would be common.	Extensive crack damage throughout the building. Crack offsets are large in many areas. Cracked wall sections are unstable; vertical support for the floor and roof framing is hazardous.
E Complete	Complete collapse or damage that is not economically repairable. Life-threatening situations in every building of this category.	Very extensive damage. Collapse or partial collapse of much of the structure. Repair of the building requires reconstruction of many of the walls.

Webster, 2009

(a)

(b)

19.35 (a) Modern well reinforced adobe house, built 1997, very close to the fault trace of the M_w 7.1 Sept 2010 Darfield Earthquake in New Zealand. (b) Damage to a reinforced adobe house near Darfield New Zealand – opening of gap at the end of doorway lintel with minor cracks propagating along line of diagonal tension.

(a)

(b)

(c)

19.36 (a) Modern adobe house that suffered modest damage, including outward movement of the chimney, minor damaging ground movement due to ground liquefaction, during the Darfield earthquake in New Zealand. (b) Damage to reinforced adobe house – typical modest cracking observed around windows and openings. (c) Earthquake damage to both leaves of pressed stabilised earth infill walls of a post and beam house. Typical steel links between leaves were ineffective.

in another publication. (Morris *et al.*, 2011) and highlighted the fact that untied double skin walls of stabilised earth were most unsatisfactory (Fig. 19.36c).

19.4   Wind and storms

Wind is a major source of damage to buildings in all areas except within 5° of the equator. Gales and tropical storms are major widespread climatic effects, and strong localised winds occur on mountain downslopes, with thunderstorm squalls, downbursts and tornadoes. Wind is particularly damaging for poorly constructed lightweight buildings, but needs to be considered for earth buildings. Wind loads on low roof pitches will create uplift, and roof tie-downs need to resist significant forces in areas that have high winds. Wind loading is the dominant design concern for lateral loads in non-seismic areas. In low to moderate seismic zones earthquake lateral loading will dominate unless ultimate limit state winds exceed 50 m/s for single-storey heavy masonry earth construction.

19.4.1 Global circulation

All wind is generated as part of a global pattern of circulation that moves air relative to the surface of the earth driven by the energy of the sun. Solar heating differs depending on the latitude, and the thermal differences between the equator and the poles create pressure differences, which drive convective circulation as illustrated in Fig. 19.37. The air movement is then subject to 'coriolis accelerations' due to the Earth's rotation; these accelerations also vary with latitude and deviate the air movement in opposite directions in the two hemispheres. In the northern hemisphere a wind directed north along a meridian diverts to the right of the velocity vector (towards east) and becomes a westerly.

19.4.2 Pressure system winds

In the mid-latitudes between 40 and 60° the largest winds are gales generated by large deep depressions. These are generated by very large weather systems, which can extend over 1000 km or more. In Europe these can last for days, while peak wind gusts occur over a period of several hours.

19.4.3 Thunderstorms

Thunderstorms occur due to warm moist air rising into cumulus cloud towers as updraughts, moving from the storm flank into the storm core as shown

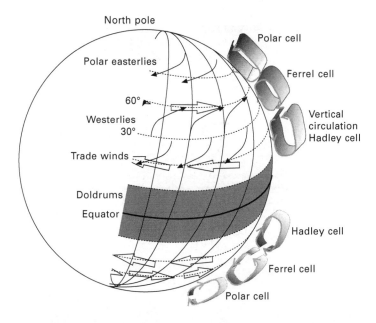

19.37 Global circulation results in bands of air flow giving predominant wind directions.

in Fig. 19.38. These storms are initiated by cold fronts, troughs or regions of low pressure further enhanced by hills. Within the core the moist air condenses into rain or hail and falls within a downdraft that spreads as it approaches the ground. The cool wind surges outward along the gust front (Australian Government Weather Bureau, 2010).

The most severe winds occur in spring and summer and are caused by the downward rush of air accelerated by rain and hail that spreads at the ground. In Australia a severe storm is defined as one that causes hailstones with a diameter of 2 cm or more, wind gusts of 90 km/h or greater, intense rainfall resulting in flash flooding or that generates tornadoes.

In temperate regions thunderstorms can be the cause of the maximum design level winds.

19.4.4 Tornadoes

Tornadoes develop out of severe thunderstorms when a rapidly rotating column of air extends from the updraft base of a cloud to the ground. Fully developed tornadoes have an interaction with the downdraft that focuses the updraft into a very tight rotating motion. Tornadoes are categorised on the Fujita scale: an F3 tornado has wind speeds in excess of 250 km/h, an

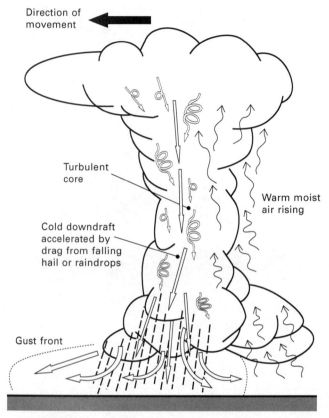

Direction of movement

Turbulent core

Warm moist air rising

Cold downdraft accelerated by drag from falling hail or raindrops

Gust front

19.38 Mature thunderstorm with warm air rising primarily at the rear flank with cold downdrafts spreading as they interact with the ground to create strong gusty wind.

F4 is in excess of 334 km/h and F5 in excess of 420 km/h. Tornadoes are localised and cause damage along a narrow path.

Major tornadoes occur in central USA where the cold Canadian air meets the warm Mexican flows. Wind loads increase proportionally to the square of the velocity so these are highly destructive extreme events. It is not cost-effective to design houses and medium importance structures to resist tornadoes, but in high-risk areas a place of refuge should be considered in the design.

19.4.5 Tropical cyclones, typhoons and hurricanes

Tropical cyclones cause very severe winds driven by latent heat of the oceans, and develop in some parts of the tropics. Water temperatures must exceed 26°C for the tropical storms to originate, and they follow tracks away

Hundreds to greater than 1000 kilometres

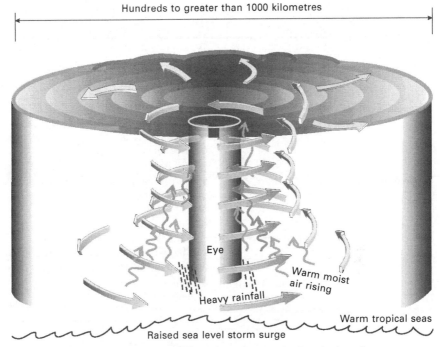

19.39 Tropical cyclone formation – warm moist air rises into a very large rotating air mass generating very high winds, heavy rain and low pressures that cause the sea level to rise.

from the equator between 10° and 30° latitude both north and south. The strongest tropical storms occur in the Caribbean and southern USA, where they are called hurricanes, in the South China Sea around south east Asia, where they are called typhoons, and off the north west of Australia, where, because of clockwise rotation in the southern hemisphere, they are called tropical cyclones.

As shown in Fig. 19.39 these are the most severe large storm systems and are frequently hundreds of kilometres. They cause very high winds, high rainfall in diameter, and the very low pressures within the storm raise the sea level, which causes storm surges. Tropical cyclones, hurricanes and typhoons cause major damage when they make landfall, but the intensity attenuates once they are over land without the warm ocean providing the energy. The storm categories, wind speeds and typical damage are listed in Table 19.5.

19.5   Earth building design for wind resistance

In each region maximum wind speeds for each particular direction are determined from meteorological data from each applicable type of wind

Table 19.5 Classification of tropical cyclones (Australian Government Bureau of Meteorology, 2010)

Category	Strongest gust (km/h)	Typical effects
1 Tropical cyclone	Less than 125 km/h Gales	Minimal house damage. Damage to some crops, trees and caravans. Boats may drag moorings.
2 Tropical cyclone	125–164 km/h Destructive winds	Minor house damage. Significant damage to signs, trees and caravans. Heavy damage to some crops. Risk of power failure. Small boats may break moorings.
3 Severe tropical cyclone	165–224 km/h Very destructive winds	Some roof and structural damage. Some caravans destroyed. Power failure likely.
4 Severe tropical cyclone	225–279 km/h Very destructive winds	Significant roofing and structural damage. Many caravans destroyed and blown away. Dangerous airborne debris. Widespread power failures.
5 Severe tropical cyclone	More than 280 km/h Extremely destructive winds	Extremely dangerous with widespread destruction

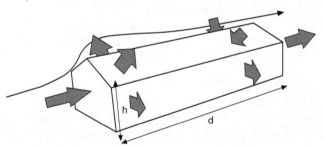

19.40 External wind pressures for wind along the ridgeline of a building. The windward end wall has positive pressure and all other exterior surfaces have suction. In long buildings the downwind roof can also have positive pressure.

event as outlined above. Design wind speeds are statistically determined and published in loadings standards with wind speed increasing as a function of height depending on the approach ground roughness. Factors are then applied to account for local topography such as ridges and valleys, shielding from other buildings, and the proportions and shape of the building.

For simple wind design the wind loading is treated as internal and external pressures, taking account of the wind direction and direction of any openings in the building. As shown in Fig. 19.40, wind along a building, parallel to the ridgeline, will cause pressure on the windward wall, negative pressure (suction) on the downwind end wall and negative pressure on the side walls.

On the upwind roof there will be suction, and in long buildings with a d/h ratio greater than three a positive pressure can apply to the downwind part of the roof.

As shown in Fig. 19.41a wind across the building, perpendicular to the ridge, will cause uplift on a low pitch building. Like an aerofoil the wind velocity increases as it crosses the building roof creating uplift. If there is a significant opening or broken window on the windward side of the building, a positive internal pressure will create upward force on the inside surface. Roof sheeting uplift or the uplift of large sections of a roof are the most common structural failures in wind storms.

On roofs steeper than 20° there can be a negative or positive pressure on the upwind surface. For roof slopes of 45° or greater there will be a positive inward roof pressure on the upwind surface as shown in Fig. 19.41b.

19.5.1 Fastening of roof cladding

Roofs for all buildings need to be well fastened to avoid uplift. In tropical storm zones the fixings must cope with many cycles of loading without

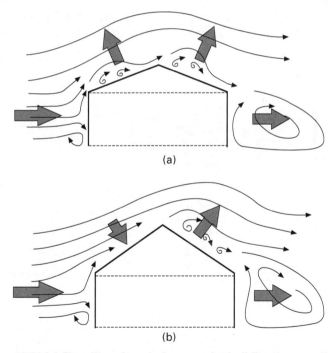

(a)

(b)

19.41 (a) Flow lines for wind across the building (perpendicular to the ridge line) and external pressure direction on a building with a low pitch roof. (b) Flow lines for wind across the building and external pressure direction on a building with a steep roof.

metal fatigue. Lightweight roofs with a timber roof structure cannot rely on direct nail pullout of standard nails. In high-risk areas, tiles are unsuitable and special load spreading connectors are nailed into the side of purlins or screwed to hold the steel sheet cladding.

19.5.2 Anchorage of roofs to walls

For earth wall buildings the tensile strength of the wall materials cannot be relied on for roof tie-down. The depth of anchorage into the wall must allow for the full wind uplift by anchoring deep enough into the wall that the self weight of the wall materials will resist the uplift. In high-strength stabilised earth with adequate tensile strength the bond anchorage strength will govern. Under the most severe New Zealand conditions this requires 1200 mm rods to be embedded at maximum 900 m centres as shown in Fig. 19.42. In tropical storm zones, anchoring rods should be full height from the roof to the foundation.

19.5.3 Out-of-plane wall strength

The important strength considerations are resisting wind for tall earth buildings, free-standing earth walls and buildings with walls made of lightweight

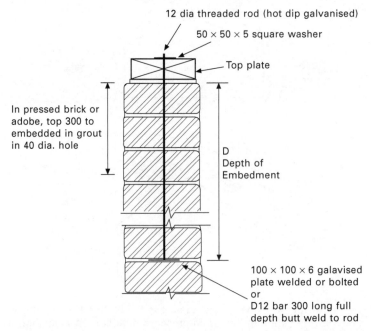

19.42 Wall anchorage to avoid roof uplift (Standards New Zealand, 1998c).

earth materials. For low-rise heavy earth walls failure of the wall itself is unlikely unless there are long unrestrained spans. A similar methodology used for out-of-plane earthquakes can be applied for wind as well, after first cracking then stability criteria will determine whether the wall collapses in an extreme event. Ytrrup proposed a set of equations for wind based on this approach (Yttrup, 1985).

19.6   Flood hazards and earth buildings

Some stabilised earth wall materials will survive inundation for days, but unstabilised walls are vulnerable to severe material degradation during flood inundation, erosion in flood waves, and extended periods of rain strike. These failures can be catastrophic and cause full wall collapses. This has serious economic and social impacts, but the structural failures have gradual onset and are not normally life threatening.

19.6.1 Causes of flooding

Temperate zone rainstorms, monsoon rains, extreme tides, tropical storm rainfalls and tropical storm surges can causes significant water level rises near rivers, streams, estuaries or along the coastal fringe. The history of these events needs to be considered in looking at potential construction sites while taking into account the changes in river catchments and any silting up or backwater effects due to downstream narrowing or obstructions. In marginal zones, consider if flood events may be worsened by climate change.

The most vulnerable areas for flooding due to rising water are flood plains, which traditionally form fertile areas and are often built on. Frequently urban development also spreads adjacent to rivers, streams and harbours. In rural areas past flood levels should be identified, in urban areas the hydrological records and flood risk assessments of local authorities should be checked to determine flood levels as illustrated in Fig. 19.43a.

Flash floods occur in steep catchments during intense short-duration rain storms and can reach a peak within minutes (see Fig. 19.43b). These floods are more localised than typical large river flooding and occur very rapidly in creeks, drains and natural watercourses. Some topographical features focus thunderstorm activity into areas that are more prone to flash flooding.

The worst cases of overland flow are due to major rainfall into catchments above the building site and can be caused by overtopped drainage systems. Minor flooding due to inadequate watercourse size and poorly maintained drainage can also have very localised severe effects: all feasible drainage paths and overland flow paths need to be considered when investigating a potential construction site.

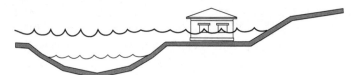

Near-coast storm surges or rising rivers across flood plain
(a)

Steep catchment flash floods and watercourse overflow
(b)

19.43 (a) Flooding due to rising rivers, and tidal surge. Mostly near streams, in the flood plains of rivers and near to the coast. (b) Flooding due to flash floods or overland flows due to high intensity rainfall.

Foundation above flood plain and/or on hardfill platform, uphill drainage or diversion for overland flow

19.44 Good site selection and preparation to be above highest water levels and well drained to divert or retain likely overland flow.

19.6.2 Design for flooding

The design for floods and earth buildings is to select and prepare a site to avoid being inundated by water. The site needs to be well above flood levels or, if it is marginal, then construct a building platform faced with resilient hardfill, rock or concrete well above the potential flood level as shown in Fig. 19.44.

Permanent drainage channels need to be well constructed for sites where there are small possible overland flows. When considering the environmental impact, a more sustainable source-control stormwater management option is to capture or slow such flows with stormwater retention ponds to reduce downstream flood peaks.

19.7   Volcanoes and landslides

19.7.1 Volcanoes

Volcanic eruptions can devastate wide areas suddenly with great intensity. The upward movement of magma prior to an eruption is often detected by seismic instrumentation, so warnings can be given prior to a volcanic event and minimise casualties. Around 500 million people live near volcanoes. Over 500 volcanoes are classed as being active, and between 50 and 65 of them erupt each year. Nagoya, Kyoto, Yokohama and Tokyo are all at risk in Japan. Vesuvius near Naples has tended to erupt every 30 years and is being constantly monitored. Eruptions also have major indirect effects on air traffic, crop failures and climate change (Munich Reinsurance, 2007).

The underlying mechanisms of volcano formation are related to the movement at the tectonic boundaries as described in Section 19.3.2. Buildings are vulnerable to a range of hazards as shown in Fig. 19.45. Lahars are liquid volcanic mud flows that concentrate in valleys and like pyroclastic flows can travel many kilometres with devastating force. Ash falls can apply significant roof loading on structures downwind of the plume tens to hundreds of kilometres away. The Etna eruption in 2002 caused over €800 million of ash damage.

For earth houses, rare volcanic events would not be a structural design consideration but, similar to flood hazard, the hazard should be considered and the site selected to minimise risk if it is near to a volcano that is defined

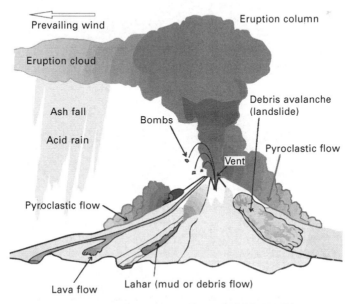

19.45 Various volcanic eruption effects (USGS 2008b).

as geologically active. In urban areas with high risk of volcanic activity there will be local authority maps of major volcanic hazards.

19.7.2 Landslides

Landslides and rockfalls mainly occur after heavy rain and also after earthquakes. Severe damage can be caused due to land sliding onto a structure. Earth houses are usually very heavy so special care is needed to avoid land subsiding or laterally spreading, especially after earthquakes. Sites at the base of slopes of soil or poor rock are most prone to landslide or pre-existing landslide sites (Fig. 19.46a). Other vulnerable sites are: in a drainage hollow; at the base or top of a fill slope, near a steep cut slope; or near a slope where there is a septic tank leach field.

Investigation should be carried out around the general area to detect warning signs: springs, seeps, or saturated ground; hummocky ground; cracks or unusual bulges in the ground, street pavements, kerbs or footpaths; sunken footpaths or street pavements; offset fence lines; or leaning telephone poles, trees, retaining walls or fences. If there are other houses nearby, then they should

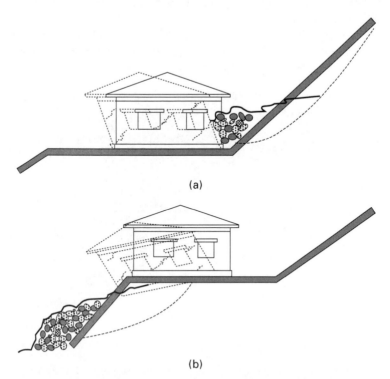

(a)

(b)

19.46 Landslide dropping or slumping earth or rock onto a house; (b) land slumping away beneath the foundations of a building.

be investigated for tilting or cracking of concrete floors and foundations, or soil that has moved away from foundations for decks and patios that have tilted or moved relative to the main house. Expert geotechnical investigation should be undertaken for any sites where possible hazards are identified.

Sites that are typically considered safe from landslides are: stable, hard, non-jointed bedrock; flat or gently sloping areas away from sudden changes in slope angle; the top of a ridge set back from the edge of the slope (USGS, 2007).

19.8   Future trends

For modern earth buildings understanding earthquake performance still requires more advanced understanding, particularly to define reliable analytical procedures. The overall principles of seismic resistant configurations are understood with continuing research around the world investigating specific details of seismic performance (Tipler *et al.*, 2010). This work is often undertaken in small research programmes with modest budgets.

Larger research programmes in burnt brick and concrete masonry will provide high quality research inputs into better understanding of seismic performance, particularly for out-of-plane walls. Some of this leading work is being undertaken at the University of Adelaide (Griffith *et al.*, 2004) and at the University of Auckland (Dizhur *et al.*, 2010; Derakhshan *et al.*, 2012). Developments of assessment procedures for unreinforced masonry are continuing as the retrofit of such buildings requires an understanding of existing structural performance. The major economic drivers for economical retrofit of large numbers of building will provide in-depth research.

19.9   Sources of further information

Books

Disaster Reduction – Living in Harmony with Nature by Kuriawa (2004) is a comprehensive volume that covers a wide range of disasters and includes social, economic and planning aspects of disaster reduction with an emphasis on South America.

Earthquakes by Bolt (2004) is the classic reference on the basics of seismology.

Earthquake Resistant Design and Risk Reduction by Dowrick (2009) provides in-depth discussion of earthquakes including economic impact, seismology and seismic design with many New Zealand examples.

Booth and Key (2006) wrote *Earthquake Design and Practice for Buildings*, which provides information aimed at designers and includes Europe.

Seismic Engineering (Betbeder-Matibet, 2008) is comprehensive with a theoretical emphasis and is focused on France.

Seismic Design for Architects – Outwitting the Quake (Charleson, 2008) gives a conceptual overview with extensive graphics that provides an easily understood introduction to seismic design for engineers and architects.

Wind sources, wind loads and design are well covered by Holmes (2007). The well-known earth building texts by Houben and Guillard (1994) and Minke (2009) have good sections that provide an overview of natural disasters and seismic resistance. Bruce King wrote a simple design approach (1996) and a chapter is also incorporated in *The Rammed Earth House* by Easton (1996).

Standards

Any serious engineering design will require familiarity with the seismic provisions of the local loadings standards such as Eurocode 8 in Europe (CEN, 2004) or NZS 1170.5 in New Zealand (Standards New Zealand, 2004).

The New Zealand Earth Building Standards (New Zealand Standards, 1998abc) are the most comprehensive available, and cover engineering design, materials and detailed requirements for non-engineered construction, and provide the detail necessary to build a house. The ASTM published a design guide in 2010 (ASTM International, 2010).

Websites

The US Geological Survey natural hazards website has an excellent overview of the various hazards and internally links to in-depth information on a full range of hazards. It is constantly updated with data on significant earthquakes worldwide. www.usgs.gov/natural_hazards/

The Earthquake Hazard Centre (2010) has a set of newsletters that provide general guidance on small-scale seismic-resistant construction and the Cyclone Testing Station (2009) in Australia has guidelines for house owners to prepare for major wind storms.

19.10 References

Ansell R and Taber J J (1996) *Caught in the Crunch: Earthquakes and Volcanoes in New Zealand*, Auckland, Harper Collins Publishers, 188pp

Australian Government Weather Bureau (2010) 'Severe thunderstorms in New South Wales and the Australian Capital Territory', available from www.bom.gov.au/nsw/sevwx/index.shtml [accessed November 2010]

ASTM International (2010) *ASTM E2392/E2392M – 10e1 Standard Guide for Design of Earthen Wall Building Systems*, West Conshohocken PA, ASTM International.

Betbeder-Matibet J (2008) *Seismic Engineering*, London, ISTE; Hoboken, NJ, John Wiley & Sons

Bolt B (2004) *Earthquakes*, New York, W H Freeman

Booth E D and Key D (2006), *Earthquake Design and Practice for Buildings* 2nd ed, London, Thomas Telford Publishing

Charleson A (2008) *Seismic Design for Architects: Outwitting the Quake*, Amsterdam, Boston, London, Elsevier/Architectural Press

CEN (2004) 'Eurocode 8: design of structures for earthquake resistance', Brussels, European Committee for Standardization

Cyclone Testing Station (2009) 'Cyclones – is your house ready? A home owner's guide', available from www.jcu.edu.au/cts/idc/groups/public/documents/other/jcuprd_053810. pdf

Derakhshan H, Griffith M C and Ingham J M (2011) 'Out-of-plane behaviour of one-way spanning URM walls', *ASCE Journal of Engineering Mechanics* (submitted May 2010, accepted Oct 2010)

Dizhur D, Derakhshan H, Lumantarna R and Ingham J (2010) 'Earthquake-damaged unreinforced masonry building tested in-situ', *Journal of the Structural Engineering Society of New Zealand*, 23, 2, 76–89

Dowrick D (2009) *Earthquake Resistant Design and Risk Reduction*, 2nd ed, Chichester, John Wiley and Sons Ltd

Earthquake Hazard Centre (2010) 'Earthquake Hazard Centre Newsletter', 10, 3, and many others available from www.victoria.ac.nz/architecture/centres/earthquake-hazard-centre.aspx [accessed November 2010]

Easton D (2007) *The Rammed Earth House*, Chealsea Green Publishing, White River Junction, Vt

EERI (2010a) 'The Mw 8.8 Chile earthquake of February 27, 2010', *Earthquake Engineering Research Institute Newsletter*, June 2010, available from www.eeri.org/site/images/ eeri_newsletter/2010_pdf/Chile10_insert.pdf [accessed 11 November 2010]

EERI (2010b) 'Learning from earthquakes', Earthquake Engineering Research Institute, available from: www.eeri.org/site/projects/learning-from-earthquakes [accessed 5 December 2010]

European Seismological Commission (1998) *EMS-98, European Macroseismic Scale*, Volume 15, European Seismological Commission, Luxemburg, 1998 available from www.gfz-potsdam.de/portal/gfz/Struktur/Departments/Department+2/sec26/ projects/04_seismic_vulnerability_scales_risk/EMS-98/EMS-98_language_versions [accessed November 2010]

Griffith M G, Lam N T K, Wilson J L and Doherty K (2004) 'Experimental investigation of unreinforced brick masonry walls in flexure', *Journal of Structural Engineering*, 130, 3, 423–432

Holmes J D (2007) *Wind Loading of Structures*, 2nd ed, Abingdon, Taylor and Francis

Houben H and Guillard H (1994) *Earth Construction: a Comprehensive Guide*, London, Intermediate Technology Publications

King B (1996) *Buildings of Earth and Straw: Structural Design for Rammed Earth and Straw-bale Architecture*, Sausalito, Cal., Ecological Design Press

Kuroiwa J (2004) *Disaster Reduction – Living in Harmony with Nature*, Self published, Lima.

Minke G (2009) *Building with Earth: Design and Technology of a Sustainable Architecture*, Basel, Birkhäuser.

Morris H (2009) 'New Zealand: aseismic performance-based standards, earth construction, research and opportunities', *Proceedings of the Getty Seismic Adobe Project 2006 Colloquium, Los Angeles*, 11–13 April 2006, Editors Mary Hardy, Claudia Cancino, Gail Ostergen, pp 52–66, available from www.getty.edu/conservation/publications/ pdf_publications/gsap.pdf [accessed November 2010]

Morris H W, Walker R, Drupsteen T (2010) 'Observations of the performance of earth buildings following the September 2010 Darfield earthquake', *Bulletin of the New Zealand Society for Earthquake Engineering*, 43, 4, 393–404

Morris H W, Walker R and Drupsteen T (2011) 'Observed effects of the 4th September 2010 Darfield earthquake on modern and historic earth buildings', Pacific Conference on Earthquake Engineering, Auckland, 14–16 April 2011

Munich Reinsurance Company (2007) 'Schadenspiegel special feature issue risk factor of earth', *Losses and Loss Prevention*, 1, Munich Re

Munich Reinsurance Company (2011) 'NATHAN world map of natural hazards', www.munichre.com/publications/302-05972_en.pdf (accessed October 2011)

NZSEE (1996) *The Assessment and Improvement of the Structural Performance of Earthquake Risk Buildings*, Wellington, New Zealand National Society for Earthquake Engineering

Priestley M J N (1985) 'Seismic behaviour of unreinforced masonry walls', *Bulletin of the New Zealand National Society for Earthquake Engineering*, 18, 2, 191–206

Standards Australia (2001) *AS 3700 Masonry Structures*, Sydney, Standards Australia

Standards New Zealand (1998a) *NZS 4297, Engineering Design of Earth Buildings*, Wellington, Standards New Zealand.

Standards New Zealand (1998b) *NZS 4298, Materials and Workmanship for Earth Buildings*, Wellington, Standards New Zealand

Standards New Zealand (1998c) *NZS 4299, Earth Buildings not Requiring Specific Design*, Wellington, Standards New Zealand. Standards New Zealand has permitted the use of content from *NZS 4297:1998* and *NZS 4299:1998* under license number 000835

Standards New Zealand (2004) *NZS 1170.5:2004 Structural Design Actions Part 5: Earthquake Actions New Zealand*, Wellington, Standards New Zealand

Tipler J F, Worth M L, Morris H W and Ma QT (2010) 'Shake table testing of scaled geogrid-reinforced adobe wall models', *Proceedings of New Zealand Society of Earthquake Engineering Conference*, March 2010, Wellington

Tolles E L, Webster FA and Kimbro E E (1996) 'Survey of damage to historic adobe buildings after the January 1994 Northridge earthquake', GCI Scientific Program Reports, Los Angeles, Getty Conservation Institute, available from www.getty.edu/conservation/publications/pdf_publications/adobe_northridge.pdf [accessed December 2010]

UNEP, UNSDR (2011) 'Global risk data platform', UNEP/GRID-Europe and UNSDR, http://preview.grid.unep.ch/ available from [accessed October 2011]

United Nations International Strategy for Disaster Reduction (2006) 'Disaster statistics 1991–2005', available from www.unisdr.org/disaster-statistics/occurrence-trends-century.htm [accessed November 2010]

USGS (2007) 'Landslide warning signs', US Geological Survey, available from http://landslides.usgs.gov/learning/prepare/ [accessed November 2010]

USGS (2008a), 2008 'United States national seismic hazard maps', US Geological Survey Fact Sheet 2008–3018, 2 pp

USGS (2008b) 'What are volcanic hazards', US Geological Survey Fact Sheet 002-97, available from http://pubs.usgs.gov/fs/fs002-97/ [accessed November 2010]

USGS (2009) 'The Modified Mercalli Intensity Scale', United States Geological Survey, available from http://earthquake.usgs.gov/learn/topics/mercalli.php [accessed 28 November 2010]

USGS (2010) Tarr A C, Villaseñor A, Furlong K P, Rhea S and Benz H M (2010) *Seismicity of the Earth 1900–2007*, US Geological Survey Scientific Investigations

Map 3064, 1 sheet, scale 1:25,000,000, available from http://pubs.usgs.gov/sim/3064/ [accessed November 2010]

Webster F A and Tolles E L (2000) 'Earthquake damage to historic and older adobe buildings during the 1994 Northridge, California earthquake', 12th World Conference on Earthquake Engineering, paper no 0628

Yttrup P (1985) 'Strength of earth masonry (adobe) walls subjected to lateral wind forces', *Proceedings, 7th International Brick Masonry Conference*, Melbourne, February

20

Embankments and dams

W. WU, T. G. BERHE and T. ASHOUR, University of
Natural Resources and Applied Life Sciences, Austria

Abstract: Based on the purposes and types of construction materials used,
embankments and dams are classified into several categories. Emphasis
is given to those embankments used for hydropower, irrigation and flood
control. These types of embankments are called embankment dams. This
chapter gives an overview of the common types of embankment dams,
construction materials, construction control, maintenance, failures and causes
of failures, analysis and design, and future trends. Moreover, the use of
natural fibers as reinforcement for embankments is briefly discussed.

Key words: embankments, filter, hydropower, stability analysis, free board,
failures, fibre reinforcement.

20.1 Introduction

Embankments are among the most ancient forms of civil engineering structures
but are still among the most relevant ones. They are widely used, e.g. as
embankment dams for reservoirs, as dikes for flood control along river banks
and as road, railway and airport runway embankments in transportation.
The requirements on performance of embankments depend mainly on their
purposes. In hydraulic engineering and flood control, embankments are used
to hold water back and for flood control respectively. Therefore, the seepage
behavior is of primary importance. In road and railway construction, the
settlements, particularly differential settlements of embankments are of major
concern. In any case, the stability of embankments must be guaranteed since
failure of embankments and dams can have serious consequences.

Embankments may vary significantly in size (height and length). To date
the tallest earth dam in the world, the Nurek Dam of Tajikistan in Central
Asia, is about 300 m high, while most embankments along river banks and
road/railway embankments are only a few meters above ground. Whereas
most embankment dams of reservoirs are only several hundred meters long,
the river and road/railway embankments can reach lengths of hundreds of
kilometres. Despite the striking difference in size and length, however,
some basic principles must be observed in design and construction of
embankments.

Based on the types of construction materials used, embankments are

538

classified into several categories. These include reinforced embankment, earthfill embankment and rockfill embankment. In this chapter particular attention is given to those embankments which are used for the hydroelectric generation schemes, and irrigation and flood control works. These types of embankments are called embankment dams.

20.2   Types and selection of embankment dams

There are several types of embankment dams. The designs have variability relating to the degree to which seepage within the dam is controlled by provision of filters and drains, the use of free draining rockfill in the embankment, and the control of foundation seepage by grouting, drainage and cutoff construction.

Based on the types of construction materials used, embankment dams are classified into two groups. These are earthfill embankment dams and rockfill embankment dams. The selection of the type of embankment dam to be used at a particular site is affected by many factors. These include availability of construction materials, foundation and/or site conditions, climate, topography and relations to other structures and time of construction. For example, an inclined core can be used when the dam site is located in a high seismic area (Tensay and Wu, 2010). An embankment dam should be designed to satisfy the specific topographic and foundation conditions at the site and to the use of the available construction materials, there are no real standard designs of embankment dams. Hence the embankment dam geometry as well as the type of construction materials varies.

20.2.1 Earthfill embankment dams

The embankment body fulfills both structural and hydraulic requirement. Depending on the height and hydraulic requirement of the dam, several construction forms are feasible. These include homogeneous embankments (Fig. 20.1), with toe drain (Fig. 20.2), with horizontal drain, with horizontal and vertical drains, and zoned earthfill (heterogeneous) with either central core or inclined core. Homogeneous embankments with small height are usually used as levees along riverbanks and in road and railway construction. Whereas the settlement of embankments is a crucial issue for highways and railways, the slope stability and hydraulic properties (seepage and internal erosion) are of central interest for flood protection.

20.2.2 Rockfill dams

For dams of great height, rockfill material is often used for the dam body. Firstly, fine grain soil is usually not available in the vicinity of the dam

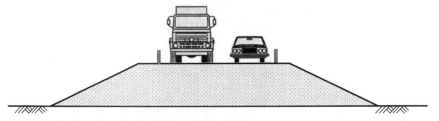

20.1 Homogeneous earthfill embankment.

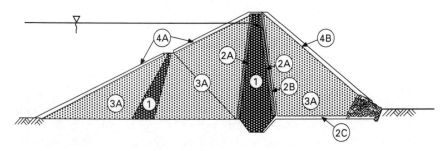

20.2 Zoned (heterogeneous) earthfill dam.

site. Secondly, the shear strength of rockfill material is much higher than earthfill. Heavy machines have given rise to better compaction and made the use of very coarse rockfill (boulder diameter of about half a meter) for dam construction feasible. The tallest rockfill in the world is the 233 m Shuibuya Dam in China completed in 2008. The large boulders and the voids among them give rise to very high permeability. In hydraulic engineering (for electricity or irrigation), water loss is important. Therefore, a water barrier must be provided to minimize the water loss. The water barrier can be located either in the dam centre or on the dam surface.

Rockfill dams can be classified into several groups according to the arrangement of the dam sections, e.g. with a centrally located core, with an inclined core (Fig. 20.3a), with bituminous concrete or asphalt facing (Fig. 20.3b). The water barrier of the Shuibuya Dam in China is a concrete face on the water side. A bituminous facing is said to show higher ductility, which is of advantage during earthquakes. The core material could be made of clayey soil, mixed soil or asphalt concrete. Frequently, it is difficult to get fine materials for the core construction. The use of mixed core and to some extent asphalt membrane as a substitute for the fine core materials has become quite common in the construction of embankment dams. The main dam body consists of rockfill material and transition zones for structural resistance, and core and facing zones shall minimize seepage. A filter zone should be provided for a rockfill dam to prevent erosion, i.e. migration of fine soil particles due to seepage force. Often a transition zone consisting of

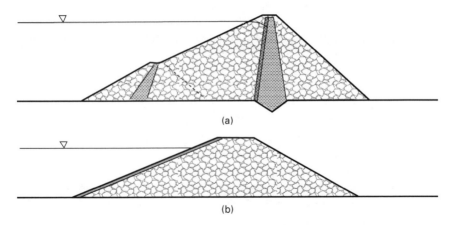

20.3 Rockfill dams: (a) with core barrier, (b) with surface barrier.

granular soil (gravel and sand) is provided between the dam shell (rockfill) and the dam core (fine grained soil).

20.3   Zoning of embankment dams and construction materials

Typical zoning of embankment dams and numbering of these zones is shown in Fig. 20.2. Accordingly the construction materials used for the embankment dam are determined by the function of each zone. The description of the construction materials (Fell *et al.*, 2005), will be provided with reference to these zoning and zone numbers.

(a) Zone 1 is described as core. The function of this zone is to control seepage through the dam. The construction materials used for this zone could be clay, sandy clay, clayey sand, or silty sand. These materials could be used with some gravel. Usually the percentage passing through the 75 μm is more than 15%.

(b) Zone 2A is described as fine filter. This prevents erosion of zone 1 and the dam foundation, and prevents build up of pore pressure in the downstream face. Construction materials used are sand or gravelly sand with less than 5% fines passing 75 μm. The fines should be non-plastic.

(c) Zone 2B is described as coarse filter. This prevents erosion of zone 2A into rockfill. Moreover, it discharges seeped water. The materials used are gravelly sand or sandy gravel.

(d) Zone 2C is described as upstream filter and filter under riprap. The upstream filter prevents erosion of zone 1 into rockfill upstream of dam core. Where as the filter under riprap prevents erosion of the zone

1 through riprap. The construction materials used include well graded sandy gravel or gravely sand with 100% passing 75 mm but not greater than 8% passing 75 μm and the fines must be non-plastic.

(e) Zone 3A is described as earthfill / rockfill. This zone provides the stability, and free drainage to allow discharge of seepage through and under the dam. The construction materials are quarry run rockfill with the oversize removed in quarry or on dam. They are dense, strong, and free draining after compaction. Based on how coarse or fine the rockfill materials are, they are compacted in layers ranging from less than 1 m to 1.5 m with maximum particle size equal to compacted layer thickness.

(f) Zone 4 is described as riprap. The materials used for this zone are selected dense durable rockfill sized to prevent erosion by wave action and sudden draw down.

20.3.1 Design and specification of filters

The basic requirements of filters are that they are sufficiently fine grained to prevent erosion of the soil they are protecting and they are sufficiently permeable to allow drainage of seepage water. These basic requirements are best met by designing the particle size of sand and gravel filters in relation to the particle size of the soil being protected. Criteria developed through many years of experience are used to design filters which will prevent migration of protected soil into the filter. This criterion is based on the grain-size relationship between the protected soil and the filter as described in Tables 20.1 and 20.2 (EM 1110-2-2300, 2004).

20.3.2 Permeability of soils

The permeability coefficient, also called hydraulic conductivity of the soil in a dam embankment or foundation, is often measured by permeability tests in a laboratory. This property of soils depends on a number of factors. Among these factors are the particle size distribution, particle shape and texture, mineralogical composition, void ratio, degree of saturation, soil fabric, nature

Table 20.1 Categories of Base Soil Materials to be used for the filter design criterion, (*US Army engineer manual, 2004*)

Category	Percent finer than No. 200 (75 mm) sieve
1	85
2	40–85
3	15–39
4	15

Table 20.2 Criteria for filters, (*US Army engineer manual*, 2004)

Base soil category	Base soil description, & percent finer than No. 200 (75 μm) sieve[a]	Filter criteria interms of maximum D_{15} size[b]	Remark
1	Fine silts and clays; More than 85% finer	$D_{15} \leq 9 \times d_{85}$	If $9 \times d_{85} < 0.2$ mm, then use 0.2 mm.
2	Sands, silts, clays and silty and clayey sands; 40 to 85% finer.	$D_{15} \leq 0.7$ mm	
3	Silty and clayey sands and gravels; 15 to 39% finer	$D_{15} \leq \dfrac{40-A}{40-15} \times \{(4 \times d_{85}) - 0.7 \text{ mm}\} + 0.7$ mm	A is percent passing the No. 200 sieve after regrading. If $4 \times d_{85} < 0.7$ mm, then use 0.7 mm.
4	Sands and gravels; less than 15% finer.	$D_{15} \leq 4$ to $5 \times d_{85}$	d_{85} can be based on the total soil before regrading.[c]

[a]Category designation for soil containing particles larger than 4.75 mm is determined from a gradation curve of the base soil which has been adjusted to 100% passing the No. 4 (4.75 mm) sieve.
[b]Filters are to have a maximum particle size of 3 in. (75 mm) and a maximum of 5% passing the No. 200 (0.075 mm) sieve with the plasticity index (PI) of the fines equal to zero. PI is determined on the material passing the No. 40 (0.425 mm) sieve in accordance with EM 1110-2-1906. To ensure sufficient permeability, filters need to have a D_{15} size equal to or greater than $4 \times d_{15}$ but no smaller than 0.1 mm.
[c]$D_{15} \leq 4 \times d_{85}$ should be used in the case of filters beneath riprap subject to wave action and drains which may be subject to violent surging and/ or vibration.

of fluid, type of flow and temperature. Table 20.3 shows in general terms the range of permeability which can be encountered.

20.4   Embankment dam construction specifications

Embankment construction methods and testing specifications are intended to ensure that compacted soils are placed uniformly and meet required engineering properties. It is common in embankment construction to prescribe a combination of method and end-result specifications to ensure quality construction. For example, method specifications typically consist of source of construction material, type of rock and degree of weathering, particle size and distribution, upper limit of fines, Atterberg limits, maximum lift thickness requirements, roller type and minimum roller passes, selection and placement of riprap. End-result specifications include specifying 95% of

Table 20.3 Permeability and testing methods for the main soil types (Head 1982)

	k = 1	10^{-1}	10^{-2}	10^{-3}	10^{-4}	10^{-5}	10^{-6}	10^{-7}	10^{-8}	10^{-9}	10^{-10}	10^{-11}	10^{-12}
Drainage characteristics		Good					Poor			Practically impervious			
Permeability classification		High			Medium		Low		Very low		Practically impermeable		
General soil type		Gravels		Clean sands		Fissured & weathered clays				Intact clays			
							Very fine or silty sand						
Test methods: direct		Large CH cell		Standard CH cell				FH cell			FH in oedometer		
indirect		✕		Computation from PSD			✕			From consolidation data			

k, coefficient of permeability (m/s),
CH-constant head,
FH-falling head,
PSD-particle size distribution analysis.

standard Proctor, description of the required soil properties or embankment performance requirements.

Moreover, the question of quality control of earth, earthfill and rockfill dams has been addressed by ICOLD and reported in *Bulletin 56* (ICOLD, 1986), which includes inspection, testing and reporting.

20.4.1 Compaction of fill materials during construction

Compaction of soils is one of the methods of soil stabilization. Through compaction, the stiffness and strength properties of the soil are improved and the permeability reduced. The grading of the fill materials is characterized by the grain size distribution curve. Figure 20.4 shows the variation of grain size distribution curves of fill materials from 12 representative earth and rockfill dams in Japan (Narita, 2000). Soil particles of fairly large grain size are used even as impervious core material because of the recent development of heavy equipment and techniques for construction control.

Consider soil compaction under given energy input and with a relatively low water content. By increasing water content step by step, the dry density of soil increases first monotonically because the lubrication effect of water allows particle movement. The dry density of soil decreases however, as the moisture content increases beyond a certain value called the optimum moisture content, w_{opt}. A compaction curve can be plotted, where the peak point is characterized by the maximum dry density, and the optimum water content, w_{opt}. The compaction test was introduced by American engineer R.R. Proctor in 1933 and became widely used around the world. The Proctor test in geotechnical laboratory is used for two purposes: first to find out the suitability of soil as fill with the maximum dry density and the optimum water content; and second to control the compaction quality during construction. In a standard Proctor test, some stipulations are made in terms of energy input

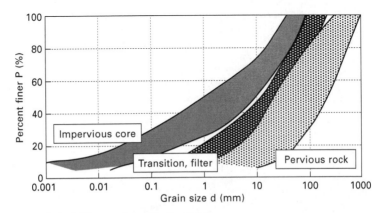

20.4 Typical fill materials (Narita, 2000).

and sample size. By using higher energy input, the dry density in construction can be higher than the Proctor density of the same material (Fig. 20.5).

20.4.2 Compaction control

To ensure the required strength, stiffness and permeability of the materials in the field, compaction conditions should be specified in the design stage. The relationship between strength and permeability of compacted soils and dry density are investigated through the use of laboratory test results. The specified after compaction field dry density ρ_{df} is determined so as to satisfy design values of strength and permeability. The field density is determined by either the conventional method (weight and volume) or with a nuclear gauge. This is usually expressed as the relative density $D = \gamma_{df}/\gamma_{dmax}$. Consider allowable field water content, compaction quality is controlled to maintain the dry density greater than the specified D-value. Usually the relative density is required to be 90% or 95%. Obviously, the field water content should be within the range between the lower bound w_l and the upper bound w_u (Fig. 20.6).

20.4.3 Particle breakage

An important issue in the design and construction of rockfill dams is the breakage of the solid particles. Marsal (1967) developed a measure of particle

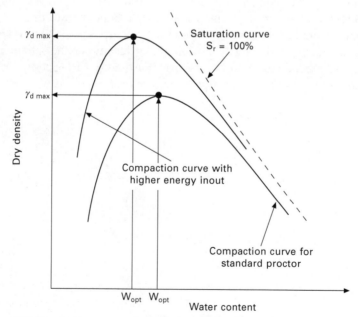

20.5 Compaction curves from Proctor tests.

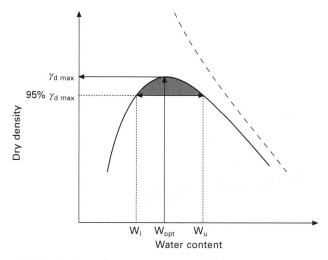

20.6 Proctor tests for compaction control.

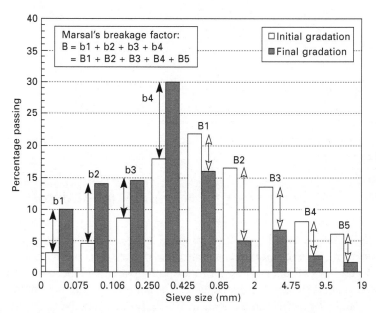

20.7 Definition of the grain breakage factor (Wang and Sassa, 2000)

breakage factor, B_g, to quantify particle breakage. His method involves the change in individual particle sizes between the initial (prior to compaction) and final (after compaction) grain size distributions. The difference in the percentage retained is computed for each sieve size. This difference will be either positive or negative. The breakage factor is defined by the sum of the difference having the same sign. Figure 20.7, illustrates the definition

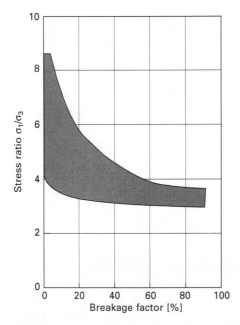

20.8 Dependence of the breakage factor on stress ratio.

of the index *Bg* (Wang and Sassa, 2000). The breakage index depends on several factors, the stress level and the stress ratio between the major and the minor principal stresses have significant influence on the breakage factor. The dependence of the breakage factor on the stress ratio is shown in Fig. 20.8.

20.5   Stability analysis of embankment dams

Slope stability analyses of high earth and rockfill dams are very important to ascertain the stability of the structure. As the geotechnical and hydraulic properties of the material in different zones of the earth and rockfill dam differ to a great extent, the static stability of both upstream slope and downstream slope under different conditions have to be analyzed. Moreover, dams located in seismically active areas need to be analyzed extensively to investigate the safety of the structure under dynamic conditions. In fact, the flexibility of dam materials provides excellent seismic stability.

The purpose of slope stability analysis is to provide a quantitative measure of the stability of the slope or a part of the slope. Traditionally it is expressed as a factor of safety against failure of that slope, where the factor of safety is defined as the ratio of the restoring force to the disturbing force, such that factor of safety greater than one shows stability and the factor of safety less than one denotes failure.

The analysis of stability of dams is almost always carried out using limit equilibrium methods. However, in some cases it is necessary to analyze dams using modern numerical analysis tools like the finite element methods and finite difference methods. This may be necessary to predict deformations, and generation and dissipation of pore pressures.

In the modern stability analysis, the finite element and finite difference methods are dominant. In particular the nonlinear effective stress and time domain approaches are leading the seismic design and analysis methods of embankment dams. This is because these methods account for the time varying intensity of ground shaking, and provide insight as to where, when and in what order elements of a dam or foundation respond to excitation. With these modern methods of design and analysis, it is possible to model the problems in three dimensions, to include the entire sequence of events like construction stages, reservoir impounding stages, operation, liquefaction and final configuration of the dam-foundation-reservoir system. Among the nonlinear computer codes used in engineering practice to evaluate the seismic response and deformation of embankment dams are the commercial software DYNAFLOW DIANA, DSAGE, DYNARD, FLAC, DYSAC2 and SWANDYNE4. Another widely used commercial software is GEO-SLOPE for the analysis of slope stability in two dimensions.

20.6   Dam freeboard requirement

Freeboard protects embankment dams from overflow caused by wind-induced tides, waves, landslide and seismic effects, settlement, malfunction of structures and other uncertainties in design, construction and operation. It is defined as the vertical distance between the crest of a dam and a specified pool level, usually the normal operating level or the maximum flood level. Depending on the importance of the structure, the minimum freeboard will vary in order to maintain structural integrity and to reduce the cost of repairing damages resulting from overtopping.

20.7   Failure mechanisms

Failure of embankments and dams may assume different forms (Fig. 20.9). Excessive settlement is the most frequently encountered problem for road embankments (Fig. 20.9a). Often poor construction quality is to blame. The loss of slope stability is another oft observed failure form (Fig. 20.9b). Most catastrophic failures of embankment dams are caused by overtopping of the reservoir water due to flooding or loss of freeboard (Fig. 20.9c). Other main factors to cause embankment failures are hydraulic internal erosion (Fig. 20.9c), generation of high pore pressure and earthquake forces. It is interesting to notice that more than 50% of embankment failures in hydraulic engineering are due to hydraulic erosion.

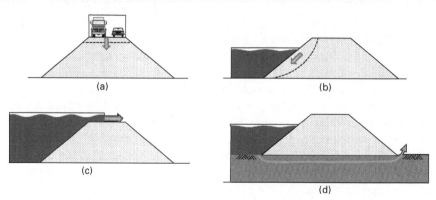

20.9 Typical failure forms of embankment dams.

The soil fractions that are considered as most susceptible to erosion are relatively uniform coarse silt and fine sand. Cohesive soils such as clays are more resistant to erosion as long as the chemical bonds are not destroyed (Srbulov, 1988).

20.7.1 Failure due high pore pressure

When soil is saturated and drainage is not provided, the external load is known to be sustained by the inter-granular stress (effective stress) and the pore water pressure. Note that both strength and stiffness of soil are related to the effective stress. This simple but important assumption is known as the effective stress principle by Terzaghi. High water pressure may reduce the effective stress to such an extent that embankments and dams fail. A good example is the construction of embankments on soil as depicted in Fig. 20.10. Along with increasing embankment height, excessive water pressure will generate in the soft soil under the embankment. The dissipation of the excessive water pressure depends mainly on the permeability. As the permeability of soft soil is extremely low, it takes a very long time until the excessive water pressure is completely dissipated. If the embankment construction is too fast, the water pressure generation will exceed the dissipation by a long way, which may lead to ground failure. As a consequence, embankments on soft soil are often constructed in stages. The construction process will be adjusted to the speed of the dissipation of water pressure. There are several techniques to enhance the consolidation of soft soil, e.g. pre-loading, drains and dewatering. Some of these techniques have been successfully used in land reclamation, e.g. the new airport in Bangkok.

The pore pressure also plays an important role in sand. Whereas the low permeability of soft soil is responsible for the slow dissipation of pre-pressure, it is the short loading time for sandy soil which is of particular relevance for earthquakes. During earthquakes the ground experiences fast

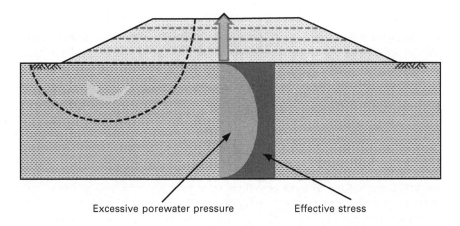

Excessive porewater pressure   Effective stress

20.10 Pore water pressure during embankment construction on soft soil.

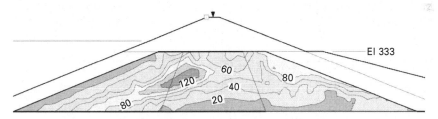

20.11 Distribution of excess pore pressure after shaking (www.geo-slope. com)

and cyclical loading. The pore pressure generation may exceed the pore pressure dissipation giving rise to excessive pore pressure and reduction of effective stress. If the effective stress vanishes, the soil liquefies and behaves like fluid. Soil liquefaction is responsible for disastrous damage during earthquakes. An example is the Low San Fernando Dam in Southern California, USA, which was damaged by liquefaction during an earthquake in 1971. This failure case alarmed the community and sparked intensive research on liquefaction worldwide. Figure 20.11 shows the numerical simulation of the distribution of pore pressure after the shakes of Low San Fernando Dam (www.geo-slope.com).

20.7.2 Seepage failure

The movement of water in soil is known to obey the law of Darcy, i.e. $v = k \cdot i$, where v is the flow speed, k the permeability and i the hydraulic gradient. The flowing water also exerts force on the soil, which can be written out readily $j = i \cdot \gamma_w$, where j is the hydraulic force and γ_w the unit weight of water. For increasing hydraulic gradient, the force j can be so large that the soil grains are suspended and transported through the voids in soil. Such phenomena are known as soil erosion, which can happen at free soil surface, at interface between soil and structure and inside soil.

Most embankments exhibit seepage to some extent. However, the rate and quantity of this seepage must be controlled. If uncontrolled, it can erode fine soil material from the downstream slope or foundation and continue moving towards the upstream slope to form a pipe or cavity to the pond or lake, often leading to a complete failure of the embankment (Fig. 20.12). Such internal soil erosions can be often observed in the form of craters formed by the fine soil grains (Fig. 20.12). Seepage failures account for approximately 40% of all embankments or dike failures.

In dam design, filter zones and drains between dissimilar zones, e.g. between rockfill and core, are provided to avoid internal soil erosion. The assessment of the soil erosion potential remains difficult because of the heterogeneous nature of geomaterials. A further important issue is the construction quality. Obviously, a poorly compacted embankment is more prone to piping.

20.7.3 Differential settlement

Differential settlement in embankments may be ascribed to the difference in ground conditions, loading conditions and material properties. The latter is particularly true for rockfill dams with an impervious core. Rockfill possesses much higher stiffness than the soft core. Moreover, some arching can be observed even for homogeneous embankments. The differential settlement may lead to internal cracks in the impervious zone and foundation. The differential settlement shows different patterns and the associated deformations appear

20.12 Internal erosion in embankment dams.

in the dam body, base foundation and abutments. Figure 20.13 shows the arching effect along the dam axis and perpendicular to the dam axis. Further settlements are caused by first impoundment of the reservoir. Although the rising water table and the buoyancy reduce the effective stress, large settlements are observed during the first impoundment. The mechanism for this settlement is not yet fully understood.

20.7.4 Earthquake damage

Earthquakes are a major geo-risk for embankments and dams. Embankment failures due to earthquake excitation can be classified into two groups; damage caused by the liquefaction of the dam body and/ or foundation, and the sliding and cracking of the dam body. In the former case, high excess pore pressure is generated during earthquake shakings by the application of cyclic shear stresses. As a consequence, large lateral and vertical deformation displacements are observed. These deformations can lead to catastrophic damages and lead to overtopping and dam break. The damaged Low San Fernando Dam after the earthquake is shown in Fig. 20.14. The dam body

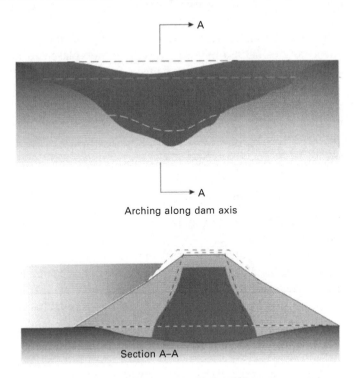

Arching along dam axis

Section A–A

Arching perpendicular to dam axis

20.13 Arching effect and differential settlement.

20.14 Liquefaction damaged dam during earthquake.

literally flew apart. Fortunately, the water table was low and there were no casualties.

According to the investigation reports on earthquake damage of embankment dams, embankment failures caused by strong excitation are classified into several patterns according to their failure mechanism. Some distinct patterns of embankment failures due to earthquake excitation are schematically illustrated in Fig. 20.15. These failure patterns were obtained from numerical simulations. However, fatal cases of dam failures during earthquakes are rare, which is probably due to the larger safety reserve compared to other engineering structures.

20.8   Maintenance of embankment dams

Due to the long service time, the maintenance of dams and embankments presents an important issue for owners and operators. The dams are to be instrumented and inspected regularly. If problems are identified during an inspection, they should be handled swiftly. Localizing and handling potential problems in the early stage reduces risk and saves time and money. Annual or long-term maintenance programs for earthfill structures may include:

- regular control of vegetation and burrowing animals
- riprap and crest maintenance and repair
- slope stabilization
- drainage system maintenance
- removal of upstream debris
- maintenance of instrumentation.

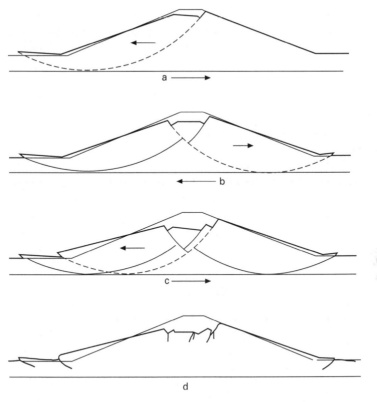

20.15 Failure patterns of embankments.

20.9   Future trends

Even though embankment dams have been constructed since ancient times, the mounting demand for renewable energy makes hydropower and therefore dam construction even more appealing. Modern dams can be constructed with impervious membranes of manufactured materials such as concrete, steel, asphalt concrete and geo-synthetics. Dams constructed with such reinforcements are safer against shear failure than conventional earth- or rockfill dams. For a given safety factor, the embankment slopes can be made steeper, the construction time shortened and construction cost reduced.

The market for reinforced soil has developed to maturity, as identified by the increasing number of construction systems.

Embankments are made of unsaturated compacted soil and their behavior is strongly affected by climatic or environmental conditions. When clayey soils are involved, significant ground movement can occur, leading to possible damage to buildings and geotechnical constructions (Arbizzi *et al.*, 2008; Borde and Desprès, 2008). More recently, fiber reinforced composites have been widely used as construction materials in various fields from civil engineering to aerospace engineering (Fu *et al.*, 2006; Baklanova *et al.*, 2006; Brownie *et al.*, 1993; Shi *et al.*, 2000). The mechanical properties of soil can be significantly improved by adding fibers to the earth matrix. Fibers of various plant residues present a cost effective way of reinforcing soil. Straw fiber helps to increase strength, control shrinkage cracks and improve toughness (King *et al.*, 2006).

On the other hand, faced with the worldwide shortage of forest resources, industry has shown increasing interest in using agricultural residues for construction purposes (Sampathrajan *et al.*, 1992). Ashour *et al.* (2010) reported on the effect of straw fibers on the strength and ductility of soils. The effect of plant fibers on the erosion resistance of cohesive soil was investigated by Ashour and Wu (2010). The fiber content has by far the largest influence on the erosion resistance of earth plaster. These and other studies show the great potential of agriculture residues as construction additives.

20.10 Norms and standards

The International Commission on Large Dams (ICOLD at http://www.icold-cigb.net/) has issued numerous recommendations on design, construction, commissioning, operation and maintenance of dams. Usually geotechnical tests for the construction of dams and embankments are carried out in compliance with the national standards. The following list of ASTM standards (American Society of Testing and Materials) applicable to the construction of earthen dams serves as example.

1. ASTM D 421: Practice for Dry Preparation of Soil Samples for Particle Size Analysis
2. ASTM D 422: Method for Particle-Size Analysis of Soils
3. ASTM D 1140: Test Method for Amount of Material in Soils Finer than the No. 200 Sieve
4. ASTM D 1557: Test Methods for Moisture-Density Relations of Soils and Soil Aggregate
5. ASTM D 2216: Method for Laboratory Determination of Moisture Content.
6. ASTM D 2434: Test Method for Permeability of Granular Soils (Constant Head)
7. ASTM D 2487: Test Method for Description and Identification of Soils
8. ASTM D 4318: Test Method for Liquid Limit, Plastic Limit, and Plasticity Index of Soils
9. ASTM D 4718: Practice for Correction of Unit Weight and Moisture content for Soils Containing Oversize Particles

10. ASTM D 5080: Test Method for Rapid Determination of Soil Density (3-Point Compaction)
11. ASTM D 5084: Test Method for Permeability of Fine Grained Soils (Falling Head)
12. Earth Manual, Third Edition 1998, by the United States Department of the Interior, Bureau of Reclamation (USBR).
13. Engineering Regulations, ER 1110-1-8155, October 2003, Department of the Army, U.S. Army Corps of Engineer (USACE).
14. Design of Small Dams, Third Edition 1987, United States Department of the Interior, Bureau of Reclamation (USBR).

20.11   References

Arbizzi S., Kreziak C., Barraud D., Larrère F., Souvignet S. and Nagel B. 2008, 'Analyse d'une base de données 'pathologies liées à la sécheresse' et mise en relation avec les sols support', Int. Symposium on Drought and Constructions – SEC2008. LCPC, Paris, pp. 385–391.

Ashour T. and Wu W. 2010, 'An experimental study on shrinkage of earth plaster with natural fibres for straw bale buildings', *Int. J. Sustainable Eng.*, **3**: 299–304.

Ashour T. and Wu W. 2010, 'The influence of natural reinforcement fibers on erosion properties of earth plaster materials for straw bale buildings', *J. Building Appraisal*, **5**: 329–340.

Ashour T., Wieland H., Georg H., Bockisch F.J. and Wu W. 2010, 'The influence of natural reinforcement fibres on insulation values of earth plaster for straw bale buildings', *Mater. Design*, **31**: 4676–4685.

Baklanova N.I., Zima T.M., Boronin A.I., Kosheev S.V., Titov A.T. and Isaeva N.V. 2006, 'Protective ceramic multilayer coatings for carbon fibers', *Surf. Coat. Tech.,* **201**: 2313–9.

Borde M. and Desprès R. 2008, 'Exemples concrets et illustrés des effets des périodes de sécheresse prolongée sur des constructions situées sur des sols argileux à fort aléa', Int. Symposium on Drought and Constructions – SEC2008. LCPC, Paris, pp. 367–375.

Brownie P.M., Ponton C.B., Marquis P.M. and Butler E.G. 1993, 'Zirconia sol coating of single-crystal ceramic fibres', *Mater. Design*, **14**: 49–51.

Craig R.F. 1992, *Soil Mechanics*, Chapman & Hall, London, UK.

EM 1110-2-2300, *US Army Engineer Manual*, 2004, Engineering and design – General design and construction considerations for earth and rock-fill dams.

Fell R., MacGregor P., Stapledon D. and Bell G. 2005, 'Geotechnical engineering of dams', Balkema, Leiden.

Fu Y.C., Shi N.L., Zhang D.Z. and Yang R. 2006, 'Effect of C coating on the interfacial microstructure and properties of SiC fiber-reinforced Ti matrix composites', *Mat. Sci. Eng. A-Struct.* **426**: 278–282.

GEO-SLOPE International Ltd, Calgary, Alberta, Canada (www.geo-slope.com)

Head K.H. 1982, Manual of Soil Laboratory Testing, Vol. 2: Permeability, Shear Strength, and Compressibility Tests, Pentech Press, London

ICOLD (International Commission on Large Dams), *Bulletin* **56**: Quality control for fill dams, 1986.

King B., Aschheim M., Dalmeijer R., Donahue K., Hammer M. and Smith D., 2006, in:

Stone N., Straube J., Summers M. and Theis B. (eds.), *Design of Straw Bale Buildings: the State of the Art*, pp. 19–55. Green Building Press.

Marsal R.J. 1967, 'Large scale testing of rockfill materials', *J. Soil Mech.* Found. Eng. Division, ASCE, **93**: 27–43.

Narita K. 2000, 'Design and construction of embankment dams', Research Report, Deptartment of Civil Engineering, Aichi Institute of Technology, Japan

Sampathrajan A., Vijayaraghavan N.C. and Swaminathan K.R. 1992, ' Mechanical and thermal properties of particleboards made from farm residues', *Bioresour. Technol.*, **40**: 249–251.

Shi Z.F. and Zhou L.M. 2000, 'Interfacial debonding of coated-fiber-reinforced composites under tension-tension cyclic load', *Acta Mech. Sinica*, **16**: 347–56.

Srbulov M. 1988, 'Estimation of soil internal erosion potential', *Computers and Geotechnics*, **6**: 265–276.

Tensay G.B., Wang X.T. and Wu W. 2010, 'Numerical Investigation into the Arrangement of Clay Core on the Seismic Performance of Earth Dams', Geotechnical Special Publication, Soil Dynamics and Earthquake Engineering, ASCE, pp. 375–384.

Wang F.W. and Sassa K. 2000, 'Relationship between grain crushing and excessive pore water pressure generation by sandy soil in ring shear tests', *Journal of Natural Disaster Science*, **22**: 87–96.

Part V

Application of modern earth construction: international case studies

21

North American modern earth construction

M. KRAYENHOFF, SIREWALL Inc., Canada

Abstract: One of the most significant strides forward toward broad implementation of modern earth building is the new ability of rammed earth to perform well in very wet and very cold climates. This paper explores the nuances of heat flow in *insulated* rammed earth for very hot climates as well as the very cold climates. Techniques to gain the required strength to withstand the forces of erosion (wet/dry and freeze/thaw) are examined. Finally, practical applications of *insulated* rammed earth in a variety of natural settings are presented to show the scope of what is now possible.

Key words: rammed earth, *insulated* rammed earth, stabilized *insulated* rammed earth, SIREWALL, thermal flywheel, humidity flywheel, thermal mass, earth blending.

21.1   Introduction

This chapter focuses on *insulated* rammed earth. Because adding insulation in a rammed earth wall creates a profound difference in performance and means of manufacture, you will notice that both '*insulated*' and '*uninsulated*' words are italicized for clarity.

Walls are the significant boundary between the indoor and outdoor experience. How well they function largely determines our level of comfort, health and safety. This chapter explores the factors that help rammed earth walls perform well.

We begin by distinguishing the philosophy that underpins the inception and development of *insulated* rammed earth. As Einstein so famously said, 'You cannot solve a problem from the same consciousness that created it. You must learn to see the world anew.' Short-term, self-interested problem solving is not the answer. Consider the world that our seventh generation descendents will inherit and what they might like to see. For them the most important shift in construction will be from toxic, disposable buildings to sustainable buildings. Because the climate is changing, the requirements for our buildings to be truly sustainable will far outstrip those at present.

Energy will play an increasing role in how we build. The level of attention to thermal detailing in rammed earth walls can result in anything from a significant ongoing energy sink to buildings that require no heating or cooling at all. Attention to detail is paramount.

561

Putting energy efficiency into action requires an understanding of how heat moves in (and out of) a building. Conductive, convective and radiant heat loss all affect a building's energy performance. Performance will largely be determined by the most active of the three modes of heat loss. Therefore, a holistic approach to heat loss is recommended.

The North American climate can be very demanding of wall assemblies due to freeze/thaw and wet/dry cycles. Only if rammed earth is sufficiently strong can it be viable in very cold and very wet climates. Compressive strength is the single biggest determinant of durability. A hydrophobic admixture can also be helpful. Examples of rammed earth work in extremely cold climates, very wet and very dry climates show the versatility of rammed earth done well.

Rammed earth and *insulated* rammed earth in particular, opens the door to a future where our buildings are truly sustainable and a gift to our descendents. First, however, rammed earth must move from a boutique cottage industry to an efficient industry with greater capacity. With that capacity and efficiency, rammed earth offers what could well be the most holistic and sustainable building option available on the planet.

In Chapter 15, David Easton describes the process of building with *uninsulated* rammed earth. Most of the practitioners of this style of North American rammed earth, or a variation thereof, are in the American southwest where the advantages of thermal mass allow for comfort in climates that are moderate or where there is enough solar insolation to allow diurnal comfort. Many of the components of *insulated* rammed earth are similar to North American *uninsulated* rammed earth as we both blend, mix, form, ram, strip and cure. This chapter will highlight the differences as well as bring to light some of the concepts that have been distinguished in the unique work of making rammed earth a functional, Building Code-approved technology, viable in very cold and/or very wet climates.

The rammed earth industry in North America is very small. There are approximately 50 practitioners producing rammed earth structures with half of them building with *insulated* rammed earth and half building with *uninsulated* rammed earth. There are very few full-time rammed earth businesses. The vast majority do their rammed earth as a passionate sideline to a related line of business that keeps them solvent (some do only one small job every couple of years). From 1992 to 2006, SIREWALL was the first and only company in North America building with *insulated* rammed earth. Many approaches were tried and had successes and failures. Slowly, with much trial and error, a system was evolved and an approach that has generated consistent quality and produces high strength walls. Through teaching courses a strong team has developed and interest has spread. To my knowledge, all *insulated* rammed earth builders in North America got their start with SIREWALL.

In the early 1990s I pledged my working life to making rammed earth a viable building technology that excelled in every way. My mentor was David Easton. Through him and his work I saw what was possible and that still motivates me today. Buildings that are healthy, beautiful, don't require heating or cooling, are durable, environmentally friendly and, most importantly, are a gift to our descendents are what I see as a significant contribution to our communities and to the planet.

We are only at the very beginning of realizing the potential of rammed earth building. In my second year of working with rammed earth I called David and asked him what the considerations are for ramming in the snow. He replied, 'Let me know what you find out ...' and I realized it was up to me to make rammed earth work in my cold, wet climate and high seismic zone. My work to make modern rammed earth a viable option in cold and wet climates stands on the shoulders of David Easton and his work stands on the shoulders of David and Lydia Miller. Although a tiny niche industry, we are developing a rich history and I am grateful to be part of it.

21.2 Seventh generation thinking and earth construction

The best place to start a conversation about building is to look at how we think about building. Otherwise, no change is possible (as the context is decisive). No problem can be solved by the same thinking that created it.

Chief Seattle is widely known for introducing the ancient aboriginal 'seventh generation' thinking, which is to consider the impact of current decisions and actions on seventh generation descendents. With a new generation occurring every 25 years or so, seventh generation thinking is 200 years in the future. For most people, it is difficult to imagine 20 years into the future. We have significant education and training to help imagine the past, but little for the future. We are ill-equipped for this vital task of creating a viable, long-term future. It is also something that cannot be done in isolation. It is difficult to design and build for a 200-year future when governments are only required to plan 5 years ahead. With the rezoning of land based on short-term thinking, some of the most robust buildings may be torn down in 10 years and replaced with parking lots and less durable and cheaper buildings. Buildings designed for passive solar can be blocked by future developments that deny access to the sun. It is essential that everyone believes, plans and is in action to create a sustainable future for there to be a viable, bright, long-term future.

Our current disposable building paradigm ensures continued and expanding erosion of the natural environment through resource extraction and land filling. We cannot continue much longer on this path. We are already experiencing the impacts of the disposable paradigm in our energy costs, our health, species extinction and environmental degradation. We also have a responsibility to our descendents to pass on durable and comfortable buildings, not toxic waste.

Buildings are typically designed and constructed to meet the programmatic, financial and technical requirements and demands of today. With the current design, funding and approval processes, this renders them out of date prior to tender and before construction is complete. If we apply seventh generation thinking to how we design and build, then we create buildings that last 200 years and those buildings are conceived of as meeting the needs at 100 years. They are then ahead of their time for the first 100 years and becoming out of date for the last 100 years. This will be a big stretch for most people and projects. However, the dysfunction of the current practice and process requires that we begin to seriously anticipate the world as it will be in the future. How much longer will we be willing *and able* to embrace the toxic, disposable building paradigm?

So, what does the future look like? It is anticipated that there will be more extreme weather events. Ocean levels and overall global temperatures are expected to rise accompanied by large-scale winds in some areas. Other areas will experience extreme rains. Some places will get much hotter and others much colder. Raw materials will become more expensive as more people consume resources that are sourced in more difficult places. Labor will be much cheaper than commodities, which will need to be reused/recycled for economic reasons. The disposable economy will lose its inertia. Energy resources will dominate decision-making. Are any of these prognostications certain to come into being? No, they are like a weather forecast. If the forecast is for heavy rain, one carries an umbrella or raincoat. It would be foolish not to, even though rain is not inevitable. We act on our best information or (likely) suffer the consequences.

For some, seventh generation thinking is an ethereal exercise. It is trying to care for the unborn, way into the future. That is well and good, but the real value in seventh generation thinking is applying it to all decisions in the building process, from design through material selection. It is above all, a practical tool.

21.3   The interplay of indoor and outdoor weather

When we see an eagle flying on thermals above a cliff, we know that warm air is rising. As the warm air rises, a lower pressure area is created beneath it and the cooler air in the valley moves to the bottom of the cliff. Nature is constantly trying to equalize air pressures.

The dew on the morning grass evaporates as the air heats up and then condenses in the evening. The existence of dew implies 100% relative humidity (RH) between the grass blades. In the morning the temperature increases and the capability of that warmer air to hold moisture also increases so the dew evaporates. We see nature equalizing water vapor pressure as well. We park our car and the engine is hot but it doesn't take long for it to lose its heat and reach the temperature of its surroundings. This can be seen as the second law of thermodynamics at work (one consequence of the *second law of thermodynamics* is that differences of temperature, pressure and chemical potential equilibrate over time in an isolated physical system). The 'weather' in the context of building walls is a force imposed on the building by the surrounding environmental conditions, mediated by the interplay of three pulls of nature (equalizing air temperature, air pressure and vapor pressure). The physical conditions inside a building may be quite different to those outside but the same dynamics of equalization are at play. The greater the variation between inside and outside conditions, the greater the demands on the building envelope.

Let's look at extreme temperatures and what rammed earth (or other) walls might be required to cope with. If the temperature is a record low of −62.2°C (−80°F) in Mayo, Yukon and the desired indoor temperature is 20°C (68°F), then there is a very large temperature difference or ΔT (82.2°C or 148°F) across the thickness of the wall. There is a very large pull (tendency to equilibrium, or entropy) for the interior heat to pass across the wall to the outside. If the temperature is a record high of 56.7°C (134°F) in Death Valley, CA, and the desired indoor temperature is 22°C (72°F), the ΔT is 34.7°C (62°F). The ΔT is positively correlated with the strength of the pull of nature to equalize temperature. The greater the pull, the greater the demand on the ability of the wall to resist heat flows. The dynamic is the same whether desert heat flows into a building or the building's heat leaves to warm the Arctic. Heat is travelling across the wall material's resistance to heat flow (expressed as R, U or k values). Larger ΔT exists in cold climates and hence technology for managing heat flows across walls is more developed in colder climates. Ironically, more energy and money is spent cooling buildings in warmer climates where they don't begin to take advantage of the cold climate technologies. It is difficult to keep walls dry because water vapor is pulled across wall assemblies. If, for example, the ambient humidity in Miami is 100% and the interior humidity is 40% due to air conditioning (AC), water vapor is pulled across the wall from outside to inside (people are most comfortable at 40–60% RH). If the humidity in Winnipeg or Albuquerque is 10% and the interior humidity is 40% there is a pull in the opposite direction, but only half as strong. Note that the magnitude of the pull will determine the demand on the resistance to water vapor flow across the wall. The mechanism that resists the water vapor transport across the envelope is the vapor barrier (or a combination of materials that has sufficient vapor resistivity), and in areas where the pull is great and the detailing is poor, the walls will always be susceptible to dampness (unless the drying capability is greater). This is a problem for wall assemblies that include organic materials because organic materials are designed by nature to decompose (due to mold, rot, insects, rodents, etc.). For modern earth buildings if they are well built, this should not be a problem as there are usually no organic materials in the wall to decay.

Air enters and leaves buildings through the building envelope. Differences in air pressure across a wall are dynamic or static. Static differential air pressure in buildings is known as the 'stack effect' and is an expression of the building acting as a chimney. Taller buildings have more stack effect than shorter buildings. Given that warm air rises, there is a higher air pressure at the top of a building than there is at the bottom. Let's look at this pressure gradient in detail. Say that the air pressure outside is 1.0 Atmospheres (all around the building) and inside at the top of the building it is 1.1 Atmospheres. At the bottom of the building it is 0.9 Atmospheres. There is a pull from inside to outside at the top of the building and at the bottom of the building, the pull is from outside to inside. Somewhere between the top and the bottom of the building, there is a neutral plane where the indoor air pressure is 1.0 Atmospheres and there is no pull across the wall. If the air tightness of the building was uniform, then that neutral plane would be halfway up, but that seldom happens. The mechanism that resists air flow across the envelope is the air barrier and it is both difficult and critical to maintain integrity and continuity of the air barrier. The quality of the air barrier is tested with a blower door and the difference between a tight wood house and a leaky wood house can be 100-fold. What happens in a leaky wood house is most air comes into the building at the lowest elevation (crawl space or basement) and exits at the top. Occupants breathe poorer quality air from the crawlspace, or other lowest entry point, throughout the house. Moisture and heat can be carried on the airstream into the wall cavity upon entry or exit from the building, potentially causing rot and compromising the thermal envelope. Rammed earth buildings have walls that are impermeable to passage of an airstream and if the detailing is well done around the windows and doors, the building is well on its way to being airtight. There is broad agreement in building science that a tight building is essential to managing temperature and humidity. As the building gets tighter it becomes essential to include an air-to-air heat exchanger to supply a continuous stream of fresh air. Typical systems used in air-tight buildings are heat recovery ventilation systems (HRV) or energy recovery ventilations systems (ERV).

Dynamic forces such as hurricanes or cyclones create a large demand on structures. A wood frame structure might have 30 tons of mass above ground level and a similar size rammed earth structure could well have 200 tons of mass above ground level. It is well known that the weight of a structure will oppose the high wind forces but the geometry and composition is also important. A tall heavy wall at right angles to a strong wind can be blown over fairly easily. Buttressing and shear walls need to be part of the design. It is much easier for a tornado to lift or tear apart a 30-ton multiple material building assembly (e.g. wood) than a consolidated 200-ton assembly (e.g. rammed earth). How the roof is secured to the rammed earth and limiting the overhang uplift are also key design considerations.

The requirements of a building envelope will correlate with the differences between the indoor and outdoor weather. The more extreme the differences in temperature, humidity or air pressure, the more robust the response required from the building envelope. The correlation need not be exact and it is best to err on the side of a higher performance response when using seventh generation thinking. Too little response, with insufficient insulation and non-airtight detailing, will cause the building to be uncomfortable and expensive to operate, whereas too much response will have the impact of an added incremental initial cost and the benefit of reduced operating costs over time.

21.3.1 Thermal mass

Passive solar design makes reference to the necessity of extra mass in the building to moderate temperature swings associated with sunshade transitions at windows. Too often the mass is insufficient to absorb the incoming solar energy and the tight passive solar house overheats while the sun's out and then slowly cools down when the sun stops shining in through the windows. Part of the difficulty of designing passive solar in the northern hemisphere is that the hottest day of the year is not when the angle of the sun is the highest. The hottest day is often several weeks after the longest day and by that time the angle of the sun is already 20–30% of its way to its lowest angle of incidence. Further, the hottest time of the day is often a few hours after noon, at which time the sun is no longer at its peak elevation. This compromises the effectiveness of the carefully calculated overhangs, which are intended to allow solar entry in winter but not in summer. Alternative designs and additional thermal mass help guard against overheating in late summer.

The relationship between thermal mass and insulation is often not well understood. That is further complicated by the varying effectiveness of that thermal mass due to climate and latitude. Conductive heat loss is a simple evaluation of a steady-state condition, whereas the dynamic thermal performance of high mass walls is often not considered because the analysis is so complicated and site specific. That does not mean there isn't significant benefit to high mass walls. Jan Kosny of Oak Ridge National Laboratory (ORNL) has done a lot of very valuable work of relevance to the *insulated* rammed earth industry. He has quantified the effects of mass and insulation in different configurations in a wall assembly (see Fig. 21.1) and he has

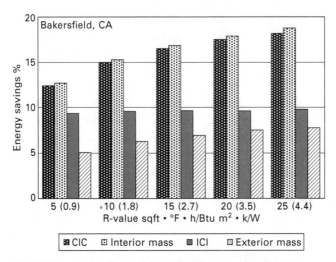

21.1 Different configurations of mass and insulation.

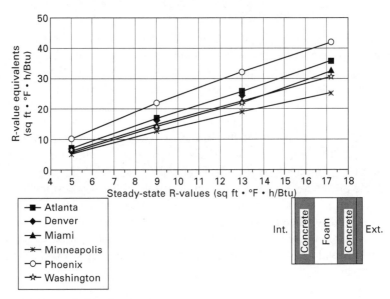

21.2 Dynamic R-value equivalents.

graphed the Dynamic R-value equivalents to steady-state R-values for six different American cities (see Fig. 21.2).

It is interesting to note (from Fig. 21.2) that the dynamic R-value equivalents range from ~1.5 to 2.25 times the steady-state R-values for mass-insulation mass assemblies.

Probably the easiest way to understand the difference between insulation and thermal mass is that insulation slows down heat loss and mass acts to minimize temperature swings. At times mass may appear to be doing the job of insulating by minimizing the temperature swings from sunny day to cold night. However, an *uninsulated* massive house in a cold climate like Winnipeg in the winter will be extremely cold and difficult to heat. It will be most affected by the average daily outdoor temperatures (combined with other contributing factors such as the temperature of the earth under the building, the effects of solar gain, thermal characteristics of the rest of the building envelope and occupant contributions). In Winnipeg the outdoor winter temperatures trump all the other contributing factors. However in San Francisco, where the climate is mild, *uninsulated* rammed earth is dependent on the other contributing factors. For example, a passive solar *uninsulated* rammed earth house may be very comfortable, but an identical house in the same general location on the north side of a hill, having no solar access, would provide a decidedly different thermal experience. When the average outdoor temperature combined with the other contributing factors results in an uncomfortable interior space, *insulated* rammed earth can almost always be a solution if the projected discomfort is identified beforehand through energy analysis.

By far the most popular configuration of *insulated* rammed earth is the 24 in. wall. While it is possible to make *insulated* rammed earth walls as thin as 18 in., the extra cost for the extra mass (usually placed on the inside of the thermal envelope) is good value. Eighteen-inch walls have 7 in. wythes on either side of 4 in. insulation. Most 24 in. walls have 12 in. of earth inside the envelope (over 70% more mass than the 18 in. configuration). The added bonus to the thicker interior wythe configuration is that one can fairly easily create window seating (or window beds) by having the bottom of the window at 14 in. to 16 in. elevation (Fig. 21.3).

Can there be too much mass? In most circumstances the answer would be no. There are many studies that calculate the optimal balance between storage and passage of heat through an *uninsulated* massive wall. These studies are not useful for the well-*insulated* rammed earth wall where the function of the mass is purely storage. We once built a large home in coastal British Columbia (BC) with over 300 tons of thermal mass (in addition to the 100 tons for its normal construction), enhanced insulation (up to R-60 in the walls), low 'e' glazing and airtight detailing. Our hope was that we would be able to move from diurnal heat storage to seasonal heat storage. In other words, the heat from the summer sun could be stored and slowly released to delay the beginning of the heating season to well after Christmas, instead of September. What we didn't account for were the impacts of comfort on lifestyle. The occupants would leave doors open in November without apparent penalty (the house did not cool down uncomfortably as most houses would) and it was certainly a nice indoor/outdoor experience, but the heating season did begin before Christmas. Moving from diurnal

21.3 Inner wythe window seat.

heat storage (gathering the day's heat for release at night) to weekly heat storage allows for periods of cloudy weather. More mass allows for this. In the end the amount of thermal mass is usually an architectural decision. Certainly the exterior of the walls should all be *insulated* rammed earth, but beyond that there are diminishing thermal benefits to adding interior rammed earth walls (though there are acoustic benefits) and the architecture typically takes precedence. It is a value in design decision – net zero energy usage vs. additional square footage in a tight floorplan.

21.3.2 Humidity flywheel

In the same way that the stabilizing effects of mass on temperature are sometimes described as a thermal flywheel, the humidity flywheel describes the inertia created by having mass that absorbs and releases humidity contained inside the vapor barrier of a building. The hygroscopic ability of rammed earth to moderate the changes in relative humidity of the surrounding air can clearly be seen in the British Columbia Institute of Technology (BCIT) study done of three SIREWALL buildings. Temperature, humidity and dew point were monitored over a one-month period on Salt Spring Island near Vancouver, BC. One of the buildings (Fig. 21.6) was unoccupied for that period, the windows and doors were shut and the power was shut off. Figure

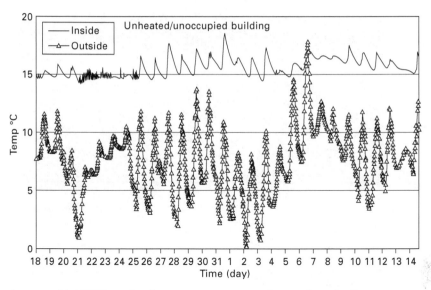

21.4 BCIT study of temperature across the north wall of an unoccupied SIREWALL home.

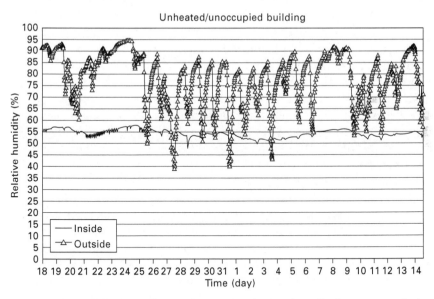

21.5 BCIT study of humidity across the north wall of an unoccupied SIREWALL home.

21.4 shows a graph of exterior and interior conditions of that house, with the rammed earth walls of the building doing the work. Notice that the exterior temperature is fluctuating widely and having only a minor effect on the temperature inside the building.

21.6 Unoccupied SIREWALL home.

Note also that the solar heat gain is around 9°C (16°F) of free stable heat. The widely vacillating exterior humidity is dramatically moderated by the stabilized *insulated* rammed earth to a very stable 50–57% RH. As discussed earlier, people are most comfortable when the RH is 40–60%. It is important and useful to know that no mold can grow when the RH is under 65%. Therefore the ability of rammed earth to moderate humidity can be an excellent tool against mold in construction. As can be seen in the graph, the RH never approaches 65% when no people are in the house and no Heat Recovery Ventilator (HRV) is operating. When people occupy buildings they contribute humidity through breathing (1 liter of water per person per day). Other contributions include pets, cooking, showering, laundry, etc. The HRV tends to counteract this effect by having a drying effect on the building's internal air quality. The volume of the space per occupant can make humidity control easy or difficult. The final factor is the rate with which the moisture passes through the rammed earth. If the mean exterior moisture content, when raised to the interior temperature, creates a desirable RH, then the perm rating of the envelope (and the insulation is the biggest variable) need not be too high. For example, fabric-coated polyisocyanurate foam, could be used without penalty. If there is an undesirable exterior mean moisture content, an aluminum-coated polyisocyanurate is the best choice to manage humidity by balancing the occupants' respiration and activity inputs with the HRV drying effect and the volume of contained space per occupant. In summary, *insulated* rammed earth is very effective at moderating interior RH, and the overall humidity regimen can be influenced by the selection of a climate-specific insulating material. Further information on this behavior is also provided in Chapter 3.

21.3.3 Heat in modern earth buildings

Heat moves by conduction, convection and radiation. Understanding heat flows in rammed earth buildings is essential in the design phase as a poorly designed building can lose or gain many times the heat of a well-designed one. There are many common misunderstandings about thermal detailing in rammed earth construction.

Conductive heat loss

Conduction is the ability of a material to transfer heat. A highly conductive material is copper (i.e. 401 W/m K) and the worst conductor is no material at all, i.e. a perfect vacuum. We evaluate the ability of materials to conduct heat and ascribe R-values or U-values to different materials. Table 21.1 lists R-values and U-values per inch for common materials used in earth building.

To determine the (Imperial) R-value of 4 in. of polyisocyanurate foam, for example, one simply multiplies the 7.0 per inch R-value by 4 in. to get R-28. For a 24 in. *insulated* rammed earth wall there are 20 in. of rammed earth (excluding the 4 in. of rigid foam) so one simply multiplies the 0.25 per inch to get R-5 for the rammed earth. To determine the steady state R-value of a 24 in. *insulated* rammed earth wall (that uses polyisocyanurate foam) we add the R-28 of the foam to the R-5 of the rammed earth to get R-33.

Table 21.1 R-values and U-values for common rammed earth building components

	R (h*sq ft*°F/BTU)	U (1/R)	RSI (m²K/W)	U (W/m²K)
Concrete	0.11	9.09	0.077	12.987
Steel/rebar	0.0032	312.5	0.0225	44.444
Rammed Earth (CSIRO)	0.25	4.0	0.1761	5.679
Roxul insulation	4.0	0.25	2.818	0.355
Fiberglass insulation	3.5	0.286	2.465	0.406
Expanded polystyrene	3.5	0.286	2.465	0.406
Extruded polystyrene	5.0	0.2	3.522	0.284
Polyisocyanurate foam	7.0	0.143	4.931	0.203
Trex recycled sawdust/plastic lumber	0.5*	2.0	0.3522	2.839
Framing lumber	1.0	1.0	0.7044	1.42

*approximate, actual value not found.
The Imperial values are per inch thickness
The metric values are per 100 mm thickness

However, we know there are significant benefits to R-values provided by thermal mass (see earlier section on thermal mass), which are quantified as dynamic R-values. From Jan Kosny's work at Oakridge National Laboratory we see that thermal mass increases the dynamic R-value by 1.5 × to 2.25 × dependent on location. We will use the most conservative multiplier of 1.5 × and apply it to the R-33 of the *insulated* rammed earth to get a dynamic R-value of R-50.

Every so often we encounter someone who adamantly believes that earth is warm in and of itself and proceeds to build a rammed earth house in a northerly climate (6000 DD) without insulation. Degree days (DD) are a measure of how much heat is required to achieve comfort in different climates. For example, Vancouver has 5660 DD and colder Toronto has 6756 DD. Looking at the heat loss characteristics of *uninsulated* rammed earth in a 6000 DD climate vs. *insulated* (with 4 in. of polyisocyanurate) we find for an 8 ft × 10 ft wall, using H = 24 h/day × DD × 1/R × area:

1 **12 in. of *uninsulated* RE;** H = 24 × 6000 DD × 0.3333 × 80 sq ft = 3,840,000 BTU/year or 1125 kWh/year for 12 in. *uninsulated*
2 **18 in. of *uninsulated* RE;** H = 24 × 6000 DD × 0.222 × 80 sq ft = 2,560,000 BTU/year or 750 kWh/year for 18 in. *uninsulated*
3 **24 in. of *uninsulated* RE;** H = 24 × 6000 DD × 0.1666 × 80 sq ft = 1,920,000 BTU/year or 562 kWh/year for 24 in. *uninsulated*
4 **18 in. of *insulated* RE;** H = 24 × 6000 DD × 0.0206 × 80 sq ft = 237,312 BTU/year or 70 kWh/year for 18 in. *insulated*
5. **24 in. of *insulated* RE;** H = 24 × 6000 DD × 0.0200 × 80 sq ft = 230,400 BTU/year or 68 kWh/year for 24 in. *insulated*

One can see in Fig. 21.7, the dramatic reduction in heat loss from *uninsulated* to *insulated* rammed earth. Inevitably and unfortunately, the result of *uninsulated* rammed earth building in a northerly climate is a very cold and difficult-to-heat home during the first winter, followed by a summer application of insulation and stucco to the exterior and subsequently a less-than-comfortable second winter due to difficulties in properly upgrading the thermal envelope post-construction (e.g. the detailing).

According to the National Building Code of Canada, the minimum insulation level in a wall assembly is to be no less than R 20. Most wood houses are *insulated* with fiber glass insulation fitted between 2 × 6 framing studs. The 2 × 6 studs measure 5.5 in. × 1.5 in.. Therefore the insulation is 5.5 in. thick, or 5.5 × 3.5 = R19.25, which when added to the drywall and sheathing totals at least R20. However, there is thermal bridging across the studs. If we look at the short length of 8 ft tall wall that has a 30 in. wide × 42 in. tall window shown in Fig. 21.8, we can begin to get an idea of the impacts of thermal bridging. For further details on thermal bridging theory refer to Chapter 3.

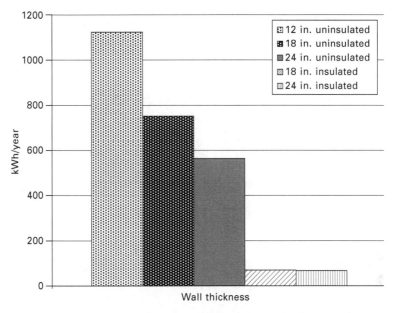

21.7 Insulated and uninsulated rammed earth conductive heat loss.

Figure 21.8 shows a 30 in. × 42 in. window with the thermal bridging areas of the surrounding woodframe structure in a (6000 DD climate);

A   **Through R-3 window;** H = 24 × 6000 DD × 0.3333 × 8.75 sq ft = 420,000 BTU/year or 123 kWh/year through the window

B   **Through the wood framing around the window;** H = 24 × 6000 DD × 0.143 × 7.125 sq ft = 146,718 BTU/year or 43 kWh/year through the framing

C   **Comparing SIREWALL detail 3 in. around window;** H = 24 × 6000 DD × 0.0333 × 3.0 sq ft = 14,400 BTU/year or 4 kWh/year through SIREWALL opening detail.

D   **Comparing RE detail with 3 in. solid RE around window;** H = 24 × 6000 DD × 1.0 × 3.0 sq ft = 432,000 BTU/year or 127 kWh/year through non-SIREWALL opening detail

E   **Through 30 in. length of concrete bond beam above window;** H = 24 × 6000 DD × 0.5555 × 2.5 sq ft = 176,000 BTU/year or 52 kWh/year though 12 in. tall by 18 in. deep concrete bond beam.

One can see from this example that more heat can be lost (or gained) through a poor window opening detail than through the window itself. Note also the heat loss through a very short length of bond beam. The bond beam typically goes around the circumference of the house or building and thereby creates a 12 in. thermal bridge that typically loses (or gains) more heat than

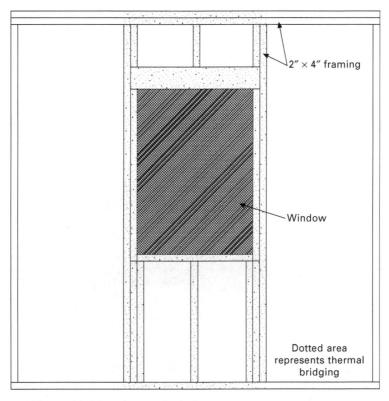

21.8 Thermal bridging around window.

the total of all the windows beneath it. Bond beams and poor window and door details can undermine an otherwise well-*insulated* building. A well-*insulated* cooler with a lid that doesn't close completely is a good analogy. *Continuity of a thermal envelope is more important than the R-value of the insulation.* What does that mean for a rammed earth building? Replacing a non-SIREWALL window detail with a SIREWALL detail cuts wall heat loss in half, or it allows for the doubling of the window area, with a similar overall conductive heat loss.

Another way to look at it is, if an 8 ft × 10 ft *insulated* wall was built with the SIREWALL detail of uninterrupted insulation edge to edge all around and that wall is compared with a identical size *insulated* wall with a 3 in. thermal bridge of solid rammed earth on two sides (perhaps for full height windows), the overall heat loss of the wall goes from 68 kWh/year to 314 kWh/year, see Figure 21.9. That seemingly innocuous detail increases the heat loss of the 8 ft × 10 ft *insulated* rammed earth wall more than four-fold (462%)! Once again, continuity of the thermal envelope is more important than the R-value of the insulation.

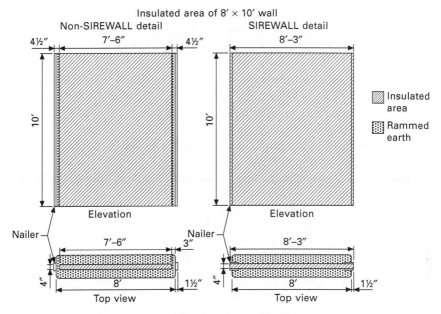

21.9 Rammed earth wall edge thermal bridging.

The previous two examples highlight some of the opportunities and impacts of decision-making around thermal envelopes. In the wood frame construction industry, conversations center on efficiency gains of 10–20%, whereas the above two examples demonstrate that 500–1000% improvements are available in *insulated* rammed earth. Every building offers multiple opportunities to make significant heat loss reductions. There is an enormous difference between *insulated* rammed earth done well and done poorly.

A review of the conductive heat loss characteristics of different wall assemblies in a 2100 sq ft bungalow allows us to assess the relative impacts of wall and window choices. For wood frame housing we use R-11.7 as the whole-wall R-value for a 2 in. × 6 in. wall with typical installation (ORNL). Assuming a 20% glazed area and 10 ft tall walls in a 30 ft × 70 ft rectangle (resulting in 400 sq ft of window and 1600 sq ft of wall) we find (still in our 6000 DD climate) that:

A **Wood frame + R-1 windows;** 5767 kWh/year + 16,869 kWh/year = 22,636 kWh/year

B **Wood frame + R-2 windows;** 5767 kWh/year + 8435 kWh/year = 14,202 kWh/year

C **Wood frame + R-5 windows;** 5767 kWh/year + 3374 kWh/year = 9141 kWh/year

D **18 in. of *uninsulated* rammed earth + R-1 windows;** 14,994 kWh/year + 16,869 kWh/year = 31,863 kWh/year

E   **18 in. of *uninsulated* rammed earth + R-2 windows;** 14,994 kWh/year + 8435 kWh/year = 23,429 kWh/year

F   **18 in. of *uninsulated* rammed earth + R-5 windows;** 14,994 kWh/year + 3374 kWh/year = 18,368 kWh/year

G   **18 in. of *insulated* rammed earth + R-2 windows;** 1400 kWh/year + 8435 kWh/year = 9835 kWh/year

J   **18 in. of *insulated* rammed earth + R-5 windows;** 1400 kWh/year +3374 kWh/year = 4774 kWh/year

K   **24 in. of *insulated* rammed earth + R-8 + shutters;** 1360 kWh/year + 937 kWh/year = 2297 kWh/year

L   **24 in. of 6 in. *insulated* rammed earth + R-8 + shutters;** 975 kWh/year + 937 kWh/year = 1,912 kWh/year

This is a very simple analysis and only considers conductive heat loss. It does not factor in solar heat gain, convective heat gains and losses, moisture content of the rammed earth or infrared heat gains and losses. Similar to most

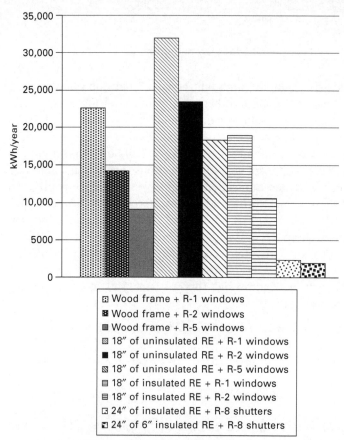

□ Wood frame + R-1 windows
■ Wood frame + R-2 windows
■ Wood frame + R-5 windows
▨ 18″ of uninsulated RE + R-1 windows
■ 18″ of uninsulated RE + R-2 windows
▨ 18″ of uninsulated RE + R-5 windows
☰ 18″ of insulated RE + R-1 windows
☰ 18″ of insulated RE + R-2 windows
▣ 24″ of insulated RE + R-8 shutters
▣ 24″ of 6″ insulated RE + R-8 shutters

21.10 Relative impacts of wall and window choices.

heat loss analysis, it only considers the heat loss that travels horizontally, where in fact, heat also leaves walls at their top and bottom.

Four points to consider when reviewing the above data:

1   Notice that in wall assemblies 'K' and 'L' (the best cases presented), there are 'shutters'. In this example we use R-20 window covering (e.g. Kalwall) at night and assume that it's in place for 12 h/day. That provides R-8 for half the day and R-28 for the other half for an average of R-18, which is the number used for this heat loss calculation. This is the kind of technology that needs to be considered to achieve net zero

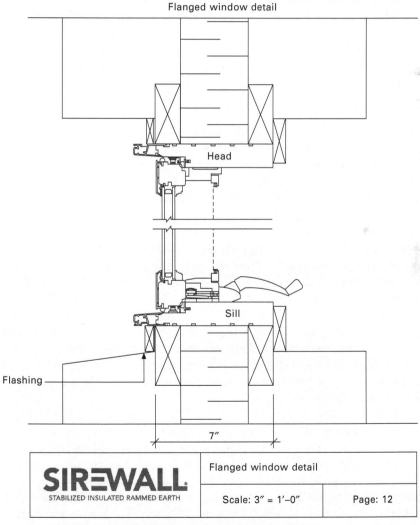

Flanged window detail

Head

Sill

Flashing

7″

SIREWALL
STABILIZED INSULATED RAMMED EARTH

Flanged window detail

Scale: 3″ = 1′–0″   Page: 12

21.11 Typical SIREWALL window and door sections.

(overall energy consumption does not exceed energy production on an annual basis)

2   There will be upgrades to windows in the future and they are fairly easily replaced whereas walls are not. If we apply seventh generation thinking then it is important to excel in those parts of the exterior envelope that cannot adjust easily over time, e.g. the wall structure

3   While energy costs in North America are typically 5 to 25 cents per kWh at present, rates will most probably rise significantly over the next generation or seven

4.   Consider internally generated heat. A person sitting down generates 60 W or, including all other higher energy activities, around 2400 W or 876 kWh/year. Two people and a dog might generate 2000 kWh/year. Add in the heat from pilot lights, electrical lights, hot water tank heat loss and phantom loads (standby power consumption) and option L above may have the heat loss through the walls fully offset by internal heat gain.

Convective heat loss

Little discussion remains as to whether or not houses should be built with an airtight envelope. There are differences of opinions as to how much makeup air is required and how to provide it, but there is broad agreement that houses ought to be airtight. This correlates with the idea that buildings are intended to create a different temperature, humidity and/or air pressure regimen than exists outdoor at a given moment. The ability to control the indoor environment largely depends on a building's airtightness.

The typical percentage of convective heat loss is 25–40% of the total heat loss in a house/building (Natural Resources Canada). The pull across the envelope will vary at different elevations inside the building due to the stack effect. SIREWALL detailing around doors and windows significantly reduces convective heat loss (see Fig. 21.11). The window flange is caulked to recycled composite boards that were previously rammed into place. Our blower door tests show that there is no measurable infiltration/exfiltration of air around the windows or doors with this detail. Once again attention to detail makes all the difference.

It is indeed easier to make an earth building tight compared to a building envelope comprised of many different materials (e.g. wood frame construction), but the areas of concern remain the same. Plumbing and electrical runs, electrical plugs, switches and lights, window and door detailing and sill plates are the primary culprits for loss of airtightness. In an *insulated* rammed earth house the plumbing runs are usually done in interior wood frame walls, but there is inevitably some plumbing requiring installation in the exterior walls. In these instances, piping can be placed in the interior wythe to maintain freeze protection. Whenever possible, the water lines are kept inside the plane of the foam (which is obviously not possible with hose bibs) and are run inside conduits. In other words, the conduit is located inside the thermal envelope (embedded rigid insulation) and then after the rammed earth walls are built, the cross-linked polyethylene (PEX) flexible water tubing is fed through that conduit to the interior surfaces where it is desired. This technique keeps water lines inside the thermal envelope and offers seventh generation flexibility in terms of water line replacements. The same is done for hose bibs and other penetrations that can weaken the thermal envelope, but in these cases the conduit will necessarily cross the thermal envelope and the conduit is spray foamed after the PEX is installed (this detail is not for extremely cold climates). The routing of the plumbing and electrical conduits needs to be planned carefully prior to construction of the rammed earth walls and clearly shown on drawings.

As the building becomes tighter the usefulness of an HRV becomes greater. Presently, there is a strong lobby in Canada to modify the Building Code such that all new houses have an (HRV) to ensure indoor air quality.

Radiant heat loss

Warm surfaces radiate heat to colder surfaces, mostly independent of the air through which the radiant infrared heat travels (for short distances). A good example of this is standing by a bonfire on a cold night. The side of one's body that is facing the fire experiences a net gain of infrared energy as the fire radiates far more heat to the cooler surface – your body – than vice versa. On the other side of your body, you are radiating your warm body's heat to cooler surfaces – land, trees and deep space – which in turn radiate less heat back to you. You experience the temperature as being dramatically different from one side of your body to the other, but the air temperature on both sides of your body is about the same. This shows the ability of infrared radiation to transport significant amounts of energy. Radiative energy from the sun travels in the same manner, but at shorter wavelengths.

Roughly one quarter of a building's heat loss has traditionally been attributed to warm interior surfaces radiating to cooler exterior surfaces through windows. The use of drapes and low emissivity (low 'e') coatings on window glazing can minimize these losses.

The ability to emit and absorb radiant energy tends to be correlated with the color of the material. Flat black is typically the best at both emitting and absorbing radiant energy and bright white has the opposite characteristics.

The Trombe wall, popular many decades ago, was a 12 in. thick panel of concrete the same size as the south-facing window it was placed directly inside of. It made use of differing absorption and emission coefficients combined with the heat storage capacity in the Trombe wall material. The black face of the Trombe wall that faced the sun absorbed the day's heat and at day's end a white-painted surface, shutter or drape minimized re-radiation back out through the glazing. The captured day's heat slowly moved to the inside surface of the Trombe wall where it radiated to the cooler surfaces in the room. In a nutshell, the Trombe wall captured infrared energy, contained it and then re-radiated it to the living space at a time more useful to the occupant.

There were three main drawbacks to the system.

1   An interior concrete wall was placed in the living room, entirely blocking the view and natural lighting from the window
2   The mechanisms used to contain the heat (drapes, shutters) were difficult to access and maintain
3   The barriers to heat loss were minimal – typically R-2 glazing and drapes. The warmest surface in the system was exposed to the poorest insulating area of the building envelope

While Trombe walls are rare these days, the principle of taking the day's heat and storing it in a thermal flywheel (significant mass) is effective and continues in different forms. Dark surfaces can be used to absorb the sun's energy. Large water containers in the living space can store solar energy. Concrete slabs can be thickened and darkened. But with *insulated* rammed earth and low 'e' coatings on the glazing, a new interior environment has become possible that has all the advantages of the Trombe wall and none of the disadvantages. We discovered this quite by chance. In a newly completed SIREWALL project we were exploring the thermal performance in the largest room which had a 22 ft tall vaulted ceiling and rammed earth on all walls. We were using a remote thermometer to measure the temperature of the wall surfaces at different elevations. We expected significant temperature stratification as is normal in any lightweight structure, but instead we found that the surface temperatures of all of the walls, ceilings and concrete floors were within 1°C! This is characteristic of our buildings. The only exception is when the floor radiant heat is on, in which case the floor is the warmest surface.

When people are heated by thermal radiation from the building interior surfaces instead of by convection (air temperature), less energy is required to create comfort. Not only is there minimal air temperature stratification and resultant heat loss exacerbation, but comfort is achieved at a lower ambient air temperature. In effect we have recreated the Trombe wall, by capturing infrared energy, containing it and then re-radiating it to the living space at a time more useful to the occupant. The dynamic is less concentrated as heat storage has greater surface area and occurs inside the thermal envelope around the whole exterior of the building. The three Trombe wall problems are solved. The window now allows light and view, there are no drapes or shutters to control manually and the mass is placed directly beside the best-*insulated* parts of the thermal envelope. There are two additional benefits. Unlike the non-structural Trombe wall this new thermal mass is load-bearing and suitable to support the roof. There is no added mass that does nothing but store heat and take space. Due to much larger mass than the Trombe wall, the *insulated* rammed earth walls needn't be black or any color in particular. They collaborate very nicely with the radiant heat in the slab and the low 'e' coating on the windows. As a result most of the infrared energy continues to be exchanged between inside surfaces, rather than exiting through the windows more rapidly as with traditional construction.

21.3.4 Holistic heat loss prevention

The three means of heat loss are interrelated. If the radiant heat is escaping through the windows then the room cools unevenly, drafts occur and there is a call for heat. The forced air furnace or baseboard heaters jump into action and heat the air, which rises creating stratification. The ΔT at the ceiling might be twice that at the floor and the rate of conductive heat loss therefore doubles at the ceiling. The stack effect exhausts the hottest air at the high elevations while taking in cooler air at lower elevations causing the occupants discomfort. There is another call for heat.

Thermal stability is best achieved through the flywheel effect of abundant *insulated* mass with substantial heat storage capacity. However, if that mass is radiating its heat to the outdoors through windows that are not low 'e', or if there is a constant bleed of heat on the airstream to the outdoors due to poor airtight envelope detailing, or if the construction detailing around windows and doors creates large thermal bridges, then the benefits of the *insulated* rammed earth are lost or greatly diminished. Heat retention in a building is only as strong as the weakest link(s).

21.4   Applications of earth construction in hot climates

Most of this chapter has been concerned with heat loss and an understanding of how heat moves, such that we can control or interrupt that movement. This understanding has its roots in colder climates. It is ironic then that the largest global opportunity to reduce energy consumption in buildings lies in the hot climates. If we liken the buildings in hot climates to refrigerators, then typically the refrigerator door is left ajar, it is poorly *insulated* (if at all) and the compressor (in the air conditioning) is working overtime to combat these structural shortcomings. There is commonly a generous selection of weakest links causing poor thermal performance. The technique currently being used to create human comfort is air conditioning. The cost of running the air conditioning accrues for the life of the building and air conditioning requires maintenance and replacement. And yet, poor building design is the root cause of the need for air conditioning in the first place. As peak oil and climate warming require large reductions in fossil fuel use, it is imperative that we shift from the first cost, disposable paradigm to long-term sustainable building. The walls of most buildings, when implemented using a holistic thermal approach, can partially or sometimes fully replace the need for air conditioning.

21.4.1 Conductive heat gain

In most circumstances the same principles used to minimize heat loss apply in reverse to hot climates where the concern is heat gain. Minor adjustments might be to lighten the color of the exterior wall surfaces and provide biological or other external shading devices particularly on the western orientation. There may be an opportunity to enhance the conductive connection to the ground and lose heat that way. Contained geothermal coupling could make use of the frost protected shallow foundation (FPSF) system used in cold climates. This technology isolates the ground under and around the building from the outside cold or heat. Once this is done, then the thermal stability or flywheel effect of the temperature of the earth under the building can be tapped. What's important with this system is the thermal continuity either of the slabs to the ground or the inner wythe of the rammed earth walls to the ground.

21.4.2 Convective heat gain

Control the indoor environment by making a rigorous distinction between indoor air and outdoor air via airtight detailing. Bring the HRV supply air into the building after pre-cooling it by running it under the ground for some distance. Carefully consider the locations of the supply and exhaust HRV vents to optimize efficiency and health without disturbing human comfort.

21.4.3 Radiant heat gain

The solar orientation of the glazing is an obvious method to control radiant heat gain. Also well known are the external shading devices such as trellises, pergolas, extended roof overhangs, deciduous trees, etc. Internally, abundant mass, taller ceilings, the reduction of inefficient lighting fixtures and the reduction of phantom loads (the electric power consumed by electronic appliances while they are switched off or in a standby mode) will reduce the cooling requirements. Collecting and venting the heat generated by heat-producing devices such as control motors, pumps and audio/visual components will reduce the cooling load. Deep space radiators are a promising technology at the cutting edge of passive radiant heat loss. With this system (and there are a number of variations), flat black surfaces emit the heat of the building to deep space. Water cooled in this way at night can be circulated in the day either in the inner wythe of *insulated* walls or in ceilings.

21.5 Applications of earth construction in wet and cold climates

The vast majority of traditional earth building has been done in warm, dry climates. The few exceptions to this principle demonstrate that it takes special care and repeated, ongoing maintenance for the rammed earth to endure. For example, the Shanty Bay church in Ontario (built 1842) has its rammed earth covered with plaster or siding to resist the forces of a very cold environment.

In order to be able to construct modern rammed earth in wet and cold climates within the constraints of Building Codes of Canada or the United States, there are specific details that are typically scrutinized by the approving authorities. First, is the insulating requirement, which is strictly based on conductive heat loss and has no allowances for convective heat loss, radiant heat loss, or the benefits of thermal mass. That objection is easily surpassed. The perm rating issue is also easily resolved with the appropriate rigid insulation product selection.

The issue of water and rammed earth continues to come up despite David Suzuki pressure-washing a 10-year-old SIREWALL with a 2750 psi (19 MPa) pressure washer on film (CBC's 'Nature of Things'). It is true that not all rammed earth can withstand that extreme level of testing. Water can come into rammed earth horizontally via rain (or sprinklers) or vertically via rising damp. Tongue and groove detailing of wall panels such that there is no uninterrupted path of rammed earth from the outside to the inside has never failed to prevent passage of water. Insurance to prevent water coming through the rammed earth is to have a hydrophobic admixture mixed in just prior to ramming such that the rammed earth is hydrophobic throughout. A surface coating is used by some. The rising damp is prevented by the hydrophobic admixture. A belt and suspenders approach is to prevent rising damp by applying a crystalline slurry to the top of the footing prior to ramming the first lift. With regards to concerns about water entry through vertical seams between panels, the SIREWALL system has the panels tongue and grooved such that any water that tries to squeeze into the vertical seam (which has been rammed together) comes up against a recycled plastic lumber face (which has complete water repellence) that has also been rammed into place. Horizontal driving rain in the West Coast of Canada has not found its way past this configuration.

The structural integrity of stabilized *insulated* rammed earth may also be a concern to the building official and the Equivalency Provision of the Code should be exercised. The Equivalency Provision, which is near the beginning of most Code books, states that if the proposed material or method meets or exceeds the requirements in the Code, that material or method shall be deemed to comply. All *insulated* rammed earth buildings that I know of in North America have been structurally engineered, and this engineering is always submitted to the Building Official for approval as per Equivalency.

The first place that engineers typically look to, when engineering modern rammed earth for cold and wet climates, is to determine the unconfined compressive strength of the rammed earth. In the early 1990s, when we started out, we did single-storey, thick-wall structures and these worked with 28-day compressive strengths of 6 MPa or 870 psi. As the engineers continually wanted more strength, there seemed no way to know how much was too much or not enough. The Portland Cement Association has done research (see Fig. 21.12) showing the relationship between strength and durability in soil cement samples and, as soil cement is a close cousin of modern rammed earth, the results of that research have become very useful.

Note that on the graph, all results are based on 7-day compressive strength. If the 7-day strength is 600 psi (4.1 MPa), then 87% of samples passed the ASTM freeze/thaw and wet/dry tests. If the 7-day strength is 750 psi (5.2 MPa), then 97% of samples passed, and at 850 psi (6 MPa), 100% of the samples passed. Now, as the result of much R&D involving diamond-drilled cores and thousands of crush tests, the average compressive strength of

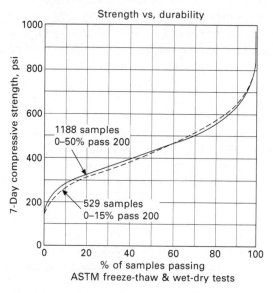

21.12 Strength vs. durability from Portland Cement Association.

SIREWALL is at least double that (at 28 days). On a couple of occasions we've had results over 40 MPa (over 6000 psi) with 9.5% cement. More typically our crush results come back between 10 MPa and 20 MPa. How is that achieved?

21.6   Optimizing rammed earth compressive strength

Optimizing compressive strength (see Fig. 21.13) in modern stabilized rammed earth is a function of optimizing the soil blending, selecting admixtures and

#6 - 854 Marion Street
Winnipeg, Manitoba
R2J 0K4
eng_tech@mts.net
www.eng-tech.ca

CONCRETE TEST CYLINDER

CERTIFIED CONCRETE TESTING LABORATORY
IN ACCORDANCE WITH STD A 283

Terra Firma Builders Ltd.
212 Cusheon LK. Road
Salt Spring Island, British Columbia
V8K 2B9

File No.:   09-166-54

Attention: Jerry Fitzpatrick

Project: GRAND BEACH WASHROOM BUILDING

No. of Cylinders: 9   Date Cast: July 21/09

Reference Number	Field or Sample Number	Date Received	Date Tested	Age at Test (Days)	Compressive Strength (MPa)	Type of Fracture
166-54-1A		July 29/09	July 31/09	10	19.8	
166-54-1B		July 29/09	July 31/09	10	21.7	
166-54-1C		July 29/09	July 31/09	10	18.4	
166-54-1D		July 29/09	Aug 18/09	28	24.6	
166-54-1E		July 29/09	Aug 18/09	28	26.8	
166-54-1F		July 29/09	Aug 18/09	28	22.0	
166-54-1G		July 29/09	Oct 29/09	100	23.7	
166-54-1H		July 29/09	Oct 29/09	100	24.0	
166-54-1I		July 29/09	Oct 29/09	100	26.5	

Specification			Plastic Test Results			
Strength:	-	MPa	Mold Type:	Plastic 150mm x 300mm		☒ A
Slump:	-	mm	Slump:	-	mm	☒ B
Air:	-	%	Air:	-	%	☐ C
Cement Type:	-		Plastic Density:	-	kg/m³	☐ D
Aggregate Size:	-	mm	Hardened Density:	-	kg/m³	☐ E
Admixture:	-		Cast By:	Client		☐ F
Supplier:	-		Cast Time:	-		
Ticket Number:	-					
Mix Number:	-		Temperature			*** TYPE OF FRACTURE REQUIRED WHEN CYLINDER FAILS TO MEET SPECIFIED 28 DAY STRENGTH
Truck Number:	-		Concrete:	-	°C	
Batch Time:	-		Air:	-	°C	
Load Volume:	-	m³	Initial Curing, Maximum:	-	°C	
Cumulative Volume:	-	m³	Initial Curing, Minimum:	-	°C	

Location: Cc: Email - jerry@sirewall.com

Comments:
Ramned earth; consisting of a blend of 10% cement and soil compacted with a pneumatic tamper.

ENG-TECH Consulting Limited

per

Danny Holfeld, Principal
Ph: (204) 233-1694 Fx: (204) 235-1579

Cylinder initial cured on site.

21.13 Unconfined compressive test results.

pozzolans that augment each other and selecting the best tamper for each specific project. Also vital is the moisture content, the soil delivery method, control of lift heights and how the tamping is done.

21.6.1 Tampers and tamping

The type of tampers and approach to tamping have a notable effect on the visual sedimentary presentation. Depending on how the tamping force is applied to the earth inside the formwork, the consolidation becomes less as the distance from the tamper butt increases. The visual face at the top of the lift displays a very dense matrix with a fine texture and is therefore usually lighter in color, whereas the bottom of the lift is somewhat less dense, more textured and therefore slightly darker in color.

Is this a problem? Is there any worry about the strength of the rammed earth at the bottom of the lifts due to less compaction and slightly lower density? The short answer is no. Part of the answer lies in the selection of tampers and the lift height. The other part of the answer lies in the fact that almost all tamper butts are round in shape. Imagine a process of tamping the visual face of a wall and running the round butt along the forms (Fig. 21.14). Assuming that the 6 in. tamper impacts the earth every 2 in. over a 6 in. distance, then at the top of the visual face there will be some soil that has not been impacted. However, 3 in. back from the surface, directly under the shaft of the tamper, the earth has been impacted four times. That is a significant difference. Figure 21.14 shows the number of impacts (up to four) on the surface of the soil being rammed. The figure is intended to illustrate the principle, not define how frequent the tamping should be.

The least compacted earth is on the visual face! This is counterintuitive. If we imagine being out in nature and taking two slices through a sedimentary deposit 3 in. apart, we would expect the characteristics to be near identical on each slice. That's because there are no edges to the layup. We would expect

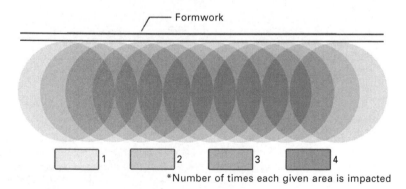

*Number of times each given area is impacted

21.14 The impact of round tamper butts.

the same to be true with rammed earth (that it is consistent throughout), but our test core drilling has revealed the result of ramming against an edge. If there is rilling and a bony appearance at the bottom of a lift, we know that it is a surface condition. We know that to see a slightly coarser texture at the bottom of a lift is a sign of strength, that the moisture content has been well controlled and the ramming well done. We know that an overly wet mix will not show texture, even at the bottom of a lift and that the strength of the material has been compromised by a poor water:cement ratio. Determining the degree of compromise takes an experienced eye and some non-textured rammed earth is very strong.

Most large countries around the world have at least one pneumatic tamper made for the forging or backfilling industries. Some manufacturers offer many models. The size and weight of those tampers varies considerably as do the size and shapes of the butts as can be seen in Fig. 21.15.

The butt at the bottom of the tamper delivers the impact to the earth. Though there are many options to choose from, two fairly common options can be compared to understand the theory behind tamper selection. If a lightweight (15#) tamper with a 3 in. butt is compared to a heavy (45#) tamper with a 6 in. butt, one would at first think that the earth would be better compacted by using the heavy tamper with the 6 in. butt. It is true

*Ingersoll Rand products

21.15 Some tamper butt options.

that the 6 in. tamper will cover a lot more 'ground' in less time than the smaller tamper, but a close look reveals more insight.

The 3 in. butt has an impact surface area of 7 sq in., whereas the 6 in. tamper has an impact area of 28 sq in. That means that, if the blows per minute and the stroke length are the same for both tampers, then the large tamper has $3 \times$ the force being applied to the earth (force = mass $\times$ acceleration) due to its extra weight, but the smaller tamper applies its load in a (4 $\times$) more concentrated manner (higher stress (stress = force/area), N/mm^2 or MPa).

Overall, the smaller tamper creates a denser matrix and has deeper penetration into the loose earth that is delivered into the forms. Empirically and from compressive testing of diamond-drilled cores in test walls, we know that the compressive strength created by smaller tampers is significantly higher than large tampers. It is also important to note that hand tampers are not equivalent when compared with pneumatic tools.

If the mass of the tampers are equal and the length of the stroke and thrust applied (in other words, the force) is equal, then the only significant variable is the impact surface area. The compressive strength of hand rammed earth, created by using a 20 sq in. impact face, is half the strength of that created using a large 45 lb tamper with a 6 in. butt.

Given the many different tampers that we've used on many different types of soils, we only have rules of thumb to offer on the most popular tampers. We believe that hand tamping creates half the compressive strength of 6 in. tampers and 3 in. tampers are 20–25% above the 6 in. tamper results. What follows from this is that the depth of lifts for large tampers should be less tall than those of small tampers and less still for hand tamping.

21.6.2 Formwork

In the close relationship of tamping and formwork, the large tamper exerts a higher force overall on the soil and on the formwork, whereas the smaller tamper exerts a higher stress on the specific areas of impact and less on the formwork. The formwork strength required to withstand the force of a 6 in. tamper is higher than the force required to withstand a 3 in. tamper. It is important to understand that most widely available formwork has been created to withstand the forces exerted by wet concrete. Wet concrete exerts hydrostatic pressure, which is to say that if 10 ft of concrete is poured at once, then the force at the bottom of the forms will be the sum of all the mass of the concrete above. With concrete at 150 lb/cu ft (2403 kg/m^3), a 10 ft tall wall of wet concrete will exert 1500 lb (680 kg) of force per square foot on the bottom of the forms. In sharp contrast, when tamping is taking place at a 10 ft height the force at the bottom of the rammed earth wall is zero. In fact, there are rammed earth forming systems where there is no formwork in place below when the tamping has moved up to the top of the wall.

There is a difference between the forces exerted on formwork by small and large tampers, but both are typically an order of magnitude lower than that required for concrete. Every different tamper will exert a different force on the formwork and those forces can be plus or minus 50%, so it is important and useful to synchronize the formwork with the tampers used. To use concrete formwork for rammed earth has the big advantage of availability, but is usually 10–20 × stronger than required (Fig. 21.16) and most of it comes with throughties, making it almost impossible to insert the insulation as the wall is rammed. Throughties also require hole patching, which can be unsightly. Unneeded strength carries unneeded weight (and cost) and thereby more time and work is needed to set up and take down the formwork on site.

We have looked at the dynamic forces on strongback and whaler forming by carefully measuring the deflections in existing rammed earth walls. Every deflection is an expression of the tamper selected for that wall, so deflection data are always correlated with the tamper used. Based on measurements that we have done we believe that the deflection in the plywood gives an indication of the local pressures and is never more than 150 kPa. The deflection in the whalers gives an indication of the global pressures. We believe that small tampers create higher local pressures and large tampers create higher global pressures. The deflection in the strong backs is dynamic in nature.

21.16 Concrete formwork used for curving rammed earth walls.

At an elevation of 1 ft, the pressure is, say 100 kPa. When the ramming is being done at an elevation of 5 ft, which might be hours later than the initial first foot of the wall, the pressure is still 100 kPa and the rammed earth at 1 ft elevation is already curing and shrinking away from the formwork, therefore, it is applying no pressure to the formwork. We know that if there is a wall that cannot be done in a day and there will be a horizontal cold joint seam as a result; the wall will have shrunk overnight by 1/16 in. to 1/8 in. on each side. This makes the following day's work appear to be a very slightly thicker wall above the cold joint. That 'overhang' cold joint can be minimized or exaggerated for visual interest. The relevance of these observations is to understand the dynamic nature of the form 'pressure' as the walls are being rammed. A similar process takes place if one places a plank across a stream and walks across it. With each step the plank bends at the place where the foot is placed and then it bounces back when the next foot is placed. Although the rammed earth formwork operating in the vertical plane is not quite so dynamic or quick in action (the spring back is not as quick), the principle remains. Understanding the close relationship between tampers and formwork appears to offer significant opportunities for optimizing efficiency, ensuring predictable quality and reducing costs.

Increasing the form panel module size also offers significant opportunities. Smaller projects with fewer repetitions are built with incremental forms. Larger projects with repetitive design elements allow for the efficiencies of modular form panels (gangforming). We have used handset form panels as small as 30 sq ft and crane-erected form panels over 300 sq ft. Removing a large form panel from a recently finished wall and immediately placing it such that it is ready for the next wall is highly efficient.

21.6.3 Delivery

The method of delivery of the earth into the forms will affect the appearance, the strength and the cost. Techniques like loading the forms with earth by dumping it in with a skidsteer, telehandler or conveyor will usually result in a certain amount of rilling and lack of control over surface texture. Rilling is where the larger aggregate will roll down the angle of repose to the visual surface and create a visual effect referred to as 'boney' texture.

The rilling only occurs on the surface and is not consistent throughout the wall. Those familiar with the concrete industry would mistakenly assume that the boney appearance runs right through the wall, as it would with honeycombed concrete. Having the surface boney does appeal to some as it creates a natural-looking effect. To avoid rilling, the soil is delivered into the formwork such that if there is rilling, it happens in the middle of the wall, not on a visual surface.

21.6.4 Moisture content

The moisture content of the soil has a large impact on the final compressive strength of the wall. We know from the concrete industry that the water/cement ratio has a big influence on the compressive strength. The more water, the less strength there is. Rammed earth usually takes full advantage of this relationship, as it is much dryer than even a zero slump concrete. What about moisture starvation? If the water/cement ratio is below 0.42 (optimum) then you will have unhydrated cement and less strength. This seldom happens. However, it is possible (and too easy) to make the rammed earth mix overwet, using the excess water as a lubricant to ease the ramming process. This technique is often used when people are hand ramming or pneumatically ramming tall lifts. Labor costs are cut at the expense of the strength of the rammed earth wall.

21.6.5 Earth blending

Insulated rammed earth requires higher strengths than its *uninsulated* cousin. As a result, *insulated* rammed earth builders largely ignore the color of the soils being considered and instead focus almost entirely on the soil's strength potential. Distinct from concrete (manufactured aggregate), earth blending uses full spectrum particle sizes sourced (when possible) either directly from nature or industrial waste, e.g. pit run, limestone tailings, recycled concrete. This allows for more strength with less cement and has a much smaller environmental footprint. Using the best of what's available requires soil analysis for every new location. Only once in 20 years have we been able to use the earth as it is available without blending. Compaction factors can vary from 0.54 to 0.97, and heavy concentrations in a specific particle size are normal. Organic materials in the soil have a deleterious effect on the strength of the wall (organic materials blow walls apart in slow motion), so they are rigorously avoided. In the early days the Fineness Modulus (FM) was the favored tool for quick evaluation of soils and soil blends. The FM is a number used to describe the mean particle size of a soil sample, e.g. an FM of 2 describes a sandy soil and an FM of 7 is a blend too coarse for rammed earth. It was developed by the US Army Corps of Engineers. As time went by we became aware of how significant it was that the silt and clay content was not factored into the FM calculation and that two soils with the same FM but different clay and silt contents can and do behave quite differently. As a result we created a rammed earth FM and a number of other metrics to quantify all the variables in earth blending, using a computer program. Now we no longer physically test different blends looking for the right soil combination with the best proportions to optimize strength. Data from all the candidate soils are sent to the SIREWALL head office and the best blend determined theoretically. Only that blend is rammed into cylinders and sent to the lab for testing. The goal with our earth blending technology is to move beyond determining the best blend, to being predictive. We see a time, when we have gathered enough data, that we receive data on all the candidate soils from anywhere in the world and are not only able to determine the optimal blend (we already do that), but also able to provide what strength one might expect to obtain from that optimal blend. To this point our success has been beyond our expectations. Only 5 years ago we were pleased with compressive strength results over 6 MPa. We now expect 10 MPa and are not surprised to see results of 20–30 MPa. Our record with 10% cement is an average of 46.0 MPa (see Fig. 21.17).

The best blend is typically weighted mostly toward optimal strength, but more and more now we are also blending for a desired finish texture, e.g. a 600 micron finish. Color is seldom a blending consideration as almost any desired color can be created by varying the iron oxides, the cement color and the pozzolans.

21.7   North American-style rammed earth

There is no one distinctive style of rammed earth in North America, as there is in Australia. Australian rammed earth is immediately recognizable due to the 2 ft × 8 ft form panels and the riveted appearance. Rammed earth builders in North America respond to a wide variety of local conditions and local economies. When there is a structural or durability requirement then stabilization is used. Insulation is used for both hot and cold climates. Unstabilized rammed earth is still prevalent in the southern USA or where walls are used as infill or for landscaping. North American rammed earth builders are rugged individualists who like to innovate, ensuring that the definition of a continental style is hard to determine.

The climates, in terms of geographical area in North America, for which stabilized *insulated* rammed earth is suitable, are many times greater than the areas of those climates for which *uninsulated* rammed earth is suitable. All of Canada and the northern half of the USA are areas where insulation in rammed earth is essential. In the southern half of the USA, there are as well, climates that are too cold, too hot or too damp for *uninsulated* stabilized rammed earth.

Where insulation in rammed earth is necessary and where it's not, have yet to be determined and that's likely to alter with climate change.

With cold-climate *insulated* rammed earth building, the necessary emphasis on strength for durability ensures a strong sedimentary appearance and controlled lift heights. In climates where insulation is not necessary in the rammed earth and where there is little seismicity, high strength is not as important as there is no freeze/thaw concern. In that situation it is possible

METRO TESTING LABORATORIES LTD.
6991 Curragh Avenue, Burnaby B.C., V5J 4V6
Tel: (604) 436-9111 Fax: (604) 436-9050

**CONCRETE
TEST REPORT**

PROJECT NO. 14143

CLIENT TERRA FIRMA BUILDERS

TO

Site TERRA FIRMA BUILDERS
212 CUSHEON LAKE RD
SALT SPRING ISLAND, BC
V8K 2B9

C.C. LEDCOR CONSTRUCTION LIMITED

ATTN: MEROR KRAYENHOFF

PROJECT VANDUSEN BOTANICAL GARDEN   OAK ST, 5251
CST   BURNABY

SET NO. 1   NO. OF SPECIMENS 9   DATE RECEIVED 2010.Aug.16   DATE CAST 2010.Aug.09

SPCM NO.	SPECIMEN TYPE	CURE CONDN	DATE TESTED	AGE AT TEST (DAYS)	AVERAGE DIAMETER (mm) OR SIDE (mm x mm)	AVERAGE LENGTH OR SPAN (mm)	MAXIMUM LOAD (kN)	COMPRESSIVE OR FLEXURAL STRENGTH (MPa)	Average	FAILURE TYPE
A	Core	Lab	Aug.19	10	150.0	285.0	632.0	35.5		
B	Core	Lab	Aug.19	10	150.0	270.0	628.0	35.0		
C	Core	Lab	Aug.19	10	150.0	290.0	756.0	42.6	37.7	
D	Core	Lab	Sep.06	28	150.0	280.0	871.0	48.8		
E	Core	Lab	Sep.06	28	150.0	300.0	852.0	48.2		
F	Core	Lab	Sep.06	28	150.0	262.0	698.0	38.7	45.2	
G	Core	Lab	Nov.17	100	150.0	283.0	649.0	36.4		
H	Core	Lab	Nov.17	100	150.0	288.0	859.0	48.3		
I	Core	Lab	Nov.17	100	150.0	262.0	963.0	53.4	46.0	

SPECIFIED STRENGTH	MPa @ DAYS	CONCRETE TEMPERATURE °C	TREND GRAPH
		AIR TEMPERATURE °C	
CEMENT CONTENT	kg/m³	SLUMP mm SPEC. ±	
CEMENT TYPE		SLUMP FLOW mm SPEC. ±	
POZZOLAN CONTENT	kg/m³	FLOW TIME sec SPEC.	
POZZOLAN TYPE		AIR % SPEC. ±	
MAXIMUM SIZE AGGREGATE	mm	PLASTIC DENSITY kg/m	
		HARDENED DENSITY kg/m³	
BATCH TIME		CAST TIME	
ADMIXTURES		CAST BY CLI	MOULD TYPE PIPE
		CURING CONDITIONS	
		INITIAL CURING TEMP:MAXIMUM °C MINIMUM °C	
		LOCATION	
SUPPLIER			
MIX NO.			
		COMMENTS	
TRUCK NO.	TICKET NO.	6 (6X12) CORES DELIVERED BY CLIENT.	
LOAD VOL.	m³ CUM. VOL. m³		
WATER ADDED	l AUTH. BY		
Page 1 of 1	2010.Nov.19	METRO TESTING LABORATORIES LTD. PER.	

Reporting of these test results constitutes a testing service only. Engineering interpretation or evaluation of test results is provided only on written request.

21.17 Record strength with 10% cement.

to be not so rigorous on soil selection, tamper selection and lift heights. The polar opposite of cold-climate rammed earth is where lift height is measured by dumping earth into forms with a skidsteer and ramming with large tampers to create thick walls that have no rebar and no foam. This approach boosts production (and economy) significantly. If the appearance of the walls is unsatisfactory, then they are plastered for a desired finish or color. It's all about getting as much earth mass rammed as quickly as possible. Another polar opposite of cold-climate rammed earth is the unstabilized or barely stabilized rammed earth that is still being built (often without permit). These projects have a very small carbon footprint but are vulnerable to erosion.

North Americans have played extensively with the visual presentation of rammed earth. The most common variables are the shape and color of the rammed earth, while less frequent are the forays into creative mix design, thin colored feature lines, creative volume displacement boxes (including fiber optics for example), carving of the formwork before ramming, carving of the rammed earth just after it's rammed, sanding the rammed earth, sandblasting, pressure washing, embeds, inserts, fabric forming, patchwork forming and enhanced overhangs. Forming techniques vary depending on whether insulation is necessary.

The stabilized *uninsulated* rammed earth builders typically use concrete formwork with through-wall tensile connectors or the California system using pipe clamps, which was popularized by David Easton, whereas the stabilized *insulated* rammed earth builders use a fully external formwork system that has no through-wall connectors as in Fig. 21.18.

Although there are many possible ways to form *insulated* rammed earth, all the stabilized *insulated* rammed earth builders in North America use the SIREWALL System or a derivative thereof. There are many ways to build fully external formworks and that topic is a book in itself. If you get a group of *insulated* rammed earth builders together, you will find many strongly held opinions about such things as strongback and waler formwork, torsion box

21.18 Common SIREWALL Cottage Forming System.

formwork, gang forming, canyon forming, form shoes, strongback options, modular options vs. full length and the plethora of materials to form with such as steel, aluminum, stainless, high density polyethylene, high density overlay, medium density overlay and fabrics or membranes.

Through-wall connectors make it extremely difficult to place rigid insulation in the center of the wall as the wall is being erected unless the formwork is being placed in very small modules, e.g. 24 in. tall. SIREWALL has developed the ability to avoid this issue altogether by devising a formwork that is fully external, with no horizontal tensile through ties. A side benefit of no through-wall connectors is the elimination of unsightly patching.

Engineering stabilized *uninsulated* rammed earth buildings typically rely on a wall-width concrete bond beam at the top of the rammed earth wall. Bond beams are terrible thermal leaks and should never be used in stabilized *insulated* rammed earth. Engineers of *insulated* rammed earth have broadly adopted the vertical cantilever for one- and two-storey walls. With this system, each individual wall is vertically cantilevered off the footing to address overturning moments. The fact that the individual walls are usually connected in the finished building is an uncalculated bonus/safety margin. This system gives the designer a lot of freedom.

For both *insulated* and *uninsulated* rammed earth, the use of wood or concrete floor and ceiling diaphragms can make the next storey of rammed earth very stable. The diaphragm configuration should be considered in an integrated design phase. A common passive solar design element in North America is to incorporate a large, uninterrupted glass expanse on the south side of the building. While that may be good for passive solar, it often creates a 'soft' wall that has little lateral or shear strength. A surprise solution to this problem, which avoids the use of expensive engineering techniques (such as steel moment frames) is the unique ability of rammed earth to extend as a buttress beyond the roofline, fully exposed to the elements. This capacity to have the rammed earth walls connected to and part of the structure, but not necessarily under the roof, creates a whole new level of design freedom along with the ability to integrate the structure into the landscape (Figure 21.19).

For green builders in North America, using either *insulated* or *uninsulated* rammed earth is an ideal way to meet the certification requirements for LEED, Living Building Challenge (LBC) and Cradle to Cradle (C2C) building perspectives. Similarly, net zero buildings, off grid buildings and Passive House designs are very compatible with rammed earth construction.

21.7.1 Insulated rammed earth business models

Most of the building of *insulated* rammed earth walls in North America takes place in the warm seven months of the year (except in the Pacific northwest).

21.19 Rammed earth buttress/retaining wall.

Building under a heated tent (heating and hoarding) in cold weather adds significant expense (when done on a one-off basis). It's a short building season for those locations that have sub-zero temperatures for 5 months per year and ironically it's those locations that stand to benefit the most from the excellent thermal performance of *insulated* rammed earth. To meet the demand in the short seasonal time frame that's available it is necessary to encourage investment in such things as custom volumetric mixers, custom conveyors, custom hoppers and custom formwork, all in the interest of getting as much done as possible in the short building season that's available. Only in the Pacific northwest is it possible to build year-round. It is therefore no surprise that the majority of *insulated* rammed earth builders are located there and that the scale of investment in equipment there is less imperative.

With a few exceptions, the demand for *insulated* rammed earth is very spread out geographically. Moving from one site to the next can and has crossed 4.5 time zones. Travel and accommodation costs can be a significant cost element for an entire crew. Transporting forms and equipment over such long distances has been problematic. The most common solution is to build *insulated* rammed earth forms on the patent pending SIREWALL System (with or without permission) using locally available wooden materials. Rather than transport an entire crew, highly trained foremen use local unskilled labor to get the job done. Only SIREWALL provides the training for these roving foremen and that training is updated every month.

Seasonal work and much travel are typical of an emerging industry. In the not-too-distant future there will be enough demand to have many centers so that there's not so much travel. The ways of dealing with seasonality will become similar to the rest of the construction industry.

These two factors currently challenge the economic viability of running an *insulated* rammed earth business. Two business responses have evolved. First is the approach of having another primary business that is viable year-round and only taking on the rammed earth work when it is local. This is the part-time approach that is favored by many. Second is the approach of training highly skilled foremen to build using local labor. This full-time approach needs sufficient winter work in the Pacific northwest or in very hot climates (like India, Mexico, Sri Lanka, Portugal, Ghana, Australia, southern USA, etc.). The challenge is to have enough trained people on the team when there is a feast and to keep them when there is a famine.

21.7.2 Costs

People understand that it costs lots of money to drive a car off the lot and that depreciation cost is immediate and ongoing. Depreciation in housing is rarely understood. For example, a house that lasts 50 years and costs $500,000 will depreciate $10,000 per year on average. The bulk of the depreciation takes place in the first 20 years and after that the repair costs begin to kick in. Most North Americans confuse land price with the cost of the house and it is common to hear 'My house has more than doubled in the last 20 years!' This is far from the truth. The value of the house has probably dropped by more than 50% while the value of the land may have tripled. The impact of this misunderstanding is detrimental in the housing market especially for homes that are designed to be sustainable. Because of this misunderstanding and despite the punitive depreciation, North Americans still base most of their decisions on first cost and appearance.

An *insulated* rammed earth building with rigorous attention to thermal envelope detailing can be net zero or have low heating/cooling bills. If not net zero the heating plant or air conditioning unit will be much smaller, hence less initial cost and less cost to run. Maintenance and repair costs of rammed earth walls (inside and out) are near zero for a very long time. Although difficult to quantify on a cost basis, there are health benefits to living inside non-toxic walls.

Payback periods depend on what is factored in (e.g. health, maintenance), the climate and the rigor of the thermal envelope. At today's energy costs, a payback of at least 10 years is probable, but would need to be calculated for each specific circumstance. North Americans select rammed earth or *insulated* rammed earth to hedge against future energy costs, for thermal and humidity comfort, for health, for low maintenance and to walk the environmental talk.

At this point in time, properly constructed rammed earth is not for the do-it-yourselfers and is not inexpensive on a first-cost basis (it's a terrific bargain in life-cycle analysis). The range of cost is affected by architectural complexity, whether the design is sympathetic to the formwork and ramming process, material selection (e.g. white cement, multi-colors, monochromatic), weather (ramming in monsoon, high heat, freezing and hoarding (building inside a tent)) and difficulty of site (little room, steep site). Other variables include location of the project, volume of work, experience of the crew, source of the material, etc. Most *insulated* rammed earth is 24 in. thick and occasionally as thin as 18 in. Most *uninsulated* rammed earth is 12 in. to 24 in. thick. Most, if not all, rammed earth built to Code in North America requires some steel rebar reinforcing. At present (2012), most *insulated* rammed earth costs start at $90/sq ft and *uninsulated* begins at $60/sq ft. Costs are widely variable as per the quality of the work and complexity of the design. Walls that are over 10 ft tall have a premium of typically 50% applied due to the necessity of fall restraint, Worker Safety Board Compliance and the 10 ft delivery height limitation of most skidsteers (assuming that is the delivery method). Mobilization costs can be significant for smaller projects and large project volumes should reduce costs.

Commercial projects carry a premium.

21.8   Case studies of North American earth construction

21.8.1   Case study 1: commercial stabilized insulated rammed earth (SIREWALL) in a cold climate

Librarian, Daphne Platte, and architect, John Carney, spearheaded the first use of *insulated* rammed earth in Wyoming. Pinedale sits at an elevation of over 7000 ft near the Continental Divide. The weather is cold and windy in the winter with lows of −30°F (−34°C) (record low −49°F (−45°C)). The construction was done in the winter inside a heated plastic tent (known as heating and hoarding). With only three summer months that have never seen sub-zero temperatures, the time available for building was to be short. The danger of a power failure (there was one) and the constant high winds, made the thermal and structural integrity of the tent a focus of concern. Also, working inside the tent with the exhaust from the skidsteer and the generator was at times difficult. On the other hand working in T-shirts when it was −20°F outside was a pleasant respite.

There were three rammed earth areas to this building: the large auditorium, the entry foyer and kitchen and the long hallway. The new construction was an addition to an older log library and is quite separate in floor plan. The stabilized *insulated* rammed earth walls are 16 ft tall in the auditorium and 11 ft tall elsewhere. They are 24 in. thick overall and have 4 in. of polyisocyanurate foam embedded in the middle. The coloration is a specific iron oxide and the proportions were slightly varied (for visual interest) as

21.20 Sublette County Public Library in Pinedale, Wyoming – AIA winner of Distinguished Building Award of Citation.

the wall was rammed. The texture of the walls is very effective at creating a good sound quality in the spaces and the building is very quiet (in downtown Pinedale) due to the acoustical attenuation of the thick, *insulated* walls. The earth was a crushed glacial till combined with a coarse angular granitic sand. The average compressive strength of the blend of these two materials combined with 10% Portland cement was 4205 psi (29 MPa) with some results exceeding 6000 psi (41 MPa).

21.8.2 Case study 2: commercial stabilized uninsulated rammed earth in a cold climate

The wall is 15 ft tall by 100 ft long and is straight with a curve towards the end. It sits on a concrete upstand stem wall. The climate is harsher than Wyoming with a range between –57°C and 42°C. The material for the walls came from a limestone quarry and was predominantly a waste product. The compressive strength at 28 days (with only 10% cement) was 24.5 MPa (3550 psi). We used a hydrophobic admixture and a crystalline admixture. There is zero erosion after two winters.

Although we applied an anti-graffiti coating, the SIREWALLs have never been 'tagged', despite the adjacent block walls being tagged repeatedly. We conclude that taggers don't think it's cool to mark rammed earth.

21.21 Grand Beach Public Washrooms north of Winnipeg, Manitoba.

21.22 Nk'Mip Desert Cultural Center, Osoyoos, BC – World Architecture Festival Winner, Governor General's Award.

21.8.3 Case study 3: commercial stabilized insulated rammed earth in a desert climate

This award-winning project is located in Canada's only desert, with temperatures that vary between –30°C and 40°C. The main wall measures

200 ft × 20 ft tall and is composed of 30 different lifts. Each lift color was specified by the architect after reviewing 40 test color samples. The intention of the building blending into its surroundings was very successfully met. The most challenging technical aspect of this project was the 100 tons of hanging rammed earth over the horizontal slot window.

There was a surprising lack of non-structural hairline cracking in these walls despite the paucity of control joints specified by the architect.

21.8.4 Case study 4: residential stabilized insulated rammed earth in a wet climate

This 2000 sq ft building constructed in 1994, has passive solar exposure to the south and is backfilled to 12 ft on the north side. Extending out the west side of the building is a buttress wall that has a cistern behind it. The buttress wall is fully exposed to the elements and shows no sign of deterioration. The acoustical attenuation is very important as the sounds from the inside shouldn't disturb neighbors, and conversely any outside sounds need to be absent when recording is underway. This is one of very few projects where soils were not blended to come up with the wall mix. This was because the local pit, at the time (lasted only a few years), had a very good particle size distribution in its glacial till, therefore soil blending wasn't necessary.

21.23 Music studio in residential neighborhood on Saltspring Island, BC.

21.9   Design elegance of modern earth buildings

Design usually considers such things as proportion, rhythm, inferred space, texture, color, massing, etc. All these are vital and valid considerations. What's missing are the three seventh generation considerations that ensure sustainability: durability, simplicity and multifunctionality. Durability answers the problems caused by our disposable approach to building and is essential if we want to leave gifts to the seventh generation. As yet few greenbuilding metrics appreciate that durability is core to sustainability.

The pursuit of simplicity, particularly in electrical/mechanical systems, is difficult to realize in isolation from the rest of the building's systems. There are net zero buildings being built, ostensibly as an expression of concern for the environment, that have mechanical rooms full of equipment sufficient for a building many times their size. The thinking is that the route to net zero is more and more photovoltaics, more and more geothermal, all of which address the supply side. Reducing the need for energy doesn't require pumps, fans, compressors, relays, anticipators, valves or all the electrical requirements. Conservation is not particularly sexy, but it has nothing to break down or maintain. If we imagine a SIREWALL in seven generations (nothing has changed) and compare it with an electrical/mechanical room in seven generations, the contrast is stark indeed. Probably nothing of the original electrical/mechanical room is original, even in two generations and if there is something original, one can be sure that parts are unavailable. Every pump, relay, compressor, alternator, electrical panel and control is a strike against sustainability. Simplicity involves using nature's tools, such as gravity, thermosyphons, radiant heat distribution, conduction and convection. Reduction of moving parts and complex manufactured components is the goal.

One material or idea serving multiple functions runs against the theme of specialization that is so prevalent today. An exterior wood frame wall requires many trades (framer, electrician, plumber, insulator, drywaller, sider, painter) and the number of materials is several times that. All that complexity can be replaced by one trade using rammed earth, rebar, conduits and insulation.

It is possible for rammed earth or *insulated* rammed earth to provide the following:

1   Conductive, convective and radiant thermal performance in one material/ process
2   Loadbearing capability
3   Toxin-free walls for healthy living
4   Stable thermal and humidity interior profiles through temperature and humidity flywheel effects
5   A radiant environment with reduced dust and other airborne undesirables (bacteriophage, viruses, molds, dander, etc.)

6 Acoustic attenuation from outdoor noises such as airplanes, lawnmowers, barking dogs, traffic, construction noise, etc.
7 Acoustic dispersion qualities for high-quality interior acoustic environments through careful attention to the mix design
8 A strong statement about concern for the environment – expressing leadership
9 Ongoing energy savings – net zero for heating and cooling
10 Ongoing energy security – not worrying about where energy costs might go
11 Durability – reduced maintenance and reduced depreciation
12 Fire safety – protection from wildfires and in-house fires
13 Beauty, drama and grace.

All of the above is achieved with *insulated* rammed earth and as one building trade.

Earlier in this chapter was the example of how the single-purpose Trombe wall was upgraded by SIREWALL into a structural element with inherent visual appeal while retaining its usefulness as a thermal flywheel (only better due to much greater heat exchange surface area). By contrast there exist a couple of SIREWALL structures where the client and architect were solely interested in the visual, visceral beauty of the material. Those structures have steel columns embedded within the SIREWALL to carry the roof loads (redundant) and unfortunately few other benefits were realized. If I were to imagine the most elegant use of the material it would be for a concert hall, museum, art gallery or winery. The *insulated* rammed earth walls would be structural, the quick-response heating, ventilation and air conditioning (HVAC) system would be one-quarter of its normal size taking advantage of the humidity and thermal flywheels, the texture of the walls would be selected for optimal acoustical dispersion, the fire-resistant capabilities of rammed earth so important for public assembly would be taken advantage of, the acoustical attenuation would allow the building to be built in almost any acoustical environment (near airports and fire or ambulance stations) and all the other benefits would come along for the ride. Note the possibility of replacing the moving parts and maintenance of HVAC with earth. Now that is design elegance!

Because the benefit most appreciated by clients is the visual appeal, let's unpack that and take a look at some of what's possible:

1 Metaphorically, the visual intent is to have the wall look as though a laser has sliced twice through a mountain and the resulting slab has been delivered to site
2 An alternative visual intent is to have the wall look as though the ground around it has been cut away
3 The intent is to have the means of construction visible in the outcome, including tie-holes, seams and any fastenings

4 Carving of the formwork to allow the rammed earth to sit proud as per the carving. Natural carving subjects are fossils and prehistoric plants, animals
5 Embedding of geodes, abalone shells, oyster shells, clam shells, etc.
6 Feature lines of contrasting, complementary or shaded coloration
7 'Soft' details like bullnosed wall ends, curved walls, bullnosed buttresses, round columns
8 'Crisp' details like hard corners, crisp chamfers of many sizes, 45° corners
9 Multi-color sedimentary layers
10 Rhythmic, undulating and 'wavy gravy', non-horizontal lifts.

It is important to note that the beauty of rammed earth is, in part, derived from its inconsistency. It is inconsistent in color and texture and may have a little efflorescence or hairline cracks. That inconsistency can make it difficult to determine what is acceptable. The danger is to require uniformity and thereby compromise the raw, natural beauty. It would be like requiring raw silk to look like finished silk. The 'flaws' are what make it interesting! What is really important is that there is no compromise on functionality (e.g. strength, water-tight integrity, thermal envelope integrity, etc.)

21.10 Future trends

Whether there is a future for rammed earth in North America is not in question. The question is when. Rammed earth and particularly *insulated* rammed earth are waiting for the market to catch up. As time goes by there is more public awareness as to the importance of durable, healthy, energy-efficient buildings that don't consume organic building materials. For our species' survival, we will need to broadly adopt seventh generation thinking and the C2C approach. As energy costs go up and our organic resources become more expensive, the first cost viability of all rammed earth will flourish (having a great life cycle cost seems to have little impact on market acceptance). The market will expand from the visionary first adopters to the mainstream.

The future potential of rammed earth is easily understood if one compares with a brick wall, where every brick is manufactured offsite, delivered to the site, delivered to the mason, picked up by hand, buttered, put in place and tapped to its final position. This process of repeatedly placing tiny modules (bricks) to create a wall is, surprisingly, economically viable. It would seem not very difficult to create rammed earth that is quicker to build than a brick wall. Rammed earth walls can be thought of as much larger modules than the brick. If rammed earth can be placed a cubic yard at a time inside of formwork that takes very little time to erect or strip and is rammed as it is now then the rammed earth labor costs will be a fraction of what they are now and the material costs not far different. However the product will be far superior, while being financially competitive. Manufacturing large modules on site has the potential to be much more efficient than the brick industry. It will take more research and development to realize that potential.

Further R&D of large rammed earth modules in North America will lead to:

- simpler, quicker forming systems, perhaps with fabric
- simpler, quicker delivery to the forms
- quicker, more reliable and accurate mixing
- offsite manufacturing of large *insulated* rammed earth modules in warehouses in winter to allow rapid erection in the summer (as a means of dealing with seasonality)
- faster and more cost-effective manufacturing processes than the current cottage industry/artisanal approaches
- less cement required but more pozzolans
- a network of architects and engineers who appreciate and implement the benefits and unique considerations for rammed earth fabrication through an integrated design process when possible.
- a network of general contractors who have learned how to work with the rammed earth process
- broad acceptance and Code approval
- established through material testing and international material standards as well as being included in national master specification systems
- recognized as a green building and regenerative building product by the most stringent green building and regenerative rating systems
- greatly expanded North American and global applications, and opportunities for *insulated* and *uninsulated* rammed earth for the full range of building and mixed-use occupancies from assembly occupancies with public and educational facilities, business, institutional and healthcare, commercial and mercantile, residential and housing and factory and high-hazard occupancies.

The future increased viability of modern rammed earth will depend on developing systems and technologies that allow scaling up and vastly increased efficiencies. As with most emerging technologies, it is difficult to afford the R&D to develop the systems and technologies that will allow rammed earth to evolve to its full potential. Until now, it is the passionate builders who, with their personal savings and/or profits, fund the incremental improvements that have made North American rammed earth so much better than it used to be.

21.11 Sources of further information

SIREWALL Introduction to *Insulated* Rammed Earth course
SIREWALL website and newsletters

21.12 Acknowledgments

For editing help my heartfelt thanks go to Catherine Green, Scott Krayenhoff, Cynthia Bennett, Fraser Krayenhoff, Matthew Hall and Diana Mulvey.

22

Australian modern earth construction

R. LINDSAY, Earth Structures Group, Australia

Abstract: This chapter describes the stabilised rammed earth (SRE) industry in Australia. It outlines the construction systems used, the design parameters that have proven successful and the technical specifications used to monitor the product. It also explains the market place of SRE within the Australian social and economic environment. Finally it looks at the future of SRE in the Australian and global context.

Key words: Australia, stabilised rammed earth, insulated stabilised rammed earth, formwork, elevated wall panels, recycled concrete, quarried aggregate, SRE, ISE, rammed earth, sustainable building.

22.1   Introduction

Thirty years ago a small team of Western Australian entrepreneurs developed new technologies to modernise an ancient building method, rammed earth. They created a successful Australian version of stabilised rammed earth (SRE) complete with new formwork systems, new business structures and rigorous building design concepts. This chapter will describe Australian soil selection procedures, construction systems, building design and market trends. SRE is the focus of this chapter as it appears to define modern earth wall construction in Australia.

In Australia, stabilised rammed earth is known by designers and engineers as the acronym SRE. SRE walls are the product of blending aggregates, sand and cement stabilisers with enough water and waterproofing admixtures to achieve a damp, optimally compactable compound. This compound is then placed within specialised in situ formwork and rammed using pneumatic tools in 150-mm layers to create a dense, water resistant and structurally useful wall panel. SRE buildings consist of a series of interlocking load-bearing monolith panels.

The appearance of the walls is an off-form sedimentary finish with the natural colours and textures of the aggregates being the defining feature. Australian designers have not embraced the coloured 'banding' of walls as often seen in North American buildings. The oxides used in these walls trump the natural colours of the base materials and the rainbow cake banding has been considered a distraction from the otherwise humble beauty of the walls.

609

22.2   Uses of stabilised rammed earth in different regions of Australia

22.2.1   The effect of regional climatic zones on earth building

For the purpose of describing Australian climate in terms of its suitability for earth building, there are two basic zones that can be identified (see Fig. 22.1).

First the humid tropical zone north of Rockhampton, Queensland, which extends in a roughly 200- to 500-km-wide strip around the northern coastline as far as Broome on the west coast. This zone is unsuitable for solid masonry building. Consistently warm to hot day and night temperatures are accompanied by high humidity. Solid masonry walls cannot release their

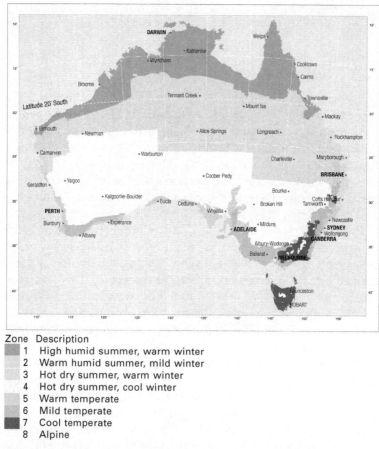

Zone Description
1   High humid summer, warm winter
2   Warm humid summer, mild winter
3   Hot dry summer, warm winter
4   Hot dry summer, cool winter
5   Warm temperate
6   Mild temperate
7   Cool temperate
8   Alpine

22.1 Climate zones as described by the Australian Building Code, 2007.

accumulated daytime heat throughout the evening. Solid masonry dwellings in this zone become uncomfortably hot and humid. Most contemporary Australian buildings in this climate area are built using lightweight timber or iron cladding, which cools down quickly unlike solid masonry.

Second the temperate zone, which extends below the tropical zone. In this zone there is low humidity, and summer night temperatures regularly drop to levels that allow solid masonry walls to adequately cool. Many contemporary earth buildings in this zone do not require air conditioning although they do need to be 'managed' by the occupants.

This management includes drawing curtains or shutters during the heat of the day and opening windows and doors at night to allow the 'dumping' of cool air through the building. To be effective, these buildings also require appropriate design with large north-facing openings to allow winter sun into interior living spaces, cross-flow ventilation for night-time cooling and protection from afternoon summer sun.

Some buildings in our alpine areas may require insulated walling, especially on sites subject to shading that would decrease sunlight on winter days.

22.2.2   Earth building in rural areas

Fewer than 15% of Australians live in rural areas.[1] However, over the past 30 years a majority of modern earth buildings have been rural or semi-rural for the following simple reasons.

Firstly, in terms of earth building, rural sites are easier to access and work on than urban sites. SRE construction requires adequate space for stockpiling and blending of materials and the storage of formwork during construction. Delivery of base materials to rural sites using cheaper truck and trailer combinations is easy. Accommodation for construction teams in rural areas is inevitably cheaper and nicer than in cities. Earth building on small, inaccessible urban sites is often cost prohibitive.

Secondly, more earth building clients have been likely to want an earth house in the country rather than the city. Culturally there has been an acceptance of earth building in rural and semi-rural districts, though this is changing.

We have no data on this, but earth building in Australian cities could traditionally be considered high risk in terms of acceptance and resale.

22.2.3   Earth building in urban areas

There are, however, an increasing number of urban buildings built using contemporary earth walling, particularly in Western Australia where the revival of rammed earth as a modern building medium has been particularly prolific.

Alan Brooks, an SRE contractor based in Perth says almost 90% of his current work is urban. He sources limestone rubble and recycled concrete from crushing plants often within the city itself so they can rely on quick deliveries of materials eliminating the need for stockpiling on small sites. Urban SRE is now their main business and they have developed tricks to streamline their production and keep costs down. In other parts of Australia this trend toward more urban earth wall construction is also growing.

Scott Kinsmore is a rammed earth contractor in Melbourne. He says he is building higher walls on smaller urban sites. The engineering for higher walls with more point-specific loads on small-site buildings is challenging and often requires more steel to be built within the wall structures. This can be frustrating and costly for the wall builder. Increasingly, Australian urban architects are meeting the challenges of sensible passive solar design and low embodied energy materials. While much of the current computer modelling that drives our '5 Star Energy Rating' programmes is 'insulation-centric', some designers are using earth walling as a way to limit the embodied energy of their buildings as well as increasing their passive solar capacity.

Seven of the recent federal government 'Building Education Revolution' schools were built using significant sections of rammed earth walling.[2]

22.3   Approaches to material type and selection

22.3.1   Regional variations of earth materials: aesthetics and use

Australia is blessed with a variety of aggregates suitable for earth building within a deliverable distance of most cities and townships. In many cases quarries will blend materials especially for the SRE industry. It is important the quarry understands the colours required and also the aggregate gradients needed to achieve ideal densities and stability of the walls.

Western Australian SRE builders have reliable supplies of reddy-brown laterite gravels, particularly in the north of the state. Importantly there has been a recent and dramatic change within the Western Australian industry to limestone materials, which, when combined with waterproofing admixtures, have offered finer textured and lighter coloured wall surfaces with minimal shrinkage as compared with the earlier laterite walls.

South Australian SRE contractors are blessed with quartzite quarries that produce beautifully textured yellow walls with minimal shrinkage. Some of these quarries have special 'rammed earth blend' stockpiles.[3]

The Victorian SRE industry has a variety of aggregates to choose from. The Mornington Peninsula south from Melbourne has a huge sandstone quarry with colour variations from strong yellow through to light grey blends.[4]

North Eastern Victoria has several decomposed granite quarries that supply naturally well graded materials that result in particularly durable walling. There are few suitable limestone materials available in Victoria except in the Geelong and Portland districts. Queensland SRE builder John Oliver sources material from two quarries located to the north of Brisbane and blends these with local washed river sand to achieve a variety of textures and colours for his clients.[5]

Sydney has predominantly yellow sandstone materials available. Up until the 1990s the high calcium levels of Sydney sandstone created problems with efflorescence leeching from SRE walls. The use of silane/siloxane waterproofing admixtures has prevented this problem recurring.[6]

More recently our own SRE company has used pink laterite materials for Sydney projects, sourced from Newcastle though the freight component is expensive.

Some Western Australian and Victorian SRE contractors have embraced the use of recycled crushed concrete and crushed brick rubble as a base material, often blended with other aggregates depending on colour preferences. These recycled materials have a lower embodied energy than quarried materials, depending on the location of the quarry and the crushing plant. Generally crushing plants are based within city fringes and are a closer source of material than aggregate quarries, which are generally further out of town.

Sourcing appropriate local quarries has been made easier with the advent of internet searching. SRE contractors needing to source materials for a job on the other side of the state should easily find a suitable material within minutes.

22.3.2   Parameters used to select materials

In Australia site soil is not considered useful for SRE construction. Often clients or architects will want to use trench or dam spoil for building their walls for sentimental reasons. Site soil is mostly highly reactive clay or loam, often with organic matter as well. SRE walls built with site soil will result in cracking, spalling and heaving within the wall structure. Most Australian SRE contractors will not entertain the idea of using site soil. The consequences are not worth it.

The rule of thumb for determining the usefulness of a quarried or recycled product is this: A road base or road topping that has minimal plastic qualities, low shrinkage and a consistent variety of particle sizes will work. Also, the closer the quarry to the site the better in terms of embodied energy demand.

Added to these requirements is the fact that most clients do not want muddy brown or black coloured walls. Thus there are also aesthetic restrictions on the quarries available. Once a suitable quarry is sourced, clients could expect a small sample of SRE rammed within a 200-mm-long section of storm-water pipe to show the sample colour and texture.

22.3.3   Soil testing methods

Typically, Australian SRE specifications and earth building codes call for testing of materials prior to construction of walls. This ensures the materials are stable, provide optimum density and a sufficient compressive strength to withstand structural loads. These tests at their most basic level include:

Particle size distribution test

In this test a dried and weighed sample is shaken through a series of ten varying sieve sizes from 25 mm down to 0.075 mm. The material retained by each sieve is weighed and then plotted on a graph to show the percentage of each particle size within the sample. The ideal material shows an even spread of each particle size, ensuring optimum density when the material is compacted. Obviously a material of similar sized particles, for example sand only, will fall apart after compaction. The test is also useful for two other important reasons. Firstly it identifies where the 'holes' are in the sample. For example, if the sample lacks any 'sand' between 1 and 2.5 mm, then the contractor can add sand of this size to the blend to create a more stable or compactable material. Secondly it identifies the percentage of clay and silt within the sample. The silt or 'pan' material of ≤ 0.075 mm is the dangerous stuff of rammed earth construction. This material can be so fine as to create heaving and cracking of walls as they dry. It also causes immense tensions within wall panels as they dry creating cracking around openings and spalling of the wall surfaces. Samples with more than 8% clay need to be either discarded or modified with added washed, sharp sand until the sample 'matrix' is suitably within the confines of the specification.

Linear shrinkage test

This is a simple test to determine the volatility of the clay materials within a sample. A small amount of the clay material, size ≤ 0.150 mm is mixed with water to form a smooth paste. This paste is cast into a smooth-surfaced semi-tubular section of steel pipe, like a trough, of a given length (normally 100 mm). The sample is left at room temperature until it is entirely dry. The cracks created along the dry sample are measured and consecutively added to provide a linear shrinkage percentage result. Samples with a result of more than 5% are discarded or more washed, sharp sand may be added to the blend at an appropriate percentage to reduce the volatile effect of the clay within the sample.

Compressive strength test

This is a common test used by the concrete industry to determine the ability of a stabilised material to withstand compressive loads. A series of samples blended with the proposed cement ratio (typically 8–10%) and optimum moisture (typically 10%) are rammed within test cylinders 90 mm wide × 180 mm high. The samples are stripped the following day and left to cure for a further 6 days after which they are subjected to a compressive load to determine their compressive strength. Compressive strength results must exceed the design load requested by the structural engineer or the specification for the project.

This test can show up weaknesses within a sample that are not obvious from previous tests. A material may test OK for clay content and particle size distribution but still have a poor compressive resistance. This is because the actual shape of the particles can allow for minimal interlocking within the matrix of the material. Imagine compacting a sample containing round glass marbles as opposed to sharp, gritty particles with interlocking sections. The marbles slide off one another. In many cases aggregates obtained from river bed quarries contain particles that are rounded from eons of being tumbled rather than, for instance, glacial aggregates that are sharp edged from being crushed. In our experience we have discarded many quarry products because of rounded or 'soft' particle shape within the sample. In some cases mechanically crushed aggregates that have sharp particle edges can be added to raise the 'interlock' of a softer material.

22.3.4   Mix designs

Given the variable nature of Australian quarried aggregates, it is often necessary to blend different materials to achieve a structurally useable outcome. For example, a local Albury NSW quarry provides a crushed aggregate material, which, although popular as a road base, is slightly too high in clay (17%) and has no significant sharp particles. It has however a lovely strong red colour, which is desirable for many clients. We blend this material with 30% 7 mm minus sharp crushed quartzite together with 30% washed white river sand, which also has a sharp particle shape. The outcome is a blend that retains a red colour (the clay content will always 'trump' other aggregates for colour) with an excellent structural density that locks together beautifully.

Despite the ability for the wall builder to blend materials in this way, we strongly advise against using site spoil as a 'base' material for SRE walling.

The risk of site soil having varying characteristics within a small distance is too great. We cannot stress the importance of this enough.

22.4   Formwork and construction techniques: the 'Stabilform system'

Almost all contemporary Australian SRE builders use a manufactured formwork system called Stabilform. Stabilform originated in Western Australia in the early 1980s and has been modified only sufficiently to allow most sets of formwork manufactured since to be compatible with each other. The outcome is a remarkable affinity between the Australian SRE construction community, whereby teams (often from different companies) can get together to construct larger commercial projects without the expense of manufacturing masses of new formwork that may lie idle until the next large project comes along.

Stabilform is a system of interlocking steel frames manufactured in 600 mm and 300 mm height increments, in lengths of 2400, 1800, 1200 and 900 mm. Each 'set' is sufficient for a three- or four-person team to build one day of walls. One set contains 18×2400 forms, 18×1800 forms, 12×1200 forms and 12×900 forms, all 600 mm height, with a smaller number of 300 mm height forms for building wall heights outside the 600 mm increments.

As well there will be six 300-mm-wide end shutters and six 400-mm-wide end shutters, all approximately 3600 mm long. These shutters retain the compacted earth at each 'end' of the wall. The faces of these steel-framed forms are made using either a high-quality 18-mm-thick marine ply coated with marine grade epoxy resin, or 18-mm-thick polyethelene or polypropylene plastic sheets, similar to plastic 'chopping board' material.

The idea of the steel frame is to provide a stiff but lightweight backing for each form, as well as a means of interlocking the forms to create a stackable system to achieve a multitude of wall lengths and heights. The formwork is stored in steel cradles, which are designed to be stacked securely onto a flat bed truck using forks attached to the skid steer loader. The truck load also includes the loader and skid-mounted diesel powered 100-cfm air compressor. The package of formwork and machinery can thus be transported in one load from site to site.

Each set of formwork will last indefinitely. The steel frames can be stripped and straightened every few years if required and the faces can be sanded and re-coated about once or twice each year, depending on useage. A set of Stabilform can produce an average 2000–3000 m^2 of walls per year, allowing for an average 200 working days in the same year. Some teams have Stabilform formwork as old as 30 years, still producing neat, straight walling.

Stabilform thus provides an extremely low embodied energy component to the CO_2 cost of each wall. There is no requirement for new formwork to be built for each new project. There are many sets of Stabilform in use in Australia, Thailand, Korea and the UK. The requirement to obtain a patent to construct the formwork has long gone, and credit must go to the system's originator Giles Hohnen in conjunction with Perth-based fabricator Boyde Metal Industries. Giles is regarded as a generous pioneer of the SRE industry in Australia, along with other industry stalwarts such as Steven Dobson and Alan Brooks from Western Australia, Ian Collet from South Australia, Oliver Petrovik from Victoria and John Oliver from Queensland. All these practitioners have advanced the SRE industry in Australia through a lifetime of persistence and ingenuity.

SRE walls are built as a series of monolithic panels. Each panel is constructed in situ. They are built allowing for control joints in longer panels at maximum 3800-mm intervals. Door and window openings are also used as control joints (see Fig. 22.2).

Formwork is erected along the very edge of the concrete raft slab or the concrete footing so water running down the face of the wall will flow over the concrete/wall joint without pooling.

22.4.1   Setting up

The 'set-up' involves selecting formwork lengths to suit the panel (forms have male–female sections for joining lengthways). Through-bolts are placed at 1000-mm intervals to restrain the compactive effort of the rammers. Shutters are clipped together using vice grip fasteners, the jaws of which can

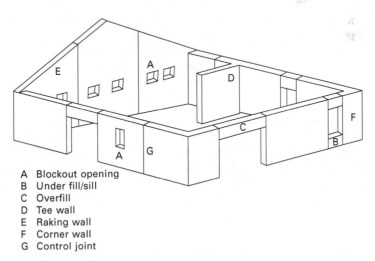

A Blockout opening
B Under fill/sill
C Overfill
D Tee wall
E Raking wall
F Corner wall
G Control joint

22.2 Typical SRE monolith panel layout describing control joint placement and the panel sections. Note: maximum length of panels between control joints is typically 4.0 m.

be opened or closed easily to secure various sections of formwork together. Thus a Stabilform set includes a bucket of up to 30 vice grips. End shutters are placed at the exact wall length position and the bolts are tightened using ratchet spanners against spacers to ensure the formwork is straight and plumb. Through-bolts are placed immediately behind each end shutter. Wedges between the bolts and end shutters keep them plumb while ramming. The end shutters are kept plumb on the sideways axis using extendable props that are secured to the head of the shutter and to the ground. These props can be wound in or out and locked off when the end shutter is true.

22.4.2   Corners

The Stabilform system uses specially fabricated 50×1500-mm-high steel sections, which interlock with the formwork to create either 90 or 45° inside or outside corners. The outside corner sections create a 50 mm chamfered edge, which matches the wall end chamfers. The inside corner sections create a square corner. Outside corners are propped to the ground to ensure they are built plumb.

22.4.3   Damp courses

An embossed plastic fabric damp course the exact width of the wall is cut to length. The damp course is then adhered to the concrete slab edge using a continuous polyurethane mastic bead. This prevents water ingress between the SRE-slab joint. Note that when detailing SRE, it is unwise to use a rebate in the concrete slab edge, otherwise common with block work or brick masonry.

A rebate takes the SRE below ground level or pavement level, thus exposing the base of the wall to potential ground water ingress (rising damp) thus compromising the effect of the damp course.

22.4.4   Key joints

The control joints at 3800-mm max intervals provide a partition line to prevent unsightly vertical shrinkage cracking, which can occur particularly in the event of prolonged hot weather or subsidence in the footing or slab. Each control joint requires a key to prevent panels from 'tipping' against each other. The key is created by nailing a 45-mm timber fillet with the hypotenuse side against the centre of the end shutter. When the end shutter is stripped away the reverse key is imprinted in the end of the wall (see Fig. 22.3).

Before the adjacent wall is rammed against the key, a 20×75 mm closed cell foam weather strip is adhered to the full length of the outside edge of

Outside of wall   A

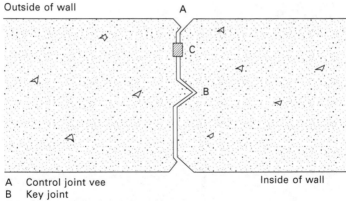

                                                        Inside of wall
A   Control joint vee
B   Key joint
C   Closed cell foam weather strip

22.3 Plan view of SRE control joints.

the wall end. This strip is extended down 50 mm beyond the base of the wall across the damp course of the adjacent wall. This ensures that rain driven into the control joint (which may have opened up to 10 mm due to shrinkage) runs down the closed cell foam strip and drains to the outside at the wall base. This detail is an important consideration especially as the walls have no rebate in the concrete to ensure water cannot penetrate to the inside.

22.4.5   Vertical chamfers

A 45-mm timber fillet is also nailed along each edge of the end shutter, providing a chamfer edge reveal to each wall end. The 45-mm fillet can be manufactured by ripping a 90 × 45-mm length of timber long-ways and then ripping the 45 × 45-mm section along the diagonal, a task most local timber joiners can do. Chamfers increase the beauty and function of an exposed wall end in a way a sharp-cornered brick wall cannot achieve.

Some practitioners use a smaller 19 mm fillet. From our experience we find the compactive effect of the round rammer head is greater using a larger fillet than a fine one.

22.4.6   Control joint appearance

In order to provide an attractive control joint, a 25-mm steel angle is placed open side out against the forms when the adjacent wall is rammed in place. This angle is taken away after stripping to reveal a vee-shaped control joint.

22.4 SRE wall end chamfers. These are typically 45 × 45 mm or 19 × 19 mm.

22.5 SRE vertical control joints. These are typically 35 mm across the face.

22.4.7 Blending of material

While the formwork is being set up, a 'mix' of materials is prepared for ramming. In Australia nearly every practitioner uses a skid-steer loader to blend the mix and then deliver the mix to the in situ formwork set-up. At the start of each project a 'mixing pad' of stabilised aggregate is laid down to provide a hard surface on which blending the material can be done without the risk of digging up surrounding ground that would contaminate the material. A section of flat ground preferably adjacent to the concrete slab and next to the aggregate stockpile is prepared for this purpose.

If the weather becomes inclement, the loader needs to be contained within the confines of the concrete slab and the mixing pad to avoid the tyres picking up surrounding clay lumps.

The loader operator batches out an amount of aggregate onto the centre of the pad. It is important the correct amount is batched to avoid blending too much stabilised earth. Mixes must be used within 1.5 hours, and less in warmer weather as the stabilising effect of the cement is drastically reduced after this time. Thus the team needs to establish how many lifts of the wall can be achieved in the following 1.5 hours and batch accordingly. The loader operator then places the required amount of cement bags on the pile and opens them carefully to ensure cement dust drifts away from them. The mix is blended by carefully turning the pile over and over using the loader until the mix is thoroughly blended (see Fig. 22.6). This normally takes 5 to 8 minutes depending on the size of the mix.

22.6 Blending and wetting the SRE material using a skidsteer loader. An average mix will take between 6 to 8 minutes to complete, including the blending of the Plasticure waterproofing admixture.

A second person then begins spraying the mix with water as it is turned using a high pressure spray nozzle. Good water pressure is essential – slow water delivery will lengthen the time taken to prepare the mix and thus shorten the time the mix is useful before it dehydrates. When the mix is partially damp – crumbly but not balling, the watering person prepares a container of diluted silane/siloxane waterproofing admixture to a volume according to the specification. This admixture is spread gradually through the mix using a watering can, all the while the material is continually turned by the loader. Finally, more water is sprayed onto the mix until the operator and the watering person are confident the correct moisture level has been achieved.

I will note here that some engineers have requested optimum moisture testing for every batch using scales to identify moisture levels by volume of material. By the time the test is satisfactorily achieved, the life of the batch is half over, especially on warmer days. The correct moisture level is such that optimum density can be achieved upon ramming the mix. Too wet and the rammers pug through the layer to be rammed. Too dry and the material fluffs around within the formwork and will not compact properly. It is as simple as that. It is experience only that guides the builders to achieve the correct moisture level. For an experienced team this important issue of correct moisture becomes second nature. Over-wet batches need to be discarded back to the stockpile and are thus a waste of material, time and money. A mix may need to be wetted down once or twice throughout its 1.5 hour life, after which it should be discarded back to the stockpile.

22.4.8 Ramming

Generally each 600 mm lift of Stabilform receives four × 200-mm-deep layers of loose mix. Each 200-mm layer is compacted down to an average 150-mm layer. The material is delivered to the leading face of the formwork using the skidsteer loader, the bucket of which is lowered carefully so the rammer team can shovel the 200-mm layer without causing spillage outside the formwork cavity.

For a smoother wall surface the material is shovelled carefully against the formwork faces ensuring stones do not roll from the apex of the delivered material onto the wall surface. For a more textured outcome the material can be 'pulled' into the formwork cavity in which case small aggregates become part of the exposed wall surface.

The rammer team lift their pneumatic rammers into the cavity and compact the layer down to a hard and consistent level (see Fig. 22.7). The rammers must be kept upright and ensure the course is compacted evenly.

Ramming a course of stabilised earth takes about two minutes per 2000-mm section of wall, normally about the time it takes for the loader to return with

22.7 A pneumatic rammer. Note the access for the operator to the formwork and material. The actual ramming is the least physically demanding aspect of an SRE builder's day.

a fresh bucket of mix. The rammers can be easily lifted over the through-bolts (there is one bolt per 1000-mm length of 600-mm high formwork). Care is taken not to ram immediately on top of the through-bolts as bent bolts are difficult to retrieve the following day.

Good quality pneumatic rammers are comfortable to use. Most Australian teams use Ingersol Rand 341 model rammers or Chicago Pneumatic model CP 0003. Generally there are two rammers used consecutively to compact walls longer than 1800 mm or one rammer for walls shorter than 1800 mm. Corners need two rammers, one for each leg of the corner.

The final (fourth) course is achieved by heaping the mix 75 mm higher than the top edge of the formwork. The compacted final course is then level with the top of the formwork. The residue mix is swept away to reveal the female rebate ready to receive the next lift of formwork. The next 600-mm forms are lifted in place ensuring the vertical form-lines are staggered (see Fig. 22.8).

22.8 Placement of SRE formwork. Note the two consecutive 'set-ups' under construction by the one team. A team will typically build one or two complete panels per day.

The formwork is bolted together, the corners and end shutters are checked for level and adjusted if need be.

22.4.9   Scaffold

All wall building requires an efficient scaffold system to provide a safe and comfortable working platform at heights above 1800 mm. The Stabilform scaffold is integral to the formwork system and consists of support brackets that locate onto the 'ramming' side of the formwork. Handrails slot onto these brackets and two 250-mm-wide aluminium planks sit across the horizontal arms of the brackets. The system ensures the ramming team have close and safe access to the loader bucket, which approaches from the other side of the wall. At the same time they are not standing on top of the formwork or within the formwork, which would be unacceptable practice within the Australian building industry.

22.4.10 Electrical and hydraulic services

In most cases all electrical services are placed within the walls using 25-mm conduits that feed wiring to the roof space (see Fig. 22.9). Wiring and services fed from the base of the walls cause difficulties getting adequate and

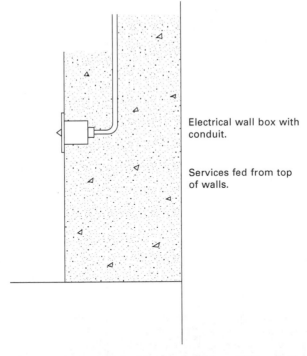

Electrical wall box with conduit.

Services fed from top of walls.

22.9 Electrical or data placement within a SRE wall. Cabling is generally fed from the top of the wall. Faces of the switches or power-points are screwed to the 'wall boxes' after the walls are completed and sealed.

neat compaction to the underside of the service box. 25-mm PVC conduits are bent into shape using conduit springs (these avoid the ridged joins on manufactured corners bends that can frustrate electricians trying to push down their wiring). The conduits are attached to wall surface boxes onto which the light switches or power points can be attached at a later time. Shop drawings of each wall show the exact location of where services need to be located. The wall box with conduit attached is placed against the formwork face and fresh mix is placed immediately behind it. The mix is then compacted gently (to avoid dislodging the level of the box) using a small hand-operated rammer affectionately called 'big foot'. Once the box is secure, the balance of the course of mix is placed and rammed. The rammers are designed to compact around the conduits.

Similarly, hydraulic pipes are placed and rammed within the wall. All hydraulic pipes must be lagged using a minimum of 20 mm foam lagging to prevent the potential for expansion cracking and piercing of the pipes by sharp aggregates during ramming.

In the event of meter boxes being required on an external wall, a polystyrene shape of the meter box is placed on top of the feeder conduits and more conduits rammed in place to take live wires to the rest of the building within the roof space. The polystyrene is later extracted and the meter box secured in its place.

22.4.11 Openings in walls

There are two ways to build window or door openings using the Stabilform system.

Block-out openings

In the event that a smaller opening is specified to be within the middle of a large panel (for example a bathroom window) it is better to use a 'block-out' form to create the opening shape. This block-out is made using 18-mm flooring sheet, which is fabricated in a way that while being strong enough to withstand the compaction of the rammers, it can be collapsed easily and

22.10 A typical SRE 'block-out' opening. Note the adjacent opening has a control joint alongside it.

re-used for other windows of the same size. Block-out openings need a minimum 500 mm SRE on either side of them to provide support.

Control joint openings

Openings using a control joint system require a wall panel to be built up to each side of the doorway or window. Assuming the head height of the opening is 2400 mm, then key joints are cast into the wall heads on either side of the opening. A tee lintel or lintel rods are then bolted or sleeved into the walls on either side of the opening (see Fig. 22.11). A horizontal timber soffit is then positioned at the head height required between the opposing walls and held securely in place using wind-up props below. Formwork is then erected and bolted in place using the same chamfers, key joints and control joint details as described for standard panels. (For a full description of lintels and reinforced masonry lintels systems, see Chapter 18.)

22.4.12 Sills

Window sills are created by setting up formwork across the opening, using all the same waterproofing key-joint and control joint details as per a standard wall. The actual sill shape can be square edged, or a 45-mm chamfer, a 75-mm chamfer or a combination. Window frames can be set at any place within the opening.

We suggest using generous window sills on the inside (useful for leaving things on) and also ensuring the external sill is protected as much as possible by an overhanging window sill.

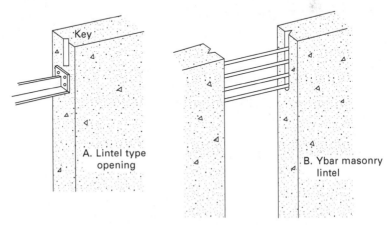

22.11 The steel tee-lintel system described on the left. The steel Y-bar masonry lintel system described on the right. For further descriptions of lintels and other structural steel elements, see Chapter 18 of this book.

22.4.13 Stripping

Freshly rammed walls are left overnight to allow the stabilisers to harden sufficiently for the safe removal of the formwork. Stabilform through-bolts have a 19-mm hexagonal head on which 0.5 in. drive pneumatic rattle guns can be attached to 'rattle' out each bolt. Form lifts are removed layer by layer using straps wielded by the stripping team standing on the scaffold planks, or on high walls standing on a scissor-lift which is levelled at the top of the wall. The last four lifts can be stripped from the ground. An average 3000-mm-long × 3000-mm-high wall takes 45 minutes to strip with forms stacked neatly on pallets ready for the next wall.

22.4.14 Plugging and detailing

The Stabilform system reveals bolt holes every 1000 mm or so at each 600-mm lift. These bolt holes are plugged using fresh sieved mix. The sieved mix can be pushed into the holes using small hand-held plugging devices and the convex mound left at the face of each hole is smacked flat to the wall surface with a hammer. Properly plugged holes cannot be identified on a wall surface among the regular bolt-hole 'dots' that appear at 100-mm intervals. The appearance of these plugged holes has never been an issue for Australian architects.

In some cases there may be small fins of SRE that stay on the wall surface between form joints. These are removed immediately after stripping using a small trowel.

22.4.15 Visual features of stabilform

An SRE wall built using Stabilform is easily identified by the 600-mm horizontal form lines and the bolt-hole 'dots'. These features can be minimised by dressing the walls carefully following stripping. However, many Australian, UK and South Korean architects enjoy the 'signature' of these lines and believe that leaving the features expressed adds to the honesty of the structure.

22.4.16 Internal sealing

Once the wall surfaces have properly dried, usually a month or so after the last wall is completed depending on the air temperature, a surface sealer is applied to bind any loose particles that may brush away over the years into the wall substrate. Diluted acrylic binder is applied to the internal wall surfaces using thick-nap roller heads. Care is taken not to seal any external wall surfaces as the binder will prevent moisture from evaporating properly

22.12 A series of monolith SRE panels showing horizontal form lines and plugged bolt holes. Note the folded steel capping along the wall heads. Pitruzello olive processing plant, Sunbury, Victoria.

from the external wall surface over the ensuing years. This can result in the face of the wall 'spalling' away as the moisture trapped within the wall pushes the bound surface off as it tries to evaporate. The surface hardness of the walls is more than sufficient to resist driving rain and the normal wear and tear of a wall surface.

22.4.17 Capping of the parapet and landscape walls

As with all masonry including brickwork, all parapet walls subject to weather require protection with either a folded metal or a cement render capping (see Fig. 22.13).

Note: We discourage using pavers or tiles, as the joints between these can grow mould during warmer wetter seasons, which then runs down the face of the wall causing unsightly marks.

22.4.18 Fixing to SRE

Light loads such as timber wall frames attached to the SRE can be held in place and either nailed using nail guns or bolted using 10-mm expansion bolts at 900-mm centres. Heavy point loads such as verandah beams, roof plates, floor beams, etc. need to be fastened using a polyester chemical injection anchor system (e.g. Ramset Chemset 101) with 12-mm bolts embedded a minimum of 200 mm at 900-mm centres. Ensure bolts are a minimum of 200 mm from wall ends, control joints or wall heads.

22.13 Cement render wall capping. Note the 18 mm overhang to disperse rain water to prevent streaking. Mansfield Golf Club entry, Victoria.

22.4.19 Machinery

The Stabilform system requires two items of heavy machinery. Firstly a skidsteer loader (minimum 50 horse power) with a forklift attachment and standard earth bucket. Secondly a 100 cubic feet per minute air-compressor, usually mounted on skids so it can be transported easily as part of a single flat-bed truck load. The skid-steer loader has three main tasks. One is to batch and blend the base materials. The second is to deliver and elevate the fresh mix to the wall under construction. The third is to load and unload the formwork cages from the delivery truck and to then move pallets of formwork about the site.

The air-compressor has two tasks. One is to power the pneumatic rammers and the second is to power the pneumatic rattle gun used for stripping the through-bolts out of the walls each morning.

Both machines use on average 25 litres of fuel each day combined.

For walls over 3.6 m height, it is necessary to elevate both formwork and the blended mix safely and efficiently. In most cases an all-terrain scissor-lift is used to elevate formwork and staff to one side of each wall (see Fig. 22.14). In some cases telescopic handlers can deliver material to the wall face (see Fig. 22.15). Material delivery to the other wall face is best achieved using a telescopic handler with a standard earth bucket. The skid-steer loader fills this bucket which is then lifted quickly by the telescopic handler to be

22.14 A scissor lift is used to elevate formwork and blended material to walls over 3.6 m height.

22.15 A telescopic handler delivers material to a 4.8-m-high wall. Telescopic handlers have been successfully used to deliver material to walls as high as 13 m.

placed within the formwork cavity. In some cases walls as high as 12 m have been built using this system. Telescopic handlers provide a surprisingly stable material delivery system.

22.5   Stabilised rammed earth (SRE) walls

22.5.1   Insulated walls

The Victorian 5 star energy rating (ref Building Commission of Victoria practice notes, 2007, 55) required all external walling for residences to have a minimal thermal resistance performance (R value i.e. m^2 k/w) rating of 2.0. The performance of a 93-mm-thick SRE panel, with a density of 2118 kg^2 m^3 submitted for testing by Earth Structures Australia Pty Ltd in 2004 gave a thermal resistance result of 0.084 m^2 k/w. Thus a 300-mm SRE wall had an R value rate of 0.252 assuming the rate of temperature passage through 300 mm of wall would be sustained at the same rate as for the 93-mm test sample.

Earth Structures Europe had previously developed an insulation system for SRE walling in the UK where long cold winters dictated high R value ratings for all external walling. This system was adapted by Earth Structures for Australian conditions using a 50-mm Styrofoam barrier sandwiched between two layers of 175-mm-thick SRE giving an R value result of 2.46 m^2 k/w[7] (see Fig. 22.16).

22.16 An insulated wall set up and ready for ramming. Note the material used for this project was a blend of recycled concrete and local aggregate resulting in a very low embodied energy. Berkeley residence, Thornton Victoria.

The system (commonly referred to as ISE – insulated stabilised earth) uses 280-mm-long 8-mm stainless steel bridging pins with 50-mm cob ends installed at 600-mm vertical and horizontal spacings. These pins provide the wall with tensile load resistance both under compression and tension, sufficient for walls up to 4.2 m heights stand-alone, and 8.4 m heights using embedded steel portal frame structures.

The ISE system, in conjunction with Stabilform formwork keeps a 150-mm solid section of wall adjacent to door and window openings to provide secure fixing. In the colder UK, however, the ISE system continues right to the wall ends where inverted fixing channels are incorporated into the end shutters.

22.5.2   Landscape walls

Landscape or garden walls require special design considerations to ensure they remain dry and stable. If used as a retaining wall, the face of any SRE subject to groundwater ingress requires thorough tanking with polyurethane or bituminous paint, then properly jointed membranes to resist scoring by sharp aggregates.

Finally the base of retaining walls requires proper drainage to ensure water-logged back-fill does not overload the designed tipping resistance of the wall (see Fig. 22.18).

22.17 Landscape SRE walls used to structure a lawn area. Flinders, Victoria.

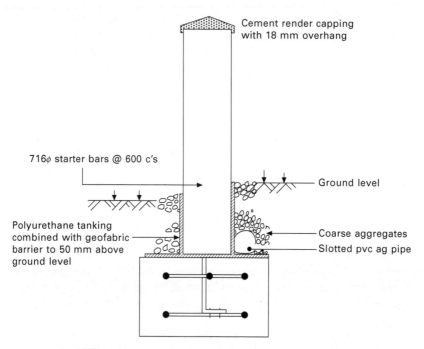

Cement render capping
with 18 mm overhang

716φ starter bars @ 600 c's

Ground level

Polyurethane tanking
combined with geofabric
barrier to 50 mm above
ground level

Coarse aggregates
Slotted pvc ag pipe

22.18 This diagram describes the minimum protection required for landscape walling. Note ground barriers must extend 50 mm above the surrounding lawn or garden bed. Inevitably, beds and lawns 'grow' and walls must remain protected from the effects of rising damp.

To provide further resistance to tipping there may be a requirement to incorporate vertical steel starter bars within the wall. Ensure these bars do not exceed one-third of the wall height, as this leads to horizontal shrinkage cracking caused by the drying walls lifting themselves up along the length of the steel bars. Any vertical steel reinforcement must be galvanised to prevent the steel from rusting and causing pressure spalling decades later. All landscape walls require a cement render capping system that ensures rainwater is cast well away from the top edge of the wall to prevent staining marks down the surface.

22.5.3   Elevated walls

In the event an architect requires SRE walls to clad an otherwise inaccessible building face, Earth Structures have designed systems of 'pre-cast' walling. These SRE panels are built off site or nearby with steel lifting lugs incorporated into the wall heads so they can be elevated in place using overhead cranes. The largest panels built using this system were 4.8 m high × 3.2 m long and 300 mm thick (see Fig. 22.19).

22.19 Placing an elevated SRE panel. Flinders, Victoria.

We have also built large corner-shaped panels, which are lifted using specially designed spreader bars to resist the lifting loads being transferred into the wall structure. Ferrules can be cast into the wall face to allow the bases or sides of the walls to be fastened onto a portal frame. Small uses of these walls also include entry walls for housing estates, entry walls for cities, entry gates for rural properties and advisory sign walls for municipal shires. These smaller walls can be rammed directly onto a cast concrete footing to the entire footing/SRE wall package. They are then lifted and transported using a crane truck then lowered into an excavated trench and backfilled (see Fig. 22.20). This practice saves the cost and inconvenience of transporting an entire SRE team to a remote place for one or two small walls.

22.6   Designing for thermal comfort

22.6.1   Passive solar design

The Australian temperate climate zone provides an ideal environment for the sensible use of SRE. We have long summer high day-time temperatures followed by cool night-time temperatures. We also experience relatively short winters with enough cloudless days to provide warming day-time sunlight. Good design can harness these climatic circumstances using solid masonry walling. Most Australian architects and designers are aware of passive solar

22.20 Transporting an elevated SRE panel. Note the concrete footing in this case is an integral part of the monolith. Mansfield, Victoria.

design. I don't intend to describe the theories of passive solar design in this chapter except to outline some peculiarly Australian design ideas that have proven over the decades to provide comfortable and low operational cost living conditions.

22.6.2   Orientation of the building and layout of the rooms

Australian winter sun tracks low across the northern sky. To ensure maximum use of this warming winter sun well-designed buildings have large openings to the north. It is preferable to have daytime-occupied rooms benefit from this winter sun. Thus kitchens, living rooms, offices and children's play rooms are best positioned along the north side of the house.

Bedrooms can, in our humble opinion, remain cold. That's where one goes at night, and that's what good thick eiderdowns are for.

To maximise the effect of winter sun beating into these north-orientated rooms, concrete or masonry floors serve to retain the heat at night as well as reflecting sunlight well into the room to warm the interior masonry wall surfaces. Window and door openings need to be covered during winter nights with curtains or shutters thick enough to provide an insulative effect. Double glazed windows are now required in most temperate zones in Australia.

22.6.3   Shading of windows in summer

Australian summer sun tracks virtually directly overhead. However it is wise to have narrow eaves along the northern openings to prevent hot sun from shining into the rooms in the heat of the day. These eaves need only be 900 mm wide to shade these openings. West facing windows-which are prone to long hours of afternoon sun, need to be either omitted or kept as small as possible. Ideally a good house design includes substantial shading from leafy trees to the west of the building or, failing that, substantial verandahs on the west that provide deep shade throughout the afternoon.

22.6.4 Natural ventilation

Sitting in a room on a hot day with a gentle breeze flowing through the room provides a welcome cooling sensation. The ability of openings to capture breezes to flush air throughout a building is a vital part of good design. Rooms that have openings along opposing walls allow for a direct flow of air across the room and out the other side. Buildings built as one large square with openings on one side of each room only have less chance for through-ventilation to happen. Casement windows have the ability to catch and redirect breezes into a room more than sash windows do. For this reason casement windows are increasingly popular as tools for passive ventilation.

22.6.5 Managing the thermal benefits

There are passive solar benefits of living in solid masonry buildings. However they must be managed in order to make them work. For summer conditions, the building must be opened in the evening to allow as much cool night air to flow through the rooms as possible. During summer days, if the house is unoccupied, these same openings need to be closed at daybreak (or as soon as the daytime temperature starts to rise) and curtains/shutters drawn to keep the night time 'coolth' within the confines of the building. If the house is occupied during a summer day the occupant may want to open strategic windows to allow breezes to flow across the room.

The reverse needs to happen in winter. Openings must allow winter sun to shine in from the north and should be closed in the evenings to keep as much of the trapped thermal heat in during the night. This system of managing openings also suits the mental comfort of most occupants. On fiercely sunny summer days, it is pleasant to be within the cool and shaded interior of a solid building. Conversely, in winter one wants as much light about the house as possible. A badly managed solid masonry house that is not given the chance to cool down from inside during summer will eventually heat up and become uncomfortable, especially during our Australian summers. All buildings, lightweight or solid, will become cold if they are mismanaged during winter.

22.6.6 Building codes and thermal comfort

Contemporary Australians have far higher demands of thermal comfort than previous generations. There is an expectation that room temperatures remain at constantly cool levels in summer and warm levels in winter to a degree that mechanical air conditioning is a given staple of modern Australian housing.

Most building design codes in Australia consequently assume that mechanical heating and cooling will be used for most of the day throughout summer and winter.[8] Due to this assumption, our current building energy rating systems do not take into account much in the way of passive heating or cooling. Therefore most recent Australian building energy rating systems (5 Star energy ratings) require all walls to have high levels of insulation to keep this manufactured air temperature intact within the building envelope for as long as possible.

For some time the new energy rating system did not allow for un-insulated solid masonry walling at all. The earth wall industry in Australia was facing closure, and in many cases the effect was severe downturns in the work for contractors, especially in the mud brick industry where insulation systems have been difficult to achieve.

Some recent adjustments to the computer modelling that drive the energy rating systems have allowed certain variables (such as increased roofing insulation in lieu of reduced wall insulation) to occur. Upgraded energy rating systems that take into account passive solar benefits are being developed and will help alleviate what has been simplistic computer modelling based on insulation alone as a means of reducing energy use in a building's life cycle.

22.6.7 House size

Perhaps the greatest failing of Australian housing over the past five decades has been the four-fold increase in house occupancy space per individual.[9] This is due to an actual increase in house size (up 200% since 1960) and the decrease in the number of occupants within individual houses (down 200% since 1960). The environmental consequence of this trend is staggering given increasing population levels. Apart from the increase in heating and cooling that big houses create, they also inevitably fill with consumables, which would not be required with a smaller house. The financial debt required for larger houses and increased volume of consumables then tends to add to the workload and anxiety of the occupants, which decreases the value of family life considerably. No end of using low embodied energy materials such as SRE walls can balance the environmental impact of designing large houses.

In some cases the environmental benefits of using earth walls and other green technologies can be used by designers and clients as a rationale to justify building larger houses. For our society to get serious about the environmental impact of humanity on all levels, one simple, easy, but overlooked way to reduce this impact is to reduce the size of our buildings.

22.6.8 SRE as fire protection

The February 2009 devastating bush fires in Victoria brought immediate and thorough upgrades to the Victorian building codes to deal with rebuilding in bushfire-prone areas. These codes are based on an assessment of the building site to ascertain what degree of exposure the new building will have in the event of a bushfire. The assessments are called BALs, Bushfire Attack Levels.[10] The most severe cases require buildings to be clad using fire-proof materials that have in excess of a four-hour fire rating including steel sheet, fired masonry and SRE.

22.7   Standards and specifications for modern earth construction in Australia

22.7.1 Building standards

The Australian building industry is regulated by the Building Code of Australia, which applies standards for the various building trades. These standards are updated on a regular basis as new technologies arise or building conditions change. For example, an environmental expectation may emerge that requires a trade to upgrade its performance. The Building Code of Australia has until recently assumed as its earth building standard a reference to 'CSIRO EBS Bulletin 5', which was derived from Middleton's experimental work in the 1940s. This reference has become increasingly redundant in the light of testing and experience available from the last 30 years' work.[11]

In 2008 the CSIRO removed Bulletin 5 as an obsolete reference document. The EBAA (Earth Building Association of Australia) is currently developing a replacement document, though it is not yet suitable for use in the Building Code of Australia.[12] In the meantime the earth wall industry has been required to use alternative means of self regulation, more often than not reverting to CSIRO Bulletin 5 as a fallback 'deemed to satisfy' document.

22.7.2 Specifications

Most Australian designers and engineers also use specifications as part of their documentation for drawings submitted for tendering by builders. These specifications ensure buildings are constructed and finished in a manner and to a level of acceptability that will satisfy the designer and engineer. Many

Australian designers and engineers currently use the Natspec Specification for Monolithic Stabilised Earth Walling Section 02350 (2004).[13] This specification is advised by the results of testing and decades of experience by ASEG (Affiliated Stabilised Earth Group) of which over 14 SRE companies in Australia, the UK and South Korea are members.[14] This specification is attached to the subcontract tender documentation and ensures elements of the trade such as materials selection, sample testing, tolerances and finishes are upheld by the SRE operator.

22.8   The cost of stabilised rammed earth (SRE) construction in Australia

The cost of SRE in Australia varies enormously according to the design of the building, the location, access and the industrial conditions of the site.

The cost for most solid masonry including double brick, core-filled and rendered concrete blockwork, double-sided stone work and mud brick are all in the same cost ball-park.

Generally the cost of 300-mm-thick walls for domestic building is AUS \$350 to \$450 per m^2, increasing if walls are over 3 m is height. Four hundred-millimetre-thick insulated walls are approximately 20% more expensive than the standard 300 mm walls. Simple shaped buildings with long runs of straight walls and minimal returns or 'nibs' can be built more cheaply than complicated structures. Commercial sites attract a far higher premium due to site issues such as higher wage agreements, portal steel reinforcement for high walls, stringent safety procedures and industrial delays. A typical cost allowance for walls exceeding 6-m heights on a commercial site can exceed \$850 per m^2 for 400-mm-thick SRE walls.[15]

22.8.1   Business models

The SRE industry in Australia is best described as an amalgam of small individually owned and operated companies. In some cases an operator may end up with two or more sets of equipment, however, it is patently obvious that the industry works best as smaller owner-operator companies. Competition is strong between these companies, however, in some cases a truce flag may be raised in order to combine forces to construct projects otherwise too large for one operator.

An interesting affiliation of small operators was formed by Giles Hohnen in the early 1990s in an attempt to enable serious self regulation and the sharing of ideas. Giles quickly advanced the industry to a high level of acceptance by the market and regulators. From this affiliation (Affiliated Stabilised Earth Group – or ASEG) over 14 companies were launched and still operate today. The group collated information and testing results for hundreds of issues relating to formwork, structural engineering, soil mechanics and marketing.

Australia is a vast country and an SRE operator inevitably travels far and wide to maintain a steady flow of work. The actual work is labour intensive but immensely satisfying. Few occupations bear such immediate and gratifying results as turning piles of earth into beautiful structures, often in remote and lovely settings. There is also the argument that the sort of people wanting to live in earth houses are better to deal with than most. This is an absurd assumption but we believe it!

The Earth Structures Group is a further extension of the ASEG model with companies sharing ideas, staff, marketing costs and group buying. The group combines on occasions where projects are too large for one company to deal with.

22.9 Case studies of modern earth buildings in Victoria, Australia

22.9.1 Contemporary insulated residential buildings

Swaney House, Flinders

Simon Swaney, a Sydney-based architect, has designed and built a simple SRE beach house for his family near Flinders on the Mornington Peninsula, one hour's drive south from Melbourne. The house comprises two significant facet-curving earth walls. The first is a 400-mm-thick 2100-mm-high solid SRE wall that acts as protection screen between the driveway and the entry of the house. The second is a 400-mm-thick 3300-mm-high ISE blade wall, which dissects the house at various intervals but acts as a west-facing external wall (thus the ISE). This same wall runs out into a coastal landscape at either end of the house.

To quote an article by Francesca Black in Issue 46 of the residential houses magazine *Houses*:

> On arrival the visitor is met with the sweeping curve of a massive rammed earth wall in the lee of the building. Presented in a refreshingly unpretentious manner, the austere beauty of the material speaks for itself. This wall is offset by an even larger rammed earth wall behind, the space in between forming a secluded courtyard. The second wall forms a central spine to the building. This weighty element serves to ground the house. With its sculptural qualities, striations of texture and rich, honeyed tones, rammed earth has been used here to great effect.

The great effect is also due to the simplicity of the wall shape. These are long, elegant and massive walls, which the designer has chosen not to play around with.

22.21 Entry to the Swaney House, Flinders, Victoria.

22.9.2   Large commercial and public buildings

Port Phillip Estate Winery, Red Hill[16]

Wood Marsh Architects are a Melbourne-based firm specialising in urban design and major public infrastructure. The Port Phillip Estate Winery is a large wine production, restaurant and wine museum facility near Red Hill, one hour's drive south from Melbourne. The building was constructed using a portal steel frame clad with 400-mm-thick SRE facet-curving panels that flow from one end of the building to the other. There are few openings in the walls which add to the dramatic effect of their mass. The staircase from the restaurant to the barrel store has visitors descending between two rammed earth walls as high as 12.6 m.

The material for the walls was sourced from a local sandstone base quarry producing a soft grey effect. The wall colour alters with the changing light of the day to produce vibrantly warm early morning and evening hues to stark grey walls at the height of a summer day.

To quote from the Australian Institute of Architecture jury who awarded the building the Sir Osborn McCutcheon Award for commercial architecture (2010):

> The building is like an archeological artifact revealed by drifting sands. The rammed earth walls of Port Phillip Estate Winery spiral from the earth and heighten anticipation of what lies below.

22.22 Port Phillip Estate, Red Hill, Victoria. Note 12-m-high faceted SRE walls with elongated block-out windows.

Tarrawarra Museum of Modern Art, Healesville

Alan Powell is a Melbourne-based architect who won the competition to design this museum for Australia's largest private modern art collection. The building required strong natural elements. SRE was chosen for its simple, rugged form in what is an elegantly shaped structure. The museum has become a much loved icon of Australian modern art and a popular weekend retreat for the Melbourne art crowd.

The SRE walls are a blend of a rocky over-burden from a quarry near Seymour, 100 km from the site. The base material was blended with sand to bring the mix within the confines of the specification. To retain the boney appearance the SRE teams 'pulled' the material into the formwork thus allowing larger aggregates to fall towards the face of the formwork prior to ramming. The long curves were created using specialised adjustable-radius formworks which were clipped to curving steel templates to achieve the uniform shape. Large 'masonry lintel' openings were achieved using steel hoops attached to 16-mm dowels bolted to the SRE panels on either side.

22.9.3   Precast and elevated modern earth structures

Lauriston Science and Resource Centre, Armadale

This building was designed by Melbourne firm Swaney Draper Architects with a brief to use a huge effort towards environmentally sustainable materials. The rammed earth walling used was a blend of recycled crushed brick rubble and locally quarried red sand. Seven SRE panels were required

22.23 Entry to Tarrawarra Art Museum, Healesville, Victoria.

22.24 Dining area at Tarrawarra Art Museum, Healesville, Victoria.

to be suspended between steel posts along the front of the building. These 4.8-m-high, 3.2-m-long panels were constructed nearby within the building site on temporary concrete plinths and later elevated using the overhead crane on site. Once located the panels were bolted to cleats extending from the steel posts and to the concrete slab structure immediately behind the walls. The walls contained steel 'lifting rod' U-shaped sections, which were rammed within the panels. The spreader bars were located over threaded bolts welded to the tops of these steel sections. The steel 'lifting rods' were housed within plastic conduits to prevent the walls shrinking along the length of the rods.

Mansfield Shire information wall, Mansfield

This small 300-mm-thick sign wall to a public area was made affordable by pre-ramming the wall onto a pre-cast concrete footing at a nearby building site where local rammed earth contractors were building a house. The entire wall/footing was then lifted onto the bed of a crane truck and transported to the site where a level based trench had been dug. The wall was lowered into the trench and settled on a bed of concrete to seat the structure. The sides of the trench were backfilled and tamped to ensure the wall remained plumb. This system saved delivering soil and formwork to the remote site, making a mess with the mixing pad for blending the material and then freighting out again.

The wall was lifted using a 16 mm U bar cast initially within 100-mm of the base of the concrete footing. Once the wall was in place the tops of

22.25 Lauriston Science and Resource Centre. Melbourne, Victoria. Note the elevated panels above the entry. The material used for this building comprised recycled brick rubble blended with locally quarried aggregates.

22.26 Lowering an elevated SRE panel into an excavated trench.
The concrete footing sits 50 mm above the surrounding ground.
Mansfield Shire sign wall, Mansfield, Victoria.

the lifting bars were cut off and a protective cement cap cast along the top
of the wall.

22.10   Future trends

The progress made so far in modern earth building in Australasia has been
due to a combination of architects and designers willing to advance sensible
eduction for sustainable development (ESD) using earth walling together
with a committed SRE construction industry.

Since the advent of SRE in Australia as a serious industry in the early
1980s there have been in excess of 4500 domestic and 300 commercial
projects completed. This is a conservative estimate as many early contractors
have not kept decent records of their achievements.

As a product, SRE is becoming more mainstream and considered less
'herbal' than it may have two decades ago. Most people in southern Australia
have seen a building made using SRE, usually a public school, winery or
municipal structure, if not in real life then in some glossy architectural
magazine, which seem to feature SRE buildings by the week.

No one company has trumped the industry in terms of patents or market
dominance, which all Australian SRE operators believe will lead to a better
future for the industry.

There is a strong commitment within the Australian industry to embrace
competition, which inevitably results in higher standards and a faster
improvement of construction systems.

22.27 SRE walls on display at a Construction Trade Expo. Seoul, South Korea.

Larger commercial projects push the boundaries of the product in terms of construction speed, safety and structural difficulty, all of which result in technical advances that trickle down into the mainstream building of SRE walls.

Globally there is an increasing demand for SRE technology, particularly in countries with enough capital to fund the gear and machinery to get started. The Earth Structures Group as an example work as consultants for projects in Thailand, South Korea, the UK, the USA and Africa. In cases where the technologies are useful the group will establish long-term affiliations with overseas companies in order to share and develop ideas and systems.

22.11   Sources of further information

Domestic

1   Merricks North House, Wood Marsh Architects, Melbourne
     www.woodmarsh.com.au/projects/detail/merricks_north_house/
2   Dromana, Vine House, John Wardle Architects, Melbourne
     www.johnwardlearchitects.com/projects/default.htm?i_PageNo=1&Proj
     ectId=3&pageNo=1&ProjectCategoryId=0&ProjectKindId=0&Featured
     =0&Archived=0&AllProjects=0
3   Red Hill House, Gregory Burgess Architects, Melbourne
     www.gregoryburgessarchitects.com.au/projects/1994/earth-house/

4   Injidup Residence, Wright Feldhusen Architects, Perth
    www.wrightfeldhusen.com/index.html
5   Shenton Park Residence, Wright Feldhusen Architects
    www.wrightfeldhusen.com/index.html
6   Dunsborough Residence, Wright Feldhusen Architects, Perth
    www.wrightfeldhusen.com/index.html
7   Yallingup House, Hofman and Brown Architects, Perth
    www.hofmanandbrown.com.au/index.php?id=112.

Commercial

1   Mansfield High Country Visitor Centre, Gregory Burgess Architects,
    Melbourne
    www.gregoryburgessarchitects.com.au/projects/2003/high-country-visitor-
    centre/
2   Hillview Quarry Offices H$_2$O Architects, Melbourne
    www.h2oarchitects.com.au/
3   Twelve Apostles Centre, Gregory Burgess Architects, Melbourne
    www.gregoryburgessarchitects.com.au/projects/1999/twelve-apostles-
    visitor-centre/
4   Charles Sturt University, Marci Webster-Mannison Architect,
    Brisbane
    www.rivertime.org/lindsay/ar_articles/ar_73.pdf
5   Tarrawarra Museum of Modern Art, Alan Powell Architect
    www.probuild.com.au/projects/tarrawarra-museum-of-art/.

22.12   Acknowledgements

The author would like to thank the following for their help and ideas in preparing this chapter. Brad Overson, Scott Kinsmore and Ben Taylor of the Earth Structures Group, Simon Swaney of Bates Smart Architects, Giles Hohnen, Alan Brooks, Steven Dobson, Howard Marshall and Charlotte Lindsay.

22.13   References

1   Australian Bureau of Statistics 2009 (abs.gov.au)
2   Kneeler Design Architects (kneelerdesign.com.au)
3   Boral Stoneyfell Quarry, Adelaide (boral.com.au)
4   Aidan Graham Quarry, Langwarrin
5   Rammed Earth Constructions (rammedearthconstructions.com.au)
6   Techdry Plasticure (techdry.com.au)
7   CSIRO test MSQ – 1530 March 2004
8   Sustainability Victoria – 5 Star Rating notes (sustainability.vic.gov.au)

9   Australian Bureau of Statistics 2009 (abs.com.au)
10 CFA Bushfire notes (cfa.org.au)
11 ASEG notes (aseg.net)
12 South Australian Govt earth building regulation notes (sa.gov.au) Also (ebaa.org. au)
13 Natspec 02350 (natspec.gov.au)
14 Affiliated Earth Structures Group (aseg.net)
15 For detailed QS notes (rawlinsons.com.au)
16 Port Phillip Estate Winery (portphillipestate.com.au)

23

European modern earth construction

M. R. HALL, University of Nottingham, UK and
W. SWANEY, Earth Structures (Europe) Ltd, UK

Abstract: This chapter presents an overview of modern (post-1970) earth building technology and selected contemporary case studies throughout Europe. It includes an appraisal of the conservation and revival of traditional earth building techniques, as well as modern construction techniques that have been adapted for commercialisation and compatibility with the modern day construction industry. Details are included for formwork types, approaches to soil stabilisation, structural detailing and fabric insulation.

Key words: European construction, rammed earth, compressed earth block, cob, adobe, case study.

23.1   Introduction

Europe has an extensive history of earth construction and a wide variety of techniques. Many of these have, at some point, been exported to other parts of the world. This chapter reviews the current state-of-the-art in modern earth construction within Europe. It summarises the techniques and products currently available for deployment within the construction industry whilst providing further links to useful resources, further reading, organisations and specialist contractors. The chapter also provides highlights of modern earth buildings throughout the continent.

Europe has a history of earth building that dates from Roman times and perhaps earlier. Examples exist in many areas and, according to where the building has been located, a wide variety of techniques have been employed (Easton, 1996). Traditional names for these techniques vary according to area but recognised names include: adobe, clay dabbin, clay lump, clom, clunch, cob, mud wall, mud and stud, piled earth, puddled clay, rammed earth, rammed chalk, wattle and daub and wytchert. The techniques have developed to suit the locally available materials. The type and proportion of clay in the soil being the most important factor, but perhaps also the proportion of sand and gravel in the soil or locally available materials that can be blended to give the desired soil grading. The types of buildings constructed in earth varied according to where and when they were built. They varied from, commonly, simple walls, barns and farmers' houses to, occasionally, fine country houses and public buildings. *Terra Britannica:*

A Celebration of Earthen Structures in Great Britain and Ireland (Gourley and Hurd, 2000) gives a good summary of the different techniques that were used historically and the areas where they were used.

23.2   Conservation and revival of traditional techniques

Considerable efforts have been directed towards understanding and conserving heritage earthen buildings and structures. Previous examples include the well-known *Project Terra* organised by leading experts from the Getty Conservation Institute (USA), International Centre for the Study of the Preservation and Restoration of Cultural Property (ICCROM) (Italy) and Centre International de la Construction en Terre (CRATerre) (France). Academic research into the preservation and restoration of traditional earth buildings has naturally developed into a desire to apply the techniques to new build projects. Within the UK context, the book *Earth Building: Methods, Materials, Repair and Conservation* (Keefe, 2005) is an excellent technical guide to traditional vernacular earth building (particularly in Devon) techniques and includes detailed references as to how compliance with modern building regulations could be sought. More recently, Alan Stokes produced an advisory document for the Devon Earth Building Association (DEBA) that advises on how proposed new buildings made from traditional mud, cob or other unfired earth buildings would be treated by the Building Regulations for England and Wales. As there are many examples of heritage traditional earth buildings in this county, it is possible to use these to demonstrate structural stability and material behaviour under the local climate, which in some cases can satisfy these aspects of the Building Regulations. However, the lack of such traditions in other regions of the UK may preclude this, and the very poor fabric *U*-value means that acceptable energy efficiency ratings can only be met by offsetting the high heat energy loss with supplementary renewable energy technologies (Stokes, 2008).

Lately, enthusiasts for environmental building have joined the cause suggesting that traditional earth construction (e.g. cob, adobe) could present a solution for low carbon buildings of the future. Throughout Europe there is a great level of interest, both academically and amongst a small number of dedicated practitioners, in the study and conservation of earthen architecture heritage buildings and structures. This has also inspired several modern builders to attempt to introduce these traditional techniques of earth building and integrate them within modern buildings, as discussed in the following sections.

The CULTURA 2000 European Programme funded a research project entitled 'Houses and Cities Built with Earth'. The main aim was to give technical training in the conservation of heritage earthen buildings. This included intensive workshops with experts and involved the Università di Cagliari (Italy), ESG/Escola Superior Gallaecia (Portugal) and Universitat de València – Dept Historia (Spain). The project led to the publication of a book entitled *Houses and Cities Built with Earth: Conservation, Significance and Urban Quality* (Achenga *et al.*, 2006). The 'Terra Incognita' project produced a state-of-the-art review of earthen architecture in Italy, France, Spain and Portugal, in terms of its heritage and the associated education, training and research. More recently, this was expanded to include all 27 European Union member state countries in the 'Terra Incognita II' project, creating an atlas of regional earth building heritage for Europe and establishing a network of expertise. Details of publications relating to all of these projects are given in Section 23.8 'Sources of further information'. The Terra Incognita projects also led to a series of awards being given in 2011 for 'Outstanding Earthen Architecture in Europe', in which 42 buildings were shortlisted across three separate categories:

Category 1: Buildings with archaeological, historical or architectural interest
Category 2: Buildings subject to remarkable and relevant intervention (restoration, rehabilitation or extension)
Category 3: Buildings constructed after 1970 (i.e. 'modern').

23.2.1 Cob

Devon has the largest population of earth buildings out of all the British counties. More than 20,000 houses and an equal number of barns, outbuildings and boundary walls may exist (Rael, 2009). This population provides a steady workload for specialist builders to maintain and repair these old buildings. Kevin McCabe is probably the best known exponent and most prolific cob builder in Britain. He claims to have built or repaired something in cob every year for the last 30 years. Starting by repairing old walls he moved on to building extensions on existing cob houses and, since 1994, he has completed four new-build cob houses. Generally these are traditional in style and have thick walls (600–900 mm), which are often curved as no formwork is used. Foundations are concrete with a stone or concrete block plinth rising about 600 mm above ground level. Windows and doorways are generally small as the cob is heavy and lintels are required due to the low tensile strength of the mud material. As a result of the uneven shape of the walls, all structural and internal woodwork is bespoke for each building. The exterior is rendered with a traditional coarse sand and lime putty mix, followed by several coats of lime wash that require periodic maintenance by recoating. To an admirer of vernacular building these are attractive houses, but are slow to build and require specialist skills. Cob in Cornwall is a small company that was started in 1994 by two enthusiasts after completing a cob building course in America. The company is led by Katie Bryce and Adam Weismann who started by repairing historic cob walls in Cornwall, then progressed to new-build cob houses and have now developed a range of natural clay plasters through the company Clayworks Ltd.

23.2.2 Clay lump/adobe

Clay lump is an English form of adobe, or unfired mud brick construction. It was relatively popular in Norfolk and Suffolk during the 18th century and was probably introduced from abroad. It seems to have suited the local glacial boulder clay, which is malleable but prone to shrinkage. The bricks would be air dried before use, thus avoiding shrinkage in the wall. Up to 20,000 earth buildings are thought to still exist in East Anglia (Rael, 2009), the majority of which were made using this technique. Often the clay lump is hidden behind a layer of render, facing bricks or flint and thus is commonly not recognised. In 1994, East Anglian Rural Telluric Houses Association (EARTHA) was formed to promote the study, maintenance and promotion of the earth buildings in the east of England and is led by Dirk Bouwens. They have studied and repaired many historic earth buildings, organised training courses in clay lump building and endeavoured to create the opportunity for new-build houses. Whilst walls built in clay lump seem to have a good record of structural performance within the existing stock of historic buildings, the thermal performance of these traditional buildings is not compatible with the ever-increasing UK building regulations. The answer being proposed to improve this is to add a layer of insulation and then a layer of weather protective material on the outside of external walls. This is not entirely straightforward because some air circulation will still be needed under the insulation to prevent moisture build up in the wall. Becky Little (Little and Davie Construction, Scotland) constructed an adobe/clay lump visitor centre at Rowardennan alongside Loch Lomond, Scotland. It was constructed using materials that were mainly sourced locally including a peg-jointed oak frame, using trees felled from the surrounding forest. The timber roof is covered with re-used slates and the natural stone plinth walls were laid using lime mortar. The load-bearing external walls were constructed using straw reinforced adobe blocks, which are externally rendered with traditional sand/lime and white lime wash, and internally plastered with clay. Currently Dirk Bouwens and a local builder are seeking planning permission to build nine bungalows in clay lump. If they are successful and the buildings are completed satisfactorily, this will be a landmark project for this traditional technique.

23.3   Modern earth construction techniques

This principally includes traditional (or unstabilised) rammed earth, stabilised rammed earth (SRE), stabilised compressed earth blocks (CEBs) and unfired clay bricks. Sourcing and testing of suitable raw materials is an add-on cost that standard building materials rarely require, and currently presents a problem since there is no mainstream requirement for earth building materials. If earth building is to become more widely accepted within Europe, and used for more mainstream building projects as in Australia and North America, then it is essential they become as compatible with the requirements of construction professionals, contractors and building regulations as other masonry materials. Wooley argues that over-enthusiasm for eco-materials can widen the gap between those seeking to be environmentally friendly and green building policies at government level (Wooley, 2000). Unstabilised rammed earth (with or without the inclusion of sodium silicate surface treatment) is recognised as an A+ rated material (highest rating) under the BREEAM Green Guide 2008 (BRE, 2008), which is an internationally recognised system of assessment for building materials. By means of comparison, an A+ rating is also achieved under the BREEAM Green Guide using, for example, medium density concrete blockwork with sand/cement mortar plus external surface render and internal gypsum plaster. Since the cement content of concrete blocks (~ 7%wt) is the same as that used in SRE, it is reasonable to assume that SRE would also achieve the same A+ rating, especially since it does not include sand/cement mortar joints, external sand/cement render or internal gypsum plaster.

A number of key projects have been built recently in the UK, almost all of which have been constructed using unstabilised rammed earth. The earth building enthusiasts who promoted these projects indicate that the use of Portland cement as a stabiliser, as used by the Americans and Australians, was against their environmental aspirations due to the increase in embodied energy of the wall elements. The soil can sometimes be sourced from the building site, but almost invariably this requires amendment through the addition/subtraction of imported soils or quarried aggregates to provide the appropriate particle size distribution. A more controlled form of material selection is to obtain overburden from a local quarry, which is often available at very low cost. If the material is to be unstabilised, then the %wt clay and silt (i.e. fines) content needs to be substantially higher than for stabilised material, and hence material selection and testing of the linear shrinkage coefficient is of paramount importance. As an unbound material, unstabilised earth often requires the addition of surface treatments, sealants or a protective screen. The surface of the rammed earth walls at the Eden Project (see Fig. 23.1), for example, are sealed with sodium silicate solution (known as 'water glass') as a dust suppressant. The earth walls at the Autonomous Environmental

23.1 Unstabilised rammed earth external walls (coated with sodium silicate) at the Eden Project Visitor Centre, Cornwall, UK (© Paul Jaquin, 2004).

Information Centre (AtEIC), as shown in Fig. 23.2, are covered by protective sheets of transparent plastic to allow visitors to see the walls without being able to touch or damage them. More recently, unstabilised rammed earth walls have been constructed inside a building (either load-bearing or non-load-bearing) and protected from weather by an external façade, such as the WISE lecture theatre, Centre for Alternative Technology (CAT) in Wales (see Section 23.5).

SRE materials typically have 7%wt white Portland cement stabilisation with encapsulated insulation (cavity) and hydrophobic admixtures, which provide capillary water resistance (see Chapter 10) whilst maintaining vapour permeability. The material is almost always sourced as crushed quarry waste or as raised natural ballast, which is ubiquitous and abundant; it can be blended to achieve the desired particle grading at the quarry in batches to suit the size of the project. Due to the lower silt and clay content requirements of stabilised soil mixes, they have low cohesion prior to stabilisation and so can easily be dry mixed to ensure consistency throughout the batch regardless of total quantity required. The exterior surfaces do not need coatings or additional surface treatments as they are a bound material. SRE is highly suitable for being recycled into new buildings or as engineering rock fill (e.g. for road sub-base, which has identical grading), and unlike unstabilised earth materials the raw material can be partially or wholly substituted with

23.2 Unstabilised rammed earth internal wall at the Autonomous Environmental Information Centre (AtEIC), Centre for Alternative Technology (CAT), Wales (© M R Hall, 2004).

crushed masonry and recycled concrete aggregate (RCA) from demolished buildings. This avoids primary aggregate extraction, thus minimising the environmental impact of quarrying, and also the costs associated with the aggregate levy (taxation) charges, which are not applied to alternative (i.e. recycled or by-product) aggregates in the UK. Although the particle grading of alternative aggregates (e.g. RCA) is highly suited to dynamic compaction, they do not contain clay minerals and so require stabilisation using a binder, e.g. Portland cement. The total content of cement required can be significantly reduced depending on the age of the demolished concrete and the amount of unhydrated cement it already contains.

Dynamic compaction is typically achieved using pneumatic tampers powered by an air compressor, typically requiring at least 20 cubic feet per minute air flow rate per tamper. Common makes available in Europe include Atlas Copco, Ingersoll Rand, Jet and others. Some contractors use lightweight, proprietary formwork systems such as Doka Frami™ or Stabilform™, as shown in Fig. 23.3. These modularised systems allow architects and engineers to design wall structures in multiples of 300 mm, which minimises construction costs/waste, and maximises efficiency. The outer panels are 600 mm high and can be assembled in horizontal lengths of, for example, 1200 mm, 1500 mm, and so on. Vertical bracing allows wall heights in excess of 9 m and is accessible by conventional scissor lifts, scaffold systems and plant.

23.3 Stabilform™ in use on a 5.2-m stabilised cavity-insulated rammed earth wall at Honiton, Devon (© Earth Structures Europe Ltd, 2007).

Other contractors have used similar lightweight systems, such as the one shown in Fig. 23.4 that was used to construct the 6 m unstabilised rammed earth internal wall at the Rivergreen Centre, Aykley Heads. In contrast, another example was the use of the California-style formwork system pioneered by David Easton (see Chapter 15). This allows the formwork to be used as a modularised system, based on standard plywood and stock timber dimensions (e.g. 1200 × 2400 mm), whilst offering the advantage of being easily customised on-site by carpenters. The timber formwork can be re-used for future projects or it can be recycled and integrated into the building for use in internal partition walls, as was the case for the example shown in Fig. 23.5.

The base of SRE walls is typically engineering brick or a concrete stem wall integrated with the foundations, either of which can be rated frost-resistant as appropriate. This allows for positioning of the formwork system and also the inclusion of a physical damp-proof layer to prevent capillary rise of salt-laden groundwater, as shown by the example in Fig. 23.6.

For small openings (typically < 900 mm) SRE materials can either be self supporting or reinforced with steel bars to increase flexural strength and resistance to cracking. Small openings such as windows can therefore be cast directly into wall sections, which often include a chamfer detail both to increase the amount of light entering the building and to eliminate the potential for accidental damage to the edges (see Fig. 23.7).

23.4 Lightweight formwork in use during construction of the 6-m unstabilised internal rammed earth wall at the Rivergreen centre, Aykley Heads, near Durham, UK (© Paul Jaquin, 2005).

23.5 California-style timber formwork system in use during the construction of a stabilised rammed earth pavilion, Chesterfield, UK (© M R Hall, 2003).

Note the two vertical rebates and chamfered edges cast into the edge of the wall section in Fig. 23.7. This enables the adjacent wall section to form a mechanical interlock, whilst just below the chamfer two thin neoprene

23.6 Engineering brick base course with damp-proof layer, at 150 mm above ground level, and SRE wall at the stables building, The Manor, Ashley, UK (© M R Hall, 2002).

23.7 Stabilised rammed earth wall with extruded polystyrene cavity insulation and small opening with chamfered edge detail and the Stabilform™ formwork system, at Straightways Farm, Devon, UK (© Earth Structures Europe Ltd, 2007).

strips are inserted such that the joint is sealed against air leakage or water penetration but allows thermal expansion/contraction of walls panels, as shown by the example in Fig. 23.8. The exact spacing between vertical

23.8 Close-up of a chamfered thermal expansion joint detail on an SRE wall at the stables, The Manor, Ashley, UK (© M R Hall, 2002).

expansion joints varies according to the linear expansion coefficient of the material, but is commonly 4 m with up to 6 m possible for lower expansion soils. The thermal expansion of cement SRE is low and equivalent to that of low modulus concrete (of the order 10 μs/K). Unstabilised rammed earth typically has a much lower modulus of elasticity, and as an unbound soil is naturally more susceptible to thermal and hygric expansion/contraction, although this can be minimised with correct material selection. Figure 23.9 shows a similar format of expansion joint detailing on the unstabilised rammed earth walls at the Rivergreen Centre, Aykley Heads, UK.

For large openings, stabilised rammed earth can support lintels in one of several ways. For shorter spans these can be formed by designing a reinforcing cage for inclusion in the SRE material, as shown in Fig. 23.10. Due to the significantly lower cement content of stabilised earth compared to concrete, and the presence of clays that can further lower the pH, carbon steel is generally not recommended. It is commonly used in concrete where the cement content is high providing sufficient alkalinity (~13.5 pH) for the steel to form a passivation layer and be protected from corrosion. The problem of reinforcement bar corrosion could be further exacerbated, especially in the presence of chlorides (e.g. on coastal locations exposed to sea spray), if the earth materials have a higher level of liquid/vapour

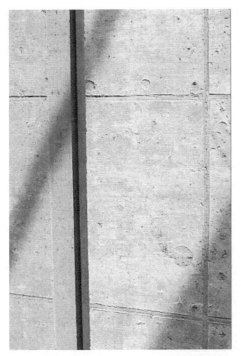

23.9 Chamfered expansion joint detail on an unstabilised rammed earth wall at the Rivergreen Centre, Aykley Heads, Durham, UK (© Rivergreen Developments Plc – Developer/Contractor, photograph by Allan Mushen, 2006).

23.10 Stainless steel reinforcement cage used in conjunction with Stabilform™ formwork to produce an integrated SRE lintel, at Straightways Farm, Devon, UK (© Earth Structures Europe Ltd, 2007).

permeability. The example in Fig. 23.10 uses stainless steel reinforcement but glass fibre reinforced polyester (GFRP) could be an equally suitable and high performance alternative.

For cage reinforced integrated lintels, the underside of the material is supported by formwork and telescopic props. The latter remains in place until the SRE has cured for several days to ensure accurate tolerances in its dimensions are maintained, as shown by the example in Fig. 23.11. Note the continuity of the cavity insulation in order to prevent thermal bridging, and the vertical expansion joint to the right of the opening.

Lintels can also be pre-cast reinforced concrete, timber or equivalent and can be simply supported bearing directly onto the earth walls. The minimum bearing depth is usually 300 mm and the wall can be cast accordingly during placement and ramming. Alternatively, lintels can be steel and fixed into adjacent SRE walls using steel bolts with an epoxy resin bonding agent (see Fig. 23.12). The same technique can be used to support upper floor joints in a multi-storey building but has not yet been applied in Europe. The steel lintel is then supported by telescopic props and formwork during compaction, as for the integrated lintel shown in Fig. 23.11.

Rammed earth walls can be built to the same level of accuracy and tolerances as masonry concrete, enabling the use of standard sized window and door frames as well as being able to install effective edge seal detailing.

23.11 SRE wall with cage reinforced integrated lintel temporarily supported by telescopic props during curing, at Straightways Farm, Devon, UK (© Earth Structures Europe Ltd, 2007).

23.12 Steel lintel attached to the adjacent SRE walls using resin-bonded steel bolts on the stables building at The Manor, Ashley, UK (© M R Hall, 2002).

This is shown by the example in Fig. 23.13 where an internal window frame has been installed. Roof connections can take the form of conventional timber wall plates, as shown by the example in Fig. 23.14, which can either be retained by metal wall ties and masonry nails or by resin-bonded bolts fastened vertically through the wall plate into the top of the wall.

23.4   Case studies of modern earth buildings throughout Europe

The number of modern earth buildings (post-1970) throughout Europe is significantly lower than in Australia, North America and India. However, there is quite a diverse portfolio of buildings and a range of innovative techniques and products available, as summarised in the following section.

23.4.1 France

CRATerre-EAG helped to ensure the conception and realisation of two significant modern earth building projects using stabilised compressed earth blocks in the 1980s. These were Villefontaine and the Domaine de la Terre housing complex at l'Isle d'Abeau (Fig. 23.15) which includes modern houses (Fig. 23.16) and apartments (Fig. 23.17).

23.13 Glazed timber window frame attached and sealed to an unstabilised rammed earth wall at the Rivergreen Centre, Aykley Heads, UK (© Rivergreen Developments Plc – Developer/Contractor, photograph by Allan Mushen, 2006).

23.14 Timber wall plate and roof truss connection on the 6-m unstabilised rammed earth wall at the Rivergreen centre, Aykley Heads, UK (© Rivergreen Developments Plc – Developer/Contractor, photograph by Allan Mushen, 2006).

23.15 The Domaine de la Terre housing complex at l'Isle d'Abeau (© Teresa Kelm, 2011).

23.16 Stabilised compressed earth block house at Domaine de la Terre, l'Isle d'Abeau (© Teresa Kelm, 2011).

Nicolas Meunier is a specialist earth building contractor, established in 1981, and member of the Ecobati Network. He typically builds between one and two modern rammed earth buildings every year, although the majority of his activity is in the restoration of traditional adobe buildings.

23.17 Stabilised compressed earth block apartment block at Domaine de la Terre, l'Isle d'Abeau (© Teresa Kelm, 2011).

Examples of his work include the community building at Valloire Rhône in the Drôme, and also an eco museum, which includes a 60-m-long rammed earth wall at Vinay en Isère in south-eastern France. In 1986, after five years of experience in earth construction, Nicolas Meunier invented a technique of rammed earth prefabrication aimed at adapting the technique to suit the economic context of the modern European construction industry. In 1987, he began experimenting with lifting equipment and earth blocks, and in the following year he built his first prefabricated earth house. The rammed earth blocks are produced in large moulds on the ground before being lifted by crane and placed on a bed of lime mortar. The maximum size of the blocks is 2.2 m long by 1.0 m high with a thickness of 0.5 m. In 1995 he successfully constructed a building comprising three units with a height of 9.4 m and a total wall area of 202 m² in Montbrison, Loire. Figure 23.18 shows the partially constructed earth walls with pre-fabricated sections being lifted into place. Figure 23.19 shows the present-day completed building with its external render and façade in keeping with the surrounding vernacular. The soil was extracted and delivered to site from Sainte-Agathe-la-Bouteresse, which is approximately 19 km from the site. The building was designed by a successful collaboration between Antoine Morand, Dorat Architect and Nicolas Meunier. Prefabrication is only used for buildings of significant height, since it requires heavier site installation and plant.

Meunier always utilises local soils, sourced from within 20 km of the construction site, and generally avoids stabilisation with the exception of 2%wt hydrated lime to assist drying of soils that are excessively wet. For research

23.18 Pre-fabricated earth walls being craned into position during construction at Montbrison, Loire (© Nicolas Meunier, 1995).

23.19 Completed 9.4-m-high unstabilised earth wall building at Montbrison, Loire with external render and traditional appearance (© Nicolas Meunier, 2011).

on thermal and paraseismic behaviour of earth walls, Nicolas Meunier has successfully collaborated with Ecole Nationale des Travaux Publics de l'Etat (ENTPE), Lyon. Load-bearing earth walls are typically 500 mm thick. One example is the house at Chasselay (69) north of Lyon, with a living area of 160 m² and non-insulated 4-m-high rammed earth corner walls designed to absorb, store and release heat energy whilst maintaining sufficient space to allow direct solar gain (see Figs 23.20 and 23.21).

23.20 Use of vertical sliding formwork system during construction of the rammed earth walls at Chasselay (© Nicolas Meunier, 2011).

23.21 Completed house at Chasselay with rammed earth corner walls to provide thermal buffering, and lightweight timber frame upper floor construction with increased glazing area to allow direct solar gain (© Nicolas Meunier, 2011).

AsTerre is The National Association of Earth Building Professionals in France and was founded in 2006. It brings together artisans and entrepreneurs, materials producers, architects, engineers and vocational training bodies. It also hosts representatives from regional organisations and others involved in developing activities within the field of earthen architecture, development of national heritage, contemporary architecture, materials research and technology. AsTerre provide a wealth of information and details of events including conferences, further details of which can be obtained from Section 23.8. In 2009, a school building at Ecole de Veyrins-Thuellin, Isère that was designed by Vincent Rigassi, was successfully built using SRE in a region classified as 'low seismicity'. The building includes classrooms, reading and computer facilities, a canteen, office space and outdoor garden activities.

Le Village in Cavaillon, Vaucluse in southern France is a rammed earth housing development scheme built between 1998 and 2009, designed by Jamal Boudchiche. They are a non-profit organisation whose main aim is to help the unemployed, or those facing other social difficulties, by helping them to find employment whilst providing accommodation in one of the individual houses within the development (25 m^2 floor area per house). There are 11 houses in total and they were built by the residents using compressed earth blocks (CEBs) made from locally sourced soil for the load-bearing walls. In a bid to achieve greater financial independence, Le Village began an entrepreneurial venture by producing and selling CEBs to the general public.

The rammed earth leisure centre in Ramatuelle, on the Saint-Tropez peninsula, was designed by Ann Guillec and built in 2005. Primarily a children's facility, the requirement was to have an environmentally friendly building, which offered a high degree of thermal comfort and indoor air quality for its users. The building is made up of wooden boxes corresponding to different spaces.

The large central hall is made from rammed earth and is adjoined by a series of small timber-frame annexes including activity rooms, a library, a dining hall, a kitchen and a sick bay. The building has a green roof and the rammed earth walls are decorated with colour striations for aesthetic appeal. The earth walls were included with the intention of providing good hygrothermal performance, for indoor temperature and humidity buffering, along with suitable acoustic and thermal insulation.

The student halls of residence at the Institut Agronomique Méditerranéen de Montpellier (IAMM) consists of three residential buildings and a cafeteria. It houses around 200 students in 82 separate studio flats, eight of which are accessible to those with disabilities. The buildings have green roofs, the ground floor walls are built from Pont du Gard natural stone, whilst the upper floors are built using rammed earth and terracotta bricks, which provides a high level of thermal performance throughout the year. It was designed by

Portal-Thomas-Teissier and built between 2007 and 2008. The passageways are generally outside of the building in order to limit the volume of access/service space requiring heating and air conditioning. The overall operational energy consumption of the building is approximately 60 kWh/m^2/year.

23.4.2 Germany

In Europe, Germany possibly has the most successful market in Europe for earth construction products, with an annual turnover of £60 million and a sustained growth of 20% per annum at a time when the rest of the construction industry experienced no growth (Schroeder, 2000). The Berlin Chapel of Reconciliation was constructed in 2000, on the foundations of the previous chapel that had been destroyed, and was designed by the architects Reitermann and Sassenroth. It was the first sacred building to be built from load-bearing rammed earth in Germany, and the altar and bells from the original chapel were preserved and are now housed in the new chapel. The 7.2-m-high internal load-bearing walls are curved and were constructed by Lehm Ton Erde, led by Martin Rauch, using unstabilised rammed earth. They are 600 mm thick with a clay content of 4%wt, optimum moisture content of 8.1%, linear shrinkage coefficient of 0.15% and compressive strength of 3.2 kN/m^2 (Perone, 2008). The building has no heating or air conditioning system and the earth walls are protected from the effects of rain and weathering by an outer wall structure that is clad using vertical wooden slats. As the earth building technique is not well known in Germany, a special case construction permit was needed that required detailed analysis of the material during manufacture and installation as well as careful supervision of the building process. Various trial mixtures were tested for their compressive, tensile and shear strength. A mixture with similar grading characteristics to concrete ballast was selected consisting of clay and stony aggregates to which a small amount of flax fibres were added.

Professor Gernot Minke is an architect and academic based at the University of Kassel. He has been active in the design, construction and research of earth building (including the development of many novel techniques) for over 30 years. His book entitled *Earth Construction Handbook* (Minke, 2000) summarises many of his activities and small-scale experimental building projects during this time. More recently, the Spandau Youth Centre near Berlin, built in 2005, contains a large internal rammed earth wall that runs for 32.5 m down the middle of the building. The south wall is glazed to maximise solar gain and the main purpose of the high density earth wall is to absorb, store and release heat as well as to regulate ambient relative humidity fluctuations. The Chapel of the Central Clinic in Suhl was completed in 2005, and also contains internal unstabilised rammed earth walls of varying thickness.

CLAYTEC is a manufacturer of earth construction products trading throughout Germany and Europe. They produce clay-based plasters in a variety of different colours including finishing plaster and coarse base coats with fibre reinforcement. They are available as a ready-mixed plaster or as a dry bagged powder that can be mixed with water. In addition, they produce a board product called Lehmbauplatten, which consists of woven fibres and clay binder and offers a rapid solution to interior wall finishing. CLAYTEC also manufacture compressed earth blocks in different dimensions and densities including 700, 1200 and 1500 kg/m^3.

The Wangeliner Gardens House, located in Buchberg, is a passive solar design building constructed in 2001 with dense rammed earth internal walls surrounded by a lightweight protective shell of timber frame construction incorporating cellulose-based insulation and with external timber cladding. Clay plasters are used on the walls and ceilings, and rammed earth floors covered with tiles are used in the adjacent glasshouse with a clay stove to provide additional heating. The building also incorporates renewable energy technology including an active solar water heating system and semi-transparent solar panels for the glass house roof covering.

23.4.3 Austria

Kapfinger and Rauch's book *Rammed Earth/Lehm und Architektur/Terra Cruda* (Kapfinger and Rauch, 2001) is an excellent case study manual of the structures built by Austria-based rammed earth contractor Martin Rauch, of Lehm Ton Erde Baukunst GmbH. The book provides technical drawings, design details, images and details of Rauch's significant contribution to modern earth construction including residential, public and commercial buildings in addition to feature walls and structures. Directions to further details of his work and current activities are given in Section 23.8. Well-known examples of his buildings include the Chapel of Reconciliation in Berlin (Germany), as detailed in the previous section. In 2001, a cemetery extension and chapel of rest at Batschuns Voralberg (designed by Marte Architects) was completed. The cube-shaped rammed earth chapel has two earth walls extending out parallel to the sides of the building creating a small courtyard. The soil used for construction was taken directly from the adjacent church grounds. The building foundations are steel reinforced concrete and a concrete bond beam is embedded in the walls of the chapel. The unstabilised earth walls were constructed under the supervision of Martin Rauch and are intentionally over-sized (approximately 450 mm thick) to allow for 'calculated erosion', which, theoretically, gives the walls a 100-year life span (Rael, 2009). The interior floor is also rammed earth, which has been stabilised by sealing with a polished wax surface to enhance its resistance to water damage.

Roland Meingast, of Natur und Lehm GmbH, has significant experience of rammed earth construction. Directions for further details about his building projects and research activities are provided in Section 23.8. More recently, Meingast has been involved in the development of LOPAS clay panels, which are a modularised pre-fabricated building system, as shown in Fig. 23.22.

The maximum panel dimensions are 10×3 m and they are assembled in a way that enables vapour diffusion through both sides, which allows drying transport processes within the multi-layer assembly and moisture equilibration for humidity buffering applications. The LOPAS system was successfully deployed within the clay passive office building 'Lehm – Passiv Bürohaus' in Tattendorf; a prototype that was built in 2004 as part of the 'Haus der Zukunft' demonstration programme organised by the Austrian Federal Ministry for Transport, Innovation and Technology (BMVIT; Bundesministerium für Verkehr, Innovation und Technologie). The biofibre clay plasters, used as part of these panels, were developed and patented by Meingast as part of the EU-funded 'Plaster+' project that ran from 2006 to 2008 (Meingast, 2005).

The clay-plaster layers (layers 1 and 2) have multiple functions including humidity buffering and thermal storage (without chemical stabilisation, as specified by the Dachverband Lehm), whilst providing space for installation of electric conduits or wall heating/cooling pipe systems, or firewall. The U-value of the walls is 0.1 W/m^2 K, and Lopas AG (Austria) produce these prefabricated passive house panels at the wood industry plant of their stockholder Holzbau Longin GmbH.

23.22 An example of the LOPAS pre-fabricated clay panel building panel system (© R Meingast, 2010).

Table 23.1 The composition of an exterior LOPAS panel (from inside to outside)

1	10 mm	Biofaserlehm™ fine clay render (reinforced with cattail – fibres)
		Note: applied on building site after mounting of the panel
2	35 mm	Biofaserlehm™ base coat clay plaster (reinforced with hemp fibres)
3	–	Clay-fleece (incorporated within the base layer)
4	22 mm	Wooden boards or OSB
5	400 mm	Insulation from chipped straw/ wall posts
6	22 mm	Wooden boards or OSB
7	5 mm	Clay-fleece, clay plastered
8	–	External cladding (selectable): timber cladding or exterior render

23.4.4 Switzerland

The Sihlholzli Sports Facility in Zurich was designed by Roger Boltshauser and includes two small storage facilities and a chronometry tower to time the track events. Built in 2002, the storage buildings were constructed from 450-mm-thick unstabilised rammed earth, with a 9–12-mm-thick layer of cement mortar for every 450 mm of height to help prevent surface erosion. The chronometry tower is two stories high and constructed from 600-mm-thick unstabilised rammed earth. The foundations for slab and first floor are constructed from concrete.

23.4.5 United Kingdom

At the Eden Project in Cornwall, the walls of the visitors' centre (see Fig. 23.1) were built by In Situ Rammed Earth Ltd in 1999 using unstabilised soil that had been excavated on site. The soil was mixed then placed using skidsteer loaders and compacted to form 40 panels, each 2.5 m high and with a mass of approximately 10 tonnes. In order to protect the wall surfaces from weathering and erosion they were painted with a chemical sealant comprising sodium silicate. Mount Pleasant Ecological Park at Porthtowan, Cornwall, provides a workspace for local businesses, as well as training and education in traditional and sustainable skills for schools, colleges, universities and the general public. The main building was constructed in 2003 from unstabilised rammed earth by Tim Stirrup, with advice from Rowland Keable at In Situ Rammed Earth Ltd. An 'off grid' dormitory and meeting space at Bradwell on Sea, Essex was built in 2010 for the Othona community by volunteers. They had a minimal budget and took technical advice on the rammed earth construction from Ram Cast CIC. The unstabilised rammed earth walls are built on a raised plinth using traditional formwork comprising scaffold planks and tubes with tourniquet bindings. The College Lake Visitor Centre at Tring, built for the Berks, Bucks and Oxon Wildlife Trust in 2009, was constructed using unstabilised rammed chalk with technical advice from Ram Cast CIC.

Pines Calyx is an eco conference centre located at St Margaret's Bay near Dover in Kent, UK (see Fig. 23.23). It was designed by Conker Conservation and built by In Situ Rammed Earth Ltd in 2005. It also acts as the centre point of an emergent Centre for Sustainable Living and a hub for sustainable enterprise in Kent, with activities and projects centred on the emerging low carbon economy. The main centre comprises two circular buildings at separate levels and constructed from unstabilised rammed chalk. Following the success of this building, Conker Conservation went on to design the Ivylands classroom based at Plumpton College for Sustainable Food and Farming in Battle, East Sussex. It was constructed in 2007 with technical advice from Ram Cast CIC. It was built using unstabilised rammed earth internal walls protected from the weather by an external timber façade. Rammed chalk walls were also used in the construction of the Sheepdrove Eco Conference Centre in Berkshire.

The Rivergreen Centre at Aykley Heads near Durham (see Fig. 23.24) was designed by Jane Darbyshire and David Kendall (JDDK) architects and built by Rivergreen Developments in 2005. It is a two-storey office building that achieved a BREEAM 'excellent' rating. The structure uses a post-tensioned concrete slab at first floor level, supported on cast in situ, reinforced concrete columns that are full height from ground floor to first floor. The bow string

23.23 Unstabilised rammed chalk walls at the Pines Calyx Conference Centre, St Margaret's Bay, Kent, UK (© A Hall, 2008).

23.24 Atrium at the Rivergreen Centre housing the non-load bearing, 6-m-high unstabilised rammed earth wall (© Rivergreen Developments Plc – Developer/Contractor, photo by Allan Mushen, 2006).

truss roof and supporting columns are made from douglas fir timber. The central double-height atrium houses a non-structural unstabilised rammed earth wall along the northern side. The height of the space allows solar irradiation to heat the earth wall, which is intended to act as a temperature and humidity buffer. The wall is 6 m tall and 600 mm thick, with 80% of the soil used in its construction being fine sand excavated from the site during construction of the building's basement. The sand was mixed with gravel and clay extracted from local quarries. JDDK Architects commissioned the Department of Architecture and Civil Engineering, University of Bath, to develop the optimum soil blend and perform mechanical testing of rammed earth test specimens. The wall itself was constructed in six separate vertical sections with expansion joints in between to limit the effects of linear shrinkage. Technical advice on rammed earth construction was provided by Simmonds Mills Architects and also In Situ Rammed Earth Ltd.

In 2003, a stabilised rammed earth pavilion was constructed at the Brimington bowling club in Derbyshire, to be used as a centre for the local residents (see Fig. 23.25). The project was initiated and led by Dr Matthew Hall in close collaboration with Chesterfield Borough Council. Limitations were imposed on the entire project by the addition of a challenge to construct the earth wall and roof structures within a continuous 24-hour period. The challenge, which was subsequently met and completed in just 21.5 hours,

23.25 Brimington bowls pavilion, built in just 21.5 hours including the 17.5 m^3 of load-bearing stabilised rammed earth walls and the timber truss roof structure plus roof covering (© M R Hall, 2003).

was proposed by a television company that had been commissioned by the BBC to produce a programme about the project called 'Home wasn't built in a day'. The resultant building is a medium-sized bungalow approximately 12 m long by 5 m wide. The purpose of the challenge was to illustrate to the public that alternative, eco-friendly construction techniques can provide a rapid and effective solution to the national shortage of housing in the UK. This project also served to demonstrate how full compliance with the UK Building Regulations could be demonstrated using SRE for external load-bearing walls. Further details of this project can be found in the article 'Stabilised Rammed Earth (SRE) and the Building Regulations (2000): Part A – Structural Stability' (Hall *et al.*, 2004).

In 2007 Straightway Farmhouse was constructed at Honiton in Devon (see Fig. 23.26). The architect was Roderick James Architects, the consulting engineers were Mark Lovell Design Engineers and the contractor was Earth Structures (Europe) Ltd. The rear external walls of the accommodation were made of solid cavity SRE, 530 mm thick including 100 mm EPS solid cavity insulation at the centre, which achieved a *U*-value of 0.19 W/m^2 K. This amounted to 89 m^2 of walling with the maximum height being 5.2 m. The building was constructed in three wings joined together at the centre of a three-pointed star. The rammed earth formed the back wall of each wing, with green oak frames and insulated panels forming the other walls of the rooms.

The internal walls and garden walls were of solid SRE 300 mm thick. Lintels over windows and doorways were made by including reinforcing steel

23.26 Straightways Farm House constructed using cavity insulated SRE in Honiton, Devon (© Earth Structures Europe Ltd, 2007).

within the rammed earth. The largest openings were for two garage doors, each being 2.5 m wide. A specially designed cage of stainless reinforcing steel was included within the insulated rammed earth walls to span these openings. The soil used for this building was a dark red naturally occurring mix of gravel, sand, silt and a small proportion of clay. It was used in the as-raised condition and extracted from natural banks just a few miles from the building site, previously offered to local firms as a stable road base. Test cores were made of this material, which achieved a compressive strength of between 5.1 and 7.7 N/mm^2. SRE was also selected for the central wall of the proposed Construction Industry Training College in the Thames gateway, but unfortunately the project was cancelled when Government funds for tertiary colleges were cut.

Earth construction materials are often used for artistic (non-building) purposes within Europe, as exemplified by the temporary display of an electricity pylon in Birmingham city centre, UK. Pre-cast SRE panels were constructed and used to represent a cross-section through the ground beneath the pylon complete with wheat to represent field crops growing on the surface of the soil.

The Birmingham Climate Change Festival was planned by Commission for Architecture and the Built Environment (CABE) and was erected in front of the Town Hall, Victoria Square, Birmingham from 2nd–7th June 2008. The Festival was intended to inspire the public to think about the creation and transmission of energy. The display was designed by Block Architecture and consisted of a 35-m-high steel electricity pylon that had been nickel plated and polished, standing in a square field of wheat that was 12 × 12 m. The field was raised 1.2 m above the pavement with exposed 'earth' sides consisting of 24 panels of pre-fabricated SRE. Each panel measured 1.2 × 2 m and weighed approximately one tonne. The SRE panels were made in Ashley, Market Harborough and transported on a low loader to Birmingham a day before the festival started and erected around the base of the pylon with the aid of a telehandler (see Fig. 23.27).

The Autonomous Environmental Information Centre (AtEIC) at the Centre for Alternative Technology (CAT) in Machynlleth, Wales, was designed by architect Pat Borer with advice from Simmonds Mills Architects who worked as the specialist design/construction team on the rammed earth components. The building atrium includes a large unstabilised rammed earth wall covered by transparent plastic sheets to prevent accidental damage from the large number of visitors. More recently at CAT, Pat Borer designed the WISE education centre, which includes rammed earth walls and also uses a series of other materials with low embodied energy including a glulam timber frame, hemcrete walls, lime render, natural slate, cork, home-grown timber flooring, and natural paints and stains. The building is intended to have high levels of fabric insulation and airtightness combined with heat recovery in some spaces. The circular 200-seat rammed earth lecture theatre is at the centre of the building. The 7.2-m-high loadbearing walls are 500 mm thick,

23.27 Pre-cast SRE wall sections with internal steel reinforcement being delivered and offloaded from a flatbed truck in Birmingham city centre (© Earth Structures Europe Ltd, 2008).

23.28 Model electricity pylon in Birmingham city centre with pre-cast SRE wall panels at the base (© Earth Structures Europe Ltd, 2008).

23.29 Load-bearing, curved unstabilised rammed earth walls forming the structure of the WISE Lecture Theatre, Centre for Alternative Technology (CAT), Wales (© Pat Borer, 2011).

curved and built from unstabilised rammed earth (see Fig. 23.29). The soil was blended to a specified mix using quarry waste taken from nearby Llynclys Quarry. The lower 1.8 m of the walls had a colourless silicate-based sealant (supplied by Keim) applied to prevent dusting of occupants' clothing. Pneumatic tampers were used for the compaction along with a proprietary shuttering system. Figure 23.30 shows the interior of the lecture theatre with the large exposed rammed earth walls providing passive temperature and humidity buffering to the occupied space (often referred to as 'mobilised mass'). This exemplifies one of the key advantages of earth materials in that when a building is periodically occupied, the increase in internal loads can be significant when a large group of people enter the building. The thermal and hygric mass of the earth walls helps to counter any reductions in thermal comfort by buffering the variations in dry bulb air temperature and relative humidity through absorption/desorption of heat and water vapour.

In 2002, Arc Architects based in Fife, Scotland, led a project to develop unfired clay bricks and mortar for the Errol Brick Company Ltd. This was followed by a three-year UK Government funded (by the former Department for Trade and Industry, DTI) 'Partners in Innovation' project to assess the commercial feasibility of low-cost unfired clay bricks, clay mortars and clay renders for modern construction. The final report is available from Arc Architects (see Section 23.8) and includes the findings from a detailed case study that included design, procurement, construction, performance monitoring (thermal, acoustic and hygric) and post-occupancy evaluation of an unfired clay brick house. The benefits of unfired clay in buffering indoor relative humidity fluctuations were clearly demonstrated: external air relative humidity fluctuated considerably between 24.9 and 96.1% but the mean internal value was around 45% and only fluctuated beyond the thermal comfort range (40–60%) for relatively short intervals (Morton *et al.*, 2005).

23.30 Interior of the WISE Lecture Theatre, Centre for Alternative Technology (CAT), Wales (© Pat Borer, 2011).

23.4.6 Spain

Although Spain has a vast history of traditional adobe building and extensive examples of heritage structures, modern examples are starting to become more frequent. Recent examples include Amayuelas de Abajo, which is a group of ten detached houses built in 2001 in Palencia, northern Spain. They were designed for a group of local ecologists by the architects María Jesús González Díaz, Jorge Silva Uribarri and Farncisco Valbuena García. The houses were intentionally designed to blend into the surrounding landscape and incorporate renewable energy technologies such as solar thermal and solar photovoltaic (PV). The walls are constructed using traditional unstabilised adobe bricks, both to provide indoor air temperature buffering (thermal mass effect) and to re-introduce a traditional method of construction to the area. The lack of skilled adobe craftsmen, no industrial-scale production and no specifications for adobe in current building standards presented problems during construction, but these were overcome. More recently, a public swimming bath building in Toro was designed by Vier Architects and built in 2010. The use of unstabilised load-bearing rammed earth plays a crucial role in the construction and intended appearance of the building. The solid walls have a U-value of 1.26 W/m^2 K and the acoustic damping of aerial noise has been stated as 69.21 dBA.

23.4.7 Portugal

Dr Arch Mariana Correia has been involved in research, mainly in the conservation of heritage earth construction but also some modern earth construction and European Union-funded projects. In terms of recent examples of modern earth construction, a monitoring centre at a water treatment station in Évora, designed by João Alberto Correia, was built in 2010. It has load-bearing rammed earth exterior walls designed to provide passive temperature buffering in the hot and dry region of Alentejo. A silicone-based surface treatment was sprayed onto the external earth wall to provide a water-resistant coating that maintains vapour permeability.

23.4.8 Poland

Amongst the range of eco buildings being constructed in Eastern Europe is an experimental house in the Ecological Park of the city of Pasłęk. It was designed by Teresa Kelm, Jerzy Gorski and Marek Kollataj from the Politechnika Warszawska (Technical University of Warsaw) and was built in 2008. Laboratory tests and investigations were conducted to determine the physical and mechanical properties of the SRE material that was used for construction. The walls were constructed using panelled timber formwork (see Fig. 23.31) to create a textured outer wall surface whilst maintaining accurate dimensions required for window and door openings. The load-bearing walls were stabilised with 8%wt Portland cement and compacted using pneumatic tamping equipment and include a large curved section, as shown in Fig. 23.32.

The purpose of the project was partly demonstration but also to make observations and conduct research during the design and construction process. The building is energy efficient and has been designed for a final wall U-value of 0.33 W/m^2 K, and utilises passive solar design by harnessing the direct irradiation gains entering via a greenhouse space on the south side of the building. The next phase will see the interior of the building completed including the installation of mechanical services and fabric insulation. The planned research aims to monitor indoor air temperature and humidity fluctuations to assess the performance of the earth walls. The completed building is shown in Fig. 23.33 and was one of the winners of the 'Outstanding Earthen Architecture in Europe' award 2011 by an international jury as part of the EU-funded 'Terra Incognita' project, in recognition of its innovative design and construction methodologies.

23.5   Future trends

People often ask why modern earth is not as widely adopted in Europe as it is elsewhere in the world. Speed, cost effectiveness, and high levels of material and structural performance are being demonstrated in North America

23.31 Panelled timber rammed earth formwork used at the Pasłęk experimental research house (© Teresa Kelm, 2008).

23.32 Pneumatic compaction of rammed earth walls, including a large curved section, at the Pasłęk experimental research house (© Teresa Kelm, 2008).

23.33 The award-winning Pasłęk experimental research house including rammed earth walls and passive solar design features (© Teresa Kelm, 2008).

and Australia, which is now starting to spread into Asia. These skills are transferrable and can be imported across the world. Regulations are in place in some European countries and not others, but this seems to make little difference to the motivation for large numbers of new modern earth buildings. Perhaps the irony exists in this being perceived as a non-conventional material, and yet by definition it is the most widely used construction material in the world; in the true sense it is the most conventional and with the longest proven track record.

Unlike North America, India and Australia where modern earth building is comparatively prolific, modern European examples have favoured a larger proportion of unstabilised or traditional earth building techniques including rammed earth, adobe, compressed earth block and unfired clay brick. However, most if not all of these examples have still used cement in other ways chiefly in concrete foundations or concrete bond beams, whilst some have used traditional lime binders as a Portland cement substitute. In northern European countries such as the UK, Austria and Germany, unstabilised earth walls are generally deemed to be suitable for internal wall applications and, although they can be load-bearing, must be protected from exposure to the weather (e.g. with wooden cladding) unless erosion of the wall surface is desired. The use of unfired clay materials, including renders, appears to be quite well established in Germany and perhaps currently represents the largest market for modern earth building materials in Europe.

In Europe, there are many individual examples of modern earth construction being used for experimental houses, commercial buildings, public centres, and for iconic centrepieces in larger buildings. However, for modern earth construction to be more widely accepted, utilised and commercially viable it may be prudent to see what lessons can be learned from other regions of the world where it has been much more prolific and has become a recognised member of the construction industry, e.g. North America, India and Australia. One of the areas that clearly separates European modern earth building from these regions is the move towards a higher level of mix design specification, standardised quality control and levels of physical/mechanical performance that are compatible with those of alternative techniques such as concrete block and/or fired clay brick masonry, timber frame, and pre-cast or in situ cast concrete including established steel reinforcement techniques. Also, in North America and Australia several supporting industries have emerged and developed specific products to compliment stabilised earth materials including proprietary additives and vapour-permeable hydrophobic admixtures to significantly enhance material performance and longevity. Significant experience already exists in these other regions of the world for mix design to control mechanical properties, durability and appearance, as well as building in very cold and/or wet climates (e.g. Canada), which would have significant benefit for use in northern Europe including Scandinavia.

23.6   Acknowledgements

The authors would like to express their gratitude to the following individuals whose helpful discussion and generosity contributed to the production of this chapter: Paul Jaquin, Hubert Guillaud, Nancy Happe, Maddalena Achenza, Mariana Correia, Jean-Claude Morel, Roland Meingast, Nicolas Meunier, Claire Moreau, Pat Borer, Teresa Kelm, Peter Candler, Dirk Bouwens, Adele Mills, Kevin McCabe, Roger Hendry, Anthony Goode, Rubén Lagunas, Mariette Montanier and Allan Mushen.

23.7   Sources of further information

Relevant publications

Terra Incognita: Discovering and Preserving European Earth Architecture (2 volumes). Published in 2006 by Culture Lab Editions and Editora Argumentum, GHHI. Editors: CRAterre-ENSAG, École d'Avignon, ESG/Escola Superior Gallaecia, Universidad Politécnica de Valencia and University of Florence.

Terra Britannica: A Celebration of Earthen Structures in Great Britain and Ireland, by Gourley B and Hurd J (eds), published in 2002 by Maney Publishing.

Terra 2000: 8th International Conference on the Study and Conservation of Earthen Architecture, Torquay, Devon, UK, May 2000, published by James & James (Science Publishing) Ltd., London.

MEDITERRA 2009: 1st Mediterranean Conference on Earth Architecture, published in 2009 by EDICOM, Universitat di Cagliari, ESG/Escola Superior Gallaecia, CRATerre-ENSAG.

Rammed Earth: Design and Construction Guidelines, by Walker P, Keable R, Martin J and Maniatidis V, published in 2005 by BRE Bookshop, Watford.

Earth Building: Methods and Materials, Repair and Conservation, by Keefe L, published in 2005 by Taylor and Francis, London

Rammed Earth/Lehm und Architektur/Terra Cruda, by Kapfinger O and Rauch M, published in 2001 by Birkhauser

Earth Building, by Jaquin P and Augarde P, published in 2011 by HIS – BRE Press

Houses and Cities Built with Earth: Conservation, Significance and Urban Quality, published in 2006 by Editora Argumentum with the support of the 'Culture 2000' programme of the European

Union, Lisbon, by Achenza M, Correia M, Cadimu M and Serra A (eds).

Earth Construction Handbook: The Building Material Earth in Modern Architecture – Advances in Architecture vol 10, by Minke G, Published in 2000 by WIT Press.

Terra Europae, published in 2011 by Edizioni ETS and Culture Lab Editions, Italy. Scientific coordination: University of Florence, École d'Avignon, ESG/Escola Superior Gallaecia, Universidad Politécnica de Valencia, CAUE.

Earth Architecture in Portugal, published in 2005 by Argumentum, with the support of ESG/Escola Superior Gallaecia; Lisbon, by Fernandes M and Correia M (eds).

European earth building contractors and specialist advisers

The following is a selection of relevant and active bodies, and not an exhaustive list:

Association of Earth Building Professionals, 67 Rue Pierre Tal Coat 27000 Evreux, website: www.asterre.org

CRAterre-ENSAG (l'Ecole Nationale Supérieure d'Architecture de Grenoble), 60 avenue de Constantine – BP 2636 38036 Grenoble, Cedex 2, France (contact: Prof Hubert Guillaud), website: http://craterre.org/

Le Pisé (contact: Nicolas Meunier), France, website: www.construction-pise.fr

Lehm Ton Erde Baukunst GmbH (contact: Martin Rauch), Austria, website: www.lehmtonerde.at

Historic Rammed Earth (contact: Dr Paul Jaquin), UK, website: www.historicrammedearth.com

Lehm Und Natur GmbH (contact: Roland Meingast), Austria, website: www.lehm.at

Earth Structures (Europe) Ltd (contact: Bill Swaney), UK, website: www.earthstructures.co.uk

Ram Cast CIC (contact: Rowland Keable), UK, website: www.rammed-earth.info/

Arc Chartered Architects (contact: Tom Morton) – specialising in ecological architecture, research and conservation – 31a Bonnygate, Cupar, Fife KY15 4BU, Scotland, UK website: www.arc-architects.com/

Little and Davie Construction (contact: Becky Little), Stone, Earth and Lime Builders, Ash Cottage, Monimail, Cupar, Fife KY15 7RJ, UK website: www.littleanddavie.co.uk

AsTerre (l'Association Nationale des Professionnels de la Terre Crue)

Clayworks Ltd (contact: Adam Weismann), Cott Farm, Constantine, Cornwall TR11 5RP, UK, website: www.clay-works.com

23.8   References

Achenza M, Correia M, Cadimu M and Serra A, 2006, *Houses and Cities Built with Earth: Conservation, Significance and Urban Quality*, Editora Argumentum, Culture 2000 Programme, EU, Brussels

BRE, 2008, *Green Guide to Specification*, Building Research Establishment, available at: www.bre.co.uk/greenguide

Easton D, 1996, *The Rammed Earth House*, Chelsea Green Publishing Company, White River Junction, Vt.

Gourley B and Hurd J, 2002, *Terra Britannica: A Celebration of Earthen Structures in Great Britain and Ireland*, Maney Publishing, Leeds

Hall M, Damms P and Djerbib Y, 2004, 'Stabilised rammed earth (SRE) and the building regulations (2000): Part A – structural stability', *Building Engineer*, 79 [6], pp. 18–21

Kapfinger O and Rauch M, 2001, *Rammed Earth/Lehn und Architecture/Terra Cruda*, Birkhäuser, Basel

Keefe L, 2005, *Earth Building: Methods, Materials Repair and Conservation*, Taylor and Francis, Abingdon

Meingast R, 2005, 'Lehm-Passiv Bürohaus Tattendorf, Berichte aus Energie- und Umweltforschung 29/2005', Hrsg.: BMVIT Wien 2005, available at: www.HAUSderzukunft.at

Minke G, 2000, *Earth Construction Handbook: The Building Material Earth Architecture – Advances in Architecture Vol 10*, WIT Press, Southampton

Morton T, Stevenson F, Taylor B and Smith NC, 2005, 'Low cost earth brick construction 2 Kirk Park, Dalguise: monitoring and evaluation', final report – DTi Partners in Innovation project, available at: www.arc-architects.com/downloads/Low-Cost-Earth-Masonry-Monitoring-Evaluation-Report-2005.pdf

Perone C, 2008, 'Prefabricated earth constructions in the UK and Europe', *The Structural Engineer* 86, pp. 32–39

Rael R, 2009, *Earth Architecture*, Princeton Architectural Press, New York

Schroeder H, 2000, 'Proceedings of Lehm 2000: Beitrage Zur 3', Internationalen Fachtagung Lehmbau Des Dachverbands Lehm E.V., 17th – 19th November, Berlin

Stokes A, 2008, 'Cob dwellings: compliance with the Building Regulations 2000 (as amended) – the 2008 Devon model', guidance document – Devon Earth Building Association, available at: www.devonearthbuilding.com/leaflets/building_regs_pamphlet_08.pdf

Wooley T, 2000, 'Natural materials, "zero emissions" and sustainable construction', Proceedings of Terra 2000: 8th International Conference on the Study and Conservation of Earthen Architecture, James & James (Science Publishing) Ltd, London

24
Modern rammed earth construction in China

R. K. WALLIS, SIREWALL China, GIGA and
A00 Architecture, China

Abstract: Although China has one of the longest rammed earth histories of any country in the world, the traditional building method has virtually died out and has even been made illegal. However, the explosion in construction combined with the lack of natural resources is creating renewed interest for rammed earth, in particular modern rammed earth. This chapter will explore the challenges and opportunities that face the growth of modern rammed earth construction in China. It will also touch on the selection and engineering of local earth, construction techniques and future trends. As the field begins to see renewed interest, it is turning away from its roots and looking abroad for technical know-how. Without a doubt, if the opportunities come to outweigh the challenges, China could well become the global source of innovation for rammed earth construction.

Key words: China, stabilized insulated rammed earth, multistorey structures, waste material, alluvial soils.

24.1   Introduction

From the famous communal Hakka round houses in Fujian to the desert portions of the Great Wall in Gansu province, passing by the hundreds of thousands of traditional structures that still dot the countryside, rammed earth construction needs no introduction in China (see Fig. 24.1). However, in its traditional form it is artisanal, largely unpredictable and more akin to cooking than building science, with mixes that usually contain egg whites, brown sugar and sticky rice. It is viewed as a technique of the past and consequently has been relegated to the annals of history, particularly as the country continues its giant shift to implementing science-based standards.

The latter has led China to favor reinforced concrete and structural steel. However, the ferocity and scale of development in China has placed enormous pressure on these natural resources globally, as well as local energy sources. With much development still needing to be done, China has begun to look into systems that use renewable, indigenous resources. The choices are few, and by far the most prominent and promising is inorganic earth.

688

24.1 Traditional rammed earth in China: Hakka round houses.

24.2   Challenges for modern rammed earth construction in China

Virtually all of the traditional rammed earth that has survived the test of time is found in areas that are mountainous and/or dry. In other words, they are found in areas of low population density. For example, the Hakka houses are found in the isolated Wuyi mountains, which is entirely due to the quality of the soil in these areas. The Wuyi mountains are dominated by craton and punctuated by zones of granite (National Geological Archives of China, China Geological Survey, 2002). Here, the relatively sharp and strong base soil allows for the creation of strong walls. In turn, the desert portions of the Great Wall are characterized by zones of granulite, greenschist and magmatic granite. Not as sharp or strong as in other locations, these soils have been adequate and the resulting walls have endured in large part due to the dryness of the areas they were built in.

By contrast, a geographic look at population density (World Trade Press, China Population, 2007) shows that China's great cities are found in some of the most challenging areas to build with rammed earth: large alluvial basins with round, weak and highly segregated soil types. Without a doubt, the greatest challenge for China is to achieve strength from the soils that are local to the great urban centers.

The second biggest challenge is legal: the lack of a building code for modern rammed earth. Unlike other countries, China does not have a provision in the code for equivalencies. The code must first be written for a particular material at the national level before that material can be used. Not only is this a time- and cost-intensive process, it also requires that the government be willing to consider the particular material being submitted. In the interim it is at times possible – albeit very difficult – to work at a district level, which at the very least requires that the laboratory crush test reports are documented and that the structural drawings are stamped by a local design Institute. It should also be noted that if the owner of the building does not require a title deed, the building need not provide any of the above. This is often the case for buildings of low value, temporary nature and/or landscape walls. It is worth noting that with rammed earth gaining in popularity as an alternative construction method, the above-mentioned structures are becoming the source of local experimentation. Given the absence of a submittal process or the need to adhere to quality standards, these structures contribute to another significant challenge: layering a new understanding of rammed earth over a traditional one that is deeply rooted into China's cultural history.

Everyone in China is familiar with rammed earth. This is both a blessing and a curse: whereas it takes very little time to introduce the concept to the average person, a lot of time is required to get beyond the traditional associations with rammed earth, particularly seeing as the strength and durability of its modern counterpart can be so radically different.

In the mind of the average Chinese person, traditional rammed earth washes away in the rain, can be eroded with one's fingers, is uneven, impossible to clean, bug-ridden, limited to low-rise structures and is a method used by the poor in rural areas. Many Chinese still have memories of these structures, having grown up in them and being happy to have moved on. Typically the only positive quality that they remember is the ability of the walls to keep houses cool in summer and warm in winter. Their experience is obviously well founded: in virtually all cases the clay content is high (35% not being unusual, as per our own findings in crushing samples of local rammed earth from the Pearl River basin, the Yangtze River basin and the Moganshan area) and the walls often contain a large amount of organic material: from topsoil to bamboo and branches used as reinforcement, to eggs, brown sugar and glutinous rice thrown in as binders. Within this context, it is understandable that the initial reaction to building with rammed earth is not taken seriously or is viewed negatively. This reaction is further compounded by the government out-lawing traditional building techniques, including traditional rammed earth and traditional clay brick. The reasons for this are environmental: both methods contribute significantly to topsoil depletion as they require surface mining.

In terms of 'earth' content, most modern rammed earth methods avoid organic topsoil and are more similar to concrete than to traditional rammed earth. Hence, the above-mentioned laws do not apply. This is certainly true of the walls that we have been building, where the 'earth' is entirely inorganic and sourced from the same quarries as the aggregate used in concrete. However, lacking standards and a clear definition of modern rammed earth, and with the existence of modern projects that use traditional mixes and techniques, modern rammed earth will continue to be perceived as being illegal until standards are created and users are educated.

Soils can be carefully chosen, people can be educated, standards can be created and new laws can be passed, but one lasting challenge for China is that of air pollution. Although it affects all buildings – buildings that look 20–30 years old in China are often no older than 5–10 years old – it is a particularly significant problem with rammed earth.

Where acceptable limits are between 10 and 30 parts per million $\mu g/m^3$, our own tests on micro-particulates have shown Shanghai to hover between 90 and 180 $\mu g/m^3$, whereas Beijing often reaches and exceeds 250 $\mu g/m^3$. These particulates are deposited onto buildings by wind and rain, resulting in gray, streaked walls within just a few months.

In areas of low air pollution, one of the key ecological advantages of modern rammed earth is that it is one of the few structural systems that does not require an outside finishing material such as tiles or paint. However, this feature is crippled by air pollution. The surface of modern rammed earth walls are porous and not easily cleaned, creating a need to consider adding an outside finish to the walls. So far, we have mitigated this issue by building large roof overhangs, however this is only a solution for low-rise structures.

The final great challenge is that of density. Globally, there is very little modern rammed earth construction that is higher than one or two storeys. As it currently undergoes the greatest urban migration in the history of humanity,[1] China clearly has a need for high-density solutions. In other words, one- or two-storey structures are insignificant in this country, as are developments in low-strength, non-predictable, traditional rammed earth methods. Modern rammed earth can already attain strengths that make it suitable for mid-rise construction, but special forming and soil delivery techniques would need to be developed, while the energy used in moving so much mass vertically would have to be weighed against the benefits of modern rammed earth.

24.3   Opportunities for modern rammed earth construction in China

While the reality of diminishing natural resources is relatively new to most of the world, it is a reality that China has been bumping up against for hundreds of years. The country and its people have a long history of making marginal resources work for them. As we reach the limits of availability for energy and natural resources, and as their resulting costs continue to rise,

---

[1] In 2007, 42.2% of the Chinese population lived in urban centers. This percentage is estimated to increase to 56.9% in 2025 and 72.9% in 2050, representing 260,958,000 people moving to urban centers between 2007 and 2025, and 205,085,000 more people between 2025 and 2050 (United Nations, Department of Economic and Social Affairs, Population Division, 2008).

it is simply a matter of time until the country turns its attention to modern rammed earth.

China is extremely cost conscious, and while this has been an almost insurmountable obstacle in the past, the stakes are beginning to even out. As the cost of conventional construction and operations continue to rise, the economics of higher performance systems are becoming more convincing, particularly when the added cost is marginal compared to other systems. This is particularly true for insulated rammed earth.

It is also significant to note that unlike North America and Europe, Chinese politicians emerge from science and engineering backgrounds, as opposed to law. Although this may seem trivial to the western reader, it is of significant importance as Chinese government officials are for more effective and proactive at recognizing the physical source of issues and evaluating the potential solutions. Their political structure also gives them the power to implement changes almost overnight. Moreover, if modern rammed earth is to make inroads in China it will have to be supported by an unshakeable scientific foundation: something that Chinese political leaders greatly value and prioritize.

In short, the combination of political structure, magnitude of construction that is left to take place and the double constraints of natural resources and energy, could set the stage for significant developments in China, which could then benefit the rest of the world. Examples include:

- The opportunity to develop competitive forming and delivery techniques for multi-storey buildings
- The opportunity to develop an economy for construction waste used to build rammed earth walls, eventually building walls with virtually no virgin material and yielding a Cradle-to-Cradle building system
- Leveraging the scale and speed of development to accelerate research in the field, from replacing cement in stabilized walls to additives for weather-proofing
- Developing high-performance hybrid systems designed to solve the challenges presented by modern high-performance buildings.

24.4   Approaches to material type and selection

As with most other areas in the world, the spread of traditional rammed earth and approaches to soil selection in China have been through trial and error. The term 'Hakka' derives from the word 'guest'. The people who lent their name to the large multi-structures in the Fujian province serve as an example of how the system spread. As they moved from province to province they brought their earth building technique with them. In places where the soils couldn't be formed or washed away in the rain after one season, the system never took root. In other areas where the soil was 'just right' the system endured for hundreds of years. Traditional rammed earth evolved as an art where soils with high clay contents were favored, particularly when they contained fragments of weathered gravel-sized rock. The old builders who still remember the traditional techniques also mention the preference for soils that contain sharp fragments rather than round ones.

For the most part, the new rammed earth projects in China have been inspired by Korean, Japanese, Australian and German techniques. While most forming techniques have been monolithic and modern in nature, the choice of soils has remained mostly traditional. In some cases, a percentage of cement was added as a stabilizer, but nonetheless, the selection of soils has been more of an art than a science. In most cases the walls are non-load-bearing. In all cases they are uninsulated. Our own interest in rammed earth was to create a system that used a scientific approach and hence could be predictable and quantifiable. It also needed to have high thermal performance not just in terms of the fly-wheel effect but also in terms of insulation. Without these qualities, modern rammed earth would not be possible to scale up in China and couldn't have a significant impact. Further investigation led us to adopting the research and construction methodology developed by SIREWALL. SIREWALL is an insulated rammed earth system that has been developed with a strict scientific and quality control process that is critical to its application in China. It allowed us to begin adapting the process to China almost immediately, focusing on multi-storey construction and, most importantly, understanding local soils.

In simplified geographic terms, China consists of one large tilted plane sloping from the jagged Tibetan plateau in the west to the alluvial plains in the east. The historical center of the country, home to Xi'an and the terracotta warriors, lies on the world's largest quaternary loess deposit. The Sichuan basin, home to Chengdu and Chongqing is alluvial and rests on unmetamorphosed strata, which is marginal for rammed earth construction due to its friable nature. The Sichuan basin is however surrounded by mountains while the unmetamorphosed strata lies on craton, which brings potential to the area. In the south, Guangzhou and Shenzhen fare somewhat better. Although they lie in the alluvial basin of the Pearl River, they border areas of greenschist and magmatic granite.

However, without a doubt, the most challenging area for rammed earth is in the east China alluvial plain, home to both the Yellow and Yangzi Rivers. Partially interrupted by the igneous Shandong peninsula, this zone spans the area from the south of Beijing all the way to Shanghai. Not only is it alluvial, it also consists of highly friable unmetamorphosed strata. In plain English, Shanghai is one of the most challenging areas in the world to build using rammed earth, and this is precisely where we have been doing most of our research.

In the words of the Chinese elders: there is no traditional rammed earth in Shanghai for a reason. However, we have come to realize that if we can build here, we can build anywhere. Working with alluvial soils leaves no room for error seeing as they make extremely poor material for rammed earth. Blending them with sharp aggregate can improve their strength, but only if the sharp aggregate is itself strong, which is not the case in the Shanghai area.

Rivers are phenomenally effective at rounding off soil particles. After having been carried downstream for hundreds of kilometers the particles found in river deltas are almost perfectly spherical. One only need to imagine building a wall out of stacked marbles to understand that round particles do not make strong walls.

Rivers are also extremely effective at segregating particle sizes. Each particle size precipitates at the point at which it can no longer be carried by the speed of the water. Hence, river deltas are characterized by areas of pure sand, silt and clay. For example, at one of our sites 150 km upstream from the mouth of the Yangzi river, 85.8% of the river sand was 0.15 mm in diameter, as per our particle size distribution analyses, and was extremely well-rounded. On land, the strata immediately below the organic topsoil was composed of pure clay, 100% of which passed through the 0.075-mm sieve. In a world where sharp, well-graded soil is critical for creating strong walls, this was not good news. Furthermore, the local quarry produced a sedimentary stone that could be crushed between one's fingers, and was consequently extremely high in fines. In a world where a wall can only be as strong as its weakest component, this was also not good news.

Engineering a wall in these conditions is extremely difficult. Blending soils to create a well-graded mix becomes absolutely critical and leaves no room for error. Relying on alluvial content only provides a narrow range of particles, typically 0.3 mm and below. Hence, one must rely mostly on quarried (crushed) content, supplemented by alluvial content where necessary. However, in an area of friable, unmetamorphosed strata, the science of grading soils becomes somewhat elusive as the ramming process partially pulverizes the soil and changes the percentage of particle sizes in the blend. For example, in another one of our projects (Moganshan, just outside of Hangzhou) we found a mix that contained 10% fines (clay and silt) became one that contained over 30% fines after ramming.

In the river conditions we found that a well-graded mix, compacted with a mechanical rammer could only achieve unconfined compressive strengths of about 1–2 MPa. No wonder the area had little to no history of rammed earth. Once stabilized with 10% cement, we still struggled to get the strength up to 11 MPa. Interestingly, the Moganshan area does have a history of rammed earth, being rammed sequentially upwards, block by block. Lime is added to the mix, with the final result being strong enough for a two-storey house, but weak enough to wear away with one's fingers. The scientific approach to the local technique, which involved blending soils to optimize particle size distribution, using the sharpest particles available with fines limited to about 5% and 10% cement, ramming mechanically and having the samples properly cured (submerged in water), yielded samples of 15 MPa strength.

It has been enough of a struggle to obtain adequate strengths that if we were not building projects in different parts of the world, I would be led to believe that our approach is wrong. Particularly seeing as our SIREWALL counterparts in Canada and the US were obtaining strengths that rivaled and exceeded concrete with exactly the same technique. It is thus interesting to note that the exact same approach used in southern Sri Lanka yielded strengths of 28 MPa after only two weeks of curing. In this particular case, the parent rock was gneiss (mostly granitic), which is both very sharp and strong. Here, it is worth saying a few words about Sri Lanka. Almost the entire island consists of metamorphic rock, including gneisses, quartzite, granites and marbles (Cooray, 1984). Only the northern tip and north-western edge of the island are alluvial and lagoonal, consisting of clays, silts and sands. Furthermore, the Sri Lankan climate makes insulation virtually unnecessary. However, in the tropical plains area and coastal belt there is a definite need to protect walls from direct sunlight in order to avoid thermal gain and the resulting radiant heat that would be given off by the walls. In fact, the Sri Lankans refer to the necessity of 'wetting' traditional earth walls so that they may cool by evaporation. Albeit, they are mostly referring to earth walls consisting of wattle and daub techniques, or sun-baked clay bricks. Of course, wetting the walls is not something we would favor given the efflorescence that could occur.

Climactically, the area best suited for modern rammed earth is the Central Highlands, where daily temperatures swing from about 20°C during the day to 15°C at night (going as low as 5°C in some areas). Here, radiant heat given off by the walls at night would be very welcome. Truly, the combination of parent rock and climate make modern rammed earth the material of choice for this particular area.

At the time of writing, Sri Lanka has emerged from civil war and is entering a period of rapid growth, particularly in the field of hospitality and tourism. This will put a definite strain on natural resources and building materials. Although it is still too early to come to conclusions, Sri Lanka may well be ripe for rapid developments and innovation in the field of modern rammed earth.

Speaking broadly in terms of material selection, we have learned that a sharp, well-graded soil derived from a strong parent rock is key to achieving strength. Fines such as clay and silt should be kept below 10%, and preferably in the 5% range. The specification for water content has not changed in centuries, requiring just enough for a handful of earth to form a ball when compressed in one's hand, and to shatter when dropped from waist height. Deviating from these five basic points creates mixes that rely increasingly on cement as the strengthening agent.

In this field, one of our primary goals is to eliminate the use of cement as a stabilizer from the blend. However, in areas such as the east China alluvial plain it is clear that using a binder is absolutely necessary. We have yet to find one that matches or exceeds the strength of cement while being benign for the environment.

Conversely, removing cement as a binder in an area like Sri Lanka should prove to be far more realistic. As mentioned above, the suitability of Sri Lankan soils and the resulting strength will allow us to obtain strengths of 10–12 MPa with far less cement. This opens the door of possibilities in terms of replacing cement with a lower strength binder.

Another key goal is to close the loop on the source of soil. Currently, the main source of inorganic soil comes from quarries, where rock is dynamited and crushed yielding coarse gravel, fine gravel and tailings. Rather than quarrying rock, waste concrete could be used as the source material, crushed to produce the same gravel-to-tailings soil gradient. In this way the ideal soil gradient could be directly produced as opposed to researched and blended, saving time, costs, transportation and enabling predictable results. However, for this to be possible, the supply chain would first need to be established.

The current state of demolition rubble is that it contains everything from plaster, brick, gypsum board and ceiling tiles, to pieces of cabinetry, posters and doll arms. In order to obtain clean material, the residual value of concrete rubble would have to be significant enough for demolition crews to first strip buildings down to the concrete structure and then demolish them. At the other end of the price spectrum, it is also obvious that the market price of concrete rubble could not exceed the price of local quarry material. Our preliminary research has shown that this could be possible on a per project basis, but is influenced heavily by local conditions such as the distance between the source material (be it quarry or rubble) and the construction site, the cost of the quarry material in the particular locality, etc. Of course, this presupposes that the quarry material is good. In areas such as Shanghai, the price of concrete rubble could exceed that of quarry material on the basis of quality. Here, the market cap would be that of modern rammed earth vs. other structural solutions. The current cost of labor, cost of transportation and amount of construction currently taking place in China make sourcing waste rubble a potentially viable solution within the areas we have researched. This will be the subject of our next project. The initial demand will have to be generated on a per project basis, requiring us to locate demolition sites and negotiate prices for clean rubble. However, to serve as a long-term solution the obviousness of market direction should still be stated. In an expanding construction market such as China's, the demand for waste material could outstrip the supply. In other words, in an expanding market, sourcing from waste material would not eliminate the need for quarrying.

The topic of sourcing waste material generates one more relevant thread: hybrid systems where modern rammed earth is but one of several components in a wall. Currently, modern rammed earth is still being used around the world in a very traditional assembly. While other structural materials such as steel, reinforced concrete, brick and wood have all evolved to become components of a total wall assembly, stabilized rammed earth (SRE) is still being used as a stand-alone material. This fact serves as a testimony to just how niche and new modern rammed earth is: unlike other materials, it has barely begun to include other system components in an effort to adapt to various climates and other constraints. Globally, SIREWALL is perhaps the only example of a modern rammed earth system that has begun to move along this path.

As modern rammed earth evolves to meet the needs of high-performance buildings around the world and as it enters the palette of more architects, developers and builders, it will evolve from being a stand-alone material to a system of components. It is just a matter of time before an architect in northern China specifies modern rammed earth as a structural material and interior finish, with 20–30 cm of insulation mounted to the outside, protected by external cladding.

One of the challenges of these new systems will be ensuring that they can be easily disassembled for recycling or reuse. For instance, the assembly mentioned above would require a modern rammed earth wall with a demountable substructure on the outside, insulation that can be easily stripped out and an exterior rain screen that can be unclipped. As modern rammed earth-related building systems begin to emerge, it will be critical that those who design these systems are thinking in reverse (in terms of disassembly) so that modern rammed earth does not follow the same path as brick, concrete, steel, stud frame, insulated masonry units, etc., which have evolved into hybrid systems that are almost impossible to disassemble cleanly.

24.5   Construction techniques and formwork

The traditional forming system in China consisted of sliding forms: essentially two wood side boards held together by through-pins and end panels. Earth was rammed into the resulting box, the panels taken down and reassembled alongside for the next block to be rammed in place. In this way, walls of any length and height could be rammed incrementally. Conversely, modern systems tend to be monolithic, with entire walls being formed at one time.

As mentioned above, our approach is founded on the Canadian SIREWALL system, where the formwork springs from the end-panels and is completely braced from the outside. In other words, long plywood side panels are horizontally supported by whalers spaced vertically every 300 mm. In turn, the horizontal whalers are supported by vertical end panels and vertical strong-backs if the distance between the end panels exceed 3 m. On the outside, the result is a forming system that also serves as scaffolding for the construction crew to climb up and down. On the inside, the result is a box that is unobstructed, which is ideal for running electrical conduits, rebar (steel reinforcement) and insulation (placed in the middle of the wall). The result is a monolithic wall with no vertical cold joints and, in many cases, no horizontal cold joints either. The monolithic system also creates a finer visual finish – an important consideration for a material if it is to become mainstream. As a side note on the quality of finish as related to form-ply, we have found that lower to mid-grade plywood produces a more consistent and nicer finish than high-grade laminated plywood. The ability of lower grade plywood to wick up a slight excess of moisture in the earth mix helps create a dry sandstone-like finish, as opposed to a smoother wet finish more akin to concrete.

Our challenge in China is that of forming multi-storey construction. These challenges include precision forming and bracing.

Precision forming is required because we use rammed earth as a final finish material, making the demands for forming more akin to fine finish structural concrete. Unlike regular structural concrete, which gets covered with a finish material, modern rammed earth leaves no room for error. Whereas these standards are fairly easy to attain at ground level, they become more difficult in multi-storey construction. This is mostly due to the external bracing.

In concrete multi-storey construction, formwork can be entirely braced from the inside. This is thanks to the horizontal ties that are used within the formwork. For our work we have opted not to use form ties for a number of reasons. One important reason is esthetics: a fine smooth surface appeals to a wider audience than one characterized by tie holes. Another reason is ease of assembly and workability: the use of form ties clutters up the space in the formwork and slows the ramming, particularly in combination with the central insulation plane.

However, our decision not to use form-ties does add constraints to the way we brace the formwork. Currently our forming and bracing techniques require one side of the formwork to be built up to full height, with the other side being raised as the layers of the wall get rammed into place. In a multi-storey building the ramming would happen from the inside, making the outside-facing portion of the formwork the full-height supporting one. Suddenly what is a simple proposition at ground level becomes a complex one several storeys up. On the one hand, climbing around the outside of a mid- or high-rise building to place, brace and remove formwork is a potentially dangerous and time-consuming proposition. On the other hand, the external formwork lacks a firm base to rest upon. It either requires supports that cantilever from the internal slab or requires an outside plinth to rest upon.

Without a doubt, multi-storey construction will require modular or pre-assembled form sections as well as external scaffolding. The impetus will be to do as much work as possible from the inside. Pre-assembled/modular forms where the actual spandrel panel, external horizontal and vertical bracing are monolithic would enable the outside facing formwork to be placed and cross-braced from the inside. However, this would not solve the greater issue of formwork removal. Once again, handling a heavy piece of formwork from the outside of a multi-storey building is a potentially dangerous proposition.

For this reason, scaffolding on the outside of the building will most likely be a necessity, either rising from the ground up or cantilevering from the finished floor slab. In China, both techniques are common for mid- and high-rise buildings.

A final proposition would be to mount and brace the outside facing formwork entirely from the inside, running it proud in all directions. This piece of formwork could be released and craned from above. This would potentially eliminate the need for scaffolding.

There are certainly other systems and methods that could be adapted from the world of reinforced concrete or that of SRE techniques that use form-ties. However, this research still remains to be done in China.

24.6   Case studies

This section is not meant to be an exhaustive study of cases built in China. Rather, it is intended to highlight the principal projects that have served as milestones in the introduction of modern rammed earth to China.

24.6.1 Split house, Commune by the Great Wall

This project, designed by Atelier Feichang Jianzhu (FCJZ) of Beijing, is the first project in China to use traditional rammed earth in modern architecture. Designed as part of the Commune by the Great Wall project and built in 2001, the architect's motivation was to develop a modern building type that could return to the earth, thereby greatly reducing waste over the life of the entire building. The earth mix was traditional, for which the soil was sourced from the neighboring plot where another house was being built for the Commune. The earth mix was unstabilized and inspired mostly by a traditional Korean approach, which involved sieving the earth by shoveling it through a screen and then mixing it with lime. The walls were hand-tamped, 600 mm wide and reinforced laterally with raw linen laid between every few lifts. The project was unable to obtain code approval and have an engineer sign off on the earth as structure, so the final walls support only their own weight and serve as a rain-screen. The actual structure is composed of laminated wood beams and columns, crossed-braced by steel cables.

Overall, the project represents an important milestone towards China's emerging interest in modern applications of the vernacular. At the time, the use of traditional techniques in a modern context was pioneering, and consequently the project enjoyed a significant amount of press. Unfortunately, this press was not successful in terms of creating a new market, or building acceptance for rammed earth. Its true success was in identifying the challenges that lay ahead of traditional rammed earth in terms of reaching modern requirements for performance, safety, quality control and scalability.

24.6.2 Cross waters eco-retreat, Huizhou

This project is better known for its use of structural bamboo than rammed earth. Similar to rammed earth, there is no section in the Chinese building code that allows for bamboo construction. Hence, without a doubt Australian architect Paul Pholeros must have faced significant challenges even in having his incredible bamboo structures implemented, let alone the rammed earth. Perhaps it is for this reason that the rammed earth walls in this project consist only of full-height, free-standing landscape walls that frame the entrance to the various houses within the retreat. In this setting, the architect did not need to obtain significant strengths nor did he need to pass the requirements of the code. Given that some of the locals were still familiar with traditional rammed earth, the approach in this case was simply to let the locals source the earth they normally would for this purpose and then stabilize it with a percentage of cement. Also, rather than building the walls incrementally with a sliding form, Australian forming techniques were introduced, thereby forming entire walls at a time. Although the scale is modest, it highlights a technique that is often used: that of a soil, which has been selected by trial and error and stabilized with cement. This represents something of a shot in the dark because, unless there is a particle size distribution analysis and the soil mix is well researched, the opportunity to optimize the strength of the walls, reduce the amount of cement and standardize quality cannot be attained. As neither code approval nor strength were of critical concern, the method used was evidently the simplest. However, in the interest of properly differentiating SRE from traditional rammed earth, the next evolutionary step would be to ensure that all walls, be they landscape or architectural, meet minimum standards for strength and durability.

24.6.3 River house eco-retreat, Zhangjiagang

This project evolved out of a client's unusual and pioneering request: to build an experimental retreat that used the most environmentally responsible techniques. In order to achieve this, the client was willing to pay the cost of innovation for technologies that would be more expensive at the onset, as long as they had the possibility of being scalable and cost-effective thereafter. These requirements set the stage for stabilized and insulated rammed earth to enter China. Not only did the walls have to be built out of local non-toxic materials, they had to be extremely energy efficient, meet the compressive, shear and tensile strength requirements of modern structures, cost-competitive, equivalent to standard building methods in terms of construction time and 'beautiful'. This was no small order, and getting there required a multi-step process, the first of which was bringing professionally developed SIREWALL techniques to China. At the time (2006), its main weakness was never having built multiple storeys of insulated rammed earth, nor having worked with alluvial soils – the challenges of which have already been discussed above.

A00 Architecture, the architects of the retreat, started by creating smaller opportunities to train their local builder: EMCC (from Einstein's $E = MC^2$, given the company's particular interest in doing things no one else has). The first of these was Just Grapes, a wine bar located in downtown Shanghai and completed in 2006 (see Fig. 24.2). Although the walls for this project were non-load-bearing and used only as an interior finish, the learning curve was high. First off, the project was akin to building a ship in a bottle, given the very narrow and long proportions of the space, as well as the necessity to form entire walls at once. The local tampers had a side-mounted pressure valve that sprayed excess air and moisture, but also oil leaking from inside the shaft. This resulted in black oil stains on the interior of the formwork

24.2 Just Grapes wine store: first stabilized rammed earth in China.

and subsequently on the walls. The crew also struggled with the appropriate water content, this being their first time. With a tendency for more water than less, many layers have a smoother, glossy look. On their first wall, the crew also struggled with the control of the surface finish, varying between very boney and perfectly distributed. This was the result of how they shoveled earth into the forms: if the earth was 'thrown' against the front side of the form work, the fines had a tendency to be projected out of the mix and against the finished side, while the larger pieces rolled out the back. This created a smooth finish. Conversely, if the earth was thrown against the back of the forms, the large particles rolled towards the finished side and the result was a boney finish. Also, attempting to achieve the smoothest possible finish, the plywood used for the formwork was faced with a black laminate, which heavily stained the earth walls. Finally, being the first project that used alluvial soil, about 15 different mixes needed to be calculated on paper, rammed and crushed in the lab before we achieved one that reached just over 6 MPa after two weeks – and this with the addition of 10% cement. Not great, but sufficient for our purposes.

The final project received awards and extensive press for beauty and design innovation. These were the first stabilized modern rammed earth walls in China.

A00 Architecture and EMCC went on to design and build [wi:]: a small free-standing test pavilion, which consisted of two parallel walls separated into six segments (see Fig. 24.3). Four of the segments were 6 m long, while two were 3 m long. All of them were 4 m tall and were used to test something slightly different.

24.3 [wi:] Studio: six test walls with water-cooled glass roof.

The walls were all built using the same reusable formwork, including modular aluminum end panels into which whalers and spandrel panels could slot. The first wall served as the pilot, with a 9-MPa mix that included 10% cement. Part of the wall also included 0.2% crystalline admixture. Although the soil was mostly alluvial and almost identical to that of the wine store, the rounder mid-sized gravel that had been used previously was substituted for a quarried, sharper gravel. The first wall also used the same black forms as the wine store. At the time, the oil projected by the local tampers was believed to be solely responsible for staining the walls, and so the valve was covered with a gauze, acting as a filter. Finally, the 6-m length was only vertically braced at the ends and in the middle.

The wall took two days to form and ram. It was stripped on the third day. The wall was sprayed for curing and in order to accelerate efflorescence, simulating long-term exposure to rain, with the effect that the portion which contained the crystalline admixture fared marginally better. However, the efflorescence was very pronounced overall. Also, the black stains were still present on the wall meaning the tamper was not the culprit. It was only by the third wall that the plywood was identified as the problem. We formed half of the wall with the black plastic laminated plywood while the other half was formed with a cheaper and rougher unfaced plywood. The difference was astounding: the unfaced ply left the colors of the wall unaltered.

The third wall was also used to train the team on carving figures into the formwork, in this case a full scale Corbusian Modular man. The remaining walls tested variations on soil delivery and spacing of the whalers and strong-backs. In the end, the six walls were covered by a glass butterfly roof with recirculating water in an effort to control the inside temperature with the flow of water. With earth walls on either side and the water running overhead, the experiential effect was that of being at the bottom of a fast-moving river. Although many of the techniques that we tested had been taught to us by SIREWALL, it was critical to enact them all in the Chinese context with a different set of tools, materials and workers – the latter being the most important. Training a group of people who could properly understand the process, strengths and limits of the system was critical to the success of future projects. Completed in 2007, these six walls represented the first free-standing and load-bearing stabilized rammed earth walls in China. They were, however, still uninsulated.

After these two projects, the client, A00 and EMCC finally felt confident enough to undertake the sample house at the River house eco-retreat (see Fig. 24.4). It consisted of a single storey of insulated rammed earth and included some of the more structurally challenging elements of the future main house. One such example was developing a special hidden steel lintel detail that would allow us to span a 3-m-wide opening without the addition of a visible concrete beam and without creating thermal bridging through

24.4 River house, sample house: first stabilized insulated rammed – earth in China.

the insulation. This was achieved with two steel angles placed against the central insulation, with their flanges facing out so as to support the rammed earth and the vertical rebar that continued up into the wall. The final span was monitored for settlement for one month. Three years later, it still hasn't moved.

Even though lessons are taught, it is clear that mistakes need to be made to understand the value of those lessons. Although the crew understood the importance of cross-bracing and how much lateral force the formwork was subject to during ramming, that lesson was learned once again on one particular wall. Being the tallest wall to date, the amount of cross-bracing was increased but the half-buried concrete blocks used to anchor the braces turned out to be insufficient. Two-thirds of the way up the wall, the force of ramming was enough to cause the bracing blocks to slip. Overall, the total slippage was about 0.9 cm spread over about seven lifts (about 105 cm).

A further lesson had to do with the challenges of ramming in extreme hot weather. Even though the crew understood the importance of keeping the base soil moist prior to mixing and ramming, a few days of extreme hot weather created additional challenges. At this heat, the soil mix dries incredibly fast and the quantities need to be kept as small as possible. Also, the side of the formwork facing the sun bakes the interior face of the rammed earth wall, causing the cement to dry before it cures. Consequently the surface of one of our walls ended up being 'soft' up to a depth of 1–2 mm. That is to say, a sharp metal object could normally only scratch the surface, but in this case it could scrape it off. Fortunately, this did not affect the overall strength of the wall.

However, perhaps the biggest lesson was the importance of designing according to the strengths of a material and avoiding its weaknesses: the main house had been designed before rammed earth had been chosen as a structural material, and included some very ambitious cantilevers. Therefore, the sample house was designed to include and test similar details. Here, two of the four walls contained an oversized puddled-earth cap that projected beyond the ends of the walls and joined to form a cantilevered corner, on top of which a steel column supported the roof. The complexity of this detail required the crew to stop ramming for the better part of a day as they installed the rebar necessary to achieve this detail. Of course, by the time the cap was ready to be poured, the rammed earth wall had time to dry and shrink just enough to pull back from the forms. EMCC knew this would happen, and tightened up the forms by hammering wood wedges between the whalers and the form-ply. Nonetheless, some of the mix from the puddled cap leached down the surface of the wall below. Here, the problem was not the skill of the crew but rather designing outside the limits of a given material.

Overall, the sample house was extremely successful. It allowed the crew to familiarize themselves with many new insulation details, in particular those related to connections with doors, windows and roofs. Completed in October of 2007, it took the further step of being the first stabilized and insulated rammed earth structure in China. However, the challenge of insulated multi-storey structures still lay ahead.

At its tallest, the main house was three storeys in height (see Fig. 24.5). It also included a free-standing central space that went up to two storeys in height.

The first challenge came with the foundation. In the areas where the structure were to be back-filled, the engineer had insisted on using reinforced concrete.

24.5 River house – main house: multi-storey stabilized insulated rammed earth.

Whereas he had agreed to design many ambitious details for a material he had never worked with, he drew the line at the foundations. The result was stepped concrete foundation walls that rose over one storey in height. The rammed earth walls were started directly on top of these foundations, and singular hairline cracks developed in the rammed earth at many of the 'steps' in the concrete foundation.

Another lesson was with ramming in near 0°C temperatures. On one particular night the weather dropped below zero with the wind chill. The last lift that had been rammed that day suffered the most damage, up to a depth of 1 cm on the outside of the wall. Interestingly, the lift right below it suffered virtually no damage. A couple of extra hours of curing had made a significant difference. Fortunately, the walls for this house were over-sized, totaling 600 mm in width with a central insulation core of 100 mm.

The house design included many ambitious elements, such as cantilevered portions of rammed earth walls as well as terracing, which were challenging in terms of structural detailing and water-proofing, as well as creating a continuous insulation plane.

However, by far the greatest challenge was in bracing the formwork on the second and third storeys. At each level, the top of the wall was finished with a puddled cap. Within that cap, a wood strip was inserted to create a shadow joint and to hold anchor bolts. The anchor bolts served to strap on a brace to support the outside facing form-work. In turn, the inside facing formwork simply rested on the new floor slab. Once the wall was completed, the formwork was taken off and the anchor bolts cut back and patched within the space of the shadow joint, making them quite discreet. Although this approach worked quite well it was far too customized to be a cost-effective and scalable solution. The formwork system was developed to suit the design of the house and, to be effective, the order would have to be reversed: future houses would have to be designed according to the formwork.

Mixing of the soil was done with a concrete rotary drum and then turned onto the ground to be finished by hand with shovels. It was then sent up in wheelbarrows hoisted by cranes. Again, although this worked well for one project, scaling up to build larger scale projects would require volumetric mixers and faster delivery systems.

Still, further lessons were learned with this project. For instance, the crew developed a two-person system, where one rammed and the other continuously 'sprinkled' earth into the forms. The result was a continuous and monolithic wall without any visible lifts and with a constant maximum density. Although this method would almost certainly make for a stronger wall, we never cored and tested any samples. The resulting wall was not visually compelling and the technique was discontinued.

A particular ongoing challenge was also mentioned earlier in this chapter: that of air pollution and premature ageing of the walls. Although this can be largely prevented with deep roof overhangs on a single-storey structure, the overhang stops protecting the walls after a certain height. We have found no solution to this problem as of yet.

Despite being a modest three storeys, the River House is currently the first and only one of its kind in Asia, and possibly the world. It has opened the door for research into multi-storey, stabilized and insulated rammed earth structural systems.

24.6.4 Naked Stables Private Reserve

Nestled in a valley that lies at the foothills of Moganshan (near Hangzhou), naked Stables is an ambitious resort that combines traditional rammed earth construction with curved, stabilized and insulated rammed earth walls. It is one of the few areas in the world where traditional and modern rammed earth were constructed at the same time. The traditional rammed earth is local to the area and is being used in the resort for structures such as the stables and the staff quarters. In stark contrast to this lie the stabilized and insulated rammed earth walls that make up the welcome center and clubhouse, as well as the 40 single-room chalets that dot the forest (see Fig. 24.6). The walls were designed by Delphine Yip and A00 Architecture, respectively, and were built by EMCC.

Although the original plan was to use site soil, it was tested as being too weak for the purposes of the modern walls. High in clay content, the parent rock was also extremely friable. Consequently all the soil came from local, neighboring quarries. However, the local site soil was used for the traditional walls, as it has been for decades.

24.6 Naked Stables Private Reserve: one of 40 identical chalets.

The most significant part of this project was the repetitiveness of a single chalet type featuring a semi-circular SIREWALL, justifying the production of curved reusable steel forms. Being extremely hands-on, many new ideas were tested by the owner himself: Grant Horsfield. The original design of the foundation by the engineer called for a deep concrete foundation; however, Grant mixed traditional and modern techniques to minimize the need for reinforced concrete. The base of the foundation was built out of dry stone walls, laid according to a local technique that positions the stones diagonally, thereby forcing the walls to further interlock after they are loaded. On top of these walls a 20-cm half-ring beam was poured in order to support the vertical rebar for the rammed earth wall that was to be built on top of the beam. At the time of writing, all of the walls were complete and the stone foundations had not shown any movement, with the oldest being almost 2 years old.

More than anything though, naked Stables is a perfect case study for the single most important challenge that lies ahead of rammed earth in terms of becoming a mainstream building material: protection of the final product. The key issue is that rammed earth is not only a structural material, but also a finish material. These two qualities are diametrically opposed in terms of when they should be appearing on site. The finished nature of the walls require that they be carefully protected until the very end of construction.

This particular project was an excellent example of how many factors can complicate the issue of wall protection, from damage done by the elements to that done by the rest of the construction process. The latter is especially challenging when multiple teams are used and when the order of 'required attention to detail' is reversed. In this case, not only were the teams for the rammed earth walls and for the roof structure different, but the latter had no culture of attention to detail. Structural teams – especially in China – tend to be extremely rough as most of their work gets covered up. Changing this entire culture represents a titanic task. For instance, the protective covering on the walls was stripped back and left to fall apart no matter how many times the project managers followed up. Habit and inertia are more powerful than common sense. From rust paint drippings to sloppy welding burn marks, many walls were irreparably damaged in numerous ways. In order to make the thermal barrier continuous, the rammed earth crew had intentionally left the insulation plane coming from the center of the wall exposed. Unprotected, this left a path for the rain to enter the center of the wall, flow back out through the rammed earth and cause rapid efflorescence. What's more, a black stain in the form of a film and streaks began to appear on the walls. The origin of the black staining is still not completely understood and is probably from multiple sources. Judging by our experience on the River House project, we think a significant portion comes from particulates in the rain as well as from the rust-proof paint. Most of the streaks run from the top ledge of the wall downwards, particularly in the areas where the steel rafters meet the wall. Finally, a portion might be coming from the impurities that are commonly found in Chinese cement, although this would have been minimized by the fact that white cement was used. Still, the black stains seem to be coming from within the walls, as though pushed out by water. Moreover, they only appear on the walls that were rammed in winter. White cement is certainly not perfectly free of impurities and it is possible that it did not fully hydrate given the colder weather, allowing residual impurities to be pushed to the surface of the wall by water. More research is required to arrive at a full conclusion.

Three of the 40 walls were more heavily affected than the others. They were the first to be completed and were wrapped prior to leaving for a three-week national holiday. In the first few days of the holiday the wrapping failed or was removed, leaving the walls exposed to direct rain for several weeks and causing heavy efflorescence. Moreover, these three walls were rammed in near-freezing weather, which we believe made the walls even more prone to efflorescence. Having built 40 identical walls where the progressive increase in temperature (from about 0°C to 30°C) was the only variable, and where all walls were exposed to rain, it is highly probable that walls rammed in cold weather are indeed more prone to efflorescence. As mentioned above, our only theory is that the colder weather prevents the cement from fully hydrating. This leaves many of the finer particles in the walls 'free' to be pushed out by water moving through the pores.

Many cleaning techniques were used on the affected walls, from water to muriatic acid with soft to hard brushes, from sponging to using a belt sander. The muriatic acid was generally successful at removing efflorescence, although it did also end up spreading the salts around and muting the walls. However, only the belt sander was able to remove the black stains. Still, none of the solutions were effective in bringing back the original beauty of the particularly affected walls. The 'old' rule stays the same: touch the walls as little as possible. That being said, the walls are still stunning, particularly when they are experienced as whole: all 40 of them nestled into the flank of a hill and weaving in and out of the forest.

24.6.5 Vidal Sassoon Academy, Shanghai

This project most likely marks the first instance in the world where a large multinational company (Procter and Gamble) has adopted modern rammed earth as a feature element for one of its major brands. In many ways, this signals a coming of age for modern rammed earth.

Located in Shanghai, designed by A00 Architecture and built by EMCC, the Vidal Sassoon Academy is similar in size and scope to Just Grapes: the walls are used as an interior finish and are non-load-bearing. Sassoon chose rammed earth as the feature of its flagship Academy in order to signal its commitment to ecological responsibility and innovation, and because it was an evocative representation of two key concepts that are shared by the fields of hairstyling and architecture: layering and shear (see Fig. 24.7). To this end, the team experimented with the control of color in the layering as well as representing shear through angled rammed earth walls. In terms of the latter, the tip of the main wall was tilted forward by 11°, while one of the side walls was tilted laterally by 11°. The team created these two conditions to experiment with ramming tilted walls. In order to account for the added vertical pressure from the mechanical rammers, the formwork was built stronger and the walls came out beautifully. However, despite its modest size, the true significance of this project is the introduction of modern rammed earth to a high-profile and mainstream market segment.

24.7   Future trends

China is a place of extremes and tends to be all or nothing. This is certainly true of the potential that stabilized and insulated rammed earth has in this country. China's current code system does not welcome new materials, and so the introduction of the technique begins with a significant uphill battle. Of course, once the country decides to adopt a certain practice an uphill battle can become one that is won overnight. Over the next couple of years, modern rammed earth will have to continue creating precedents and accumulating science-based data. It will also have to demonstrate how it is scalable. Armed

24.7 Vidal Sassoon Academy: large multi-national (P&G) features stabilized rammed earth.

with the proof of results and within our context of growing environmental challenges, modern rammed earth could be well positioned to have the code written in its favor. If this happens, we will see the development of multi-storey rammed earth structures, the adoption of insulation in rammed earth construction, the development of walls made almost entirely of waste material and, finally, the development of hybrid structures that leverage the structural and thermal mass properties of rammed earth. Beyond this, two great challenges will remain.

I am confident that we can deepen our knowledge of soil types and the science of blending, and thereby increase the strength of rammed earth walls. Following this, I am confident that we will be able to significantly reduce – if not eliminate – the use of cement in stabilized rammed earth. I am also confident that we will soon be able to build insulated multistorey structures cost-effectively. These are all technical issues that mostly require time and research. I am less confident about the time it will take for China to clean up its skies, and hence resolve the problem of the premature ageing of the walls. I am also less confident in the ability of people to properly protect the walls during the rest of construction. Although this is not an issue with a small, dedicated team working on specialty projects, it will most certainly become an issue with large-scale projects that employ hundreds of workers and see the regular turnover of project managers. Maintaining quality will require steel-fisted project managers, or cultures with a universal sense of personal pride and respect.

Without a doubt, if the opportunities come to outweigh the challenges and China chooses to invest itself, it could well become the global source of innovation for rammed earth construction.

24.8   References

China Geological Survey (2002) A series of small scale geological maps of China, Beijing. Available from: http://old.cgs.gov.cn/ev/gs/Geomap.htm

Cooray P G (1984) An Introduction to the Geology of Sri Lanka (Ceylon), 2nd revised edition, Colombo, National Museums Department

World Trade Press (2007) China population, Petaluma, Cal., World Trade Press, www.worldtradepress.com

United Nations/Department of Economic and Social Affairs/Population Division (2008) *World Urbanization Prospects: The 2007 Revision*, New York, United Nations Publications

Appendices

Appendix 1
Techno-economic analysis and environmental assessment of stabilised rammed earth (SRE) construction

R. LINDSAY, Earth Structures Group, Australia

A1.1   Introduction

This appendix deals with a technical and economic analysis of modern stabilised rammed earth (SRE) building and an environmental assessment in the light of this analysis. The analysis deals with the technical ability of these buildings to perform realistically within the confines of current economic parameters. It also deals with how design affects the economic and environmental cost of SRE.

Interestingly, the construction industry in Australia is rapidly adopting the notion of building to reduce environmental cost as well as economic cost. Recent carbon tax impositions will further create an incentive to build using less carbon-extravagant materials and, secondly, to design buildings that result in a frugal use of carbon-extravagant heating and cooling.

As a modern SRE wall builder, I deal with solving technical, marketing, pricing and environmental challenges every day. As such I have structured the appendix to review issues that drive the commercial and environmental realities of the modern SRE construction industry.

Please see below some definitions of acronyms that are regularly used throughout this chapter:

- SRE – In the Australian 'stabilised rammed earth' industry the product name is SRE. This is not a product brand in any way associated with the author – it is shared by the entire Australian stabilised rammed earth industry
- ESD – Environmentally sensitive design.

It should be noted that the earth buildings discussed in this chapter are stabilised using Portland cement. For a detailed analysis of the difference between stabilised and unstabilised earth walls in the context of a techno-environmental impact assessment see Section A1.2.4 of this appendix. The relatively temperate Australian climate calls for vastly different thermal specifications to those in say, Canada or the northern mountain states of the USA.

It should be noted that this appendix describes an Australian approach

715

to the economic and environmental issues pertinent to the earth building industry. In extremely cold climates extensive thermal barriers and vertical steel reinforcing are sometimes required. While the Australian SRE industry uses insulated walling systems, we do not have the extended cold climates that require high insulation to achieve thermal comfort. These (North American) earth walling elements would be considered extravagant given our more temperate Australian climatic conditions.

Furthermore, modern stabilised rammed earth operators in North America use customised ply and timber formwork as opposed to our re-useable manufactured formwork systems and operate with very different business models to those in Australia. The economic and environmental differences between the Australian and North American models are substantial and for this reason we decided to offer an outline of both approaches by way of this appendix for our readers.

A1.2   The technical parameters of modern earth wall construction

A1.2.1 Materials and locality

In the context of both the economic cost and environmental sustainability, the more local the source of base materials for earth buildings the better. In almost all cases, unless the client has an approved quarry on their property, site soil is not useful for modern SRE. There is potential for incorrect materials to be volatile, to heave, to crack and to shrink excessively.

Table A1.1 shows how materials must fit within the specifications of what soil types are useable. Source: Natspec Specification – Stabilised Earth, 2005 www.natspec.gov.au

Local earth building contractors have long-standing relations with their local quarries and understand the best material blends for different construction applications. The increasing use of recycled concrete and brick rubble products is a positive outcome particularly for urban projects where recycling crushing plants are often within a few kilometres of the site.

A1.2.2 Site and access

A major challenge for modern earth builders is site access. We are often described as a mobile factory. Base materials are delivered and dumped on site and we need room to store these materials, mechanically blend them and deliver them safely and efficiently to the wall face. Very small urban sites often have little room for dump truck access, mixing pads or delivery of earth to the wall faces. Some sites have improbable prospects for the use of SRE. Similarly, steep sites with multiple levels of concrete footings and

Table A1.1 Shows how materials must fit within the specifications of what soil types are useable.

Materials		
A. Gravels, laterite soils and soil blends Soil contents:		B. Recycled crushed brick rubble
Organic content	Less than 2%	Crushed building rubble ex nominated supplier to suppliers standard.
Clay and silt content	Material below 0.075 mm to be below 20%	
Soil contents	Material between 0.075 mm and 4.75 mm to be not less than 50%	Cement content shall be no less than 8%.
Gravel content	Material between 4.75 mm and 75 mm to be above 30%	Proportion to be determined by mix design and strength evaluation test.
Not more than 5% to be retained on 37.5 mm screen size. Cement content by volume shall be 6% minimum to 12% maximum determined by mix design and strength evaluation test.		

Source: Natspec Specification – Stabilised Earth, 2005

slabs can provide enormous and expensive challenges in terms of access and safety.

We are often asked why we don't use mobile mixing plants and elevating machines on small sites instead of our current skid-steer loaders, which may appear to require more room. The answer is economic. The cost or hire of specialised machinery is prohibitive given the scale of most small urban projects. Pumps, conveyors and pug mill mixers are expensive pieces of equipment that need to be financed, serviced and freighted using heavy haulage. They make the process more capital extensive, environmentally extravagant and thus less affordable. This doesn't eliminate the use of earth building in small urban areas or steep sites, but the cost of the walls is inevitably more than on larger, level and more accessible sites. Accessing very high wall sections or walls that have steel portal frames or steel cages obstructing formwork and rammers can make an otherwise efficient system slow down to a less affordable pace. Consulting with earth building contractors early about site access in the design stage can result in positive economic and consequently environmental outcomes.

A1.2.3 Designing to suit the earth building technology available

There is a balance between designing SRE walling for sensible outcomes and for extravagant outcomes. By sensible we mean design that takes into

A1.1 Steel portal frame within cloister walls at the RACV project, Torquay.

account the building shape, construction access and structural parameters that are sympathetic to the available formwork and construction system. By extravagant we mean buildings with wow-factor budgets but little regard to the buildability of the project. These 'build at all cost' projects can have side benefits for the industry by 'pushing the envelope' with new technology, but more often than not result in very complicated one-off formwork construction, tedious use of labour and an extravagant use of materials.

The balance between frugality and extravagance in construction is a moral dilemma that should be faced by all architects and clients. Anything can be built, but often at an extraordinary financial and consequent environmental cost. The ability of clients and their architects to justify extravagant buildings in terms of 'we are using sustainable earth walls in this large project therefore it is OK' is disturbing. In reality, massive budgets allow for anything to be built, but often the outcome is an excessive use of structural steel (you can make anything stand up with steel), reinforcing, bracing and custom-made formwork.

The techno-economic-moral challenge here for designers is obvious. Glamour and an obsession with 'design status' is achievable but at what environmental cost? Small, simple and elegant earth buildings can create wonderful outcomes for clients at a fraction of the cost.

A1.2 Port Phillip Estate under construction.

A1.2.4 Stabilised vs. unstabilised earth walls

There is a long-running debate in the general earth building industry about whether earth walling can still be considered 'earth walling' if it has stabilisers added. As earth walling contractors, we weigh towards using stabilisers for reasons to follow, but quite understand the context of a purist approach to the argument. For a more detailed analysis of soil stabilisation see Chapter 9.

For centuries most of the world's solid masonry buildings were built using unstabilised earth walls either as rammed monoliths or mud-brick structures. The Industrial Revolution and consequent reliance on manufactured building systems have made traditional unstabilised earth building diminish to a tiny fraction of the building industry in developed countries. There are obvious (and often disappointing) reasons for this shift away from unstabilised earth building. Firstly, modern codes and regulations require buildings to fit within structural parameters, most of which will not allow for unstabilised earth walls to restrain roof loads, point loads or weathering capacities.

Secondly, modern humans are suckers for marketing. For decades we have been conditioned to believe and trust as consumers that 'the more refined and manufactured the better'. This applies to our food consumption, our leisure and of course our living environments. It could be argued that for 'modern folk', the idea of living in a mud brick house when everyone else is in a fired brick or cement sheet house is somehow ridiculous. Maybe like wearing a woolly jumper when everyone else is wearing polypropylene fleece. Thirdly, the warranties placed on modern building contractors, architects and engineers for the design life of buildings are such that there is an understandable fear of using any products that are perceived as 'risky' in terms of weathering and durability. Building unstabilised earth walls is risky. They are prone to weathering at the base of the wall particularly and also on wall surfaces exposed to driving weather. Unstabilised earth walls are also prone to instability and face spalling, even within our current industry where some contractors insist that walls do not require stabilising.

A structural engineer designing load-bearing earth walls for even the most modest dwelling will be unlikely to accept a load resistance of less than 3 MPa. For moderate point loads they will require a load resistance of maybe 5 MPa, given a 300- or 400-mm-thick wall. These typical requirements to resist compressive or tensile point loads are not attainable for most load-bearing unstabilised earth walls.

There is the option of using unstabilised earth walls to clad steel or timber portal frames, as is often the case with traditional mud brick construction. By using a portal frame however, the environmental benefit of using no stabilisers is trumped by the high-embodied-energy materials and labour to erect the portal frame. There is also the argument of the life cycle of the building. The effect stabilisers have on the longevity of the building may balance out the increased embodied energy of the actual stabilisers.

A stabilised earth building that lasts 300 years may be a better environmental option than an unstabilised earth building that lasts 30 years. Maintenance issues are also valid. An unstabilised earth building in Mali may require rendering as a biannual event. This, given the traditions and artistic benefits

A1.3 Bird in Bush, London, UK.

A1.4 Unstabilised earth walls, Omeo, Victoria.

A1.5 Mali house.

of the activity, is a bonus to the community and owners of the building in Mali. But applied to the modern earth walling industry in a developed country the practice of re-rendering walls each year would be economically unsustainable.

Therefore, while unstabilised earth walls will always have a place in modern earth building, the use of stabilisers is considered a necessary ingredient for the existence of a modern commercial earth building industry.

We cannot avoid the fact that stabilisers are big polluters. Our challenge is to design buildings that require minimal stabilisers by using strategies such as protecting parapet wall heads, avoiding massive point loads, sheltering weather-prone walls and designing sensibly within the parameters of the modern earth walling systems available.

A1.3   An economic analysis of modern earth wall construction

The cost of SRE building is subject to all the normal cost elements of other solid masonry construction systems with the exception of large base-depot premises – SRE construction is site-built and requires no 'factory' to produce manufactured blocks or panels. There needs to be a distinction made between a commercial cost for modern earth walling and an owner–builder cost for the same. For the purpose of describing SRE wall construction in this appendix we believe it will be more useful to focus on the construction of earth walls in a commercial setting, i.e. undertaken by a contractor. The vast majority of SRE walling is built by commercial operators rather than owner–builders. The cost of modern earth wall building falls into these categories:

1   materials 35%
2   labour 32%
3   overheads 21%
4   machinery and formwork 10%
5   freight of construction gear 2%.

This analysis was undertaken assuming 400-mm-thick insulated walling. The site was 250 km from the depot. The earth was sourced 100 km from the site. The construction speed was an average 10.68 m²/day using a three-person crew.

Of course these cost percentages vary according to how remote the jobs are, the cost of the raw materials delivered to site and the cost of accommodation, etc. (in this context 'accommodation' refers to a short-term rental of a farm house, or cabin within a caravan park).

A1.3.1 Materials

Within the materials outlay, the cost breakdown is as follows:

1   quarried base aggregate (earth) 32%
2   insulation sheets and bridge pins 29%

3   cement 19%
4   plasticure waterproof admixture 7%
5   general hardware, damp course, control-jointing, hold-down bolts 7%
6   steel lintels 6%.

Base aggregate

Base aggregate costs vary according to the locality of the quarry and the commercial supply/demand context of the material. It also depends on how much the aggregate requires processing. For instance, a decomposing granite material may simply be excavated straight from the quarry face and placed in a dump truck for delivery to site. Other materials may require washing (to reduce the clay content), crushing (to bring the particle sizes within the appropriate specification) or blending with other materials to bring the matrix within the confines of the specification.

Insulation

Fifty-millimetre Styrofoam (the trade name for an insulation panel product made by Dow Chemicals) is used by most Australian SRE contractors to achieve an insulation rating to satisfy the 5 Star rating code in the southern states. The 50-mm-thick Styrofoam panels are placed in the centre of a 400-mm-wide wall thus creating a 175–50–175 mm sandwich panel. Eight-millimetre stainless-steel bridge pins with 50-mm returns at each end hook into the centre of each masonry skin. These bridge pins are located at 600-mm centres at each 600-mm lift. The thermal bridge they create is negligible – 0.008% of the wall total.

Cement

Bagged cement is obtained from the nearest cement distributor and delivered on pallets by truck. The pallets are unloaded using the skid-steer loader on site with a forklift attachment. Cement proportions can be varied according to the structural performance of the base aggregate. There are significant economic and environmental savings in making every attempt to source or blend optimal base materials to reduce the cement requirements of the walls.

Waterproofing admixtures

Plasticure (a waterproofing admixture made by Techdry Building Products in Melbourne, Australia) or other waterproofing admixtures are usually manufactured in the nearest metropolitan city and delivered in 20-l containers and stored on site. Accurate batching methods reduce wastage of these expensive admixtures. We ask our teams to treat the stuff like whisky – then wonder why they become ill!

Hardware

Most SRE hardware can be purchased at the local hardware store. This includes damp course, control joint weather-proofing strip, plywood or chipboard for making soffits or window block-out forms, nails, screws, etc. Hardware use can be minimised with good design. For example, by using similar sized window openings as often as possible, the same soffits or plywood block-outs can be re-used time over.

Steel lintels

Finally, steel lintels can be fabricated at the nearest steel shop. They are simple to make and are generally prime coated, though galvanising is advisable within 40 km of the coast.

The economic and energy cost of materials

The economic cost of all these ingredients will vary minimally according to the distance of the site to the nearest product distributor. The embodied energy cost however increases markedly as the freight distance increases for bulky, heavy items such as earth and cement. See Section A1.4.1 for further information.

A1.3.2 Labour

The labour component of SRE building is moderate compared with say, 200-mm-thick concrete block-work walls, the costs of which are broken down in Table A1.2.

Experienced labour for modern earth walling is a scarce resource, as any commercial earth building contractor will tell you. The number of experienced operators will increase as the demand increases, however it is not as simple as looking in the local paper for a rammed earth foreman – there are very few, unlike finding say a brick-layer or concreter, of which there are many. In some cases experienced staff may be less likely to stay in the industry for long periods of time due to the prevalence of 'away' work. Another labour problem is that due to the demand fluctuations of our smaller industry there are occasions where teams are quiet for weeks on end and may require other work. 'Communities' of rammed earth contractors within one country or even internationally can and will share labour depending on demand. This

Table A1.2 The comparable cost breakdown of concrete block-work walls and ISE walls

	200 mm concrete block-work	400 mm ISE
Labour	41%	32%
Materials	39%	35%
Overheads	12%	21%
Steel lintels	5%	2%
Machinery	3%	10%

ISE = insulated stabilised earth
Source: Ange Golin Bricklaying, VIC

happens regularly within the global Earth Structures Group and is one of its greatest assets. However, there are advantages with the SRE industry from a worker's point of view. For instance, the skill set required on a SRE site is diverse. A good team has everyone being multi-skilled – machinery operation, machine maintenance, erecting and dismantling of formwork, ramming, placing services into wall cavities, reading plans, dealing with architects and clients, dealing with the vagaries of weather and site conditions, working out efficient schedules for construction, procurement of materials, maintaining a neat site and, importantly, getting on together.

A rarely identified but important factor is that the sites for most SRE buildings are beautiful and the clients and builders are generally people who enjoy the challenges of interesting structures. This makes a difference to the happiness of the work environment, unlike a site using highly manufactured and generic products with regimented construction systems and careless clients.

There are also social benefits. Most customers and onlookers are genuinely interested in the process of SRE construction. It makes for good discussion about work which would not occur otherwise. Because teams are often away from home, there is an added cost of accommodation and travel over more common systems of construction. There is no doubt that a project near to a SRE contractor's base is economically preferable to a remote one.

A1.3.3 Overheads

The overheads cost of any product or service is hard to quantify because it will vary markedly between similar businesses producing the same product. Needless to say a small owner-operated SRE contractor whose partner may do the books and has a steady turnover of local work without requiring a large marketing effort will have a much smaller overhead component that a more corporate SRE business with a wider area to manage. A corporate 'group' of contractors however can pool resources for costs such as website development and group purchasing of products. The commercial advantage of sharing ideas during challenging projects is also significant. Commercial projects require higher overhead margins than small domestic work due to the bureaucratic nature of commercial project management.

A1.3.4 Machinery and formwork

SRE building requires elements of mechanisation to remain viable in terms of construction speed and workers' well-being. In early days this author had experiences of hand mixing, hand delivering and hand ramming of earth walls. Manual production of rammed earth walling is frankly boring, repetitive and exhausting work. Mechanisation where available is imperative, so long as the cost of financing, servicing and freighting the machine is not unaffordable. Since its inception in the early 1980s, the Australian SRE industry has relied on two significant items of machinery, the skid-steer loader and the diesel-powered air compressor. Other machinery such as lifting equipment (scissor-lifts or telescopic handlers) is generally hired for the duration of specific projects. There are two expensive capital elements required to efficiently construct modern SRE walls. One is machinery (a skid-steer loader and an air compressor as mentioned). The other is manufactured formwork.

Machinery

The skid-steer loader and diesel-powered air compressor need to be reliable to enable a viable pace of production. A mechanical breakdown of machinery effectively halts the site, and remote repairs can be costly and frustrating in down time. Hence most modern operators will tend to finance newer equipment and spend considerable money maintaining their machinery to avoid mechanical breakdowns. We are often asked why we don't use various material delivery or mixing machines over and above our skid-steer mixers. The answer is cost. The more items of machinery owned or hired by the contractor, the more potential for finance stress, mechanical breakdowns and freight costs.

Formwork

In pure cost terms, we believe manufactured formwork has many advantages over site-made formwork. The capital outlay for a new set of manufactured formwork may be a daunting financial obstacle for many wanting to build their own house, indeed a prohibitive cost. However, for a contractor intending to build many buildings throughout their working life, a well-designed and fabricated set of formwork can last indefinitely.

The Earth Structures Group has several sets that between them have built

Table A1.3 The accumulative cost of maintenance and equipment

Equipment	Cost (AU$)
Stabilform formwork set	82000
New 50 horse power skid-steer loader with forks attachment	54000
New 100 cfm diesel air compressor	22000
New pneumatic rammers × 3 with hoses	9500
Specialised pneumatic and electric hand tools	6000
Total	173500

many hundreds of houses and commercial projects. Some of these formwork sets are over 25 years old and still entirely serviceable. Using manufactured formwork there are no requirements for the construction of customised shutters. This eliminates the cost of employing skilled form-workers. It also eliminates the time spent by the team making up customised forms rather than getting straight into ramming walls. There is a cost element in maintaining the formwork. Approximately once a year the plywood or polypropylene faces of the forms are sanded thoroughly and re-coated with an abrasion-resistant epoxy resin.

The cost of maintenance, finance and depreciation for a set of this equipment, including machinery, is approximately AU$74000 per year, which needs to be amortised across the contractor's annual quoting system.

A1.3.5 Freight

Freight costs vary obviously depending on the distance of each project away from the contractor's base depot. The most cost-beneficial rule regarding freight is to keep the load within the confines of one flat-bed truck only.

A1.3.6 Benefits of local SRE contractors

Obviously, the more local the SRE contractor can be to the project, the greater the potential for economic and environmental savings. This works on many levels. Locally employed SRE specialists won't need to travel so far to the project. The freight component will be far less. The accommodation costs may be minimal as each worker can return home each night with less travel time and less fuel cost. The social impact on the staff will also be less.

The advice a local SRE operator can give the architect and head builder towards the design and 'buildability' of the project will commonly be more appropriate if the operator is a local. They will understand what design issues work locally, which quarry products suit which wall outcomes, and they can appear in person for meetings without the extravagance of flying or driving long distances.

A1.4   An environmental analysis of modern earth wall construction

The environmental costs of most building materials are assessed as:

1   the embodied energy of the material
2   the life cycle energy of the material
3   the toxicity of the material throughout its lifespan
4   the lifespan of the material
5   the degree the material can be recycled
6   an assessment of all these in combination.

A1.4.1 Embodied energy

The embodied energy of SRE is the energy consumed by all the processes associated with creating the wall, not the operation and disposal of the building, which we will discuss in the life cycle section below. The embodied energy of SRE includes the analysis of the following consumables used to create the walls:

• the excavation, processing (if required) and delivery of the base earth material
• the manufacture and delivery of the cement stabilisers used
• the manufacture of the waterproofing admixtures used
• the manufacture and delivery of the formwork and machinery (calculated over the lifespan and disposal of both)
• the fuel used to power the machinery on site
• the fuel used to freight the machinery and formwork to and from the site
• the fuel used to transport the team to and from the site
• the manufacture and delivery of wall coatings used
• the embodied energy ratings are divided into two parts, one the Gross Energy Requirement (GER) and the other the Process Energy Requirement (PER).

The GER describes the energy used in the entire production of the material, including the upstream components; for example, it includes the power used for the lighting of the office of the factory where the cement was made – this is an extreme example but gives you the idea. The PER is perhaps truer to reflect the process of building SRE walls. It can be easily identified and quantified.

Naturally there will be variations between different operators as the energy used by one may differ significantly from the next depending on how frugal they are with the maintenance and replacement of machinery or their operational efficiencies, etc. The PER embodied energy of standard

SRE (uninsulated) is 0.7 MJ/kg. This compares with the MJ/kg rate of these other solid walling products:

- imported dimension granite 13.9
- in situ concrete 1.9
- precast tilt-up concrete 1.9
- clay bricks 2.5
- concrete blocks 1.5
- stabilised rammed earth 0.7.

(Source: Lawson, 1996)

It is worth noting that low embodied energy consumption is highly desirable. Embodied energy can be the equivalent of many years of operational energy. The single most important factor in reducing the impact of any embodied energy is to design long-life, durable and adaptable buildings. In this sense, using low embodied energy products is worthless if the building will have a limited lifespan. SRE rates well in the embodied energy stakes. Generally, the more highly processed a material is the higher the embodied energy. The process of building SRE is relatively simple. Raw material is quarried locally, minimally processed (crushed perhaps) and delivered to the site. The material is then blended with stabilisers (cement is the highest contributor to the 0.7 MJ/kg rate) and rammed into in situ formwork. The formwork itself has an incredibly low MJ/kg rate as each section of steel-framed form will last indefinitely and requires minimal maintenance. If the facing fabric of the formwork is made using recycled polypropylene (chopping board plastic) then the MJ/kg rate is even better.

A1.4.2 The recurrent energy use of SRE buildings

The recurrent energy use of buildings can also be referred to the life-cycle impact. The energy requirements specific to running SRE buildings can be described within two basic variables:

1 design of SRE buildings
2 management of SRE buildings.

Design of SRE buildings

Firstly design – the three defining factors for efficient design of residential SRE buildings are orientation, ventilation and size.

Massive masonry walls such as SRE have enormous thermal mass and need to be used by designers to enable the building to absorb as much winter sun as possible and to be shaded from as much summer sun as possible. Unless the building is in an Alpine environment, which may have an extra two months of cold weather tacked on, most Australian architects will design buildings for the hot months, for while summers enable some wonderful outdoor living they can make building interiors like ovens. SRE walls can serve to provide a summertime solar bank, storing night-time coolth in their massive fabric, which can be released back into the interiors of the building during the day. This cycle of storing and releasing coolness must be managed ('Management of SRE buildings') or else the reverse situation of passive heating during hot months may start to occur.

The walls also need to be able to absorb night time coolth during hot summer months and to be protected from fierce afternoon western sun throughout the same hot months. To enable these passive solar benefits good design will ensure the building has large northern openings allowing low-cast north winter sun into living areas. The correct orientation of the building to allow northern winter sun into living areas is paramount. The Australian summer is a long and relentless test for humanity. To gain optimum passive cooling, however, a well-designed building will allow the cross-ventilation of breezes to pass through living areas. The slightest zephyr will cool a human and it is not uncommon to see our clients parked nearby an open window on a hot day. If window placement can allow breezes to pass across rooms to an adjacent opening, then all the better. Casement windows (that wind out sideways) can be used to funnel passing breezes into living areas more effectively than sliding windows.

Perhaps the greatest, but least recognised aspect of environmental design is the actual size of the building. While this segment is not specific to SRE buildings we need to include it in this discussion. It is a largely hidden aspect of the Australian ESD debate.

Obviously the larger the building the more embodied energy is required. There are more materials used to build a big house than a small one. Secondly, the larger the building, more energy is required to heat and cool it. This obvious assumption applies to all buildings regardless of the materials used. Reducing the size of buildings is a sensible way to help occupants remain more thermally comfortable for less energy outlay.

Interestingly, in a society such as Australia currently dominated by environmental and financial anxiety, we still manage to overlook the effect of our housing size. Australians are, as of 2009, the occupiers of the world's largest dwellings, with the USA now second. This issue is pertinent to all building materials, not just SRE. All the efforts towards reducing both the embodied and life cycle energy of building materials are trumped by our obsession with big houses. Our residential occupancy space per individual has increased 400% since 1960 (Year Book Australia, 2005). In 1960 these were Australian home buyer expectations:

• children shared bedrooms

- families shared one bathroom
- families shared one living room
- families shared one garage car space.

In 2010 these are Australian home buyer expectations:

- Children have their own bedrooms. Master bedrooms have en-suite bathrooms. Children do not share bathrooms with parents
- There are two if not three separate living areas given names such as rumpus rooms, studies, media rooms, gymnasiums, billiard rooms and so on
- Garages have space for two cars.

The financial demands placed on Australian families due to mortgage repayments and the fluctuations of interest rates has major social implications, again rarely put in the context of house size. The 1960 cost of an average Australian house equated to ×2.5 years average gross family income. Our current ratio is ×7.25 years average gross family income (Yates and Milligan, 2007). The Australian expectation of large house size is not on the current ESD agenda but is an obvious remedy for many environmental and social challenges ahead.

Management of SRE buildings

There is no doubt that buildings with solid masonry walls require sensible management to enable the building to perform to create thermal comfort. The beauty of solid masonry walling such as SRE is that well managed buildings need not require air-conditioning as opposed to highly insulated lightweight buildings.

A1.4.3 Case study of SRE house management

As an example of effective SRE design I will describe the difference between living in in a brand new brick veneer-clad house stuffed with insulation with a high R value (my rented residence while working on a large construction project at Torquay, Victoria) and living in our own uninsulated family SRE house at Mansfield, Victoria.

Both buildings exist in a similar climate, though the Mansfield winter is colder for a few weeks longer than the coastal Torquay. It is a stinking hot summer week at Torquay. I finish work at 5pm and return to my suburban brick veneer house. I open the front door and although the interior had been tepid in the early morning when I left (most summer evenings cool down enough to sleep comfortably if you leave the windows open) the house is now warm. I close the front door quickly to stop the hot afternoon air from rushing inside. However, after going in and out a few times to do the garbage and to get things from the car, the hot outside air is now inside the house and it is really uncomfortably hot. What to do? I find the remote control for the refrigerated air conditioner and crank it up to a temptingly cool 20°C. The machine blows refrigerated air into the house and after half an hour I am almost cold. I open a window but the cold air rushes out and it's hot again in minutes. My skin is dry from being cold. Now I'm really uncomfortable. I reach for the remote again. The machine has to work even harder to regain the coolness. I am living in a mechanically conditioned environment. I check the power meter. It spins madly. I leave the machine running until I go to bed at 9pm and only then open the windows to let in some cooler night air. The plasterboard interior fabric of the walls has no hope of holding any of the air conditioned coolness and it's soon warm again ... During the day I come back to do some work on the kitchen table. The house is hot again. I have to turn the machine on to get the inside temperature down to a workable level of 26°C. When I leave again two hours later I turn it off, hoping the highly insulated walls and roof will keep the place cool enough for when I return. I must admit it is still quite OK when I get back after work.

I have learnt my lesson. The only way to manage the climate of this house is by using the machine and by keeping the place tightly closed up.

The following afternoon I am home at our rammed earth house near Mansfield. It is even hotter – 35°C at 5pm. On that same morning my wife and kids had woken and closed the windows and curtains of the house before heading off for the day.

The previous night they had opened all the windows and doors (we have fly screens to stop mosquitos and moths coming in) to allow the cooler night air to swamp the interiors of the house. By closing the house up the following morning they prevented the hot daytime air from coming in, and shaded the windows from radiant heat. Throughout the day the coolth stored in the interior fabric of the massive walls from the night before radiates back into the house. During the summer days the inside of the house is shady, almost dark. It is a pleasant refuge from the fierce heat outside.

At 5pm the house is still quite cool. We have dinner and at about 7pm the outside temperature has fallen to about 18°C. Out of habit we open all the doors and windows and head off to bed. It's a warm night but a slight breeze pushes the air around and keeps us cool enough to sleep. Apart from a fan we run occasionally on really hot, still nights beside the kids' beds, there is no mechanical air conditioning. A properly designed and managed SRE house does not require an air conditioner in hot climates, so long as the evening temperatures fall to comfortable levels.

For the five or six really hot nights we get during the year at Mansfield, we manage with bedside fans, which use very little power to run.

This scenario is typical with many hundreds of our SRE house clients. As inhabitants we have developed regular habits of managing our houses – habits which are simple and sensible and by no means arduous. Interestingly, it is refrigerated air conditioning that brings the power grid to its knees every summer in southern Australia.

During winter the process is reversed. We draw open the northern curtains of the house to allow low cast sunlight into the living areas of the house. At night we close these and light the fire in the living area. By morning the fire has gone out but the living area is still warm enough. We open the curtains to let the morning sun in. Our bedrooms stay cold. We have a warm shower and get into bed and get warm. That's what beds and duvets are for.

A1.4.4 Toxicity of SRE buildings

Depending on the binder used to seal the internal faces of the SRE walls, there is little or no toxicity released into living areas. In Australia the wall binders used are diluted acrylic resins or diluted PVA emulsion. Both are applied as soon as the walls are dry. Because the management of SRE buildings involves plenty of natural ventilation (unlike living environments dependent on refrigerated air conditioning), a good circulation of fresh air lowers potential toxicity from carpets, paint, etc.

A1.4.5 Lifespan of SRE

Two things reduce the lifespan of SRE walls. First is rising damp. The prevention of rising damp is dealt with in Chapter 22. The second is uncapped wall tops. Again, this is dealt with in Chapter 22. Otherwise, properly built SRE walls have an unlimited lifespan.

A1.4.6 SRE as a recycled product

SRE can be demolished and crushed in the same way as concrete or bricks. The fines and particle sizes within crushed concrete or crushed SRE aggregates are suitably stable for recycling as road base, pathways or indeed as an SRE base.

A1.5   Conclusions

Given the complexity of manufacture, delivery and construction of most common building products in Australia, the simplicity of SRE construction (in an overall context) appears to contribute a relatively small environmental impact.

It is apparent the more processing events in each product, the higher the embodied energy. SRE performs comparatively well in this capacity. It

Table A1.4 The comparative environmental impact in terms of embodied energy of four typical Australian masonry construction systems

Energy contributor	SRE	Double brick	Rendered block-work	Brick veneer – including timber frame, plaster, insulation and paint
Source of raw material	Equal	Equal	Equal	High
Delivery raw Material to factory*	Equal	Equal	Equal	High
Manufacture	Nil	Extreme	High	Extreme
Delivery from factory to site	Nil	High	High	Extreme
Assembly	High	Low	Low	High**
Embodied energy of assembly MJ/m²	405	860	492	906
Embodied energy of raw materials MJ/kg	0.7	2.5	1.5	2.1

* In the case of SRE the 'factory' is the building site
** There are multiple trades used contributing to multiple deliveries, etc.
Source: Embodied energy amounts from Lawson, 1996

is also apparent that given the ability of well designed and managed SRE buildings to achieve thermal comfort without mechanical air conditioning, the environmental impact of the product is comparatively low.

A1.6   References

B. Lawson (1996) *Building Materials, Energy and the Environment: Towards Ecological Sustainable Development*, RAIA, Canberra (Source: *Lawson 'Buildings, Energy and the Environment' 1996*)

Judith Yates and Vivienne Milligan (2007) *Housing Affordability: A 21st Century Problem*, National Research Venture 3. Australian Housing and Urban Research Institute, www.ahuri.edu.au/search.asp?CurrentPage=1&sitekeywords=housing+affordability+venture+2007&SearchType=And

Year Book Australia, Australian Bureau of Statistics (2005) Australian house size is growing, www.abs.gov.au/ausstats/abs@.nsf/Previousproducts/1301.0Feature%20Article262005?opendocument&tabname=Summary&prodno=1301.0&issue=2005&num=&view=

Appendix 2

Techno-economic analysis and environmental assessment of stabilized insulated rammed earth (SIREWALL) building

M. KRAYENHOFF, SIREWALL Inc., Canada

Abstract: The aim of this appendix is to provide a North American perspective of the emerging modern rammed earth industry and how it is continually affected by changes in the economy, environmental awareness and technology. Many of the widely held assumptions about the environment and greenbuilding are questioned and a way forward offered. The challenges and opportunities created by changes in the business cycle are explored. Upstream technological solutions are advocated and anticipatory design is seen as essential.

Key words: rammed earth, insulated rammed earth, stabilized insulated rammed earth, SIREWALL, environmental building metrics, embodied energy, Cradle to Cradle, rammed earth technology.

A2.1   Introduction

This Appendix is about three dynamic aspects of modern earth building and how they relate to one another. The three are constantly changing and interacting on a macro and micro level. They are the context inside of which earth building takes place.

Environmental – Increasing awareness of climate change and peak oil has resulted in an incipient shift in our societal institutions and practices, including how we build. How we measure building performance has an extraordinary impact on environmental building. Furthermore, how we build with earth is constantly evolving. The challenge is to maintain a small footprint while redefining what it means to achieve health, durability and energy efficiency in our buildings.

Economic – All sectors of human endeavor are impacted by business cycles and earth building is no exception. Due to the small scale of the emerging earth building industry, downsizing due to a sluggish economy will gut many companies. Conversely, when there is a burst of demand it is difficult to accommodate the need. At the same time as this feast/famine economic phenomenon, there is a strong desire among earth builders to make their services and products more affordable. At present it is primarily the cost of marketing, educating the public and doing R&D that hinder affordable

735

earth construction. There is some sharing of these costs in the earth building community, but the highly competitive environment limits collaboration.

Technological – Earth building technology improves as modern earth building increases in popularity and as its principles and techniques are shared around the world (e.g. this book). Its integration into standard construction is progressing as each side learns the needs of the other. There is no precedent for the capabilities and scale of today's modern earth buildings. What the market demands of buildings is ever changing and yet sustainability requires a long-term perspective. Anticipatory design is a key tool to get us there.

A2.2   The environmental impact of stabilized insulated rammed earth building

A2.2.1 Building for tomorrow's environment

Rammed earth building offers the possibility of building or rebuilding with significantly increased resilience in the wake of the more frequent and devastating natural disasters that are occurring. For example, with the rebuilding of housing in New Orleans post-Katrina, the wisdom and validity of rebuilding in the same location with designs, methods and materials that did not withstand previous disasters should be questioned. Rebuilding with earth containing hydrophobic admixtures is a sustainable, healthy solution.

There are frequent and destructive fires on the coast of California, the inland portions of Washington, Oregon, Idaho, Wyoming and the Canadian Okanogan. Small communities are repeatedly evacuated and consumed by fire with only the inorganic parts of the buildings such as chimneys and foundation walls remaining. When they rebuild, they rebuild with wood. Earth building is non-combustible.

The big earthquake in Haiti left an abundance of rubble and thousands of people without homes and work. The chosen solution was to clear the rubble and dispose of it in order to build new buildings from imported materials. An opportunity missed was using modern rammed earth to create seismically resistant housing that is more comfortable and healthy using local labour and materials (rubble run through a concrete recycler), both of which were in abundance.

Every year twisters touch down in Kansas hurling roofs and entire houses into the air. Every year in the southeastern US, people board up their houses and evacuate in fear of hurricanes. According to a local demolition specialist, mobile homes (trailer park houses) may weigh as little as a couple of tons, while stick frame houses weigh 50 (2.4 kilo newtons/m^2)/sq ft lb or 50 tons for a 2000 sq ft unit (45 tonnes for a 186-m^2 unit). If that 2000 sq ft house was rebuilt with modern insulated rammed earth, its weight would be 150 tons, which is considerably more difficult to blow over. Every year trailers and houses are destroyed and replaced with identical lightweight units. We can do better. Many houses are destroyed by termites, carpenter ants, powder post beetles, rot and mold, all of which attack the wood framing. Organic materials by their nature decompose. Houses destroyed this way are typically rebuilt with wood framing with an added host of toxins to make the wood less appealing to nature's decomposition agents.

In all of these circumstances the opportunity for modern rammed earth construction is to provide resilient, durable, healthy and sustainable environments that have a better chance of weathering and surviving extreme weather and pests.

A2.2.2 Environmental building metrics

How we measure something largely determines the outcome. The impact of the methods used to measure something has a significant impact governing perception, values, choices and actions. It is difficult to underplay the importance of metrics. One universally recognized metric is the gross domestic product (GDP), which measures economic activity regardless of what the activity is. Economies are measured and policies created based on GDP statistics that encourage busy-ness, regardless of whether that busy-ness improves people's well-being. When the earthquake, tidal wave and nuclear meltdown impacted northeast Japan, there necessarily was a lot of activity that wouldn't have occurred otherwise. Controlling and cleaning up the Fukushima disaster was (and will continue to be) an addition to the Japanese GDP. Lawsuits contribute to the GDP. Building disposable housing (or anything else disposable for that matter) is typically good for GDP. The environmental building community has been searching for a metric that produces the desired results, although there is little agreement on what those desired results are. There is no agreement on a destination, target or where we want to get to. Few metrics recognize that disposability (short lifespans of materials or buildings) can be addressed by durability. There is considerable agreement on what we don't want and very little agreement on what we do want. We don't want unhealthy, energy intensive, transportation intensive, material intensive, carbon intensive, ugly buildings that are disconnected from nature. Even the definition of sustainability (i.e. not diminishing the prospects of our descendants) states what we don't want. Reactive metrics that reward less-ness are the result. Buildings can be awarded points for more insulation (less heat loss), smaller carbon footprint, use of recycled materials (less new materials), harvesting and recycling water (less potable water demand), reducing the need to drive (less oil consumed), etc. It is difficult for people to get really excited about doing with less. Less is not the most alluring prospect. Reductionist metrics embody an intrinsic acceptance of the current paradigm and context that have created the circumstances we are dealing with as well as conventional and habitual resolutions, leaving little room for imaginative approaches and creating progressive opportunities.

Most building metrics ignore the impact of disposability. The result is 'green' buildings that use technologies that stop working and are expensive (or impossible) to repair in 20 or 30 years. In our pursuit of green-ness we have unintentionally created metrics that embrace (or at best ignore) our worst habit, disposability. Ignoring disposability is tantamount to ignoring the needs of our grandchildren and their grandchildren.

Alternatively, we might envision a bright and ambitious 200-year future and backcast to define the steps we need to take this year to get there. It is imperative that we start to rally around a great vision of the future. The absence of a positive, agreed-upon, collective future is probably the largest single failing of the environmental movement, including environmental building. We believe that SIREWALL offers a small component of a future that can be proudly handed down through the generations. The more SIREWALL buildings replace disposable buildings, the better. A future where buildings are not a burden in terms of maintenance, health, safety, environment and a sustainable economy yields space for other creative and positive uses of our energies.

Collective agreement about what we want over the long term is essential, not only for the well-being of the planet and future generations, but for the present. Creating metrics that honour the wishes of our 200-year descendants is the place to start.

A2.2.3 Questioning the importance of embodied energy

In the environmental building industry a lot of time and energy is spent discussing the embodied energy of various building materials and methods. For the most part these are myopic analyses of initial conditions. A common criticism of cement-stabilized rammed earth is the amount of embodied energy contained in the cement. This type of analysis is what has landed us in the environmental situation that we find ourselves in today. What matters is the broader picture. How long will it last? Inside the context of the building and over the lifetime of that building, what scale of impact does it have? Broader yet, how does it affect health, comfort, fire safety, seismic stability, beauty and acoustics? There is a study from the University of Michigan (Blanchard and Reppe, 1998) that demonstrates where energy goes over the 50-year lifetime of a conventional house. Surprisingly, only 6.1% of the energy is consumed in the construction phase, while 93.7% is consumed in the occupancy phase and 0.2% in the deconstruction/demolition phase. Now if that building were not disposable (50 years is a very short lifespan) and instead lasted for 400 years as many buildings in Europe and around the world do, then the energy consumed in the construction phase would be amortized over eight times the lifespan, resulting in less than 1% (6.7% divided by 8) of the total energy consumed over the life of the building. Consider also if that building were to be net zero for heating and cooling over its lifespan: the energy consumption would drop by at least half during the occupancy phase as half the energy consumption in a building goes to heating and/or cooling. Both increasing the lifespan of the building and removing requirements for heating or cooling are at least an order of magnitude more beneficial than trying to reduce the initial embodied energy during the construction phase. Embodied energy, while not totally irrelevant, is a red herring that distracts us from more fruitful environmental endeavors such as eliminating disposable thinking/building and establishing net zero as a baseline for buildings.

A2.2.4 Reducing building size

It has become popular over the last decade to reduce the size of buildings. Sarah Susanka's 'not so big house' movement has captured the attention of people who are concerned about the environment. While the intention behind this movement is laudable, reducing house size is one of the least effective solutions. I grew up in a neighborhood of 33 × 120 ft lots. Every house was required to have 3 ft side yard setbacks and front and backyard setbacks that resulted in houses that were uniformly 27 ft wide by 60 ft long. While the long sides of the houses received very little light and had no views, they also had ~70% of the house's wall heat loss surface area. Extrapolated over a one-block area approximately two/thirds of the heat loss is expected through walls that faced each other only 6 ft apart. For wood frame construction there are benefits to the 6-ft gap, such as fire retardation and acoustic attenuation. However, if the gap is replaced with a 24 in. SIREWALL, then the fire safety and the acoustic attenuation improve substantially. Each building also gains 48 in. of floor space on each side resulting in an additional 20% of interior square footage, which is of significant value in urban contexts. The one drawback in the shared wall approach is a cultural stigma against attached housing, which has its roots in the context of lower economic status, less autonomy and privacy, poor acoustics and increased fire experiences, as well as unimaginative designs associated with wood frame-attached and multi-unit housing. Perhaps there is also a metaphorical aversion to rubbing shoulders with our neighbors.

Building smaller houses does not guarantee better thermal performance. Neither does it guarantee fewer materials are consumed, when viewed from a single point in time. Building eight 'not so big' (say 1000 sq ft) houses that need to be rebuilt every 50 years will have a much higher environmental footprint than one house with a 400-year lifespan that is ten times as big (10,000 sq ft). For some this might seem counterintuitive. Let's look at the wall areas per square foot of floor area. If the 'not so big' house is 25 × 40 ft with 10-ft tall walls, then the exterior wall area is 1300 sq ft containing 1000 sq ft of heated space or 1.3 sq ft of wall per sq ft of floor. Now let's look at the large house. If the dimensions are 80 × 125 ft with 10-ft tall walls, then the wall area is 4100 sq ft to contain 10,000 sq ft of heated space or 0.41 sq ft of wall per sq ft of floor. The 'not so big' house has over three times the heat loss wall surface area per sq ft of floor area! The large house is over three times better than the 'not so big' house from a wall energy efficiency point of view. Five families could split the 10,000 sq ft and have 2000 sq ft each with a third of the wall heat loss per sq ft of the 'not so big' house. Perhaps there is appeal in the 'not so big' house to consuming fewer materials? Well, the same analysis will show that one needs to build over 3 times as much area of exterior wall per square foot of floor area when building the 'not so big' house. On a per square foot of floor area basis, the 'not so big' house has tripled the energy cost and tripled the wall building cost. The reason to go through this analysis is to expose that the 'not so big' house has only one benefit, which is to reduce the overall square footage being built by the community. If the community's overall energy consumption or the community's overall building cost was paramount, then larger, connected buildings would be a far more effective course of action. It is only inside a 'single family detached' housing context that 'not so big' makes any sense at all. Actually, there may also be a socio-political use for restricting house size as a tool to restrict the wealthy from consuming more than their share of resources. That may make sense if buildings last only 50 years, but building healthy, energy-efficient buildings that last 400 years is a gift to our descendants, regardless of size. Toxic, disposable, energy-consuming buildings leave a bad legacy, regardless of size.

A2.2.5 Indoor environment

The human body today is not very different from the human body of 2000 years ago and yet the indoor environments in that same timespan have changed radically. Our bodies now have to cope, in the indoor environment, with a variety of toxins that we have never before been exposed to. We have surrounded the spaces wherein we spend 90% of our lives with materials that contain carcinogens, fungicides, volatile organic compounds, endocrine disruptors, etc. (the list of toxins is long and disturbing). It is difficult to find a single component of wood frame wall assemblies that does not have a toxic material in it. It is logical (but difficult to prove) that people living in spaces enveloped by carcinogens will have higher rates of cancer. Human health effects resulting from toxic indoor environments are generically known as Sick Building Syndrome. The most common response is to support the search for a cure (to cancer or asthma or compromised immune systems, etc.). This is a downstream approach. Conversely, an upstream approach addresses the causes, not only the effects. ('An ounce of prevention equals a pound of cure.') There are solutions that attempt to improve a building's health by prohibiting specific toxic materials (e.g. lead or volatile organic compounds in paint, asbestos in insulation, lead in solder, etc.). However, it is impractical to evaluate all chemicals in order to prohibit the toxic ones, as there are over 100,000 chemicals in production today in North America. The Precautionary Principle, which suggests staying away from products that have a chance of being toxic, is a somewhat better way forward. An even better approach is to create environments that most closely mimic those environments in which we evolved as a species. That's where earth building comes in. People have lived successfully in earth buildings for millennia. 'One half of the world's population, approximately 3 billion people on six continents, lives or works in buildings constructed of earth.' – Ronald Rael, eartharchitecture.org. Prohibiting toxic building materials is next to impossible as there are so many toxins available. Avoiding unnecessary use of materials we're unsure of is still reactionary and doesn't provide a target. Changing how we build to the extent that we can, given our modern ideas of comfort, in order to mimic the environments our bodies evolved in, is the most certain way to ensure health.

A2.2.6 Cradle to Cradle

All buildings, even stabilized rammed earth (SRE) and SIREWALL buildings, must at some point in time come to the end of their life. At that time, the construction materials will be a burden or a boon to those who undertake the deconstruction. Which they turn out to be is a function of the sustainability of the initial design. Little or no thought is given to the end-of-life deconstruction of the vast majority of today's disposable buildings. Priority is given to reduced cost per square foot along with an acceptable or impressive veneer. The result is a high level of toxicity in the construction materials and a disturbingly short lifespan, and virtually nothing can be reused at the end of the building's life. This effectively maximizes the lifecycle cost, resource extraction and its related environmental impacts, and the waste and disposal of toxic materials in open landfill sites. By contrast, the SIREWALL team has taken the Cradle to Cradle approach very seriously in its system design. At the end of a SIREWALL's life, the plumbing and wiring can be pulled out of the metal conduits, the windows and doors can be easily removed, and the wall materials can for the most part be reused. The rammed earth can be ground up in a concrete recycling plant for reuse in the next rammed earth building and the rebar, metal conduits and electrical boxes, all of which are a low grade steel can be made into more rebar, metal conduits and electrical boxes. Only the foam insulation cannot be reused for the same function (and possibly the TREX nailers if they are used). We are working on these issues. The profound difference between reuse and recycling is distinguished in the inspiring book, *Cradle to Cradle* – William McDonough and Michael Braungart (2002).

A2.3   The economic impact of stabilized insulated rammed earth building

A2.3.1 Business cycles

While it is true that the business cycle affects all businesses it does not affect them equally. When there is a downturn in building and a reduction of discretionary spending, values such as health and the environment may fall away, diminish in importance, or be deemed 'non-essential'. All that matters is the first cost. People are only interested in very short payback propositions. That is not a good climate for earth building to do well. However, it is encouraging to note that this now appears to be changing. Earth building is economically extremely attractive when a long-term point of view is used. However, in an economic downturn the disposable culture takes over.

Business downturns seem to correlate with the price of energy. When energy costs are low, business thrives, and vice versa. Building with earth is much less energy dependent than other forms of building. Around 90% of the materials are typically sourced within a half hour's drive. With regular construction, building materials get transported from around the globe, assembled into modular shapes (4×8 ft plywood or drywall, 2×6 in. lumber, etc.) in urban centers and then shipped to the builder at the site. This modality of building exacerbates our energy cost exposure. As well, it streams money away from local communities, making them more vulnerable to economic downturns. Building with rammed earth, by virtue of using so much local material and labor, acts to vitalize local economies and keep the money circulating therein.

A2.3.2 Lifecycle cost analysis

Individuals are slow to realize the usefulness of lifecycle costing. In North America houses are typically owned for less than five years. Governments foresee and know they will face the impacts of making first-cost decisions favoring disposability and poor performance, and are therefore willing to spend a little extra to construct green and durable buildings. During economic slowdowns, governments try to stimulate the economy by spending on infrastructure, which includes buildings. At these times, prices and costs are considerably lower than during a booming economy.

A2.3.3 Scheduling

Scheduling work is always difficult for the small earth building company because the lead times are so long. It is rare that a client comes to the door with a finished set of plans and says, How soon can you start? Almost always it's a matter of shepherding a design that's already in progress, or starting on a design from scratch with the client. Either way start dates are a long way from first contact, and the certainty of a specific start date is an illusion. So many things can happen in the municipal and permitting process, the engineering, the client changing their mind and circumstance, funding and contract negotiations. Operating inside this long timeframe environment in concert with the business cycle can be challenging, especially in a downturn. It takes longer to get going again, when things slow down. When business is hopping, it can be exhilarating with a full pipeline of work to the horizon.

A2.3.4 Collaboration inside the earth building community

The earth building community around the world is very small. At present, all companies are small or part-time and the culture is oriented to small-scale endeavors. There are a couple of exceptions to this but the cottage industry approach has proven to be reasonably durable, although difficult. Unlike wood, concrete and steel building, people have many questions and concerns about building with earth. These need to be addressed, over and over again. What is firmly lodged in people's minds is that earth buildings will wash away in the rain, that they are dirty, that they are dark and cave like, and that they are best described as a developing world mud hut. Modern earth builders spend an extraordinary amount of time trying to counter these erroneous preconceptions. This effort is made through marketing, answering questions over the phone and by email, and occasionally by teaching courses. All of the preceding does not make money for the small business owner. These efforts are simply necessary in order to address potential clients' preconceptions, questions and concerns. Given that this education is necessary and deleterious to the viability of the small earth building business owner, there is a clear benefit to sharing the education burden. Builders are typically autonomous, rugged individualists, so this sharing and collaboration can be difficult. Collaborating with one's competitor creates many conflicts. As the earth building community grows, there is evidence of a greater willingness to collaborate to educate the public.

R&D has generated whole new aspects to modern earth building. Now it is possible to confidently build with rammed earth in very wet and very cold climates. Now there is pise available. Big strides have been made in adobe and cob. These developments did not happen without a lot of time and money invested. R&D requires private initiatives and happens only through an individual's passion for what's possible. It does not make money. Some of those who have spent time and money on R&D are able and willing to share what they learned freely, while others try to recoup some of their costs. It's usually a matter of how much money has been invested. Typically, small dollar R&D is shared, while large dollar initiatives are not. From strictly a business perspective, it seldom makes sense to do R&D. It's far better to let someone else do the work and absorb the cost, and try to copy the results. This fact is a significant disincentive to do R&D on one's own. Once again the sharing of resources by small earth building companies is an obvious solution. The difficulty here is while the benefits of the R&D are shared among the group, clients will often want competing bids for the enhanced technology, to ensure a fair and best price. Once again the difficulty of collaborating with one's competitor arises. This difficult dynamic is an impediment to rapid progress, and people in the earth building field are continually searching for a solution.

A2.3.5 Commercial work

Taking modern rammed earth technology onto commercial building sites evokes many different reactions. The general contractor may view it as a loss of control or intrusion into their concrete business. It may be seen as an unwelcome scheduling complication. The contractor may see the advantage of building a relationship with a deep green building technology, or may embrace the simplicity of having so many subtrades replaced by one subtrade.

There has been an enormous amount of energy directed toward increasing productivity and efficiency in the mainstream building technologies such as concrete, steel and wood. These technologies have learned to work together. At this point in time, modern rammed earth is a disruptive technology to building as usual. To bring a modern rammed earth project to a successful conclusion requires strong advocacy by the architect and client. Only under those circumstances is the disruption accepted and worked with. As more and more commercial projects get built, more synchronicities will be found and the level of disruption will continue to diminish.

A2.4   Stabilized insulated rammed earth building technologies

A2.4.1 Shifting away from 'downstream solutions'

Concerns about the cost of energy over the last 35 years have resulted in, almost exclusively, downstream solutions. The conversations are about whether to go to nuclear or how to 'achieve clean' coal, or the viability of wind and solar energy. Political talk is about the instability caused by dependence on foreign oil. Very seldom is there serious discussion about reducing demand. Buildings consume 40% of the energy in North America and, of that, 50% is used in heating and cooling. We have the technology to build net zero buildings, but that initiative is left exclusively for the visionary individuals of the world to fund and put in place. If we changed our focus from downstream to upstream, and embarked on ensuring that all buildings were net zero, total energy consumption could be reduced by 20%. We could stop buying air conditioners and turning up the thermostat if we could redirect downstream funding to upstream solutions.

A2.4.2 Technological advances

Despite impatience to have rammed earth achieve its true potential, it's important to acknowledge the 'ground that has been taken' moving from rammed earth to modern rammed earth. Up until 40 years ago, rammed earth technology had improved little from the time when the Great Wall of China was built, over 2000 years ago. Forming was done in a slipform fashion presenting less than 10 sq ft to ram against, before needing to reset the forms. Mixing was done with shovels and hoes. Delivery was with hand carried pails. Mix design was by feel. Color and texture results were random and of little concern. The results were buildings that had a very small environmental footprint, took a long time to build, and were vulnerable to rain and freezing weather. That reputation for lack of durability has been difficult for modern rammed earth to shed, despite ample evidence to the contrary.

Mix design has evolved considerably from, 'Let's try this, it seems/feels good' (2 MPa) to sieve analysis and particle size distribution analysis (5–8 MPa) to using tools like Fineness Modulus (6–15 MPa), to more sophisticated proprietary systems (10–40 MPa). Results have moved from concern over load-bearing capability and durability issues to achieving large safety factors in both those areas.

Mixing of soil has moved from shovels and hoes to rototillers, to mortar mixers, to skid-steer mixing, to pug mills, to custom volumetric mixers.

Forming has moved from 10 sq ft panels to panels that can measure in the hundreds of square feet (see Figs A2.1a and b). The necessity of having form ties run through the walls for tensile strength is no longer necessary. This move began with David Easton's 'California' forming system using pipe clamps and small wall segments. The Australian technique of ramming the full height wall segments first and following with underfills and overfills was a good complement to the California system. There are now several ways to form walls that don't require through-wall tensile form ties. This advancement has benefited the visual appearance of rammed earth and has eliminated the need for patching holes.

Delivery has moved from pails to tractor buckets to skid-steers to

(a)

(b)

A2.1 (a), (b) Examples of panels used in stabilized insulated rammed earth buildings.

telehandlers, scissor lifts, conveyors and cranes. Customization of conveyors and hoppers specifically for the needs of rammed earth is now taking place. Through trial and error, efficiencies are beginning to emerge.

Ramming has gone from wood hand-tampers, to steel hand-tampers, to pneumatic tampers of various sizes and weights. Some customization of tampers has proven beneficial. Control of color and texture has developed from being constrained by the color and texture of local soils to being able to create almost any color desired with increased control over final texture.

System design has transformed from analysis of the individual elements to an integrated approach. For example, knowing the form pressures created by different rammers allows optimization of the formwork for specific rammers. Designing formwork for easy delivery, and integrating mixing and delivery in system design offers benefits if there is enough lead time and, ideally, if there is an Integrated design process.

Given the small size of the global rammed earth industry, improvements in the above areas are still in their infancy and many have not yet achieved widespread adoption.

A2.4.3 Anticipatory design

Although not common thinking at this time, anticipatory design is essential to sustainability. In the 1960s, Buckminster Fuller spoke about 'Comprehensive Anticipatory Design' where emphasis is placed on global trends and local needs, whole systems thinking and critical inquiry of nature's fundamental principles as an integral part of design science. That was many decades ago and the need for such an approach grows yearly.

Although in its infancy on a global scale, the current green building movement is beginning to address some of the issues and problems with current building approaches and practices. With a vision mandate to make green buildings mainstream and the standard level of care, there is a smorgasbord of green building rating systems, codes, policies, directives and incentives to develop and promote green building policies and best practices. In the struggle to take hold and become established within the market place, the phenomenon of 'greenwashing' has emerged, which continues and relies on narrow, short-time-frame thinking. For example, most green rating systems reward the use of styrene insulation and the more of it used the more points are received. While styrene insulation does indeed insulate well, health penalties also accrue from persistent, bioaccumulative toxins in the brominated fire retardants. The Global Warming Impact (GWI) of blowing agents in styrenes make them of questionable benefit with respect to climate change especially in disposable buildings where they end up exposed in landfill sites after demolition.

Where most green thinking and action to date has taken a 'less bad' approach to green building, the Cradle to Cradle (C2C) certification for building products assesses product safety to humans and the environment and design for future lifecycles. C2C offers a holistic paradigm and program that focuses on using safe materials that can be disassembled and recycled as technical nutrients, or composted as biological nutrients. Resource extraction and land filling almost disappear. This is the most positive approach to an anticipatory design future that I'm aware of. Although the implementation of C2C is daunting in scope, the imperative is clear and the rare possibility of a bright sustainable future is clear.

Our materials and methods of construction need to be sustainable and we need to be clear about where we are going with our design approach and building technologies. Comprehensive Anticipatory Design and C2C certification offer effective tools to create a healthy, sustainable and regenerative future.

While most businesses operate by reacting to consumer demand, it is also important to lead. Running a business by reacting guarantees an inability to be current and relevant. Anticipating market demands in a context of unworkable strategies and situations such as massive landfilling, wanton resource extraction, peak oil and climate change offers unprecedented opportunities to create and provide sustainable leadership. Balancing leadership with satisfying market demands is essential and critical at this point in history.

A2.5   Acknowledgements

My very sincere thanks go to Scott Krayenhoff and Catherine Green for their thorough and insightful editing.

A2.6   References

Steven Blanchard and Peter Reppe (1998) Life cycle analysis of a residential home in Michigan, Study from the University of Michigan
William McDonough and Michael Braungart (2002) *Cradle to Cradle*, North Point Press, New York

Index

749